GOLD PLATING TECHNOLOGY

GOLD PLATING TECHNOLOGY

by

FRANK H. REID

and

WILLIAM GOLDIE

ELECTROCHEMICAL PUBLICATIONS LIMITED

1974

ELECTROCHEMICAL PUBLICATIONS LIMITED
29 Barnes Street, Ayr, Scotland.

First published 1974
Second printing 1982

Printed in Great Britain by The Anchor Press Ltd
and bound by Wm Brendon & Son Ltd, both of Tiptree, Essex

To my wife
Elizabeth
and our children
Lorainne and Alison

W. GOLDIE

PREFACE

One of the most striking features of developments in the electroplating field during the past two decades has been the growth in the use of gold plating for industrial applications, a growth directly related to the rapid development in the electronics and telecommunications industries over the same period.

Why is gold plating used to such a wide extent? This question is answered in considerable detail in the applicational sections of the present volume. Here it will suffice to state that no answer is really necessary as relating to decorative applications since the colour and permanence of gold have appealed irresistibly in this context since the metal's first discovery. It is the same basic property of chemical nobility which has conditioned the use of gold plating as an industrial finish, pre-eminently as an electrical contact surface, in which role its immunity to tarnish or corrosive attack and freedom from surface films confers on it a low and stable contact resistance even under arduous environmental conditions, and the ability to transmit the smallest electrical signals with a minimum of distortion.

It is not surprising therefore that usage of gold plating should have paralleled the growth of industrial areas in which the successful functioning of equipment and systems depends critically on the best possible electrical contact between components, be it through the medium of permanent or semi-permanent connectors or through sliding or wiping contacts operating at very light pressures.

Intensive use of a process in critical applications, however, often tends to reveal shortcomings, and the development of gold plating has proved no exception. Problems related to process or performance have arisen in several areas due, for example, to attack of printed circuit board adhesives by conventional alkaline cyanide electrolytes, to limited wear resistance of the relatively soft pure gold coatings in sliding contact applications, to cold welding tendencies of similar coatings in semi-permanent contact and reed switch applications, and to loss of protective value due to porosity of deposits. Problems have been encountered in soldering to gold plate, and in relation to high temperature stability of coatings due to diffusional effects or to the presence of foreign matter.

In the course of meeting and overcoming such difficulties it has become increasingly apparent that whilst gold must, by virtue of its innate quality of chemical inertness, continue to play a key role in the industrial field, the properties of gold plating require to be "tailored" in a variety of ways to meet specialised requirements, whether these be in the traditional contact field, in the welding of contacts to integrated circuits, or in the use of gold as a thermal control surface or as a high vacuum bearing surface in space applications.

With the development of acid and neutral gold plating solutions, and more recently non-cyanide electrolytes, potentialities for the modification of the properties of gold plate by alloying have become very wide, and indeed, the last twenty years have seen a level of activity in the development of plating processes and in the study of deposit characteristics in relation to applications which is unparallelled for any other rare or precious metal. As a result the patent and technical literature on gold plating is, today, quite extensive. In general reference books on electroplating, however, the treatment of this important subject is inevitably superficial. The same criticism applies, though of course to a lesser extent, to the few books available dealing specifically with precious metal plating, in which the treatment accorded to gold must of necessity be limited by space considerations when silver and other precious metals of the platinum group are covered at the same time.

The present volume is the first, so far as the editors are aware, to aim at an in-depth coverage of all major aspects of gold plating on the basis of contributions from eminent specialists in the fields of electrolyte development, deposit evaluation, and industrial application. To this end, following a historical introduction, four main sections are presented dealing respectively with electrolytes (including immersion and electroless plating solutions), practical plating techniques, coating properties, and testing methods. In connection with techniques, due emphasis is given to aspects of particular current interest such as selective area plating; this section also includes an extensive review of pretreatment procedures for virtually all materials of practical interest.

These general sections find co-ordination and amplification in a final major section on applications, covering both decorative and industrial fields, from the plating of jewellery and watchcases to current uses in printed circuitry, reed switches, connectors, semiconductor and microelectronics technology and space applications.

In a book of this nature some degree of overlap between individual contributions is inevitable, particularly in the applicational section. Although this has been "edited out" to a certain extent, efforts in this direction have been deliberately restricted with a view to making each chapter as self-contained as possible, thus avoiding the need for too frequent cross-reference on the part of the reader.

The editors' aim has been to produce a book which, despite the rapid pace of development, should serve as a useful work of reference for some time to come, both to those immediately involved with any aspect of gold plating and to electroplaters and engineers in other fields who may at any time find themselves concerned with the subject in a practical way. If this object has been attained their thanks must go to the individual Authors whose experience has been so willingly contributed to the project. Acknowledgement is also due to the American Electroplaters' Society, for permission to reproduce certain material and illustrations that have previously appeared in their official journal *PLATING*.

FRANK H. REID
WILLIAM GOLDIE

CONTENTS

PART 5—**TESTING**

PART 6—**APPLICATIONS**

BIOGRAPHIES

THE EDITORS

Frank H. Reid is a well known figure in the precious metals field, with which he has been associated throughout an industrial career of over 30 years, initially as a Chemist in the Precious Metals Refinery of International Nickel Limited at Acton, London, and subsequently in research and administrative capacities in the Platinum Metals Division of the Development and Research Department of the Company, where he was mainly concerned with the development of improved processes for the electrodeposition of the metals and the study of deposit properties. During this period awards for technical papers included the Westinghouse Brake and Signal Company Prize and the Johnson Matthey Medal (twice) of the Institute of Metal Finishing.

He holds both General and Special (Chemistry) BSc degrees of the University of London and is a Fellow of the Royal Institute of Chemistry. He was elected an Inaugural Fellow of the Institute of Metal Finishing in recognition of his contribution to the precious metal plating field.

William Goldie has been associated with the electroplating industry for fifteen years during which time he has contributed a number of important research papers to the technical literature. He has also been granted a number of patents in this field. A pioneer and acknowledged authority in the field of electroplated plastics, he is the author of the accepted standard reference work on this subject, a two volume book entitled "Metallic Coating of Plastics".

Besides running a consultancy firm, William Goldie and Associates, he is also the Chairman of Electrochemical Finishes, a highly specialised company involved in precious metal deposition and situated in Alexandria, Scotland.

Mr. Goldie is a member of several professional institutes.

THE CONTRIBUTORS (*in alphabetical order*)

Morton Antler was born in 1928 in New York, N.Y. He received his BA from the University College of New York University in 1948 and his PhD in Chemistry from Cornell University in 1953. After several years in research on combustion science and on mechanical wear in the automotive-petroleum industries, he joined the International Business Machines Corporation in 1958 as Advisory Chemist where he led studies on electrical contact phenomena and materials. In 1963 Dr. Antler was appointed Deputy Director of Research of the Burndy Corporation and in 1970 he joined Bell Telephone Laboratories, where he is continuing his studies of electrical contacts, electrodeposition and tribology. He has numerous publications and many patents in these areas.

In 1968, and again in 1971, Dr. Antler received the Precious Metal Plating Award of the American Electroplaters' Society. In 1971 he also received the Captain Alfred E. Hunt Memorial Award of the American Society of Lubrication Engineers for significant contributions to the understanding of the basic mechanisms of sliding wear of metals.

Dr. Antler is a member of the American Electroplaters' Society, American Society for Testing and Materials, American Society of Lubrication Engineers, and several other professional societies. In addition, he is a member of various groups including the Committee of the Illinois Institute of Technology for the Ragnar Holm Seminars, the Books and Symposia Committee of the American Electroplaters' Society, and in 1972 was co-chairman of the Sixth International Conference on Electrical Contact Phenomena.

Kenneth J. Clews is a Principal Scientific Officer at the Welding Institute where he currently holds the position of a Senior Lecturer in the Education and Training Division. Prior to joining the then British Welding Research Association, Mr. Clews was with the Metals Division of Imperial Chemical Industries (now Imperial Metal Industries) where he specialised in the welding, brazing and soldering of non-ferrous metals and their alloys.

Walter K. A. Congreve received his BASc in Metallurgical Engineering from the University of British Columbia, Vancouver, in 1949 followed by a PhD in Metallurgy from Manchester University in 1951. Since 1961 he has been in charge of all aspects of electrodeposition at GEC, Hirst Research Centre, Wembley, Middlesex.

Bruce M. Dickinson is at present employed in the R & D laboratories of Engelhard Industries, Cinderford, as a Research Chemist. Prior to joining Engelhard, he held the post of Finishing Manager at Metal Link (Jewellery) Ltd., a large plating company involved in the jewellery plating trade.

Mr. Dickinson studied chemistry and metallurgy at Glamorgan College, Wales.

Edward F. Duffek is presently Technical Director and Vice President of Chemline Industries, Santa Clara, California. He was formerly Manager, Materials & Processes, Advanced Manufacturing, at Fairchild Camera and Instrument Corporation where his responsibilities included the direction of technical projects and development of low cost, high volume, assembly and packaging of semiconductor devices. Prior to this, Dr. Duffek was Section Head of Electrochemistry and Electroplating, R & D Laboratories, Fairchild Semiconductor Division. A previous appointment was that of Electrochemist at Stanford Research Institute, Menlo Park, California. A recognised authority in the electroplating of electronics hardware, Dr. Duffek has innumerable patents to his credit as well as having contributed widely to the technical literature on this subject.

Education attainments include a BS from Loyola University, Chicago, Illinois and an MS and PhD from Carnegie Institute of Technology, Pittsburgh, Pennsylvania.

D. Gardner Foulke Director of Research Services of Sel-Rex Company, a Division of Oxy Metal Finishing Corporation, has been associated with the metal finishing industry for thirty years. He has held lecturing posts at Rutgers University, where he received his PhD in 1942, at Beaver College and at Brooklyn Polytechnic Institute. He has also been employed as a consultant with A. K. Graham and Associates, Foster D. Snell Incorporated and as an industrial chemist with Republic Steel, Houdaille-Hershey Corporation (Manhattan Project), Hanson-Van Winkle-Munning, and Research Director of Sel-Rex, Nutley and Sel-Rex International S.A., Geneva, Switzerland.

Dr. Foulke has authored many papers on precious metal electrodeposition and has a number of patents to his credit both in precious and non-precious metals. He was co-editor of "The Electroplaters' Process Control Handbook" and has contributed several chapters to "Modern Electroplating" and "Handbuch der Galvanotechnik".

Dr. Foulke is a member of the American Electroplaters' Society, American Chemical Society, American Society for Testing and Materials, American Society of Metals, the Electrochemical Society and the Chemist's Club. He has served as Chairman of the Electrodeposition Division of the Electrochemical Society, Chairman of the National Technical Task Committee on Industrial Waste and also served as Executive Secretary and as Director of the American Electroplaters' Society.

Samuel M. Garte received his AB in chemistry from Harvard University in 1937 and his PhD from New York University in 1964. He has been active in the plating field for 23 years as a chemist, plating engineer, and in research and development with Raytheon Manufacturing Company, Coro Incorporated, the Bulova Watch Company, and Lionel Electronic

Laboratories. He joined the Burndy Corporation Research Division in 1964 as a research scientist and is now Chief Inorganic Chemist. He has written several papers on precious metal plating and electrical contact properties of metals, and was awarded the American Electroplaters' Society Gold Metal for the best paper of 1968. Dr. Garte is a member of the American Electroplaters' Society, American Chemical Society, the Electrochemical Society, Society of Sigma Xi, and ASTM Committees BO8 and BO4.

Henry A. Heller attended the University of Illinois and Tri-State College where he received the degree of BS in chemical Engineering in 1936.
His experience in the field of analytical chemistry began at the Carnegie-Illinois Steel Company in Chicago, Illinois. In 1940 he joined the US Bureau of Mines in Boulder City, Nevada where he gained his early experience in instrumental methods of analysis. In 1953 he was made Head of the Instrumental Analysis Department of the Bureau of Mines in Reno, Nevada. In 1956 he joined the National Lead Company of Ohio, as operating contractor for the US Atomic Energy Commission. Here he developed X-ray and spectrographic methods for the determination of impurities in atomic reactor feed materials and prepared the instrumental analysis sections of a comprehensive manual of analytical methods which was later published by the Atomic Energy Commission.
In 1963 he joined AMP, Incorporated in Harrisburg, Pennsylvania where he is employed at the present time. He organised the analytical department at the AMP Research Division where the laboratory activities now include work in the fields of emission spectroscopy, X-ray spectrometry and diffraction, infra-red and ultra-violet spectrophotometry, thermal methods of analysis, liquid chromatography and classical "wet" chemical analysis.
Mr. Heller is a member of the American Chemical Society, the Electron Microscope Society, the Society of Plastics Engineers and the Society for Applied Spectroscopy in which he served as President of the Cincinnati, Ohio Section. He is also a member of the American Society for Testing and Materials where he has participated in the establishment of standard methods for the measurement of plating thickness by beta-ray backscatter and X-ray spectrometry.

John F. Houlston is the Chief Chemist of FR Electronics, a division of Flight Refuelling Limited. Educated at Newport and Monmouthshire College of Technology, he obtained his HNC in chemistry in 1961. Prior to taking up his present position in 1966, he spent two years as an electroplating engineer at Texas Instruments.

John M. Leeds was awarded his PhD by London University for research into the porosity of gold electrodeposits. A porosity test method developed

during this investigation is incorporated into Gold Plating Specifications DTD 938 and BSS 4597. For this work he was co-recipient in 1968 of a scientific award from the American Electroplaters' Society for making the most outstanding universal contribution to the development of methods for the evaluation of electrodeposit properties.
Dr. Leeds who is Technical Director of Electrochemical Finishes Limited, is an Associate of the Royal Institute of Chemistry and an Associate of the Institute of Metal Finishing.

David R. Mason was educated at Dulwich College and obtained his LRIC in 1964 specialising in Radiochemistry. He was previously employed as a Plating Chemist with GEC and as a Research and Development Chemist at Ronson Products. He holds several patents on gold and platinum group metal plating and is currently the Chemical Manager of Engelhard Industries, Cinderford, supervising the production of all chemical and electrochemical products, as well as being responsible for all R & D aspects.

Michael Massin, a metallurgist by training, received his first degree and subsequent doctorate for studies on factors affecting the hardenability of carbon and high-speed steels. Since 1960 he has specialised in materials problems in micromechanics with particular reference to the watch industry, and is the author of numerous papers in the field, relating especially to precious metal electrodeposits and their thermal treatment. He is at present Head of the Metallurgy and Materials section at the Centre Technique de l'Industrie Horlogère at Besançon, France, where he is responsible for all aspects of research, control and standardisation concerning materials and applications in the watch industry and in micromechanics, including electrodeposited gold on base metals.
Dr. Massin is also in charge of the course in metallurgy for students of chemistry and micromechanics at the University of Besançon.

Raymond Mills received his technical education at the Northampton College of Technology (now the City University) London, where he studied metallurgy, with specialisation in the corrosion and surface treatment of metals. He is an Associate of the Institution of Metallurgists and a Fellow of the Institute of Metal Finishing.
He has had very wide experience in metal finishing and materials technology both in research and industry. His first association with gold plating occurred when he was employed by the Design and Research Centre for the Gold, Silver and Jewellery Industries at Goldsmiths' Hall, London, where he saw the growing use of gold plating on items of jewellery which had hitherto been manufactured from rolled gold. He was later to witness, as Group Leader of a materials technology group at

GEC (Electronics) Limited, the more spectacular rise in the use of electroplated gold coatings in the electronics industry to provide reliable contacts for printed circuits.

The author has also been employed by prominent suppliers to the metal finishing industry and as a Senior Research Officer with the Tin Research Institute. He is at present in charge of the national and international activities of the British Standards Institution in the field of metallic and related coatings.

Leo Missel is an Advisory Engineer at the International Business Machines Corporation. He has developed new functionally oriented surface treatment processes used in the fabrication of nose cones (including the original re-entry nose cone), antennae, matrices, connections, radiation devices and heat shields. He has published numerous articles on plating, etching and other surface treatments.

Mr. Missel holds a BS degree from the City College of New York.

Fred I. Nobel received his BSChE from the City College of New York and his MChE from the Polytechnic Institute, Brooklyn, New York. His experience in the metal finishing field covers a period of 28 years. He is presently Technical Director and Executive Vice-President of Lea-Ronal, Incorporated. Previously he had been with the Bureau of Aeronautics United States Navy, working on materials and processes for aircraft, then later in private industry as a plater, chemist and finishing supervisor. Mr. Nobel is a member of the New York Branch of the American Electroplaters' Society, American Association for the Advancement of Science and the Institute of Metal Finishing.

Yutaka Okinaka received his PhD in 1959 from Tohoku University, Japan. He has been a member of the technical staff at Bell Telephone Laboratories since 1963. Until the end of 1967 he was engaged in research and development of secondary batteries, and he holds the 1970 Research Award of the Battery Division of the Electrochemical Society. Since 1968 he has been working on electroless and electrolytic metal deposition processes. He received the American Electroplaters' Society Award for excellence in presentation of his paper on electroless gold at the 1970 Convention of the Society. Before joining Bell Laboratories he was active in the areas of polarography and related electroanalytical chemistry at Tohoku University and at the University of Minnesota.

Dr. Okinaka is a Life Member of the Electrochemical Society and a member of the American Chemical Society, Society of Sigma Xi, the Electrochemical Society of Japan and the Metal Finishing Society of Japan

Barnett D. Ostrow has been involved in the metal finishing field for over 30 years as a plater, chemical engineer and consultant. He received his BSChE degree from the City College of New York and conducted graduate study in metallurgy at the Polytechnic Institute of Brooklyn. He is at present Chairman of the Board and President of Lea Ronal Incorporated.

Mr. Ostrow is a member of the American Electroplaters' Society, American Chemical Society and the Electrochemical Society.

Edward A. Parker, Executive Vice President and Technical Director, Technic Incorporated, obtained his education at the University of Illinois where he received the degrees of BS in 1930, MS in Organic Chemistry in 1932 and his PhD with a Thesis on X-ray Diffraction in 1937. Following appointments as Textile Foundation Fellow (1935–1937), with Long Island Biological Association (1937–1939) and as Atlas Powder Fellow (1939–1941), he entered the plating field in 1941 when he joined the consulting firm of A. Kenneth Graham and Associates. He then became associated with Alrose Chemical Company on black oxide finishes, leaving in 1945 for his present position as Technical Director of Technic Inc. Dr. Parker is an active member of the Providence-Attleboro Branch of the AES, having served in the capacities of Secretary-Treasurer, President, Board of Managers and presently as Educational Chairman. He has been Secretary-Treasurer, New England Council of the American Electroplaters' Society since 1970.

He is a former member of the American Electroplaters' Society Research Project Sub-Committee No. 12 from inception (1948) to 1966. He is a former member of American Electroplaters' Society Research Committee, Chairman, 1969–70; the American Electroplaters' Scientific Achievement Award Committee, Editorial Board, Educational Committee, Paper Awards Committee and the Branch Education Committee. He is also a member of the American Chemical Society, American Society for Testing and Materials, American Society of Metals, American Institute of Chemists, the Electrochemical Society, Institute of Printed Circuits, Society of Sigma Xi and the Institute of Metal Finishing.

Dr. Parker received the American Electroplaters' Society Gold Medal (Heussner Award) in 1952 and was made a National Honorary Member of the American Electroplaters' Society in 1969.

Marvin Rubinstein has spent many years in the electroplating field and is well known internationally. He holds two degrees in widely divergent fields—a BS in Chemical Engineering from Cornell University in 1943 and a more recent acquisition, a JDr degree from Brooklyn Law School in 1967.

His work in electroplating has spanned a period of 30 years, and includes positions as plating chemist, shop foreman, factory manager, finishing engineer, technical writer and electroplating consultant. He has published widely and acted as a consultant in the United States, Europe and Israel. A field in which he has also received wide recognition is that of electroforming.

For the past fifteen years, Dr. Rubinstein has dedicated most of his efforts to one finishing specialty—selective (brush) plating. He is President of Selectrons Limited, of New York, one of the leading companies in this field. Starting with a hobby shop or touch-up type tool, Dr. Rubinstein has helped to expand the horizons of selective plating so that today it is finding extensive use for many industrial applications.

Dr. Rubinstein is a member of the New York Bar, practically a life-long member of the American Electroplaters' Association, and on the Board of Directors of several US corporations.

Michael Scott graduated from Trinity College, Cambridge with a degree in Natural Sciences (specialising in Metallurgy). Following appointments with the Royal Naval Scientific Service and the Central Electricity Research Laboratory, he joined the then British Welding Research Association in 1960. At Abington, where he is employed as a Principal Scientific Officer in the Metallurgy Department, he has worked on various aspects of joining most of the common non-ferrous metals and their alloys.

Gordon R. Strickland is a product of the Woolwich Arsenal School having studied under A. J. Hothersall and R. A. F. Hammond. During a career in the metal finishing industry which has spanned nearly thirty years, he has been involved in just about every conceivable aspect of this subject from the role of research and development chemist to that of production manager. For the past seven years he has concentrated his efforts in the field of printed circuit technology. He is currently the Managing Director of Electrochemical Finishes Limited.

Mr. Strickland has contributed frequently to the technical press and holds a number of patents. He is an Associate of the Institute of Metal Finishing.

Colin J. Thwaites graduated in 1948 with a BSc (Hons) in Metallurgy from the Royal School of Mines, Imperial College, London. Joining the British Non-Ferrous Metals Research Association he studied creep and fatigue properties of lead alloys and was awarded an MSc based on this investigation. In 1952 he joined the Tin Research Institute and has held the position of Chief Metallurgist since 1969. His main interests are in soft soldering, particularly for the electronics industry. Other spheres of interest are tin in ferrous materials, bearings, and the properties and uses of tin-containing coatings.

Mr. Thwaites is a member of several British Standards Institution Committees, a member of the Metallurgy Committee on the Institute of Metals, and Chairman of the Soldering Group of the British Association for Brazing and Soldering. He is an Associate of the Royal School of Mines, Member of the Institute of British Foundrymen, Fellow of the Institute of Metal Finishing and a Fellow of the Institution of Metallurgists.

Thomas Warlow joined BICC-Burndy in 1966 and was responsible for installing and supervising the precious metal plating department. He is currently involved in research and development on gold plating for the electronics connector industry.

Mr. Warlow is an Associate of the Royal Institute of Chemistry and an Associate of the Institute of Metal Finishing.

Alfred M. Weisberg graduated from the University of Harvard in 1949 with an AB in Chemistry. He is at present President of Technic Incorporated, a company which he joined in 1960 as Vice President.

Mr. Weisberg is a member of the American Chemical Society, American Electroplaters' Society, American Society for Testing and Materials, American Association for the Advancement of Science, The Electrochemical Society, Society for the History of Technology, Institute of Electrical and Electronic Engineers, Institute of Metal Finishing and the Institute of Metals.

PART 1

Introduction

Chapter 1

HISTORICAL BACKGROUND

A. M. WEISBERG

Luigi V. Brugnatelli, a professor of chemistry in Pavia, was probably the first person to electrodeposit gold from solution. But an insult from Napoleon Bonaparte which led Brugnatelli to confine publication of his work solely to his own Journal in Pavia, Italy, buried the information for some thirty eight years. His work was rediscovered after John Wright, a surgeon from Birmingham, England, found potassium cyanide to be a suitable electrolyte for gold and silver electroplating. Wright's solution was combined with some other developments of Oglethorpe Wakelin Barratt and the Elkingtons, and was subsequently patented by Henry and George Richards Elkington. This discovery and patent was the forerunner of modern gold and silver plating in late 1840.

Brugnatelli was a colleague and close friend of Allisandro Volta at the time the latter made his momentous discovery of the Voltaic Pile. The importance of a new and awesome power made a particular impression on French scientists and Napoleon Bonaparte, presumably in his position as a member of the National Academy, invited Volta to Paris to demonstrate his discovery. Brugnatelli accompanied him. Before or during the course of three lectures, Volta, after being presented to Napoleon, introduced Brugnatelli to Napoleon as "my colleague, the great Italian chemist". Napoleon turned away with the comment that there were no *great* chemists in Italy. Highly insulted, Brugnatelli returned to Pavia and never again communicated with the French Academy of Sciences. As a result, his early work with Volta using voltaic electricity on various metallic solutions was never published in Paris and therefore escaped major notice. His electro-reduction of metals, probably first performed in 1800, finally resulted in his "reviving" of gold. A letter he sent to the editor of the Belgian Journal of Physics and Chemistry in 1805, was partially reprinted in Britain. "I have lately gilt in a complete manner two large silver medals, by bringing them into communication by means of a steel wire, with the negative pole of a voltaic pile, and keeping them one after the other immersed in ammoniuret of gold newly made and well saturated." When Brugnatelli's work was rediscovered in the 1840's, several attempts were made to duplicate it, but these were unsuccessful, in part due to the considerable confusion about the exact meaning of the word "ammoniuret".

Other scientists pursuing their individual interests deposited gold and other metals, but they either did not think of any practical application to the "trades" or merely overlooked them.

By contrast gilding, and the coating of metals and non-metals with gold without the use of an electric current, was already an ancient art in 1800. The Egyptians gilded with gold leaf and with gold plates. Cleopatra's palace was reported to have had gilded beams. The Arabs gilded copper and silver by applying a coating of mercury and then burnishing on a layer of gold leaf.

During the middle 1600's, many copper coated iron items and gilded silvered copper trays were produced by chemical displacement in Hungary and sold as souvenirs. Many of these items have only recently been found although they are described in the early literature. They bear inscriptions such as "Eisen war ich, Kupfer bin ich, Silber trag ich, Gold bedecket".

In the 1700's, Baumé made the suggestion that gilding of brass or copper was best carried out in a gold chloride solution containing as little excess acid as possible. He suggested evaporating the aqua regia solution until it just began to crystallise.

By 1800 a great deal of immersion gilding from dilute gold chloride solutions, called water gilding, was carried out on the very cheapest of trinkets, but all other items from spectacle frames to candelabras were mercury amalgam gilt. This is a very dangerous process and resulted in many deaths and disabling diseases.

While the natural philosophers and scientists were "reviving" metals in their experiments, practical chemists and working men were investigating metal deposition. In Birmingham, England, Henry Elkington and his cousin George Richards Elkington were in several businesses, separately and together, as gilders, manufacturers of gilt toys and novelties, gilt spectacle frames and gilt pen nibs. It would appear that Henry, the older of the two, was the experimentalist and practical chemist, although George did some developmental work in addition to being the entrepreneur. Their work culminated in a patent being issued to George Richards Elkington for an immersion gold bath based on gold chloride, neutralised with potassium bicarbonate. The bath was vastly superior to the usual water gilding methods, produced a very even and uniform colour and could equal in thickness the thinner deposits achieved by mercury gilding. This patent was applied for in France, the United States and several other countries and served as the basis for a monopoly on gold and silver plating that the Elkingtons maintained for many years.

During the latter part of 1838 and 1839, copper electroplating, or more properly, electrotyping and electroforming, was invented more or less simultaneously by Maurice Hartmann Von Jacobi (Boris Simonovitch Jacobi) in St. Petersburg, C. J. Jordan in London and Thomas Spencer in Edinburgh. This discovery motivated the Elkingtons, their chemists Barratt and Alexander Parkes, as well as others, to try and adapt the galvanic current to the deposition of gold and silver. Although Barratt's process was not workable, G. R. Elkington applied for a patent on March 25th, 1840, under the broad title of "Improvements in Coating, Covering or Plating Certain Metals". The exact method and details of the invention did not have to be submitted until the following September 25th. However, June came and Elkington was forced to admit that they had been unsuccessful.

At the same time that Barratt and the Elkingtons were carrying out their experiments, there was another gentleman in Birmingham working toward the same end, a surgeon by the name of John Wright, who tried cyanide and was soon able to exhibit a gilded chain and a silvered plate. He was unable, simply by showing the chain and plate, to obtain any financial support in Birmingham, so he went to London to apply for a patent.

G. R. Elkington, facing the deadline for the filing of his specification, hired John Thomas Cooper as a chemical consultant and travelled to London to draw up the final specifications. From an examination of the Elkington records, it was determined that both Elkington and Wright stayed at the same hotel and consulted the same patent agent. Ethics being somewhat different in those days, the agent introduced the two men. They very generally discussed their proposed patents and returned to Birmingham for further negotiations. When Charles Askin, representing John Wright, showed Elkington some gold plated objects, their colour and lustre were such that Elkington felt they had been coloured by fusion. Askin, not certain himself, wrote to Elkington the following day "I spent yesterday evening with Mr. Wright and he informed me that the articles I showed to you yesterday are not coloured. It is a process he does not understand." Askin would not let Wright divulge his process until a purchase agreement was drawn up. Elkington, on his part, found it hard to commit himself before he knew what he was buying.

It was not until September 1st, only 25 days before Elkington's completed specification had to be submitted, that an agreement was reached between Wright and Elkington. The first agreement signed concerning the silver plating part of the patent, called for an equal partnership of Henry Elkington, George Richards Elkington and John Wright.

Finally, the Elkingtons were allowed to know what they bought and they tested Wright's invention. On the 15th of September, Cooper wrote about the difficulties and ambiguous results he was having with the Barratt part of Elkington's specification but added "the cyanides with anything that contains gold does the trick instanter".

The specification, including Wright's potassium cyanide solutions for gilding and silvering, was rushed to completion to meet the final date for filing and in the rush, the drafting made the patent vulnerable to both attack and criticism.

William Sturgeon, the famous scientist and electrician, thought very little of Elkington and the practical men of Birmingham. He wrote, "A patent for gilding various kinds of metal has been obtained by Messrs. Elkington of Birmingham, bearing the date March 25, 1840, but the specification is such a rigmarole undefined jumble as clearly to denote the entire ignorance of the parties of many things they pretend to claim." Sturgeon claimed two years later in 1842 that, "7 years before, I contrived a magnetic electrical machine . . . by means of which I coated metals with tin, copper, etc., and I have employed the same machine to great advantage in silvering, gilding, and platinising various kinds of metals of inferior value. . . ." It is interesting to speculate if Sturgeon's dislike for the Elkingtons was based on his not having been employed by them as a consultant.

The scientists derided the discovery, the silversmiths and Sheffield platers dismissed it and only an occasional word of praise for the gold and silver plating patent was heard.

Necessity for speed in completing the patent was required by the deadline for the filing of the specification. The reason, however, for the filing of the application even before the process was successful can only be surmised. Although it is possible that the Elkingtons and their chemists worked in isolation in their workshop, it is much more probable that they were in touch with the scientific writings of the time. They may have known that Joseph Shore applied for a patent on March 3rd, 1840, for "Improvements in Preserving and Covering Certain Metals and Alloys of Metals" and that the patent involved the use of galvanic batteries made by John Woolrich. (Woolrich, on good authority, was supposed to have gilded by galvanic battery as early as 1838, but never published his findings). They may also have read the March/April 1840 publication of Auguste de la Rive concerning his method of gilding without the use of mercury by electrolysing, with a very weak current, a neutral solution of gold chloride contained in a bladder diaphragm.

In any event, improvements in the art of copper plating, copper electrotyping, and copper electroetching were coming with such rapidity, and so many hundreds of people were experimenting with electrodeposition, that the need for filing a patent to obtain an early priority date must have been obvious. (It is difficult to imagine the excitement that the discovery of copper plating and electroforming caused in Britain and the Continent. Copper plating literally swept across Britain). Dr. Alfred Smee, inventor of the first practical battery for electro-metallurgy, (a phrase that he coined), wrote ". . . there is not a town in England that I have happened to visit, and scarcely a street of this metropolis, where prepared plasters are not exposed to view for the purpose of alluring persons to follow the delightful recreation by the practice of electro-metallurgy."

Auguste de la Rive, a professor of physics at the University of Geneva, had worked on the problem of depositing gold without the aid of mercury as early as 1825. He had tried to duplicate the work of Brugnatelli, but he had little success. He returned to the work in the early part of 1840 and influenced by Antoine Becquerel, whom he credits, and perhaps Elkington, whom he doesn't, he electrolysed a neutral solution of gold chloride contained in a bladder which was immersed in a weak chloride solution containing a zinc anode. Several craftsmen in Geneva made limited use of his method; Hamman did gold resist etching and the Bergeons carried out gilding of watchcases, but found the process unsatisfactory for watch parts. De la Rive's process, published in April, 1840, was modified by Boettger, a chemist from Frankfort, to incorporate a salt patented by Henry Elkington in 1837. Although this modification produced better results, the process still did not give satisfactory results.

Becquerel wrote that de la Rive's process was not practical, which was also the opinion of Thomas Spencer, who probably shares the honour of the invention of copper plating with Jacobi and Jordan, and who wrote in August, 1840, "Since my processes have been published, Mr. de la Rive has adopted an inefficient method of gilding by the same process, by putting the metal to be gilt in a bladder with the solution of gold and having a piece of zinc attached in another solution." The French Academy, in November, 1841, also confirmed this view. When awarding a small part of the prize money (for a process to replace mercury gilding) to de la Rive, they added that although the deposit from his solution and apparatus had thickness, it lacked solidity and had no adherence. Nevertheless, August de la Rive deserves credit for the first published method (after Brugnatelli) for gold electroplating.

The major portion of the prize money referred to was divided between the Elkingtons and a French scientist, de Ruolz. Henry Catherine Camille de Ruolz was an opera composer and chemist born in Lyon in 1810. He did some chemical consulting work on fabric dyeing for a neighbour and friend, William Edward Chappee and through Chappee's brother, a jeweller, was presented with the problem of how to gold coat small objects without the use of mercury. Within a few months, on December 19, 1840, he applied for a patent for a gold bath based on gold oxide dissolved in potassium hydroxide. This was followed by a series of patents of addition dating from June 1841, claiming first, potassium cyanide and later, ferrocyanide and ferricyanide. (In the same month he also patented silver plating from a solution of potassium cyanide and later additions claimed ferrocyanide and other salts).

Meanwhile the Elkington's British patent of September, 1840, was divided; Henry Elkington patented the plating of gold from potassium cyanide solutions in France, while George Richards Elkington patented the plating of silver from potassium cyanide. Both French patents precede de Ruolz's first patent.

The Elkington patents were not well publicised in France, where de Ruolz was considered the inventor of gold and silver plating. A year later, when the French Academy was about to investigate de Ruolz's patent for the possibility of awarding it a part of the prize money, Henry Elkington's patent was first brought to their attention. John Wright was sent from England to demonstrate the process in the laboratory of Dumas, one of the members of the Academy. Henry Elkington's priority was thus established, but nevertheless, he had to share the prize with de Ruolz under protest.

The Academy, in its award citation, mentioned that potassium cyanide was expensive and difficult to conserve, and they doubted whether it would replace mercury gilding in practice. Elkington, through his agent, lodged a complaint with the Academy. He pointed out that 48 centimes* of potassium cyanide would dissolve one ounce of gold. He further pointed out that de Ruolz had almost three months to look over his patent before he filed his own and that his patents of addition incorporating cyanide did not appear until much later. Nevertheless, de Ruolz's prize and honour remained and he was accorded the more favourable "press".

From this point, the history of gold electroplating divides itself into several lines of independent or only slightly related developments. The best documented were those of the Elkington's commercial development in England, Christofle's jewellery factory in Paris, Max Leuchtenberg's quasi-governmental development in Russia and academic investigations that tended to ignore the commercial facts of the new and growing industry. Less well documented are the many independent workers, such as Werner von Siemens, who patented a gold plating solution in Prussia in 1841, the growth of gold and silver plating in the United States, where the Elkingtons did not patent the cyanide baths, and the many infringers of the Elkington's patents in France and England who show up primarily in legal records. (275 seizures of infringing solutions were made in France by the end of 1850).

The Elkingtons reportedly tried to license their gold and silver plating patents rather than exploit them themselves, but were unsuccessful due to the resistance of the Sheffield plate manufacturers who ignored them and would not take out licenses. Finally, in 1842, the Elkingtons, together with Josiah Mason, were forced to set up their own factories and licensed dealers. Silver electroplating had to struggle for survival for another 9 years in Britain, but gold plating was more readily accepted.

While Elkington and Mason pursued success in silverware manufacture, George Richards Elkington continued the production of gilt novelties, spectacle frames, pen nibs, etc. By using the new gold electroplating process, he was able to produce low cost items of such uniformity and overall quality that he soon dominated the entire trade in Birmingham. Henry Elkington continued his own business of plastic art work and gilding. Undoubtedly, the new electroplating process was introduced into his works because he, too, became a dominant factor in Birmingham. Within three years some 300 people were employed and approximately 600 gallons of silver and 50 gallons of gold plating solutions were in use. In the same period their first Sheffield licensee, John Harrison, paid them royalties on plated objects that were valued at £5,311.

In France, gold and silver plating was almost immediately a commercial success. Charles Christofle, who had purchased the patents of de Ruolz, apparently gave up the jewellery business and established an electroplating company bearing his name in 1842. (Shortly after commencing business, he took a license from Elkington and eventually bought the Elkington French patent). Since France has no "Sheffield" plate industry, electroplating was

---

* 48 centimes at this period of time was equivalent to 12 ¢ (U.S.) 6d. (U.K.).

Fig. 1.1 Christofle's gold and silver plating shop about the year 1845.

readily accepted. Christofle's fame for jewellery manufacture had made him many friends in both the military and government. This helped him to introduce his electroplated flatware to various official institutions. Universities, seminaries and hospitals quickly proved his electroplate under conditions of adverse wear and general public acceptance followed. Some years later when he began the manufacture of hollow ware, his electroplated and electroformed silver and gold were readily purchased by the cash short "fashion leaders" of French society and government. Christofle supplied the table setting consisting of more than 1200 pieces for the official receptions of Napoleon III, and by 1860, he was depositing 140,000 ounces of silver and 800 ounces of gold annually.

Tsar Nicholas I in Russia bought M. H. Jacobi's copper electrodeposition patent in 1839. Later, he set up his son-in-law, Maximilian Camille Napoleon, Duke of Leuchtenberg Romanov, in business as a copper, silver and gold electroplater. The Russian scientists gave little or no credit to Elkington and dismissed de Ruolz's plating work, but their innumerable "improved" gold formulations were in reality no different from those of Elkington and de Ruolz. It was, however, in Russia that large scale gold plating was first carried out. Max's Electroforming Casting and Mechanical Plant gold plated the domes and interiors of a number of cathedrals as well as copper electroforming and gold plating of an untold number of ikons and statues. The outside dome of St. Isaac's, in St. Petersburg, had been mercury gilded with the attendant deaths of 60 workmen from mercury poisoning. The inside was done by electroplating. The bases and capitals of the interior columns necessitated gold baths of over 1,300 (US) gallons, (5,000 l), the total project requiring 280 kg of gold. The work was first struck in a solution containing 0.41 g/l of gold followed by plating in a solution containing 10.7 g/l of gold.

In the middle 1850's, they gold electroplated copper plates for a cathedral dome that required 498 kg of gold. This appears to be the first time that a modern form of specification and quality control testing was used. The specification called for 28.44 g/m^2 of gold $\pm 20\%$ (0.000182 in. $\pm$ 0.000036 in.) and required that two plates out of every 100 be checked for gold thickness by a strip and weigh technique. If the two plates failed to meet the specification, the entire 100 plates were rejected. The capacity of the tanks for processing these plates was 1,500 (US) gallons, (5,677 l).

From the 1860's, gold plating grew in size as an industry but few, if any, important new scientific developments occurred. With the passing of the Victorian influence and the demand for grand objet's d'art, gold was relegated to the jewellery workshops where it was used only for inexpensive novelties and trinkets. Much of the technology required to handle large items, to electroform massive gold and to plate heavy deposits, was lost or forgotten. Apart from the empirical development of alloy coatings for the watchcase industry, virtually no new development work was done. This scientifically quiescent period lasted for almost 80 years.

In 1942, with the emergence of the electronics industry, a renewed interest in gold electroplating was evidenced by technologists and scientists. Requirements arose for heavy deposits of gold not requiring polishing or burnishing. Furthermore, it was necessary to deposit given thicknesses with accuracy and reproducibility. The 40's marked the development of chemical and electrical controls to ensure these results. The 50's brought the rediscovery of solutions and techniques for thick, uniform and fine grained gold deposits, and the introduction of modern bright acid and neutral gold solutions. Following this, in the 60's, acid gold and gold alloy systems were developed to produce deposits with specific physical properties such as ductility, corrosion and wear resistance, purity and so forth. During this period, the non-cyanide gold electrolytes were also reintroduced in bright and hard plating modifications.

Developments in gold plating in the early 70's pointing the direction for the next decade are both chemical and technological. New equipment is being devised to perform continuous plating of wire and strip and, most important, continuous plating that selectively deposits gold on the functional surfaces only. This results in tremendous savings in the amount of gold used and in the cost per unit. Stripe plated strip from which contacts and connectors are later stamped, and coils of incompletely stamped connectors selectively plated on critical surfaces only, are rapidly replacing the gold barrel plated connectors for many applications. Continuously plated leadframes with a gold spot plated on the portion to receive the chip, have virtually eliminated rack plated leadframes. Further economies may be envisaged from the use of more highly alloyed gold deposits, which are the subject of much recent evaluation. In this case the possible disadvantage of high contact resistance may be overcome by a final coating of pure gold.

Chemical developments are primarily directed toward increasing the speed of deposition, and increasing the levelling and throwing power of the baths. In modern plating machines, spots of gold 1.25 microns thick, are deposited in under 2.5 seconds. Future progress will undoubtedly lead to baths capable of depositing gold at even higher speeds.

Chapter 2

WHY USE GOLD?

A. M. Weisberg

Gold is electroplated not only for its psychological effects, but because of its unique combination of physical properties. Mythology, folk tales, the Bible and other written and oral records from all ages attest to gold's influence on the human animal. It has inspired greed, covetousness and ostentation. It has been the instigator of war, rape and murder. At the same time, it symbolically represents the sun, life, eternity, solidity, ultimate value and love. Although many countries have restrictions on the ownership of gold, it pervades our minds and lives. Gold is mentioned over 400 times in the Bible, and nearly 300 times by both Shakespeare and Tennyson. It is a family name in many countries of both the East and West. It's effect on many was recognised by Chilon in 560 BC, "Gold is tried by a touchstone and man by gold".

It is impossible to describe what makes humans respond to the warmth and beauty of this yellow metal, but they do in virtually every country in the world. In every age, people have desired to possess gold in the form of ornaments as well as bullion, but the high and increasing cost restricts private ownership of solid gold even when legally permitted. However, electroplating allows a small amount of the metal to be thinly and evenly distributed over a very large surface area so that it may simulate bulk gold and as such be used to decorate a variety of less expensive consumer items.

There are many other ways of coating objects with decorative gold besides electroplating, but the ancient methods such as gold leaf gilding with mastic, gilding on hot metal (French plating) and "gilding with the rag" required much skill and handwork. The most popular method, fire or mercury gilding, had severe toxicity hazards. Modern methods of mechanically bonding or cladding a sheet of gold or gold alloy on to a basis metal, with or without a solder, cannot produce deposits as thin as electroplating. They require relatively expensive equipment and are applicable only to flat stock strip or wire, but not intricately shaped individual pieces. Other methods, such as physical and chemical vapour deposition, explosive cladding and aerosol plating, are generally more expensive, cannot be easily adapted to bulk handling and are limited to specialty applications.

Jewellery, trinkets and other novelties may be electroplated with as little as 0.025 microns of gold and heavier deposits of 1.75 microns and 2.5 microns can be controlled to within 0.025 microns. Such coatings over nickel are widely used as decorative finishes on functional objects such as buttons, buckles, metal picture frames as well as bathroom, luggage and furniture hardware, tableware, cosmetic containers, boutique and boudoir decorations. Recently, thin gold deposits on stainless steel flatware have had a remarkable commercial success probably due to the fact that they connote the glamour of solid gold or silver gilt, yet at the same time offer good service life.

Heavier deposits of gold and gold alloys from 2.5–40 microns are used on moderate priced jewellery, pens, eyeglasses and watchcases. Although some of these applications use gold for its corrosion resistant properties, the decorative aspect is uppermost in importance. It is estimated that 34,000,000 Troy ounces of gold are used annually for the manufacture of jewellery and other decorative items, while about 200,000 Troy ounces are used in the form of decorative coatings by electroplating.

In the technical field gold is also electrodeposited in order to make use of its excellent characteristics in terms primarily of electrical, chemical and optical properties. Generally accepted values of the more important properties of bulk gold are listed in the following compilation, together with some characteristics of electrodeposited gold and the more frequently electrodeposited gold alloys. It should be borne in mind, however, that the properties of electrodeposited metal may differ from those pertaining to the bulk material, due to differences in structure, the presence of occluded material and so on. For a similar reason properties measured for electrodeposited alloys of the same nominal composition produced from different electrolytes may also show variations, depending on the plating conditions, method of measurement, etc.

PHYSICAL PROPERTIES

Atomic No: 79
Atomic Weight: 196.97
Naturally occurring isotope: 197
Crystal Structure: Face Centered Cubic, $a = 4.07$ Å
Closest approach of atoms: 2.88 Å
Density: 19.3 g/cm^3
Atomic Volume: 10.2 cc/g-atom
Melting Point: 1063°C (1945°F)
Boiling Point: 2970°C (5380°F)
Latent Heat of Vaporisation: 41.5 cal/g

THERMAL PROPERTIES

Specific Heat: 0.031 cal/g/°C (20°C)
Coeff of Linear Expansion: 0.0000144/C° (0–100°C)
Thermal Conductivity: 0.71 cgs units (20°C)

Table 2.1
Effect of Impurities on the Thermal Conductivity of Gold

Impurity	*Thermal Conductivity (cgs units)*
1% Silver	0.61
1% Copper	0.58
1% Palladium	0.65
1% Platinum	0.67
3% Zinc	0.282
3.1% Cadmium	0.296

ELECTRICAL PROPERTIES

Table 2.2
Effect of Impurities (alloy) on the Resistivity of Gold

	Resistivity (micro-ohm/cm)	*% Increase in Resistivity*
1% Silver	2.82	28.2
1% Palladium	2.9	31.8
1% Cadmium	3.07	39.5
1% Platinum	3.3	50.0
1% Copper	3.59	63.3
1% Indium	4.5	104.0
1% Zinc	4.93	124.0
1% Nickel	5.1	132.0
1% Tin	7.6	245.0
1% Cobalt	17.8	710.0
1% Iron	26.9	1220.0

Table 2.3
Contact Resistance
(Crossed Wire Method Using Burndy Resistometer)

	Purity (%)	*Contact Resistance (milliohms)*
High purity gold		0.20
Nickel-refined gold	99.95	0.43
Cobalt-refined gold	99.9	0.65
Silver-gold	95.0	0.46
Non-cyanide gold	99.9	0.33
Non-cyanide gold	99.0	0.39

Resistivity (18°C): 2.2×10^{-6} ohm-cm.
Temp. Coeff. of Resistance (0–100C°): 0.004/C°

* Relative Attenuation (copper = 1) : 1.19
In waveguide 0.400×0.900 ID (=3.2 cm)
= 0.139 db/metre (measured)
In waveguide 0.170×0.420 ID (= 1.25 cm)
= 0.6 db/metre (measured)

† Skin Depth:

1 cm	0.000018 in.	(4.53×10^{-5} cm)
10 cm	0.000057 in.	(1.43×10^{-4} cm)
100 cm	0.000180 in.	(4.53×10^{-4} cm)

‡ RMS Noise Voltage. Gold ring—graphite brush
(Pressure—11 g/cm² Linear speed (35 cm/sec)
Noise voltage = 0.6×10^{-6} volts RMS (0.5–200 cps)

§ Contact Erosion:
Cathode glow discharge in air. Weight loss
(Platinum = 1) : 0.886
Capacitive current. Weight loss
(Platinum = 1) : 1.114

Photoelectric Work function: 4.9 volts
Specific Magnetic Susceptibility (18°C): 0.14×10^{-6} cgs units.

CHEMICAL PROPERTIES

Valency: 1 and 3
Ionisation Potential: 9 ev
18.56 ev

Heat of Formation of AuO: −12.0 Kcal/g-mole
Resistant to: Sulphuric acid (100°C), fuming sulphuric acid, persulphuric acid, 70% citric acid (100°C), fuming nitric acid, 30% hydrochloric acid (100°C), perchloric acid, hydriodic acid, phosphoric acid (100°C), acetic acid, tartaric acid, citric acid, selenic acid (room temperature)
Resistant to: fluorine, dry iodine, sulphur (100°C), moist H_2S
Attacked by: Sodium and potassium cyanide plus oxygen, aqua regia, chlorine, chlorine water, bromine and iodine in alcohol, selenic acid (above 230°C).

Shelf Life Characteristics

These are best illustrated in graphical form (Figure 2.1) where various gold and alloy gold deposits are considered in relation to nickel, copper and silver basis metals.

---

* MIT Radiation Series, Vol. 9 (McGraw-Hill, New York).
† Author's calculation.
‡ Personal communication from James Nye Company.
§ Bell Telephone Laboratories.

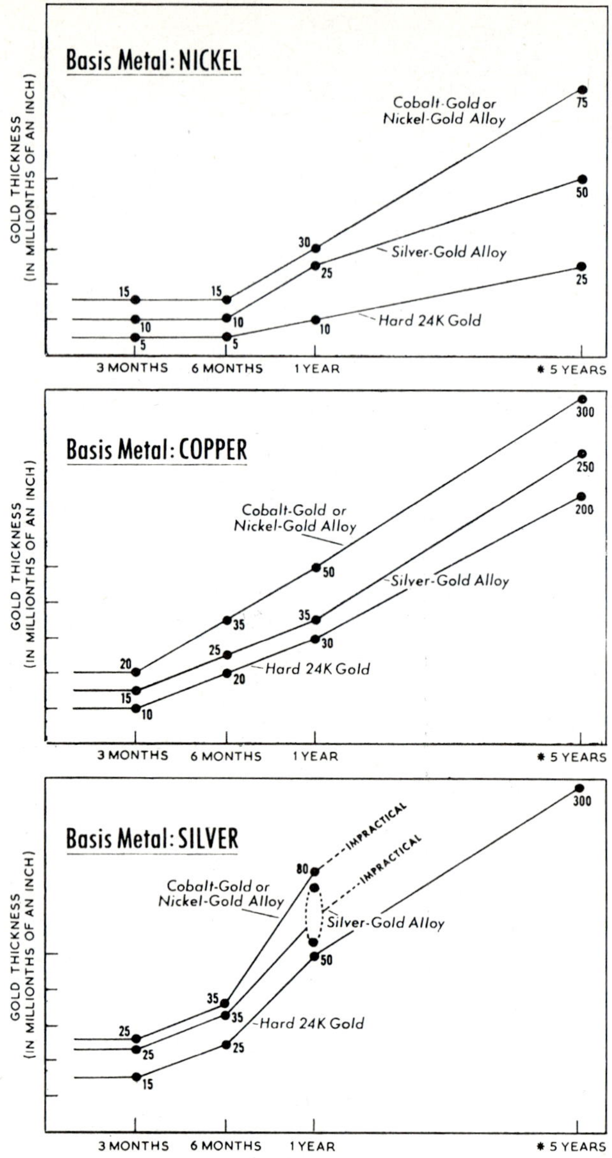

NOTE ALL VALUES APPROXIMATE.
OF GREATEST IMPORTANCE IS FINISH OF BASIS METAL ETC.
* DEPENDS ON STORAGE ATMOSPHERE.

Fig. 2.1 A guide to gold plating shelf life for three basis metals.

METALLURGICAL PROPERTIES

Hardness

Table 2.4
Hardness of Electroplated Golds

Type	*Gold Content* (%)	*Hardness* $(KHN)_{25}$
Ultra-pure	99.99	45–85
Cobalt-refined	99.9	140–160
Cobalt-Gold	99.6–99.8	200–245
Nickel-refined	99.95	110–120
Nickel-Gold	99.9	130–140
Nickel-Gold	99.8	140–160
Nickel-Gold	99.5	200–260
Non-cyanide	99.9+	130–150
Non-cyanide	99.0	210–250
Silver-Gold	95	110–130

OPTICAL PROPERTIES

Reflectivity

The optical properties (% reflection of incident radiation) are presented graphically in Figure 2.2. Silver and rhodium are included for comparison.

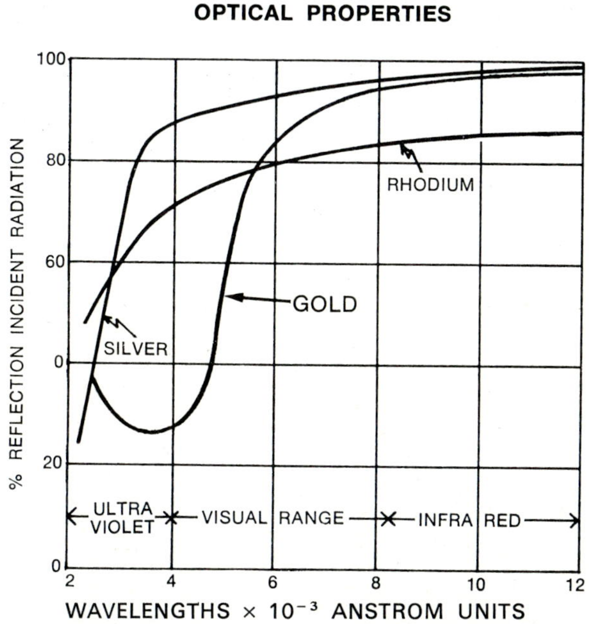

Fig. 2.2 % Reflection incident radiation of gold, silver and rhodium at various wavelengths.

SUMMARY

The high electrical conductivity of gold, the low contact resistance and the good solderability, combined with the constancy of these properties due to the chemical inertness of the metal, make gold the ideal choice for many items of electrical and electronic hardware. Contacts, terminals, connectors, conductors, chassis, printed circuit boards and shields are routinely plated with gold or gold alloys. The grain refined bright deposits of gold and its alloys or electroplated admixtures of base metals and gold, allow the electronic engineer to alter the hardness of the gold deposit to give long wearing and abrasion resistant surfaces for moving and sliding contacts. The addition of 0.1% to 1.0% or more of other metals to a gold deposit modifies other properties besides hardness such as porosity, apparent rate of diffusion of basis metals through the gold layer, and the electrical and contact resistance (refer to properties and shelf life charts in this chapter).

The good electrical properties combined with the excellent ductility of gold make it ideal for flexing or vibrating current carrying components. The resistance of gold to the formation of oxides, sulphides and other corrosion products suit it to applications on safety devices, alarms and high reliability switches. Gold is also used on reed relays, waveguides and other RF conductors, grid wires and glass seals. Transistors, headers and multipin headers make use of both pure and doped golds for eutectic die bonding, resistance welding, formation of beam leads and making *n*, *p* and ohmic contacts to semiconductors.

The excellent reflectivity of gold, particularly in the infra-red region and the mirror brightness of modern electroplating formulations that do not require polishing or buffing, determine electrodeposition as the ideal method of applying gold for all but the most special reflectors and mirrors. A monomolecular layer of oxide is thought to form on the surface of gold, but as it is transparent, the optical properties of gold are stable over a wide range of environments, including space.

For highly contoured or intricate shapes and for parts that must be treated in bulk, electroplating is the method of choice for applying gold since it is the most inexpensive and flexible. Mechanical cladding of a stripe (inlaying) previously had some economic advantage, but now continuous stripe and spot electroplating has been developed to a point where it is faster (2.5 microns in just under 5 seconds from a 12 g/l bath) and often less expensive to produce as there is no gold clad scrap to contend with. Electroplating is also clearly indicated where multilayers of different types of gold are to be applied and where precise thickness control is required, particularly for deposits less than 5 microns in thickness.

PART 2

Electrolytes

Chapter 3

INTRODUCTION TO ELECTROLYTES

D. G. FOULKE

Gold, like silver, requires only a very small negative potential for reduction from the ionic to the metallic state, hence in solutions containing free ions, non-adherent immersion deposits are rapidly formed on base metals. Deposits from electrolytes of this nature are of large grain size with a tendency to "tree". Gold is therefore invariably plated from electrolytes in which the metal ions are complexed.

HISTORICAL

Gold plating has been carried out in alkaline cyanide electrolytes ever since the inception of commercial gold plating in the 1840's[1]. The alkaline cyanide bath had limitations with respect to the deposition of hard alloy golds of constant colour because rigid control of the free cyanide was necessary to permit the continuous and uninterrupted codeposition of nickel, cobalt and other alloying constituents which were themselves complexed by the cyanide ion. Consequently, gold plating over the years became more of an art than any other plating system. Moreover, the picture became complicated due to the fact that the baths were frequently prepared from gold chloride. In this case, addition of sodium or potassium cyanide leads to the formation of auricyanide.

$$AuCl_3 + 4CN^- \rightarrow Au(CN)_4^- + 3Cl^-$$

In alkaline solutions the auricyanide is reduced, slowly or rapidly, depending on the conditions, to the aurocyanide.

$$Au(CN)_4^- + 2e \rightarrow Au(CN)_2^- + 2CN^-$$

The reaction may well be more complex than indicated, but the end result is the same, namely, that a certain amount of auricyanide is likely to be present in baths prepared from the chloride. Failure to appreciate this led to many conflicting reports in the literature. No such ambiguity is encountered when solutions are prepared by simply dissolving gold anodically in potassium or sodium cyanide solution, when the aurocyanide is formed directly.

$$Au^\circ + 2CN^- \rightarrow Au(CN)_2^- + e$$

The mechanism may well be more complicated than shown, but this is unimportant. What is important is that for many years, both $Au(CN)_2^-$ and $Au(CN)_4^-$ were present in the baths prepared from gold chloride. This led to all sorts of confusing reports in the literature.

However, utter confusion did not entirely prevail. Fritz Volk[2], for instance in 1949, recognised the importance of maintaining free cyanide at a low value. By this time it was standard practice to add gold to the bath as $KAu(CN)_2$ instead of using gold anodes (in which case free cyanide is required) and by lowering the pH to between 6.5–7.5, he was able to plate bright alloy deposits. Nevertheless, difficulties were still experienced due to build-up of free cyanide which made effective daily control troublesome.

Rinker[3] in 1951 discovered that silver in a solution with a high cyanide content was more noble than gold, and developed the first foolproof bright gold solution. The coatings had a brassy cast and, consequently, were of limited appeal to the jewellery industry. The relatively low hardness of the deposits and the high alkalinity of the electrolyte also militated against the use of this process in many applications.

A major technological advance occurred when Rinker and Duva[4] in 1957 discovered that $KAu(CN)_2$ was stable at a pH as low as 3.0. This eliminated the presence of free cyanide and, as a result, nickel, cobalt and other metals could be codeposited with gold, in an uninterrupted and controlled manner, to give pale shades and hard deposits.

More recently, the industry has returned to non-cyanide baths and there is widespread interest in gold sulphite complex solutions[5] which offer some advantages in effluent control as well as good possibilities for alloy deposition.

Although all texts on gold plating mention the acid chloride type of electrolyte, this is in fact of negligible importance in the commercial gold plating field. Its principal use is the electrolytic refining of gold by the Wohlwill[6] process, in which impurities such as the platinum group metals and silver are removed by anodic dissolution of an impure gold anode in an electrolyte containing 25–40 g/l of gold and a small excess of free hydrochloric acid. Platinum metals which dissolve remain in solution and are not deposited at low concentrations, while silver is precipitated as the chloride. Gold of 99.5 % minimum purity is deposited in coarse crystals on the cathode of fine gold sheet. This type of electrolyte has been used for electroforming, but its highly corrosive nature is a severe practical disadvantage.

ELECTROLYTE CLASSIFICATION

For purposes of the more detailed discussion of electrolytes in the following chapters, a convenient classification is as follows:

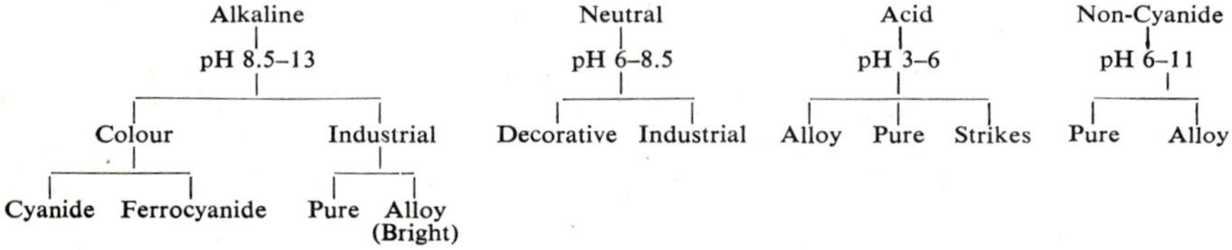

Alkaline electrolytes in general differ from the closely related silver baths in that they usually contain phosphates and a rather low free cyanide content. However, certain authors recommend 4 g/1 of potassium hydroxide and as much as 120 g/1 of free potassium cyanide.

The standard electrolyte has always contained sodium or potassium gold (I) cyanide, so that all but the few baths discussed under the subsequent non-cyanide classification, have been cyanidic in nature. Even the ferrocyanide solutions, described in a precious metal text[7] as "developed primarily for non-toxic electrolytes" were essentially $KAu(CN)_2$-$KAu(CN)_4$ electrolytes and as such, are not to be recommended as a substitute for beer or wine at lunch.

Acid gold plating solutions moved rapidly into the picture in the 1960's and are now used for a substantial proportion of gold plating throughout the world. The pH of such solutions, based almost exclusively on $KAu(CN)_2$, has been arbitrarily set at 3.0–6.0, although the alloy type of solution works best at a pH of 3.0–5.0. Besides potassium aurocyanide, these baths contain conducting salts, such as the salts of organic or inorganic acids, in particular citric and phosphoric acids. In addition to this, they also contain complexes and chelating agents of base metals to make possible the codeposition of such base metals with the gold.

The neutral gold plating baths were developed in Germany and Switzerland where there was a particularly high demand for low carat decorative alloys for watches and jewellery applications. The patent and operating manual evidence states quite clearly that these baths should be operated at pH values between 6.0–8.5, this depending upon the specific process used.

Although the non-cyanide baths have been known for some considerable time, it is only within recent years that they have achieved any commercial significance.

Table 3.1 indicates the principal use of each type of gold plating process.

Table 3.1
Applications for Various Types of Gold Electrolyte

Type	*Applications*
Alkaline	
(a) Colour	Jewellery
(b) Industrial	Electronic
Acid	
(a) Pure	Headers, flatpacks, etc.
(b) Alloy	Connectors, printed circuits
Neutral	
(a) Decorative	Jewellery, watchcases
(b) Industrial	Headers, flatpacks, etc.
Non-Cyanide	
(a) Pure	Connectors, printed circuits
(b) Alloy	Connectors, watchcases, Switches, printed circuits, etc.

Applications for gold plating, along with structure properties and methods for measuring properties, are covered in a critical bibliography (1940–1968) published in 1970 by the American Electroplaters' Society[8].

In the following chapters, each of the major types of gold plating electrolyte which have been briefly described in the preceding paragraphs, will be discussed individually and in considerably more detail.

REFERENCES

1 Foulke D. G., *Metal Finish.*, **67,** 4, 50 (1969).
2 Volk F., U.S. Pat., 2,812,299 (1957).
3 Rinker E. C., private communication (1951); Rinker E. C., U.S. Pat., Reissue 24,582 (1958).
4 Rinker E. C., Duva R., U.S. Pat., 2,905,601 (1959).
5 Smith P. T., U.S. Pat., 3,059,789 (1962).
6 Wohlwill E., *Electrochem. Metallurg. Ind.*, **2,** 221 and 251 (1904).
7 Fischer J., "Galvanische Edelmetallueberzuege", Leuze, Salgau/Wurt (1960).
8 Niehoff R., Faust C. L., Cady J., Freberg C., Am. Electropl. Soc. Res. Rep. No. 58, Jan. (1970).

Chapter 4

ALKALINE CYANIDE ELECTROLYTES

D. G. FOULKE

Alkaline cyanide electrolytes may be used for the production of pure or alloy gold deposits. The most suitable classification, however, is on the basis of applicational areas, i.e., colour gilding (jewellery) and industrial uses. Colour gilding is based entirely on alloy plating while industrial applications may call for pure or alloy gold deposits depending on the technological requirements. There is also an immense difference in control of the two types of process. The deposition of colour golds remains much of an empirical art, in contrast to the close control necessary to meet present day specifications in the industrial gold plating field.

COLOUR GOLDS

CYANIDE ELECTROLYTES

Colour golds are of somewhat limited interest in that only thin "flash" coatings are normally deposited, and the amount of gold used in this context represents only a very small percentage of the total consumed in electroplating. To reduce inventory and minimise drag-out, colour gold baths are formulated with less than 4 g/l of gold, and are invariably operated at elevated temperatures to achieve bright deposits. This applies even to the green golds, though these have recently been shown to produce bright coatings at 20–30°C.[1] High temperature operation results in lower bath voltage, which causes difficulties in obtaining a deposit of consistent carat. These are further aggravated by the increase in free cyanide concentration which normally occurs during operation.

The colour of the gold alloy deposit does not depend solely upon the composition of the electrolyte. The operating current density (voltage), temperature and the degree of agitation also play important roles. Table 4.1. shows the effect of various alloying additions on the colour of the deposit, i.e., nickel or cobalt (light yellows) silver (green) and copper (red). However, cadmium will also produce a greenish colour and nickel, under certain conditions, a rose colour. It is therefore obvious that in principle an unlimited variety of colours can be obtained, but it is difficult to maintain a specific colour on a continuous basis, so agitation and current density control (really voltage) have become an integral part of the art of the colourer. Table 4.1 is sufficient to indicate the general possibilities in the present state of the art.

Table 4.1
Typical Colour Gold Plating Electrolytes

Constituents	*Bath Composition (g/l) and Operating Conditions*			
	Green	*Yellow (pale)*	*Yellow*	*Red*
$KAu(CN)_2$	4	4	4	4
$KAg(CN)_2$	1.9–2.2			
KCN	2–7	4	1.5	3–7
$K_4Fe(CN)_6$		30		30
CuCN				1.5
$K_2Ni(CN)_4$		1.8		2.0
Temperature, °C	50–70	60–70	60–70	60–70
Current density, A/dm^2	1–2	2–4	2–3	2–4

In some instances, the type of current is also varied. Several German processes employ cyclic variation of the applied direct current voltage, while a method developed in the UK utilises an intermittent DC cycle.[2]

Because colour gilding has been so much of an art, it is difficult for the colourer to set down clearly the rules to follow, even if he wished to do so. Certainly in recent years, the metal has been added as the sodium or potassium aurocyanide instead of using gold anodes to maintain the gold content. For "flash" plating solutions the choice of alkali cyanide is not critical.

The free cyanide content is, however, important when trying to achieve specific colours. Although free cyanide improves conductivity and throwing power, it also complexes the base metal ions and must be carefully monitored in order to obtain consistency of colour.

Phosphates and carbonates improve the conductivity and throwing power. If the bath is an all-sodium solution, the limited solubility of sodium gold cyanide can adversely affect the successful operation of the bath, particularly if the gold content is high. In this instance, the use of potassium gold cyanide is recommended.

A knowledge of the relationship between the current density and the cathode potential (polarisation curves) is useful when studying the deposition of a single metal, depolarisers sometimes permitting the use of higher current densities at lower potentials. Where a metal can co-exist in more than one valency state, such relationships are of particular interest, as in the case of $Au(CN)_2^-$ and $Au(CN)_4^-$ ions which is discussed later in the text.

Raub[3] has carried out a great number of studies in this field and many of the subsequent curves are taken from his papers.

Figure 4.1 shows the variation of cathode potential with current density for a simple potassium gold cyanide—potassium cyanide bath which has a pH of about 11.0, as well as for the acid system.

The current density-cathode potential relationship may be very different when base metal complexes are present, depending principally on the type of

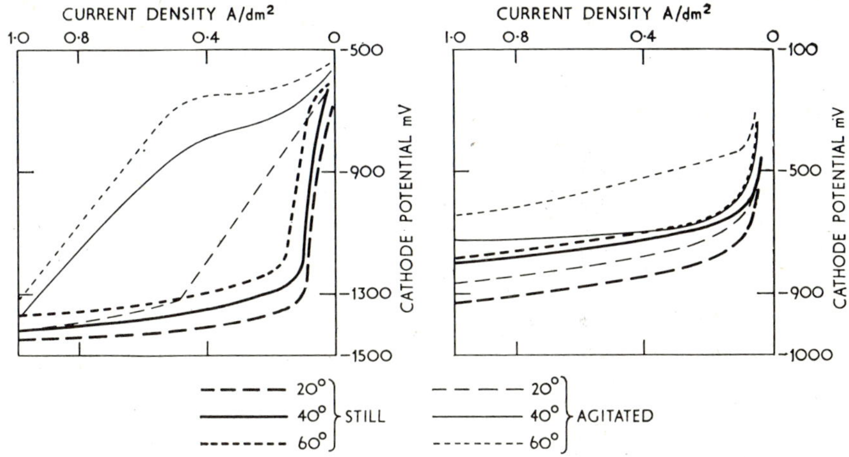

Fig. 4.1 Cathode potential—current density curves for $KAu(CN)_2$ electrolytes. (a) Au 2 g/l, KCN 8 g/l, pH 11.0 (b) Au 2 g/l, citric acid 40 g/l, sodium citrate 60 g/l, pH 4.2.

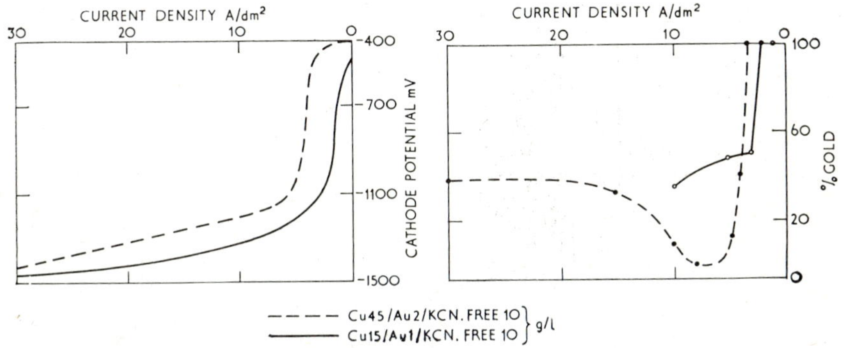

Fig. 4.2 Cathode potential—current density curves for two copper–gold cyanide electrolytes and variation of carat.

complex which forms the intermediate compound from which the discharge takes place under electrolysis, the ratio of the metals to be codeposited and their total concentration.

The current density-cathode potential curves for two simple gold-copper systems are smooth (Figure 4.2). However, the gold content of the deposit fluctuates considerably with the current density. It is reported[3] that these copper-gold alloys were not deposited as a solid solution, which accounts for the poor tarnish resistance of alloys from these baths. When the pH is almost neutral, however, a considerable proportion of the deposit is solid solution.

Gold-silver cyanide baths are much more straight forward. The argento-cyanide complex is considerably weaker than the aurocyanide and moreover, in cyanide solutions silver behaves as a more noble metal than gold. However, the codeposition of gold takes place considerably below the limiting current density for silver and solid solutions are easily formed over a wide range. The alloy composition changes with conditions in the normal manner, i.e., increased polarisation, low temperature and high current density favouring a higher gold content in the alloy, whereas increased temperature, low current density and agitation increase the amount of silver in the deposit. Although colour gold solutions are normally used at elevated temperatures, present day commercial gold-silver baths are usually operated at room temperature, with good agitation and high free cyanide. Under these conditions when the silver content of the bath is 1.0% of the gold content, a 99% alloy is formed and if the silver is ten times higher, a 90% gold-silver alloy is obtained at 0.3–0.5 A/dm^2.

Zinc and cadmium have also been used in colour golds, both as binary alloys and, in the green silver-gold baths based on Au-Ag-Zn and Au-Ag-Cd, as ternary alloys. These two metals form relatively weakly bound cyano-complexes and can be codeposited with gold in spite of being relatively much less noble. Figures 4.3 and 4.4 show the current density-potential curves for typical zinc- and cadmium- containing baths. At about 1 A/dm^2 the deposited alloys contain approximately 30% zinc and 25% cadmium.

The curves illustrated in Figures 4.3 and 4.4 cover binary alloys only. In colour gold systems, it is not uncommon for the deposit to contain two base metals. For example, the red gold solution in Table 4.1 contains both copper and nickel. It has been reported[4] that if gold and copper cyanide complexes alone are used, there is an abrupt change from yellow to pink as the copper builds-up in the solution. However, if either nickel, cadmium or silver is present, the change is gradual and better control is therefore possible.

To maintain the free cyanide at a fairly constant value for pink gold, Parker[4] suggests using as much as 15–30 g/l of free cyanide. In general, however, this is kept below 7 g/l, because at high temperatures there is considerable breakdown with the formation of cyanide polymers which contaminate the solution. The higher the free cyanide, the more accentuated this problem becomes.

The pH value of colour gold solutions is normally about 11.0. If higher, the

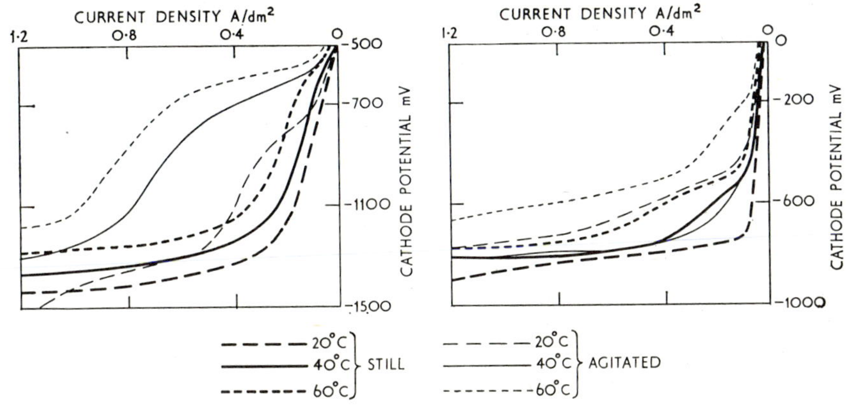

Fig. 4.3 Cathode potential—current density curves for zinc–gold electrolytes. (a) Au 2 g/l, Zn 5 g/l, total KCN 24 g/l, KOH 12 g/l, citric acid 40 g/l (b) Au 2 g/l, Zn (sulphate) 5 g/l, citric acid 40 g/l, sodium citrate 60 g/l, pH 4.2.

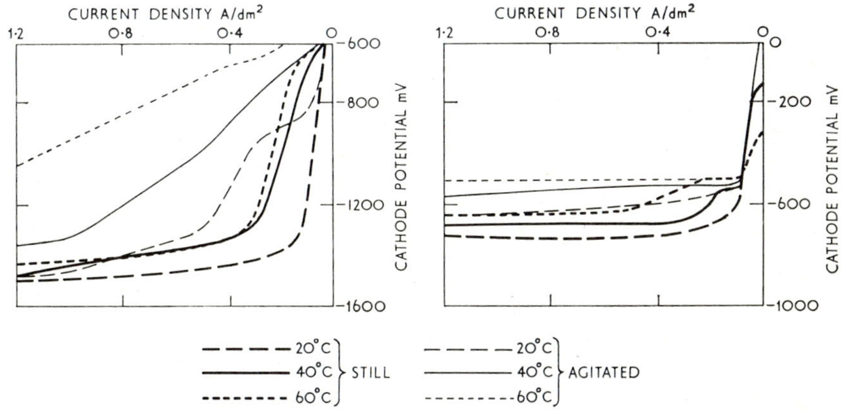

Fig. 4.4 Cathode potential—current density curves for cadmium–gold electrolytes. (a) Au 2 g/l, Cd 2.5 g/l, total KCN 18–34 g/l, KOH 4–10 g/l (b) Au 2 g/l, Cd 1.2 g/l, citric acid 40 g/l, sodium citrate 60 g/l, pH 4.2.

bright range is affected and if lower, cyanide polymers may build-up in heavily worked baths.

FERROCYANIDE ELECTROLYTES

In addition to the sodium or potassium gold cyanide electrolytes used for decorative plating, early workers employed "ferrocyanide" solutions. These were in reality potassium gold cyanide and not ferrocyanide solutions as presupposed.

The preparation of gold baths using potassium ferrocyanide as a reactant with gold chloride goes back some 150 years[5], when alkali cyanides were impure and expensive, while potassium ferrocyanide was fairly pure and cheap. This type of solution was resurrected in 1954 by Wogrinz[6] and in 1961 by Mornheim[7]. The overall equation according to the latter is:

$$6HAuCl_4 \cdot 3H_2O + 14KOH + 4K_4Fe(CN)_6 \cdot 3H_2O + O_2 \rightleftharpoons$$

$$6KAu(CN)_4 + 4Fe(H_2O)_3 \cdot (OH)_3 + 24KCl + 22H_2O$$

This might well be considered a somewhat lengthy and illogical procedure for preparing a cyanaurate solution when one considers that the same essential product can be obtained by simply adding potassium cyanide to a gold chloride solution. Some hydrogen cyanide is liberated, but this can be controlled by proper pH adjustment

$$AuCl_3 + 4CN^- \rightarrow Au(CN)_4^- + 3Cl^-$$

In this case, there is no need to remove hydrous iron oxide by filtration.

The chemistry of solutions has been somewhat confused over the years. Wogrinz and Mornheim state that potassium cyanaurate (III) is decomposed by hydrochloric acid[7] to give aurous cyanide and gold. However, potassium cyanaurate (III) is stable in mineral acid solutions, the pK of potassium cyanaurate (III) being 56.

This point has been clarified by Knödler and Raub[8]. Although $KAu(CN)_4$ is stable at a pH below 7.0, if the solution is rendered alkaline and heated, considerable $KAu(CN)_2$ is formed and then the bath is not stable when hydrochloric acid is added. Baths containing potassium auricyanide only, however, are stable at very low pH values.

Publications as recent as 1966[9,10] have not clearly stated the composition of the ferrocyanide gold electrolytes. However, as early as 1843 Graeger[11] recommended that an excess of potassium ferrocyanide and potassium hydroxide should be avoided. This led to a neutral plating bath which was superior to the alkaline solutions generally described. He noted that the neutral plating bath performed better if made strongly acid with sulphuric acid prior to use. This is the first reference to an acid gold cyanide electrolyte, in this case the gold being present as potassium auricyanide (III).

Besides the auricyanide (III) most of the ferrocyanide baths also contain the monovalent gold cyanide. Langbein[12], as early as 1842 stated that the gold chloride should be added to a boiling solution of ferrocyanide and sodium carbonate, after which it should be boiled for a further 15 minutes. This would rapidly transform the gold to the aurous state:

$$2OH^- + Au(CN)_4^- \rightarrow Au(CN)_2^- + CN^- + CNO^- + H_2O$$

Some formulators however, neglected to mention that the bath had to be heated. Consequently, the transformation to aurocyanide was slow and the baths contained gold in both valency states, plating out at two efficiencies. It is therefore not surprising that the literature is very confusing and that the colourer experienced so many difficulties.

There is relatively little published information comparing the behaviour of $Au(CN)_4^-$ and $Au(CN)_2^-$ complexes. Knödler and Raub[8] have reported that the cyanaurate (III) complex is so stable in acid solutions that the reaction which causes the precipitation of gold cyanide (AuCN) in *N* sulphuric acid solution takes place only after eight days of heating at 95°C. Boiling concentrated hydrochloric acid slowly converts $Au(CN)_4^-$ to the chloroauric complex, $AuCl_4^-$. On the other hand, it is well known that the cyanaurate (I) complex is rapidly decomposed to AuCN at a pH value below 3.0.

Figure 4.5 shows a current density-potential curve for the two complexes in acid and cyanide conditions. In still solutions there is a well defined limiting current density region above which hydrogen is evolved. The value for cyanaurate (III) is about three times that for the monovalent complex. Since three times the current is required for the reduction of $Au(CN)_4^-$ compared to $Au(CN)_2^-$, it would appear that both complexes have similar diffusion constants.

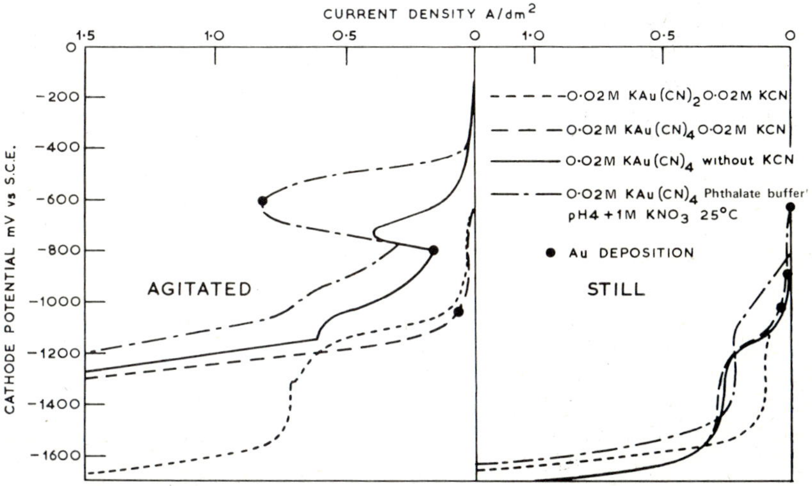

Fig. 4.5 Cathode potential—current density curves for cyanaurate (I) and cyanaurate (III) electrolytes.

It is of interest that for cyanaurate baths free of potassium cyanide, a definite potential must be reached before gold is deposited. Below this potential, $Au(CN)_4^-$ is being reduced to $Au(CN)_2^-$ and at more negative potentials direct reduction to metallic gold takes place. In actual practice the ferrocyanide solutions are operated at rather high voltages and current densities.

The ferrocyanide solutions have been used to deposit both 24 carat golds and colour golds, particularly the latter, but this system is now rarely used.

The same additions as those given in Table 4.1 are used to obtain the various colours. However, the amount of alloying constituent has to be modified because the $Au(CN)_4^-$: $Au(CN)_2^-$ ratio varies at any given time. Allowances for this can only be made on the basis of experience.

The comments on composition and function of the constituents discussed under colour golds are generally applicable to the ferrocyanide golds in that the baths, in most cases, are primarily $KAu(CN)_2$ solutions. However, the presence of auricyanide complicates matters, especially in terms of free cyanide. For each mole of gold electroplated from an auricyanide bath, four moles of cyanide are released, compared to two from an aurocyanide bath. This, of course, complicates matters with respect to the codeposited base metal and places more reliance upon the experience of the colourer. However, a properly prepared electrolyte, with almost no excess free cyanide, will function well initially. The ferrocyanide solutions are not well documented in the literature with regards to specific operating instructions. It would certainly be advisable when setting up a colour gold plating facility to use the modern bright acid gold cyanide plating baths instead of the ferrocyanide and to consider depositing more than a flash of gold onto the basis metal.

INDUSTRIAL GOLDS

These processes are of two types, (a) pure golds and (b) bright alloy golds, both of which are used by industry primarily for functional purposes as opposed to purely decorative applications. This distinction will be apparent in other chapters, but nowhere is the divergence quite so clear as in the alkaline golds. Timing, i.e., the requirements of the electrical and electronics industry, has had much to do with the fact that in recent years the more modern processes have been clearly documented. Consequently, during the last decade, industry other than the jewellery trade has used both pure and hard golds as functionally required. Previously, the electronics industry had had no choice but to use soft golds because it could not rely upon the colour golds which tended to vary considerably in quality and could not be electroplated consistently to a specified thickness. More recently the physical properties of gold deposits have become of the utmost importance.

Industry was slow to accept solutions containing large amounts of gold. By the late 1940's, however, an industrial bath containing 22.5–26 g/1 of gold was recommended.[13] The free KCN was 50–60 g/l with 5 g/l KOH also present.

In the next decade Bauer[14] recommended gold at 10 g/1, with vanillin and potassium sulphate added for grain refinement. This brought gold plating more in line with silver plating practice, except that the operating temperature was still about 80°C. In 1960, Seegmiller and Gore[15,16] increased the gold content to 20 g/1, held the KCN content at 70 g/l and added turkey red oil. The recommended operating temperature of the bath was 50°–70°C, which was more and more approaching silver plating practice.

This was the status of alkaline gold plating in the early 1960's when Rinker and Duva[17], Ehrhardt[18], and Ostrow and Nobel[19], disclosed the acid gold systems, which have now been widely accepted.

ELECTROLYTES FOR PURE GOLD DEPOSITS

The pure alkaline systems have not changed much since the early 60's except for the application of ultrasonics[20], the use of potassium hydroxide,[21] an addition agent[22] and the use of ammonium gold cyanide[23] to minimise salt built-up in the bath. Operational aspects of high purity gold plating baths in recent years have gradually more and more resembled those of high speed base metal electrolytes, i.e., use of higher metal content and temperatures, continuous purification (special absorbents to remove cyanide polymers) as well as gold recovery from the rinses to give effective recovery of metal from the higher drag-out.

The alkaline industrial gold solutions in current practice which are formulated primarily to give 24 carat, may be as low as 23.5 carat or 98% gold. 2% copper, for example, would render gold totally unacceptable for semiconductor applications. As a result, the operating conditions and composition of industrial (pure) gold solutions are much more limited than for colouring solutions.

Typical high purity gold plating solutions disclosed in the literature are given in Table 4.2. There are, of course, many possible modifications.

The gold is usually added as potassium gold cyanide and the concentration is generally strictly controlled. This is also true for the conducting salt to ensure satisfactory current distribution and efficiency. The efficiency is usually close to 100% at 60–70°C. Current densities vary with the application, a typical figure being 0.5 A/dm^2. In barrrel work about $\frac{1}{3}$ of the surface is being plated at any given time, so the current density for the full load will be 0.15–0.2 A/dm^2.

Many of the solutions in practical use for high purity deposits are operated in the alkaline pH range. A number of these are proprietary baths, whereas others are "captive" operations, so it is rather difficult to assess bath compositions except in a general way. Conducting salts may be orthophosphates, polyphosphates, salts of organic acids, borates, carbonates and so forth. Frequently, chelating agents are added to complex the impurities which tend to contaminate the deposit. Maintaining the free cyanide at 15 g/1 and higher will complex most impurities when operating at low current densities in the region of 0.1–0.3 A/dm^2.

Table 4.2
High Purity Alkaline Gold Electrolytes

Constituents	*Bath Composition (g/l) & Operating Conditions*			
	Ref. 20	*Ref. 21*	*Ref. 24*	*Ref. 32*
Au (as $KAu(CN)_2$)	56+	40	4–500	4–12
Potassium cyanide	60	7.5		30
Conducting salt*	60–125		15	30
Potassium hydroxide		40		
Potassium carbonate				30
Wetting agent†	0.1%			
pH	10.5–11.5	> 13	6.1–10.5	> 12
Temperature, °C	60	75	70	50–65
Current density, A/dm^2	1.0	4.5	0.1–0.8	< 0.5

* $(NH_4)_2HPO_4$, K_2HPO_4, sodium potassium tartrate, etc.
† Pentafluoro–octanoic acid, alkali salt.

The deposit thickness is of the order 2–3 microns, but may be very much thicker if required, hence the structure of these deposits can be studied. The high purity golds have large grains as shown in Figure 4.6 (plated from Bath No. 4., Ref. 32, Table 4.2). The hardness of the high purity golds is of the order of 65 Knoop.

ELECTROLYTES FOR BRIGHT ALLOY GOLD DEPOSITS

In the area of alkaline alloy gold deposits, the bright to semi-bright deposits of Rinker[1,25], Ostrow and Nobel[26], and Raub[27] involve the codeposition of silver, tin and antimony (Table 4.3).

The (1–2%) silver-gold deposit is particularly interesting because this represented the first commercial bright gold plating process. Although the high free cyanide was beneficial in that most elements which might plate out were complexed and not codeposited, the alkalinity of the electrolyte was a serious drawback for the plating of printed circuits.[18] A further disadvantage was that the yellow deposit had a faint greenish tinge and Hamilton shades were not possible. Fischer[28] applied for a patent for a very similar process in Germany in 1949, which included the use of an acrolein-carbon disulphide reaction product as an addition agent. Figure 4.7 shows the fine structure of deposits from this bath.

The baths containing antimony found considerable use in the electronics industry, but the application to *n* type transistors was actually much less than the literature suggests.

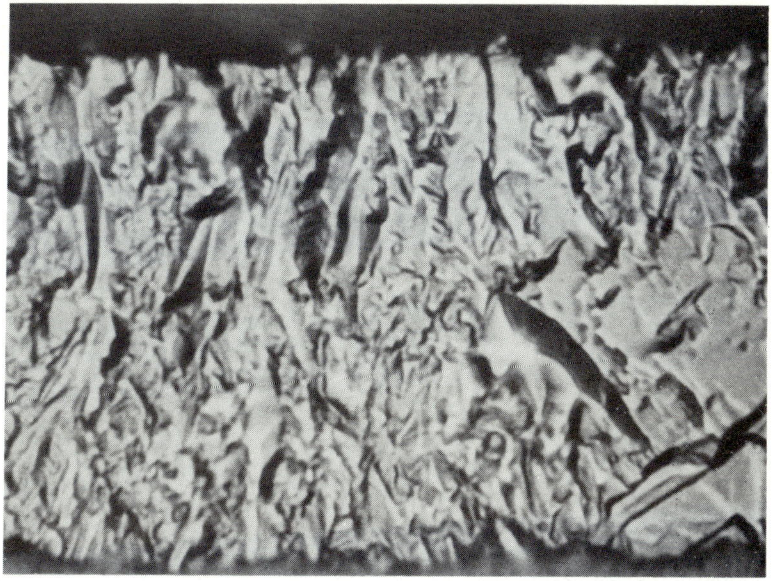

Fig. 4.6 Structure of deposits obtained from the hot gold cyanide electrolyte, Bath 4, Ref. 32, Table 4.2. (1500 ×).

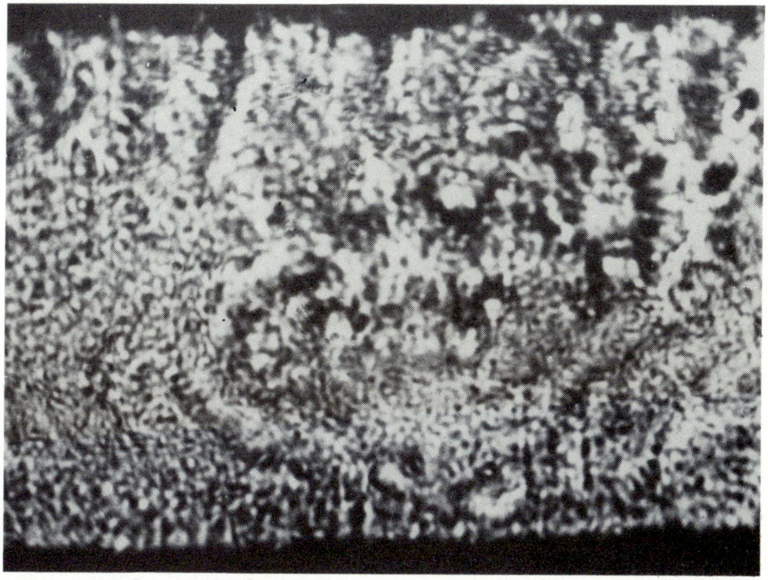

Fig. 4.7 Structure of deposits obtained from the bright silver–gold electrolyte, Bath 1, Ref. 28, Table 4.3. (1500 ×).

Other references to alloy deposits from alkaline cyanide baths include claims that uranium and molybdenum have been codeposited with gold[29] and that nickel-copper-cadmium-palladium-gold alloys can be electrodeposited[30]. Neither process has experienced any commercial success and, because palladium forms a very stable cyanide complex, Pd-Au alloys cannot be codeposited on a production basis from baths replenished with potassium gold cyanide, contrary to claims that it is possible.

Table 4.3
Alkaline Industrial Gold Electrolytes

Constituents	*Bath Composition (g/l) & Operating Conditions*			
	Ref. 1	*Ref. 26*	*Refs. 25 & 27*	*Ref. 28*
Au(as $KAu(CN)_2$)	8	7.5	20	4
Ag(as $KAg(CN)_2$)	0.125			0.4
Potassium cyanide	90	80	120	80
Potassium hydroxide			80	
Potassium carbonate				20
Potassium nickel cyanide	5–30	3		
Antimony tartrate			3	
Potassium stannate		4		
Sodium thiosulphate	5–35			
Brightener*		20^a		4^b
Wetting agent†				0.3
Temperature, °C	room	70–90		20–70
pH	< 12	< 12	< 13	< 12
Current density, A/dm²	0.3–0.5	0.1–2		1–14

* *a* = n-methyl glycine reaction product, *b* = CS_2-acrolein product.
† polyglycol esters.

In a recent summary of gold and gold alloy plating processes[31], a large number of solutions currently available are listed. The majority of solutions are either neutral or acid and it would appear that the trend for baths yielding high purity deposits is now directed towards the neutral pH range.

REFERENCES

1 Rinker E. C., U.S. Reissue 24, 582 (1958).
2 Gardam G., Tidswell N., *Trans. Inst. Metal Finish.*, **31,** 418 (1954).
3 Raub E., *Metall*, **21,** 709 (1967).
4 Parker E. A., *Plating*, **38,** 1134 (1951).
5 Frary C., *Trans. electrochem. Soc.*, **23,** 25 (1913).
6 Wogrinz R., *Metalloberfläche*, **8,** B162 (1954).
7 Mornheim A., *Plating*, **48,** 1104 (1961).
8 Knödler A., Raub E., *Plating*, **53,** 765 (1966).
9 Shenio B. S., Gowrie K., Indira K. S., *Metal Finish.*, **64,** 5, 54 (1966).
10 Lainer L., Kudryavtsev N., "Fundamentals of Electroplating", Part II, Translated U.S. Dept. Commerce (1966).

11 Graeger *J. Prakt. Chemie*, **30,** 343 (1843).
12 Langbein G., *Dinglers polytech. J.*, **88,** 30 (1842).
13 Shell J. S., Lauletta J. R., Mulligan J. J., "Metal Finishing Guidebook & Directory", pp. 123, Metals & Plastics Publications, Inc., Westwood, New Jersey (1968).
14 Bauer C. L., *Plating*, **39**, 1335 (1952).
15 Seegmiller R., Gore J. K., *Proc. Am. Electropl. Soc.*, **47,** 74 (1960).
16 Seegmiller R., Gore J. K., U.S. Pat., 2, 801, 960 (1957).
17 Rinker E. C., Duva R., U.S. Pat., 2, 905, 601 (1959).
18 Ehrhardt R. A., *Proc. Am. Electropl. Soc.*, **47,** 74 (1960).
19 Ostrow B. D., Nobel F. I., U.S. Pat., 2, 967, 135 (1961).
20 Schneider E., Lidell M., U.S. Pat., 3, 427, 231 (1969).
21 Edson G., U.S. Pat., 3, 445, 352 (1969).
22 Marlow E., Brit. Pat., 913, 356 (1962).
23 Greenspan L., U.S. Pat., 3, 637, 473 (1972).
24 Camp E. K., U.S. Pat., 3, 397, 127 (1968).
25 Rinker E. C., U.S. Pat., 3, 020, 217 (1962).
26 Ostrow B. D., Nobel F. I., U.S. Pat., 2, 765, 269 (1956).
27 Raub E., German Pat., 849, 787 (1952).
28 Fischer J., U.S. Pat., 2, 800, 439 (1957); Appl'n Germany (1949).
29 Taormina S., Marinaro A., U.S. Pat., 2, 754, 258 (1956).
30 Campana C., U.S. Pat., 2, 719, 821 (1955).
31 Anon, *Metal Finish. Plant & Proc.*, **6,** 3, 87 (1970).
32 Duva R., Korbelak A., "Metal Finishing Guidebook & Directory," pp. 284, Metals & Plastics Publications, Inc., Westwood, New Jersey (1971).

Chapter 5

NEUTRAL ELECTROLYTES

D. G. FOULKE

DECORATIVE

This type of electrolyte was originally developed to provide 14–18 carat alloy gold deposits. Typical examples are the solutions developed by Volk[1] and by Spreter and Mermillod,[2] both of which processes have been extensively used in the plating of watchcases and jewellery.

The baths of Volk, and Spreter and Mermillod have been used to plate millions of watchcases (5–20 microns) and jewellery of an even greater magnitude. Both processes, however, are suffering severe competition from newer solutions which have been developed.

VOLK GOLD ELECTROLYTE

Volk[1] recognised that the free cyanide concentration should be kept low so that base metals could be codeposited with gold. By buffering the solution with phosphates at a pH of 6.5–7.5 he was able to prevent a large build-up of cyanide in baths fed with potassium gold cyanide. Electroplating personnel familiar with this bath still operate it successfully to obtain 14–18 carat pink deposits which are bright up to a thickness exceeding 20 microns. An example cited in the patent describes a bath containing:

Gold (as $KAu(CN)_2$)	7 g/l
Sodium phosphate, Na_2HPO_4	28 g/l
Copper (as $Na_2Cu(CN)_3$)	7 g/l
Iron (as ferrocyanide)	3 g/l
pH	6.8–7.5
Temperature	65–75°C
Current density	0.5–1 A/dm^2

The baths most widely used omit the ferrocyanide and use small amounts of the double cyanide of silver and/or cadmium.

Raub[3] gives polarisation curves for copper-gold baths at a pH of about 7.0 (Figure 5.1). He found that at a neutral pH and low free cyanide concentration, the alloy deposit was a solid solution. It is known that deposits from the Volk bath possess good tarnish and nitric acid resistance, which confirm the deposit to be a solid solution and not individual crystals of gold and copper. This is apparently because of the bath pH and the presence of cadmium and/or silver cyanide complexes. The Volk process has experienced wide usage in Europe, consequently the operating conditions are well documented. The bath, when put into operation must be adjusted to a pH of 7.2 with phosphoric acid (20%), followed by a 2 g/l addition of potassium cyanide, the bath pH then being re-adjusted to 7.2. Plating commences at 2 volts and the voltage is then rapidly increased to that required to give the desired colour. The current density at this stage is about 0.5 A/dm^2. The

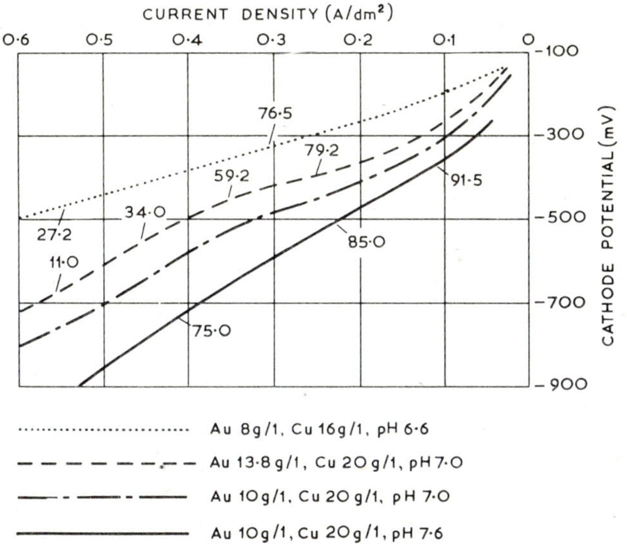

Fig. 5.1 Polarisation curves for neutral copper–gold baths (figures appearing on the curve denote percentage gold present in the alloy).

temperature is maintained at 80°C and should not exceed 85°C. The pH should never fall below 7.0 and the free cyanide content must be carefully monitored by adding about 1 g/l of potassium cyanide each morning and 0.5 g/l each afternoon, the exact amount depending upon the temperature, the area to be plated and the colour desired. In any event, the pH should be adjusted to 7.2 before the addition of the potassium cyanide.

The replenisher contains a silver complex, obviously the double cyanide, and copper. The copper addition is stated to be in a ratio of 1 : 4 with respect to the gold, whereas the silver addition is 3–5% of the gold content. The deposit is an 18 carat alloy, but by careful control of the base metal constituents, the deposit can be as low as 15 carat.

SPRETER AND MERMILLOD GOLD ELECTROLYTE

Spreter and Mermillod,[2] employed copper EDTA and potassium gold cyanide together with conducting and buffer salts to hold the pH under 8.5. A typical formulation is:

Gold (as $KAu(CN)_2$)	0.7–1.5 g/1
Copper (as copper EDTA)	8 g/1
Na_2EDTA (free)	16 g/1
pH	8.5
Temperature	53°–57°C
Current density	1.0–1.5 A/dm^2

EDTA decomposition products necessitate regular activated carbon treatment of the bath. In order to obtain consistent corrosion resistance and harder (400 HK_{25}) coatings, the deposits are often heat treated at 300–400°C to ensure the presence of Cu_3Au and CuAu with no free copper in the deposit.

Rochat[4] presents Debye-Sherrer diagrams of the Spreter-Mermillod deposit which indicate a two-phase system, with free gold and copper absent. On the other hand, these deposits could be slightly attacked by nitric acid indicating the presence of some very rich copper alloy. After heat treating at 300°C for three hours he reported an increase in crystal size. Furthermore, the lines are no longer diffused (perhaps due to stress relief) and the structure becomes polyphase. In this condition the deposits were resistant to nitric acid.

The Spreter-Mermillod gold bath of Pierre De Robert et Cie de Geneve[5] is probably the best documented gold plating process ever released, in that De Robert issued a 55 page brochure in 1957 describing the characteristic properties of the deposit, including levelling and X-ray studies. Unfortunately, little was said about the operating conditions of the process. Deposits of 16–21 carat were studied with respect to corrosion resistance, but this was not related to solution composition, so it is not possible to supply reliable operating information on these baths. It is, however, known that the bath has levelling properties and the deposit is laminar (Figure 5.2).

Other neutral alloy gold baths available for decorative plating are essentially modifications of the previous two examples. They are currently being widely displaced by acid alloy gold plating solutions, where the colour and composition of the deposit can be maintained more easily, and by the new cyanide-free electrolytes.

INDUSTRIAL

A very high proportion of high purity gold plating is carried out in barrels, neutral electrolytes generally being preferred for this purpose as best meeting the requirements of good throwing power and metal distribution, brightness or smoothness, and temperature resistance of the deposits. The superiority of neutral electrolytes in this context has been demonstrated by Nobel and Thomson[6] in a study of barrel plating of transistor headers from a variety of electrolytes, including proprietary solutions of pH 5.0, 7.0 and 8.0, an ammonium citrate solution at pH 5.0, alkaline phosphate and phosphate/cyanide baths at pH 10.0 and a potassium cyanide-potassium carbonate bath at pH 11.0.

In general, good throwing power, an almost neutral pH and an electrolyte of the correct composition are the essential requirements to satisfactorily plate transistors and integrated circuits with high purity gold.

Industrial neutral baths yielding high purity deposits can be described more conveniently on tne basis of patents. Table 5.1 lists the composition of a number of these solutions. In general, conductivity salts do not differ from those used at a higher pH. Differences, however, exist in that small quantities of addition agents such as arsenic are introduced as grain refiners and new

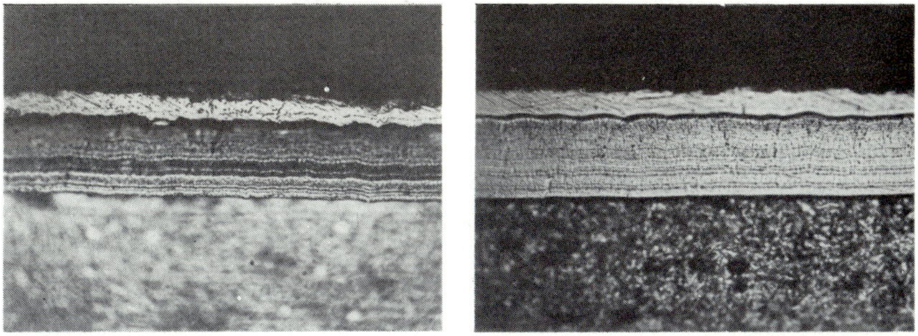

Fig. 5.2 Structure of a typical copper–gold alloy deposit from the Spreter–Mermillod bath. *Left*: before heat treatment. *Right*: after heat treatment at 300°C. (100×).

types of complexing agents such as organic phosphonic acid derivatives are used to reduce the codeposition of base metal contaminants with gold to less than several parts per million.

Zimmerman and Brennerman[7] (Table 5.1) describe a solution containing orthophophosphate and citrate at a pH of 6.0–6.5. In addition to trivalent arsenic, they also use a thio-compound such as a thiosulphate, mercaptan, etc. Deposits from this type of electrolyte contain about 0.1% arsenic which results in a small grain size.

The citrate/organic phosphate gold solution developed by Hodgson and Szkudlapski[8] based on potassium instead of ammonium salts, is claimed by the authors to have several advantages over the conventional ammonium citrate/gold cyanide type of acid gold solution described by Ehrhardt.[9] The addition of ethyl phosphates gives brighter, smoother deposits, at efficiencies of about 95% at 65°C. Furthermore, the surface tension of the solution is reduced to 52 dyne/cm, with the throwing power being comparable to that of an alkaline cyanide bath. Alkyl phosphates are also claimed to have excellent ability to complex metallic impurities such as copper. Use of potassium rather than ammonium buffer salts was stated by the authors to give better stability of bath pH, and has a practical advantage in eliminating the inconvenience of ammonia evolution when gold is recovered from spent or contaminated baths by precipitation with zinc dust under hot alkaline conditions.

Table 5.1
High Purity Neutral Gold Electrolytes

Constituents	*Bath Composition (g/l) & Operating Conditions*				
	Ref. 7	*Ref. 8*	*Ref. 9*	*Ref. 10*	*Ref. 11*
Au(as $KAu(CN)_2$)	10–31	10–20	7–18	6	8.2
Potassium dihydrogen phosphate	60		82		
Potassium pyrophosphate					150
Citrate	60*	60–125†	50–75‡	90§	
As(as $NaAsO_2$)	0.02–0.04				
Sodium thiosulphate, pentahydrate	5–10				
Ethyl dihydrogen phosphate		30–60			
Amino trimethyl phosphonic acid				80	
Benzyl alcohol (optimal)					0.05 %/wt
pH	5.5–8.0	6–8	5–6.5	6.0	7–8
Temperature, °C	60	70	45–100	65	60
Current density, A/dm²	0.1–1.5	0.1–0.3	0.1–0.4	0.1–0.5	0.1

* potassium salt † potassium or ammonium salt ‡ ammonium salt § water soluble citrate.

Addition of potassium sulphate[8] (44 g/l) has been used to increase the specific conductance of neutral solutions to 0.178 $ohm^{-1}\ cm^{-1}$ at 65°C, which is some 30% higher than that of commercial acid gold formulations, and favours improved deposit distribution.

Ostrow and Nobel[10] have divulged organic phosphonate compounds while Pokras[11] prefers to use pyrophosphates.

REFERENCES

1 Volk F., U.S. Pat., 2,812,299 (1957).
2 Spreter V., Mermillod J., U.S. Pat., 2,724,687 (1957).
3 Raub E., *Metall*, **21,** 709 (1967).
4 Rochat R., *Bull. Ann. Soc. Suisse Chronomet Lab.*, Suisse Recherches Horl., **4,** 11, 45 (1957).
5 deRobert P., Le Plaque OR Thermocompact, Geneva (1957).
6 Nobel F. I., Thompson D. W., *Plating*, **57,** 469 (1970).
7 Zimmerman R., Brennerman R., Brit. Pat., 1,275,386 (1972).
8 Hodgson R. W., Szkudlapski A. H., *Plating*, **57,** 693 (1970).
9 Ehrhardt R. A., *Proc. Am. Electropl. Soc.*, **47,** 78 (1960).
10 Ostrow B. D., Nobel F. I., Brit. Pat., 1,198,527 (1970).
11 Pokras D. S., U.S. Pat., 3,505,182 (1971).

Chapter 6

ACID ELECTROLYTES

D. G. FOULKE

Acid gold plating solutions may be classified in the following three categories, (a) alloy* gold solutions (b) pure gold solutions, and (c) strike solutions. The latter can be further sub-divided into solutions based on aurocyanide and auricyanide respectively.

It is now recognised that advantages accrue in principle from reduction in the free cyanide content and pH of gold plating solutions in permitting more readily the codeposition of base metals such as nickel and cobalt, with possible improvements in hardness and other properties of the coatings. However, this possibility was relatively late in practical realisation since it is well known that copper and silver cyanide plating baths could not be operated without decomposition at pH values below about 5.5 and 6.5 respectively and it was therefore naturally assumed that gold cyanide electrolytes would be subject to a similar limitation.

A strong impetus for the development of acid gold solutions was provided by the rapid growth of the printed circuit industry during the late 1950's. Here the problems associated with the gold plating of edge contacts, using conventional alkaline electrolytes, were essentially twofold:

1 Attack of the adhesive bond between the copper foil and laminate by high temperature, high pH gold solutions.

2 Limited wear resistance of deposits due to the relative softness of the pure gold coatings.

The bright plating process developed by Rinker[1], based on codeposition of a small amount of silver, went some way towards solving some of these difficulties by providing a harder coating and reducing the deposition temperature, but this did not provide a complete solution.

The real breakthrough came with the discovery, independently by Rinker[2] and Ehrhardt[3], that potassium cyanuarate (I) remained stable in acid media down to a pH as low as 3.0 before decomposition occurred with slow precipitation of aurous cyanide. Atwater[4] had in fact earlier suggested a solution with a pH of 1.8, but this was too low for commercial use with potassium cyanaurate (I) as replenisher.

---

*It should be noted that "alloy" golds in this context do not necessarily refer only to highly alloyed deposits, but to those produced from solutions to which base metals are deliberately added to improve the brightness and physical properties of the deposits. The amount of alloying may be quite small.

ELECTROLYTES FOR GOLD ALLOY DEPOSITION

The first acid gold solution based on potassium cyanaurate (I) was reported by Rinker and Johns[5] and details were disclosed in a subsequent patent.[6] This type of bath comprised essentially of a solution of potassium gold cyanide together with the salt of a base metal in an aqueous medium buffered to a pH of 3–5 by an organic acid/salt mixture, usually citric acid/citrate. In addition to its buffering effect on pH the citrate forms complexes with base metals such as nickel and cobalt, from which metal ions are released for codeposition in a controlled manner to give consistent colour (composition) of the deposit.

A number of other complexing agents have been claimed in the literature, probably the most commonly used being amines such as tetraethylene pentamine[7] and EDTA[8,9]. Typical formulations of this type from the patent literature are shown in Table 6.1.

Table 6.1
Acid Alloy Gold Electrolytes

Constituents	*Bath Composition (g/l) & Operating Conditions*		
	Ref. 6	*Ref. 7*	*Refs. 8 & 9*
Au(as $KAu(CN)_2$)	8	4	12
Citric acid	40	120	105
Sodium citrate	40		
Tetraethylene pentamine		20	
Phosphoric acid			12.6 ml
In(as $In_2(SO_4)_3$	5		
Ni(as $Ni_3C_6H_5O_7$)		2.5	
Co(as CoK_2EDTA)			1
pH	3–5	4.0	3–4.5
Temperature, °C	Room+	40	35
Current density, A/ft^2	5–20	20	5

Almost all of the bright acid gold processes depend on the codeposition of a small amount of base metal, e.g., nickel, cobalt, indium, which is added as a salt, complex or chelate to provide sufficient base metal ions in the cathode film[10]. The relative concentrations of base metal and free complexing agent are important. For instance, in the case of cobalt some excess of uncomplexed cobalt salt must be present for codeposition from EDTA solutions since the metal is not deposited from the cobalt-EDTA complex. Cobalt complexes with pK values lower than that for the EDTA compound will release sufficient ions to permit controlled codeposition, to an extent depending on the pK value for the complex and the amount of free complexing agent present.

In contrast, solutions without complexing agents in the acid pH range release base metal ions very rapidly. For example, acid phosphate cyanaurate (I) solutions give white deposits even when only very small amounts of cobalt or nickel are present, the amount of codeposition being hardly controllable in the absence of an effective buffer for base metal ions. On the other hand, certain complexing agents, such as tetraethylene pentamine, become more effective at a pH above 5.0 and they can therefore extend the bright plating range somewhat. However, most bright acid gold plating solutions are consistently operated in the pH range of 3.5–5.0. The advantage of the organic acid/salt solutions lies in the fact that this milieu acts as a buffer both for hydrogen ion and base metal ion activity, hence control of the process is relatively simple.

At the present time, less difference exists between decorative and industrial acid alloy plating solutions than was the case prior to 1960. It is now possible to deposit coatings of up to about 10 microns with fairly good control of colour, composition and properties to suit both types of application. The bright acid gold cyanide baths are limited to depositing thicknesses not normally exceeding 10 microns because a certain amount of cyanide-citrate polymer[11] is codeposited with the gold. The mode of formation and constitution of the latter is not clearly understood. It is present only in negligible amounts in deposits from this type of solution in the absence of metallic hardening additions, but has been isolated in the form of a fragmentary colourless transparent film from deposits produced in a proprietary electrolyte, containing 15–20% of nickel as a hardener. The presence of this polymeric material probably accounts for the abnormally low densities which characterise this type of deposit (see Chapter 24) and has also been suggested as a factor contributing to their generally good wear resistance.

ELECTROLYTES FOR DEPOSITION OF PURE GOLD DEPOSITS

In the early development of acid gold plating electrolytes, emphasis was on the production of bright deposits of enhanced hardness, which was achieved by deliberate additions of base metals. In the mid 1960's interest turned to the production of high purity gold deposits to meet the requirements of semiconductor applications, in which field the presence of metallic impurities such as copper, is deleterious to the successful functioning of devices.

The first baths developed were similar to those used in bright acid plating, but with the omission of base metal additives, i.e., organic acid/salt systems buffered in the pH range of 3.0–5.0. These solutions were also able to complex small amounts of base metals introduced into the solution as adventitious impurities, though less effectively than the high cyanide baths. The types of electrolyte in current use are best summarised from the patent literature, of which a number of formulations are presented in Table 6.2. These include three[3,14,18] modifications of the original high purity acid gold plating baths (Bath No's 2, 3 & 6, Refs 3, 14, 18) two only[14,15] including the use of phosphate as the conducting salt (Bath No's 3 & 4, Refs 14, 15).

ADDITIVES

It is obvious from the patents issued that there are three principal types of additive associated with high purity acid gold electrolytes, viz., complexing agents, grain refiners and hardening agents.

Complexing agents are added to reduce the activity of metallic impurities present in the solution by forming stable complexes and hence minimising codeposition. Typical examples include the pyrophosphate ion, organophosphorus compounds and polyphosphates. Organic chelating agents are also widely used, but these compounds, including EDTA and related compounds, are not well documented in recent patent literature because this approach has been part of the "art" for many years and as such is not generally patentable.

For grain refinement relatively small amounts of base metals may be used which exert sufficient effect to lead to smoothing and semi-brightening of the

Table 6.2
Pure Acid Gold Electrolytes

Constituents	*Bath Composition (g/l) & Operating Conditions*					
	Ref. 2	*Ref. 3*	*Ref. 14*	*Ref. 15*	*Ref. 16*	*Ref. 18*
Au(as $KAu(CN)_2$)	8	4–30	4	8	2.7	16
Potassium phosphate			45	100		
Citric acid	40	40				55
Sodium citrate	40					135
Ammonium citrate		40	7.5			
Potassium pyrophosphate					10	
Superphosphoric acid (42% pyrophosphoric)					10	
Arsenious oxide					0.2	
Thallium (as sulphate)				0.0025		
Hydrazine sulphate						6
pH	3–6	2.3–6.6	2.0	3–6	5.5–6.0	4.2
Temperature, °C	Room	30–60	—	20–90	50–65	70
Current density, A/dm^2	0.5–1.0	0.1–0.8	1.0–2.0	0.1–1.0	0.5–1.0	1.0

deposit, but are not codeposited to a significant extent, so that the response of the coating to thermal test requirements is not affected. As in the case of neutral baths, arsenic[16] can be used for this purpose (Bath No. 5, Ref. 16, Table 6.2), but the amount codeposited is not disclosed. Thallium has also been suggested[15] (Bath No. 4, Ref. 15, Table 6.2). This additive is claimed to improve metal distribution as well as the colour and brightness of the deposit.

Certain additives are claimed to harden the electrodeposited gold without being codeposited. These include alums[17] and hydrazine sulphate[18] (Bath No. 6, Ref. 18, Table 6.2), and the hardening effect may simply result from their influence on the grain size of the deposits, similar to the effect of potassium cobalticyanide additions to alkaline cyanide baths.

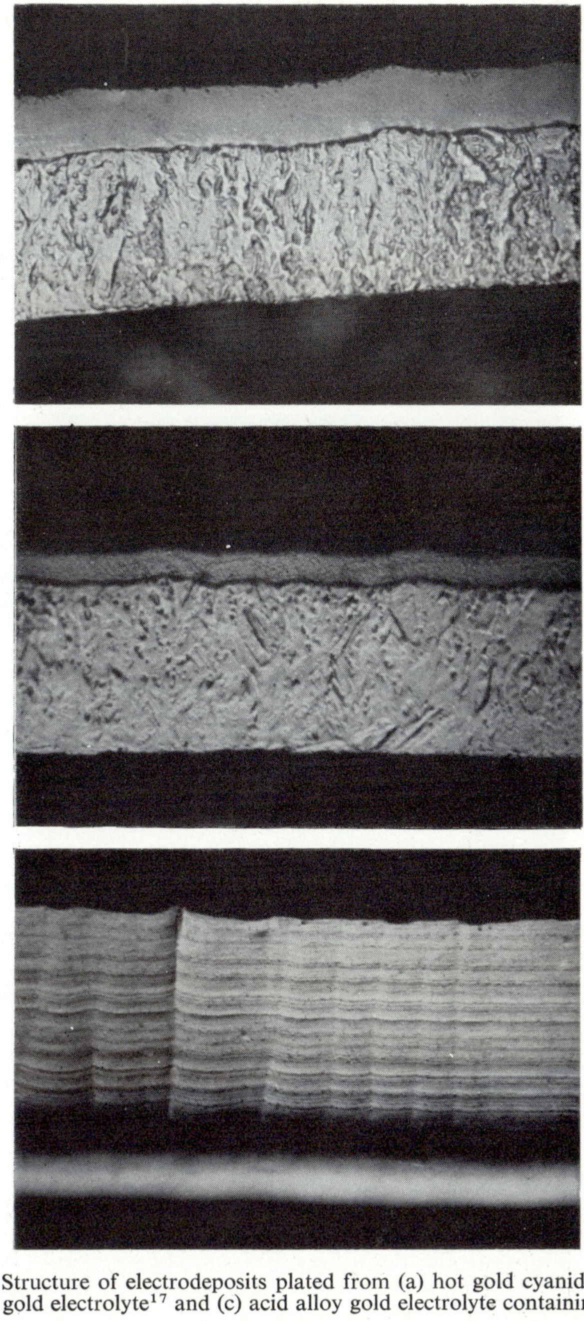

Fig. 6.1 Structure of electrodeposits plated from (a) hot gold cyanide bath (b) pure acid gold electrolyte[17] and (c) acid alloy gold electrolyte containing nickel[2]. (1250×).

STRUCTURE OF GOLD DEPOSITS FROM ACID ELECTROLYTES

The fine grained structure of deposits from electrolytes containing base metals, as compared with the columnar structure of pure gold deposits is illustrated in Figure 6.1.

X-ray data of Raub[12] and Foulke[13] show that gold-nickel deposits are single-phase, the diffuseness of lines in the diffraction pattern indicating lattice distortions which are in line with the known stresses in gold-nickel and gold-cobalt deposits. Raub[12] states that it is possible to obtain alloy deposits of any composition from weakly acid or acid gold solutions. This author worked with solutions of low gold content, but even with the more usual concentration of 8.2 g/l, deposits containing up to at least 35% nickel can be obtained which, however, are very brittle and stressed.

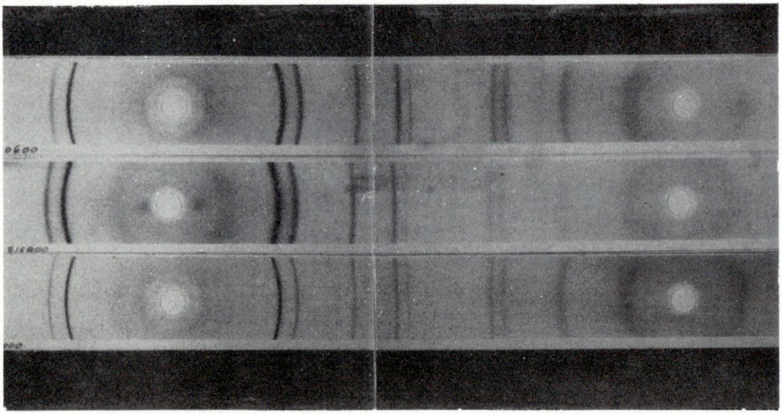

Fig. 6.2 Debye–Scherrer diagrams of deposits from acid gold electrolytes containing (a) cobalt 500 ppm (b) cobalt 2.5 g/l, indium 1 g/l (c) nickel 2.5 g/l, indium 1 g/l.

Debye-Scherrer diagrams for a number of deposits are shown in Figure 6.2.[13] Powder samples for this work were produced by dissolving the basis metal (copper) from 2 micron deposits by means of ammonium trichloroacetate and pulverising the residual coating. Copper radiation with a nickel filter was employed.

The structure of all deposits studied was face centred cubic with lattice parameters very close to that of pure gold. No evidence of a second phase or of non-alloyed base metal was apparent. Line broadening, especially of the back reflections, is a good indication of this condition and there appears to be little difference in this respect between the five gold alloy deposits studied. However, there are certain limitations to the use of this method to determine unambiguously whether isolated base metals are present in the crystal.

Craig[19] and his co-workers have reported the appearance and structure of electrodeposits from fourteen gold electrolytes, ten of which were acid solutions. A comparison was made with deposits produced from hot gold cyanide electrolytes (see Table 4.2), the surface microstructure of which showed very large grains (60–120 μm) separated by open grain boundaries which penetrated deeply into the deposit (Figure 6.3). Most of the acid gold cyanide deposits showed non-uniform grains of size from 5–50 μm with sharper grain boundaries (Figure 6.4). Deposits from baths containing a metallic brightener[16] showed further refinement of grain size (1–5 μm), as illustrated in Figure 6.5.

Craig *et al*[19] used Berg-Barrett X-ray topology[20] to determine the maximum crystallite size of deposits. Deposits from baths with no impurities present show a crystallite size of 0.1 μm or more, but solutions containing impurities or addition agents yield deposits of much smaller crystallite size (0.05 μm at the solution side of a 127 μm deposit). At the basis metal/gold interface the crystallite size in general levels off and is not very different from bath to bath ($\sim$0.04 μm).

X-ray diffraction studies showed that the pure acid gold cyanide baths produce deposits with an initial [111] orientation which, as deposition progresses, is replaced by higher order orientations [311], in agreement with the theory that close-packed orientations, forming from the slowest growing crystal face, are natural for heterogeneous nucleation. The [111] orientation is also converted to a [311] type (more random) when brighteners are added to the solution.

STRIKE SOLUTIONS

A high proportion of the acid gold strike solutions now in use are proprietary. They contain potassium gold cyanide with phosphoric acid or an organic acid and a conducting salt formed by partial neutralisation of the acid. The pH is generally within the range of 3.0–6.5 and the gold content is normally 2 g/l, but may be much bigger when the solution is used for difficult-to-cover items such as transistor headers. Base metals may also be present.

In the late 1930's an acid gold strike solution was described by Lukens[21] for the plating of stainless steel pen nibs, and this solution is still in limited use for this purpose. Essentially an acid chloride electrolyte, the formulation is as follows*:

Sodium gold cyanide	6.25 g/l
Sodium cyanide	50 g/l
Hydrochloric acid	450 ml/l

* *Editors Note: Extreme care would be necessary in the preparation and use of this solution due to the hazard of hydrogen cyanide evolution. All operations should be conducted under an effective fume hood.*

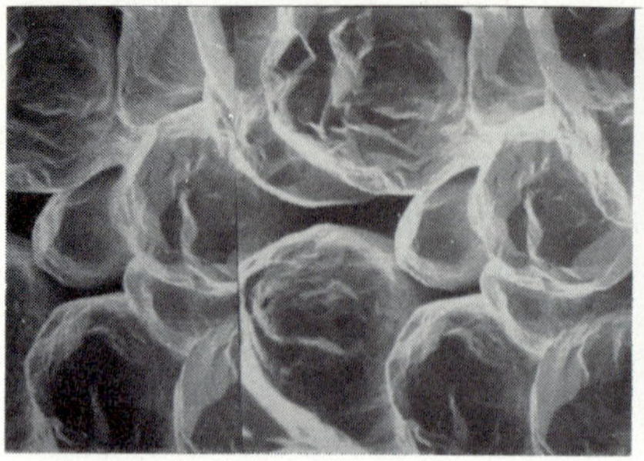

Fig. 6.3 Stereo-electron photomicrograph of a deposit from the hot cyanide electrolyte. (Ref. 32, Table 4.2).

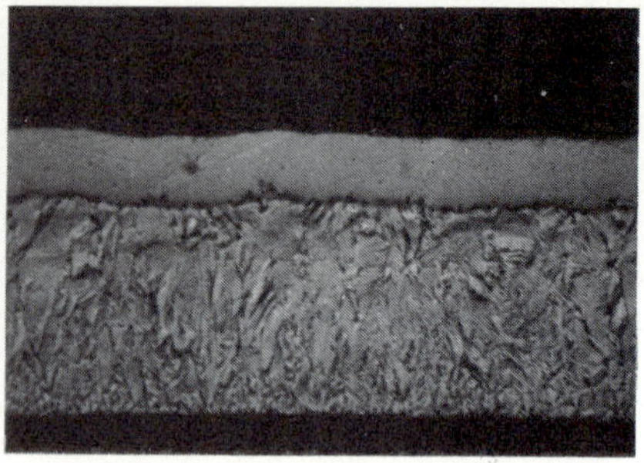

Fig. 6.4 Deposit from an acid gold cyanide electrolyte with a metallic brightener[8].

Fig. 6.5 Electron photomicrograph of typical acid gold deposits mounted on one of a conventional alkaline hot gold cyanide deposit. *Top left*: acid gold electrolyte (negative replica): *Bottom left*: same electrolyte (carbon positive). *Top right*: acid gold electrolyte containing cobalt (negative replica). *Bottom right*: acid alloy gold electrolyte containing cobalt, nickel and indium (negative replica) Mag: 20,000×.

The aim is to maintain free hydrogen cyanide in the solution at an acidity slightly acid to litmus, the strike being followed by plating in a conventional alkaline cyanide solution.

This strike has been used successfully for many years, but not for heavy deposits and, of course, good ventilation is necessary since periodic additions of sodium cyanide and hydrochloric acid are made. The tendency for precipitation of aurous cyanide in the absence of a buffer is minimised because the solution is operated under conditions which provide free cyanide and this tends to favour the reaction.

$$AuCN + HCN + K^{+} \rightarrow KAu(CN)_2 + H^{+}$$

If an alkali metal auricyanide is used instead of the aurocyanide it is possible to operate the baths at pH values well below 3.0. Such solutions differ from the Lukens type, in that the conducting media may be acids such as citric or phosphoric, or acid salts such as bisulphate. The gold content of these solutions can vary over a wide range and, of course, the cathode efficiency is only one third of that for the aurocyanide strike solutions.

REFERENCES

1 Rinker E. C., U.S. Re-issue, 24,583 (1958).
2 Rinker E. C., Duva R., U.S. Pat., 3,104,212 (1963).
3 Ehrhardt R. A., *Proc. Am. Electropl. Soc.*, **47,** 78 (1960).
4 Atwater A. R., Julich E. M., U.S. Pat., 2,978,390 (1961).
5 Rinker E. C., Johns E., *Iron Age*, **181,** 25, 118 (1958).
6 Rinker E. C., Duva R., U.S. Pat., 2,905,601 (1959).
7 Ostrow B. D., Nobel F. I., U.S. Pat., 2,967,135 (1961).
8 Parker E. A., Powers J., U.S. Pat., 3,149,057 (1964).
9 Parker E. A., Powers J., U.S. Pat., 3,149,058 (1964).
10 Foulke D. G., Brit. Pat., 928,088 (1963).
11 Munier G., *Plating*, **56,** 1151 (1969).
12 Raub E., *Metall*, **21,** 709 (1967).
13 Foulke D. G., *Plating*, **50,** 39 (1963).
14 Camp E. K., Deitz E. G., U.S. Pat., 3,303,112 (1967).
15 Duva R., Simonian A., U.S. Pat., 3,562,120 (1971).
16 Greenspan L., U.S. Pat., 3,423,295 (1969).
17 Schumpelt K., U.S. Pat., 3,367,853 (1968).
18 Duva R., Foulke D. G., U.S. Pat., 3,156,634 (1964).
19 Craig S. E., Harr R. E., Henry J., Turner P., *J. electrochem. Soc.*, **117,** 1450 (1970).
20 Barrett C., *Trans. Am. Inst. Min. Engrs.*, **161,** 15 (1945).
21 Lukens H. S., U.S. Pat., 2,133,995 (1938).

Chapter 7

NON-CYANIDE ELECTROLYTES

D. G. FOULKE

One of the reasons why the electrodeposition of gold was widely acclaimed in the 1800's was that this method of coating basis metals would eliminate the mercury poisoning hazards of fire gilding. However, what industry eventually settled down to using were solutions based upon potassium (I) gold cyanide along with free potassium cyanide. Such solutions, of course, were still poisonous. As early as 1845 Elsner[1] suggested that non-cyanide gold electrolytes (as well as copper and silver) might have advantages. He mentions a $KAuCl_4$ solution, a ferrocyanide bath (it is obvious now this was a cyanide bath) and a gold sulphite solution which he studied after von Seimens[2] reported that a similar silver bath could be used. Woolrich[3] also suggested similar formulae in 1843.

Other types of electrolytes appear mostly in the patent literature. Schloetter[4,5], for instance, suggested a thiocyanate and an iodide bath. Thiosulphate and thioacid[6] (e.g., thiomalate) solutions have also been reported, but little commercial use has been made of them.

ALKALI GOLD SULPHITE SOLUTIONS

Sodium gold sulphite solutions were first suggested in the 1840's at a time when a considerable amount of research was being carried out on gold plating. However, the alkali gold cyanide electrolytes soon replaced all others and for well over a century were the only solutions used for gold plating in industry. In 1962, Smith[7] described a sodium gold sulphite solution containing gold sulphite and a metal brightening agent which produced bright electrodeposits. Since then, the commercial development of this type of electrolyte has progressed rapidly and various modifications are now in widespread use. Deposits range from 12–24 carat alloys. Good micro- and macrothrowing power are claimed for these solutions.

The gold sulphite system contains sodium gold sulphite as a source of the metal along with free sodium sulphite. The pK for sodium gold sulphite is much lower than the pK for the gold cyanide system. Consequently, the need for free sulphite is more pronounced than for the cyanide systems, hence all solutions of the former type contain free sodium, potassium or ammonium sulphites[8]. Most commercial processes are operated at a pH above 8.0 and contain metallic brighteners, together with complexing and conducting salts.

These baths operate at almost 100% efficiency, a current density of about 0.3 A/dm^2 and at a temperature in excess of 40°C.

The first commercial electrolyte contained a small amount of copper as a brightening agent. Subsequent baths produced binary alloy deposits containing gold, together with cadmium, copper or palladium. Ternary alloys are also obtainable. Characteristically, the deposits are quite ductile, the distribution is good and the baths tend to promote brightness (smoothening or geometrical levelling) as the coating thickness increases.

The whole area of gold sulphite electrolytes has been opened up by the publication of Meyer and co-workers[9], who report that gold sulphite (alloy) solutions are stable at a pH as low as 6.5 which is just slightly higher than the pH at which sulphur dioxide is released from the bath.

As in the case of cyanaurate (I) systems, higher voltages tend to increase the base metal content of the alloy. Lower temperatures also produce a higher base metal content in the deposit (lower conductivity) and higher concentrations of base metal permit the formation of lower carat alloys. Silver does not appear to behave as it does in the cyano (I) system, but this is not well documented because of the instability of the alkali silver sulphite system.

The earlier publications were concerned with simple preparative details and information to the effect that the deposits were not so attractive as those obtained from the ferrocyanide process. Elsner[1] reported that dissolving gold fulminate in sodium sulphite gave a clear yellow solution, yielding a light brown, gold deposit which was not so attractive as the yellow gold (goldgelbe Farbé) colour of gilding achieved with the ferrocyanide and potassium cyanide process.

Other than five patents[7,8,9,10,11] since the 1840's, there have been but two literature references[9,12] to the sulphite system. Craig and his colleagues[12] investigated an electrolyte containing a single metallic addition agent and reported the deposits to have outstanding physical properties. Meyer[9], Losi and Zuntini described one capable of producing a gold-copper alloy of about 16 carat. The deposit is hard, 390–410 Kg/mm^2 (25 g load) and the plate distribution outstanding (Figure 7.1).

Three of the patents[7,10,11] are rather simple, but the recent Shoushanian[8] disclosure is quite complex. This bath must be maintained at a pH above 8.0, in contrast to the Meyer solution, and contains the gold as the sodium or potassium gold sulphite complex (1–30 g/l Au). Shoushanian claims that the bath must contain 40–150 g/l of an alkali metal sulphite and 5–150 g/l of buffering and conducting salts (phosphate, carbonate, acetate, citrate). In addition to this, there should be at least one additive selected from:

(a) brightening metal additives (Cd, Ti, Mo, W, Pb, Zn, Fe, In, Ni, Co, Sn, Cu, Mn, V), 1–5000 mg/l,
(b) brightening semi-metallic additives (Sb, As, Se, Te), 1–400 mg/l,
(c) an alloying element (as for (a), plus Sb, but excluding Ti, Mo, W and V),0.5–5 g/l.

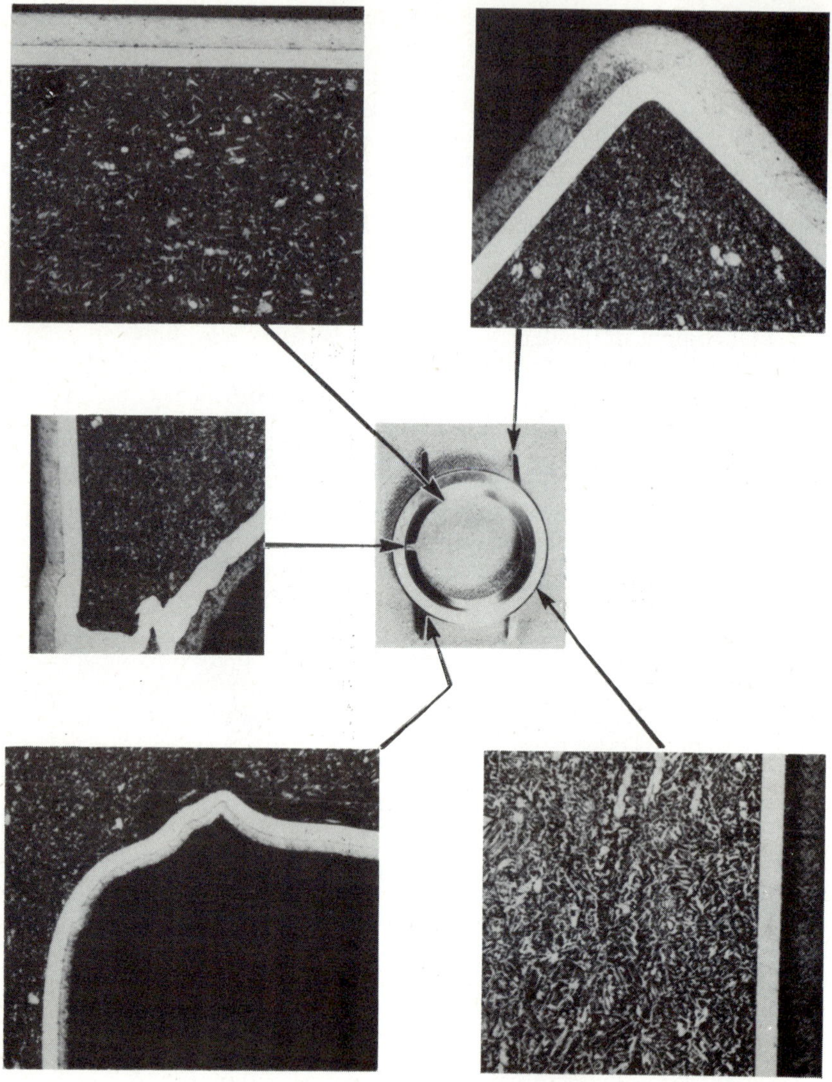

Fig. 7.1 Deposit distribution obtained from the electrolyte of Meyer, *et al.*[9].

Typical examples from the four patents[7,8,10,11] are shown in Table 7.1.

Table 7.1
Sodium Gold Sulphite Electrolytes

Constituents	*Both Composition (g/l) & Operating Conditions*				
	Ref. 7	*Ref. 8*	*Ref. 9*	*Ref. 10*	*Ref. 11*
Au(as sulphite complex)	3–4	10	10	6–8	1–8
Sodium EDTA	18	30	20		
Copper EDTA	0.075		2.5		
Cu(as diethylene-triamine penta acetate)				0.5–3	
Diethylenetriamine penta acetate				40–60	
Ethylenediamine (25%)			2.5*		40–120 ml
As(as AsO_3)		0.004			
Sodium sulphite	NR	90	25	NR	NR
Temperature, °C	40–60	50	60	40–50	40–50
pH	9–10	10	6.5	9.5–10	alkaline
Current density, A/ft^2	1–3	8	2	1–2.3	0.4–1.0

* Amine not characterised.
NR = Not revealed.

One of the advantages of the gold sulphite process is that the deposits are bright, hard and very ductile, whereas the cyanide based acid (alloy) electrolytes yield deposits of limited ductility.

Another advantage is the good throwing power of these electrolytes. Thickness ratios of near 1 : 1 can be obtained in recesses versus high points and, surprisingly, the microthrowing power ratio h_3/h_1 approaches 1, which leads to geometric levelling (see Chapter 9).

The structures of the sulphite gold electrodeposits are amorphous. Craig states, "Surfaces from this bath are completely featureless, even when viewed at 30,000 × " (Figure 7.2).

COMPARISON OF SULPHITE AND CYANIDE ELECTROLYTES

The sodium sulphite gold plating solutions were documented as far back as the 1800's and for both these and the cyanide electrolytes, no essentially new developments took place in over one hundred years though, unlike the sulphite baths, the cyanide solutions were used extensively during this time to meet the relatively undemanding requirements of gold plating.

Reappraisal of both systems was triggered off by the rapid growth of the electronics industry which created demands for specific deposit characteristics which could not possibly be met by existing electrolytes.

Since the sulphite complexes of gold were quite different from the cyano-complexes and the corresponding base metal complexes were not well documented in the literature, it is not surprising that advances in this area were slow. However, with the development of new complexing systems, it became possible to formulate baths containing gold sulphite at a relatively high pH, and to incorporate base metal ions using the new complexing agents to ensure non-precipitation of the base metal hydroxides. Moreover, since many base metals are less strongly complexed by sulphite as compared to cyanide, it became readily possible to produce heavy alloy gold deposits of low or high carat and which were found to be surprisingly ductile. In the

Fig. 7.2 Electron photomicrograph of a deposit from the gold sulphite electrolyte containing a metalloid addition agent. (60,000 ×).

cyanide system the problem of alloy deposition was partially solved by eliminating free cyanide and operating at a pH of 3.0–5.0, but under these conditions heavy deposits (10 microns or more) were not possible.

The sulphite systems are also attractive because of their non-toxicity as compared to the cyanide systems. Effluents, however, do have BOD* problems as do the majority of acid gold cyanide plating baths used for the production of bright deposits.

REFERENCES

1 Elsner *J. Prakt. Chem.*, **35,** 361 (1845).
2 Von Siemens W., "Lebenserrinerungen," Springer, Berlin (1832).
3 Woolrich J. S., *Dinglers polytech. J.*, **88,** 48 (1843).
4 Schloetter M., Ger. Pat., 608,268 (1936).
5 Schloetter M., U.S. Pat., 1,857,644 (1932).

* Biological Oxygen Demand.

6 Duva R., Raleigh J. P., U.S. Pat., 3,520,875 (1970).
7 Smith P. T., U.S. Pat., 3,059,789 (1962).
8 Shoushanian H. H., U.S. Pat., 3,475,292 (1968).
9 Meyer A., Losi S., Zuntini F., Proc. Fachtagung. Galvanotechnik, Leipzig (1970); Swiss Pat., 506,828 (1969).
10 Smagounova N. A., Gavrilova J. P., Balashova N. N., Russ. Pat., 217,167 (1968).
11 Smagounova N. A., Youdina A. K., Gavrilova J. P., Russ. Pat., 231,991 (1969).
12 Craig S. E., Harr R. E., Henry J., Turner P., *J. electrochem. Soc.*, **117,** 1450 (1970).

Chapter 8

ELECTROLYTE PARAMETERS

D. G. FOULKE

The structure and nominal composition of gold deposits obtained from alkaline, neutral, acid and non-cyanide electrolytes have already been briefly discussed under their respective headings (see Chapters 4–7). However, the effect of modifying the parameters of the more important gold plating solutions will be developed further in this present chapter insofar as is consistent with data currently available.

ALKALINE CYANIDE ELECTROLYTES

The colour golds and the alkaline ferrocyanide electrolytes have been covered in sufficient detail (see Chapter 4), inasmuch as these solutions no longer represent an important volume in our economy, nor will their use increase.

The industrial alkaline golds are still widely employed and may well continue to be of relative importance. They have already been classified in Chapter 4 as (I) pure golds, (II) alloy golds and of these, the former are less complex. Since they operate in a milieu containing an appreciable amount of free cyanide, which normally complexes impurities possibly introduced from basis metals, they tend to produce relatively pure electrodeposits. However, occurring simultaneously, and particularly at high current densities, cyanide polymers are liberated, so steps must be taken to eliminate organic impurities from such solutions during operation.

In general, higher gold concentration, more vigorous agitation and higher temperatures will permit operation at higher current densities at high efficiency (see Chapter 4). The addition of trace quantities of metals such as arsenic, which do not form cyanide complexes, is common, but such additions must be carefully controlled otherwise the heat resistance of the deposit will be impaired. Actually, if such elements are not carefully controlled the deposits fall into the category of alkaline alloy golds.

The temperature of the alkaline cyanide gold baths is generally held within the range of 50–65°C with a pH of at least 10.0. The latter appears to have little effect on the current efficiency, but the degree of agitation and the gold concentration are important. Lee[1] reports that the plating efficiency is normally 90–100%. At low gold concentrations, in the region of about 1 g/l, the efficiency can be less than 20%, but when the gold content reaches about 3 g/l, with good agitation, it is 100% at 0.2 A/dm^2. His curves show that at 1.0 A/dm^2 with good agitation the efficiency will be about 100%, whereas without agitation at 2.5 A/dm^2 it drops to below 50%, both at 10 g/l of gold. Lee suggests that for jig plating, a gold content of 8.2–10.3 g/l is advisable for maximum current density and efficiency. In more recent years the gold content has been increased drastically to permit ever increasing current densities. Parker[2], in a barrel study reported that the efficiency reached a maximum at about 4–6 g/l between 0.1–0.5 A/dm^2.

It is impossible to give standard conditions because of the effect of one variable upon another. Generally speaking, increased agitation, elevated temperatures, high gold content and the presence of conducting salts will permit higher current densities at high efficiencies.

The alloy industrial gold processes listed in Table 4.3 (see Chapter 4, Refs 1,25,26,27 & 28) were described in patent disclosures of limited extent. The only one described extensively in the technical literature is the Rinker[3] process. The temperature of this bath should be maintained below 40°C, preferably at 20–25°C, to achieve bright deposits. The gold concentration is normally held at 8 g/l, but it may vary widely as long as the brightener is maintained at approximately 1% of the gold content. The current density used is normally 0.2–0.6 A/dm^2. Obviously at 4 g/l of gold, the current density must be lowered; conversely at 30 g/l the current density can be increased appreciably. Since this process operates at high free cyanide concentrations and silver is more noble than gold under these conditions, higher current densities produce a richer alloy while low current densities (often true of barrel plating) yield an alloy lower in gold. The current efficiency is 95–96%, but with vigorous agitation and a high gold content, 100% efficiency is possible. The high free cyanide content of the bath prevents the codeposition of most metals with gold, but on the other hand, as generally for gold and silver baths containing large amounts of free cyanide, considerable organic matter is codeposited. Increasing the silver concentration makes it possible to deposit alloys containing 25% and more of silver.

More recently arsenic[4] has been added as a brightener in alkaline gold solutions. In this case, arsenic acts as a depolariser, allowing the gold to deposit more easily and improving the efficiency.

When the basis metal is copper, the free cyanide content can be very important, since copper codeposition can be disastrous in semiconductor applications. Consequently, there is a trend to use highly complexing agents as conductivity salts in such solutions.

NEUTRAL ELECTROLYTES

Neutral gold electrolytes in modern plating practice have been directed towards pure, rather than alloy baths, primarily due to the increasing importance of electronics as compared to decorative applications. There is, however, very little published data available.

Zimmerman and Brennerman[5] describe an essentially neutral solution (see Table 5.1, Ref. 7) which deposits about 0.1 % arsenic with the gold. They also recommend the addition of 5 g/l of sodium thiosulphate. The preferred temperature is 55°–72°C and the current density 0.1 A/dm^2, the current efficiency being in the region of 93–100 %.

Many of the neutral baths contain alkali metal and ammonium citrates and/or phosphates, the latter group including pyrophosphates (see Table 5.1, Ref. 11). In general, these baths are operated at elevated temperatures, with approximately 6–30 g/l of gold, good agitation and a pH of about 6.5. Under these conditions, the current efficiency is close to 100 %.

When a base metal brightener is present, higher current densities will result in more base metal plating out. This is also true for base metal impurities introduced adventitiously into the bath unless the formulation incorporates suitable chelating agents for copper, nickel, cobalt and iron, which are the most probable contaminants in general practice. There are, however, a number of exceptions to this rule (Table 8.1).

ACID ELECTROLYTES

The high purity acid gold solutions are of two types (a) baths without metallic addition agents and (b) those containing very low concentrations of metallic brighteners (see Table 6.2, Refs 15 & 16). These baths are operated at temperatures of 50°C or above in commercial practice. The gold content averages 8 g/l. Vigorous agitation is employed. The current density is normally 0.5 A/dm^2, but may vary between 0.3–1.0 A/dm^2. Higher current densities may be used by increasing the gold content. At 70°–90°C, with good agitation and a pH of 5.0–6.0, the current efficiency is essentially 100 %. This can be somewhat less, depending upon the brighteners and chelates present. Lower gold concentrations, less vigorous agitation and low pH will reduce the current efficiency. Lower temperature will also reduce the current efficiency, but the deposit will generally be harder, probably due to the inclusion of organic materials in the deposit.

The acid alloy gold plating bath parameters have been fairly well documented in the literature. Table 6.1 (see Chapter 6) summarises some of the processes in use. More recent developments include the patenting of a polyamine[6], an acid pyrophosphate along with a nickel or cobalt chelate and a polyhydroxy organic acid[7], organic phosphorus compounds, which may or may not act as chelating agents[8], and a German patent describing a similar complex organic phosphorus compound[9].

The acid alloy gold solutions are usually operated from room temperature up to about 35°C. For high speed applications the temperature may be raised. The gold concentration is maintained at approximately 8.2 g/l, although this is be no means mandatory. However, at this relatively low concentration, drag-out losses are minimised and maintained at a reasonable level.

The pH and degree of agitation affect the current efficiency considerably. Consequently, a pH as high as is consistent with good brightness is employed with good solution agitation by mechanical or other means. In the latter context, air and nitrogen agitation have been used as well as ultrasonics.

The amount of base metal present in the solution and codeposited with the gold is also a factor bearing on the current efficiency. With a low base metal concentration (100 mg/l), a temperature of 35–40°C, good agitation, a pH of 4.5–5.0 and a current density of 1.0 A/dm^2, the current efficiency will be in the order of 50%. However, if the pH is lowered to 3.5 and the temperature to 25°C, the efficiency will be much lower.

In the early days of acid gold plating the current efficiency for the sodium citrate bath was about 33% which led some workers to postulate that gold was present as the trivalent citrate compound. This is incorrect. When base metal ions are absent, this process operates at essentially 100% cathode efficiency. When they are present polarisation occurs, resulting in the evolution of hydrogen. The presence of some organic chelating or complexing agents will also lower the current efficiency, even though they extend the current density range. Consequently, it is better to consider the deposition rate in terms of mg/A-min and check the plating rate regularly.

NON-CYANIDE ELECTROLYTES

Discussion of the parameters of non-cyanide electrolytes will be limited to the gold sulphite electrolytes, the only such baths in present use by industry to any extent. These baths are of two types (a) high purity baths, usually containing metalloid additives and (b) alloy baths.

The high purity baths are generally operated at elevated temperatures in the order of 40–60°C, the gold concentration being 8–12 g/l. With vigorous agitation, the maximum permissible current density is about 0.5 A/dm^2. Increasing the gold concentration will permit the use of higher current densities, if a low voltage is maintained by using sufficient conducting salts. The efficiency is essentially 100%, calculated on monovalent gold.

The alloy electrolytes are generally operated at the same temperatures as the non-alloy baths, except that for highest brightness and levelling, a temperature below 40°C may be preferable. The gold concentration is nominally 8 g/l and moderate agitation is employed. The efficiency, based upon 100%, being 123 mg/A-min for pure gold, will vary from this figure with the composition of the deposit. A lower pH does not adversely affect the efficiency, but below a pH of 7.0 the bath is less stable.

In general, increase in the base metal salt concentration and current density (voltage) and decrease in temperature will tend to increase the base metal content of the deposit. However, this general effect may be modified in specific cases by the amount and nature of complexing agents present, and by the presence of cathodic depolarisers. Permissible current densities for the alloy electrolytes may exceed 1.5 A/dm^2.

When the bath is formulated to produce heavy deposits of ternary and quaternary alloys, for example, in the case of watchcases, it is best to agitate the work and to use moderate solution agitation. In the watchcase industry where thicknesses up to forty microns are deposited, this is carried out by using a circular rack in a circular tank, the workpieces being carefully spaced upon the rack, which is moved in a clockwise direction, then in a counter-clockwise direction at moderate speed. In barrel plating, however, it is difficult to obtain uniform colour on the parts when plating ternary or quaternary alloys because of the great variation in current density within the same load of workpieces.

HIGH SPEED PLATING

Appropriate variation of the electrolyte parameters will permit the rapid deposition of gold and gold alloys. There are, of course, cases where specially engineered equipment is used for high speed gold plating but in all cases the bath parameters are also optimised, as discussed in the following text.

METAL CONCENTRATION

As in base metal electrodeposition, to attain higher plating rates, higher metal concentrations must be used. Raising the gold concentration to 25–37.5 g/l will greatly increase the permissible operating current density. No reliable data have been published to indicate the improvement, possibly because in most cases other parameters have also been changed. A fair generalisation is that the current density can be doubled by increasing the gold concentration from the nominal 8 g/l to 32 g/l.

CONDUCTING SALTS

Increasing the amount of conducting salts present in the electrolyte will permit the use of higher current densities, but more important still, will result in better current distribution in the higher current density regions.

BRIGHTENERS

The inter-relationship between operating parameters, such as agitation, pH and temperature, and mass transfer factors affecting the incorporation of brighteners in the deposit is complicated. In general, the brightener concentration must be increased when operating at high current densities. This is not necessarily the case, however, for certain semi-metallic brighteners (for example, arsenic present as arsenite is deposited more rapidly at higher pH values), or for amine complexes of cobalt or nickel where the free metal ion concentration is dependent on the pK value for the complex and the pH of the solution.

If the brightener is an oxidisable species its effective concentration may well be reduced as a result of anodic oxidation when high current densities are employed. In this case it may be unnecessary to use a higher initial concentration, but an increased dosage rate will be needed during operation.

TEMPERATURE

In general, high speed baths are operated at higher temperatures than in normal operation. The resulting increase in mobility of the ions, and hence the conductivity of the solution is generally useful, but since less base metal will be codeposited at the higher temperatures it is necessary to increase the base metal concentration of the bath to obtain deposits of the same composition as those produced from a low temperature bath.

CURRENT DENSITY

If the conductivity is not correctly balanced, the voltage may be an important factor with respect to the rate at which brighteners and hardeners (Co, Ni, etc.) are codeposited with the gold. Increasing the current density

Table 8.1
Effect of Current Density on the Deposit Composition (Wt. % Base Metal) of "Alloy" Acid Electrolytes

Bath Type	*pH*	*Current Density, A/dm²*			
		0.1	*0.5*	*1.0*	*1.5*
Co, 100 mg/l	3.5	0.1	0.16	0.16	0.17
Co, 100 mg/l	4.0		0.11	0.12	0.12
Ni, 100 mg/l	3.5	0.94	0.26	0.21	
Ni, 100 mg/l	4.0	0.19	0.20	0.10	
Ni, 4.5 g/l	4.0	3.61	0.96	0.60	
Ni, 4.5 g/l*	4.0	4.11	2.57	2.63	

* Also contained an organic polyamine addition agent.

(voltage) does not favour the deposition of nickel with gold (Table 8.1) as has been claimed. Cobalt, on the other hand, does plate out slightly faster at a higher current density (voltage). Change in pH has little effect on the efficiency of the alkaline cyanide, non-cyanide, and neutral gold baths, but in the case of acid baths for either pure gold or alloy gold plating, efficiency decreases with pH. Hence in high speed plating these solutions are operated at higher pH values than normal. In alloy plating, the increase in pH decreases the amount of base metal present in the uncomplexed form, hence higher overall concentrations must be used to obtain the same alloy composition as that produced under normal conditions.

AGITATION

Agitation is particularly important, because gold exists in plating baths in the form of an anion. Thus, electrolytic migration tends to move the gold anion towards the anodes and this must be overcome by extremely vigorous agitation.

In general, the greater the degree of agitation, the higher the permissible current density. Hence the most vigorous agitation is recommended for high speed plating. To maximise this it may be desirable to use a combination of conventional means, e.g., work movement, solution circulation and air or nitrogen simultaneously.

FIXTURING

Much can be accomplished by proper fixturing and by good anode-cathode spacing as well as anode shielding to ensure good current distribution over the work. This, in turn, permits higher overall current densities and, consequently, higher plating rates.

Conforming anodes, such as are used to plate bumper bars at current densities exceeding 100 A/dm^2, have not been employed for gold plating.

High speed gold plating has many limitations. It is most useful when plating limited areas (selective plating), when rack plating (where jigging, anode-cathode distance and anode or cathode shielding can be optimised) and for strip and wire plating. When applied to barrel plating no significant advantages are achieved.

ULTRASONICS

Since agitation is an important parameter in all types of plating, particularly in the case of gold where the metal is present as the anion, it might be expected that ultrasonic agitation of the solution in the vicinity of the workpiece might be widely adopted, but this is not in fact the case. However, among the few applications encountered are the electroplating of slip rings and commutators,[10] where ultrasonic agitation is used to depolarise the electrodes to ensure homegeneity of the rather thick deposits produced, and the barrel plating of components, to improve uniformity of the deposit in blind holes.

From a purely investigational point of view, Vrobel[11] has investigated the influence of ultrasonics on the deposition of bright gold and claimed that an increase in brightness, microhardness, current efficiency and current density can be achieved. The effect of ultrasonics has also been examined by Weiner[12] and is also the subject of a patent[13].

MODULATED CURRENT[14–23]

The term modulated current is proposed for any current that undergoes multiple cyclic changes of value and/or of sign, in the course of an electroplating operation. There are two types of modulated current, one in which there is a reversal of polarity, and the other in which there is no reversal of polarity. A.C. superimposition, with the AC lower than the DC, is of the latter type. Many other types, such as ripple current, pulse current, etc., also fall into this group, whereas periodic reverse (PR) current, and asymmetric AC belong to the former category.

The pulsed current plating technique of Rockafellow[24], Girard[25] and Popkov[26] is now being evaluated by a number of workers with reference to gold plating. Avila and Brown[27], using a direct current source pulsing at 10–15 milliseconds ON and 115 milliseconds OFF, reported less nodular, finer grain deposits at a pulsed current density of 4 A/dm^2 than were obtained with direct current alone at 1 A/dm^2, and it was claimed that the deposition rate with pulsed current could equal that for direct current plating. Cheh[28], however, on the basis of calculations based on a diffusion model, has suggested that although the magnitude of the instantaneous current for deposition could, under suitable conditions, be considerably higher in pulsed current plating than in DC plating, the limiting overall deposition rate is, in general, lower. Some support for this conclusion was provided by results on acid citrate, neutral phosphate and alkaline cyanide electrolytes.

More recent work, not yet published has shown that a pulsed current cycle of 1 millisecond ON and 10 milliseconds OFF can indeed deposit gold at essentially the same rate as direct current plating because of the higher permissible current densities. In addition, these higher current densities do give denser deposits with finer grain structures. One point worth noting is that the pulse plating operation minimises edge build-up or treeing, and gives better overall metal distribution, especially on large flat surfaces.

REFERENCES

1 Lee W. T., *Corros. Technol.*, **10,** 4, 59 (1963).
2 Parker E. A., *Plating*, **45,** 633 (1958).
3 Rinker E. C., *Proc. Am. Electropl. Soc.*, **40,** 19 (1953).
4 Greenspan L., French Pat., 1,508,895 (1968).
5 Zimmerman R. H., Brennerman R. L., Belgian Pat., 747,808 (1969).
6 P.M.D. Chemicals Ltd., French Pat., 2,028,402 (1970).
7 Engelhard Industries Ltd., French Pat., 2,028,113 (1970).
8 Nobel F. I., Ostrow B. D., French Pat., 1,539,226 (1968).
9 Todt G., German Pat., 1,909,144 (1968).
10 *Electroplg Metal Finish..* **15,** 2, 69 (1961).
11 Vrobel L., *Trans. Inst. Metal Finish.*, **44,** 4, 161 (1966).
12 Weiner R., *Metall*, **15,** 97 (1961).
13 Schneider E., Lidell M., U.S. Pat., 3,427,231 (1969).
14 Jernstedt G. W., U.S. Pat., 2,451,341 (1945).
15 Bregman A., *Metal Prog.*, **58,** 199 (1950).
16 Jernstedt G. W., Patrick J. D., U.S. Pat., 2,678,909 (1954).
17 Yampolsky A. M., *Electroplg Metal Finish.*, **16,** 3, 76 (1963).
18 Heilmann G., U.S. Pat., 3,586,611 (1971).
19 Bandaver V. V., Krupnikova E. I., Rodionova L. I., TR Gos. Nauch-Issled Inst. Splavov Obrab Tvset Metal, 31, 93 (1970).

20 Bertorelle E., *Galvanotecnica*, **4,** 141 (1953).
21 Gardam G. E., Tidswell N. E., *Trans. Inst. Metal Finish.*, **31,** 418 (1954).
22 Shenoi B. A., Gowri S., Indira K. S., *Metal Finish.*, **64,** 5, 54 (1966).
23 Winkler J., German Pat., 576,585 (1933); 728,497 (1942).
24 Rockafellow S. C., U.S. Pat., 2,726,203 (1955).
25 Girard R., "Electrodeposition of Thin Magnetic Films," French Atomic Energy Commission, Grenoble Centre of Nuclear Studies (1965).
26 Popkov A. P., *J. Appl. Chem. USSR*, **39,** 1747 (1966).
27 Avila A. J., Brown M. J., *Plating* **57,** 1105 (1970).
28 Cheh H. Y., *J. Electrochem. Soc.*, **118,** 551 (1971).

Chapter 9

THROWING AND LEVELLING POWER

D. G. Foulke

In all applications of gold plating a major aim is to achieve maximum uniformity of thickness over functional areas. This is important from the economic viewpoint to avoid wastage of gold in meeting minimum thickness specifications, and from the technical viewpoint, in that variations in thickness may affect other characteristics of the coating such as porosity, stress, wear and corrosion resistance, solderability, etc., which are of critical applicational importance.

In any electrodeposition process, thickness distribution may be improved by a number of well known devices, such as adjustment of anode distribution and/or anode-cathode distance, current shaping by the use of conducting or insulating screens or auxiliary electrodes, jetting of electrolyte through anode perforations and so forth. However, the extent to which such devices may be necessary is determined by the basic electrolyte characteristic of throwing power. The present chapter, therefore, is concerned with the effect of deposition and electrolyte variables on this and related factors for various types of gold plating solution.

DEFINITIONS AND MEASUREMENT

Covering Power denotes the ability of an electrolyte to cover over the recessed areas of a component. It is usually measured in terms of the depth of penetration of the deposit into recesses of standard dimensions.

Throwing Power is a measure of the ability of an electrolyte to deposit coatings of uniform thickness over a component of irregular shape. This is termed macrothrowing power as distinct from microthrowing power, and a number of techniques have been devised for its quantitative expression, based for instance on local thickness measurements across a Hull Cell panel or on components of awkward shape. It is often referred to qualitatively in terms of the distribution of thickness on actual components, such as connector contacts, or across printed circuit board edge connectors.

The most commonly used technique for measuring throwing power utilises the Haring and Blum Cell, a rectangular cell with two cathodes disposed at different distances from a central anode, all the electrodes completely filling the cell cross-section. Throwing power is expressed by the empirical formula:

$$\text{T.P. } (\%) = \frac{100(K-M)}{K+M-2}$$

where M is the ratio of the weights of metal deposited on the two cathodes (secondary current distribution ratio), and K is the primary current distribution ratio, usually taken as the ratio of the distances of the two cathodes from the common anode.

The usefulness of the Haring-Blum cell in gold plating is, however, somewhat restricted. The low metal content of most gold solutions necessitates the use of vigorous agitation, but it is difficult to achieve this to a uniform degree on each side of the cell because of the difference of volumes. The metal distribution ratios may not, therefore, be strictly applicable to normal deposition conditions. In the case of gold alloy deposition a further difficulty

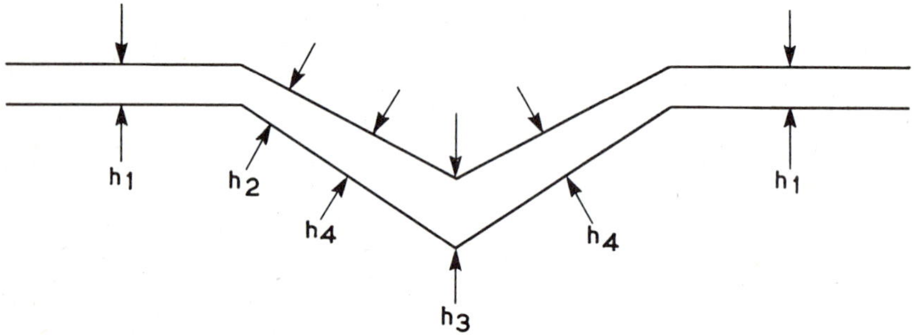

Fig. 9.1 Measurements made in scratch for thickness ratio.

arises due to the variation in deposit composition with current density, which affects the secondary current distribution value as derived from simple weight gain of the two cathodes. Ideally the cathode deposits should be analysed and an appropriate correction applied. However, despite these limitations, the Haring-Blum cell remains a valuable tool in experienced hands, and is particularly useful for comparative studies of gold plating solutions, as in the assessment reported by Foulke and Johnson.[1]

Microthrowing Power denotes the ability of an electrolyte to deposit coatings of uniform thickness over microgeometrical surface irregularities. In this case the primary current distribution over the surface profile is uniform and variations in deposit thickness occur because of differences in the thickness of the cathode diffusion layer, which in turn influences the transport of dischargeable ions to the cathode surface.

Microthrowing power is usually measured by the relative build-up of deposit thickness at high and low points of a standard surface microprofile, e.g., a microgroove record "mother". Thus, referring to Figure 9.1, the microthrowing power index is defined as h_3/h_1.

Levelling is related to microthrowing power and is used to describe a decrease in microgeometrical surface roughness brought about by electrodeposition. Measurements are usually made on the same microprofiles as are used in microthrowing power studies.

For more detailed general treatments of the above aspects the reader is referred to references 2–5.

PRACTICAL STUDIES

Although the literature abounds with qualitative statements, relatively few quantitative studies are reported. For example, the covering power of gold plating solutions is frequently stated to be excellent, although no actual measurements have been presented, to the writer's knowledge.

MACROTHROWING POWER

Comparative studies of various types of gold plating solution, using the Haring-Blum cell, are reported by Foulke and Johnson.[1] The results are summarised in Table 9.1.

Bright, high free cyanide gold baths and solutions based on the sulphite complex and containing metallic brighteners have excellent throwing power (68–90%). Acid potassium gold cyanide electrolytes containing brighteners do not, however, exhibit such good throwing power as their pure matte counterparts, particularly when the latter are operated at low current density. In practice, however, better distribution is obtained in barrel plating from the bright acid baths, since the deposits are smoother and consequently the parts tumble well without bunching.

No quantitative measurements of throwing power for modern neutral baths have been reported, but it is known that those baths that deposit fully bright coatings exhibit exceptional throw, particularly when barrel plating recessed connector parts. Foulke and Johnson[1] report measurements for a matte neutral phosphate solution operated at low temperature (40°C) and relatively high pH (8.0), which show a marked improvement in throwing power with an increase in current density.

These authors also studied the effect of current density, solution pH, specific gravity and temperature, all of the factors affecting the main electrolyte properties which influence the secondary current, and hence deposit distribution, i.e., specific resistance and polarisation characteristics. For all types of solution, throwing power was improved by an increase in the current density, pH and specific gravity of the solution. Although raising the bath temperature should increase the specific conductance and hence improve throwing power, the improvement was in fact only slight, probably being offset by a reduction in cathode polarisation.

MICROTHROWING POWER

In electrolytes of high metal ion concentration the depletion of metal ions in the cathode diffusion layer during deposition is relatively slight, so that differences in thickness of the diffusion layer have only a relatively small effect on the transport of dischargeable ions to the surface, and hence microthrowing power of such solutions is usually good since concentration polarisation is low.

Table 9.1

Macro- and Microthrowing Power of Typical Gold Electrolytes

	Reference 1				Reference 6			
					Macrothrowing Power		*Microthrowing Power Index*	
Type	*Temperature °C*	*Current Density A/ft²*	*Macro-throwing Power*	*Micro-throwing Power Index*	*Agitation*	*Still*	*Agitation*	*Still*
Alkaline								
Conventional	60	1	8					
	60	2	37					
	55	3			25±7	18±5		
Bright Cyanide	40	2	67					
		2.5		0.8				
		3			92±3	62±2		
	40	4	34					
		5		0.7			0.8	0.35
Acid								
Bright Acid								
75 g/l-Ni	35	4			50±5	38±2		
(0.50 g/l-Ni)	40	5	29					
	40	10	58	0.9			0.80	0.88
(0.1 g/l-Ni)	35	4			15±5	35±5		
	40	5	11	1.0				
		6		1.0				
	40	10	42				0.88	0.99
Matte	55	4	46		50±5	32±3		
	40	5		0.7				
	40	10	49	0.5			0.91*	0.94*
Non Cyanide †								
(1 metal)	55	3			68±2	48±2	0.91	0.93
(2 metals)	55	3			65±6	40±3	0.97	0.97

* Temperature for macrothrowing power studies was 55°C
† Brighteners, 1 and 2 metals, respectively

In electrolytes of low metal ion concentration such as most gold plating solutions, depletion of the diffusion layer occurs more readily and variations in its thickness at various parts of the surface have a considerable effect on the supply of dischargeable ions and hence the rate of growth of the deposit which tends to be reduced in recesses as compared with other parts of the profile.

Gold electrolytes would, therefore, be expected to exhibit poor microthrowing power. This is not necessarily the case, however, for in practical solutions microthrowing power observations are further complicated by the effects of preferential absorption of brighteners on metal distribution.

Two studies, both using record mothers as a standard substrate, have been reported[1,6], the results of which, as shown in Table 9.1, are in fairly good agreement.

The high cyanide bright gold bath is interesting in that it exhibits excellent microthrowing characteristics together with the highest macrothrowing power of all the baths (Figure 9.2).

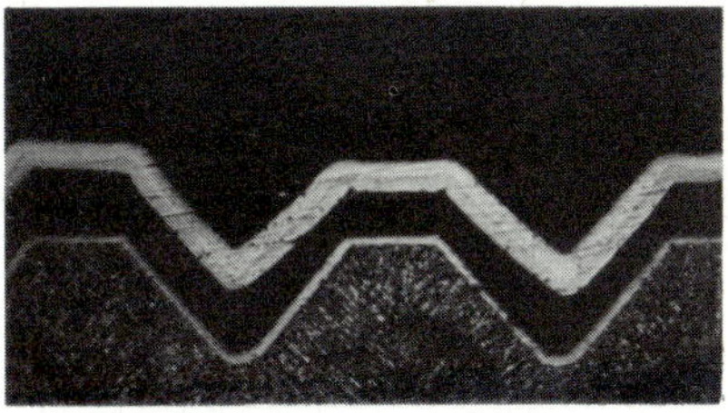

Fig. 9.2 Throwing power of the bright silver–gold electrolyte.

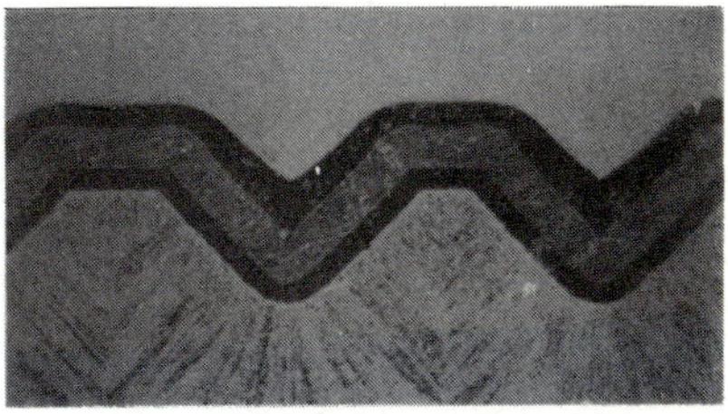

Fig. 9.3 Throwing power of the acid nickel–gold electrolyte.

The non-cyanide (gold sulphite) electrolytes also show good micro- and macrothrowing power. Otherwise the solutions appear to follow the more general rule that good macrothrowing power is associated with rather poor microthrowing characteristics. Microsections of some typical deposits produced in this work are shown in Figures 9.3–9.5, the gold concentration in all cases being 8 g/l.

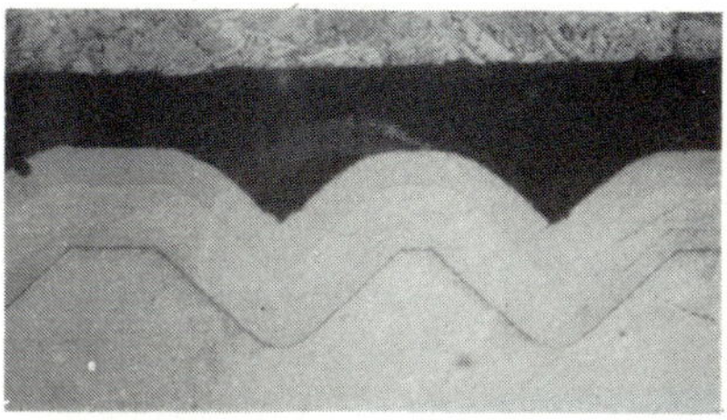

Fig. 9.4 Throwing power of the sulphite gold electrolyte containing two additives.

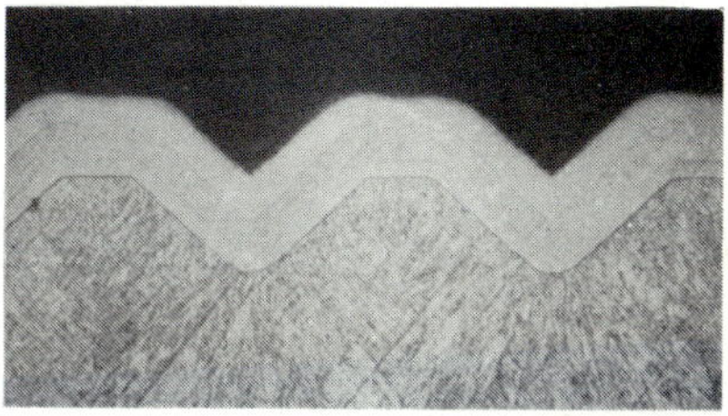

Fig. 9.5 Throwing power of the sulphite gold electrolyte containing one additive.

The addition of brighteners improves microthrowing power but reduces macrothrowing power, Vigorous agitation, increase in electrolyte temperature and the use of low current densities, all factors that reduce metal ion depletion in the diffusion layer, would be expected to improve microthrowing power, and this conclusion appears to be borne out for gold solutions by the limited results shown in Table 9.1.

LEVELLING

True levelling occurs when the deposition rate in a groove recess is greater than that on the peaks. Levelling agents are frequently added to plating solutions to induce levelling by being preferentially absorbed on peaks, thus favouring metal deposition in the recesses. Brightening agents frequently have a levelling effect, but there is little documentation on this aspect referring specifically to gold solutions.

Two types of gold electrolyte with near levelling ability are the sulphite and neutral cyanide solutions containing brighteners. With these exceptions, there is no evidence for levelling during gold plating which would compare with that obtainable for nickel or copper plating solutions with additives. Since gold is usually plated onto relatively smooth and well prepared surfaces, levelling ability is a much less important requirement in this case, which probably explains why this aspect has received little investigation.

REFERENCES

1 Foulke D. G., Johnson D. C., *Proc. Am. Electropl. Soc.*, **50,** 107 (1963).
2 Vagramyon A. T., Solor'era Z. A., "Technology of Electrodeposition", Draper, Teddington (1961).
3 "Microthrowing Power—A Literature Search", AES Research Report No. 17, Plating **43**, 388 (1957).
4 Raub E., *Plating*, **44,** 486 (1958).
5 Raub E., Muller, K., "Fundamentals of Metal Deposition", Elsevier, Amsterdam (1967).
6 Foulke D. G., Meyer A. R., Interfinish (1968).

Chapter 10

IMMERSION SOLUTIONS

E. A. PARKER

Processes by which metal coatings may be applied to base metals from aqueous solutions without the use of an externally applied EMF fall into two main categories:

1 Those based on chemical replacement of the base metal by the more noble metal coating. In this case the reaction will, in principle, either completely cease or else proceed at an immeasurably slow rate once the base metal surface is effectively masked by a thin coating. Suffice to say that solutions in this particular category are only capable of yielding deposits of limited thickness.

2 Those based on controlled chemical reduction of the noble metal to form a coating on the base metal. Although a replacement reaction may occur in the initial stages, this is rapidly superseded by the reduction process. The coating metal may or may not act as a catalyst for the reduction process; in either case the coating process can continue after the base metal is completely covered.

These two types of process differ basically in the mechanism of coating formation, a point which is thoroughly analysed by Okinaka in Chapter 11. Immersion Plating is a term that is usually restricted to processes of type 1 and will be so defined for the purpose of the present chapter. Processes of type 2, the so-called electroless or autocatalytic processes, are dealt with elsewhere by Okinaka (see Chapter 11).

THEORY OF REACTION

Probably the most familiar example of immersion plating is the formation of a copper deposit on an iron nail when it is immersed in an acid copper sulphate solution.

$$Cu^{++} + Fe \rightarrow Cu + Fe^{++}$$

Many such reactions could be listed, but quantitative consideration of possibilities depends on a knowledge of the standard electrode potentials of the metals involved, i.e., the potential developed relative to a normal hydrogen reference electrode when the metal is immersed in a solution, normal with respect to ionic concentration. A list of these values for a number of common metals in N sulphuric acid solution is given in Table 10.1[1]. Electrode potentials depend primarily on the concentration of free metal ions in solution and can therefore be changed considerably by the addition of complexing agents, as indicated by the figures shown for cyanide complexes in certain cases.

Normally a metal in this series will displace from solution any metal below it on the list. However, if the difference in potential of the two metals is too great the displacement will proceed too rapidly, with precipitation of the more noble metal occurring as a spongy, non-adherent mass, while the more active metal will continue to dissolve and in doing so undermine any tendency for

Table 10.1

Standard Electrode Potentials

Metal	*1N H_2SO_4*		*Cyanide Complex*
Aluminium	$Al \rightarrow Al^{+++}$	−1.70 V	
Zinc	$Zn \rightarrow Zn^{++}$	−0.76 V	−1.26 V
Iron	$Fe \rightarrow Fe^{++}$	−0.44 V	
Nickel	$Ni \rightarrow Ni^{++}$	−0.22 V	−0.82 V
Tin	$Sn \rightarrow Sn^{++}$	−0.14 V	
Hydrogen	$H_2 \rightarrow 2H^{+}$	0.00 V	
Copper	$Cu \rightarrow Cu^{++}$	+0.34 V	−0.43 V
Silver	$Ag \rightarrow Ag^{+}$	+0.80 V	−0.30 V
Gold	$Au \rightarrow Au^{+}$	+1.50 V	−0.60 V

The signs are those adopted by the Electrochemical Society.

the noble metal to adhere. In practice, therefore, there is only a very narrow range of conditions under which the reaction will proceed in such a way as to result in the formation of an adherent, bright and reasonably continuous coating on the more active base metal. These conditions are as follows:

1 The electrolyte must dissolve the base metal without the formation of insoluble salts, oxides or other products which might hinder contact with the surrounding ions.

2 Attack of the base metal by the electrolyte should not be too vigorous, otherwise there will be no suitable substrate or lattice on which the depositing metal can develop a continuous structure.

3 The noble metal ion must deposit at a rate such that its lattice structure is filled in as orderly a fashion as possible.

4 The driving force (EMF) must be sufficiently low to meet the requirements of (2) and (3).

Since the potential difference necessary to initiate the reaction is small, any decrease in this will slow or stop the reaction. This can occur by a number of mechanisms:

(a) The anode (base metal) area is large at the start. As soon as 60–80% of the area is covered, the ratio of anode : cathode area decreases rapidly with decrease in the rate of displacement.
(b) Any local cold working of the base metal surface, as in various forming operations, will produce a difference in the potential energy in the area concerned and thus alter reactivity to the electrolyte. Some reactions are speeded up by this effect, but most are slowed down or stopped.
(c) The presence of impurities and their segregation at the surface through manufacturing processes will also change the potential in those areas. Such impurities may be intentional additions, e.g., aluminium, lead, selenium, to improve machining characteristics, or introduced accidentally via the embedment of metal slivers, oxide scale and so forth, during the manufacturing process.
(d) Any non-conducting film on the surface arising, for example, from organic contamination, will effectively prevent immersion deposition.

The general connotation of immersion plating in the metal finishing industry is that of an evil that must be overcome in order to obtain good adhesion of deposits, hence the recommended practice of ensuring that all work is electrically alive before it enters the plating solution and the use of high voltage, low metal strikes, where necessary. However, there is a tendency, arising from this, and unjustified in the writer's view, to generally downgrade immersion processes, when poor or non-reproducible results may well be attributable in many cases to inadequate quality control of commercial starting materials, leading to the presence of impurities as mentioned in (c).

Although the theory of immersion plating is quite simple and the general effects of all operating parameters are well known, at least quantitatively, the narrow range of conditions under which useful coatings may be produced requires that any new possibility must be tested by trial and error to determine how critical each of the above factors are. Since the end effect is mainly decorative, there has been little stimulus to develop new processes for mass production.

TYPICAL SOLUTIONS

The historical development of immersion plating processes is described by Weisberg in Chapter 1. The present section will therefore be confined to processes and solutions that are relevant to current commercial practice, as applicable to individual basis metals.

COPPER

In 1931 Sizelove[2] published the following formula for an immersion gold plating solution, which is similar to that used in some of the earlier salt water gilding solutions:

Gold, (as fulminate)	1.25 g/l
Potassium ferrocyanide	180 g/l
Sodium carbonate	90 g/l
Ferric chloride	1 g/l
Sodium hydroxide	2 g/l
Temperature	80–90°C

Since the colour of the coating was very light when the solution was made up initially, a small amount of sodium cuprocyanide was sometimes added to darken the shade. Other workers[3] preferred a higher gold content with the addition of a small quantity of sodium cyanide (1 g/l). Such formulae dominated the literature until 1937 and were probably the forerunners of the following which appeared in the Platers Guidebook in that year[4], and in fact is still listed in the 1973 edition[5].

Sodium gold cyanide	3.7 g/l
Sodium cyanide	30 g/l
Sodium carbonate	45 g/l

According to the original publication, the solution was prepared by boiling in an enamelled iron vessel for 2 hours, and operated at a temperature of 85°C.

Soon after the new acid golds (pH 3.0–6) came into use in the middle to late 1950's, it was found that immersion deposits could be produced from these solutions under certain conditions, temperature and pH being the most critical factors. Typical[6] baths contain some or most of the following: potassium gold cyanide, ammonium hydroxide, citrates, chelates, amines and metallic brighteners. Modifications of these solutions for immersion plating soon appeared in the literature.

Both Vrobel [7] and Robinson[8] increased the pH to 7–8, and used an immersion time of 5–20 minutes at 80°C to deposit gold on copper. The former author further improved his process by adding ammonium chloride[9], while Robinson developed a process in which copper (i.e., printed circuits) was immersion plated with palladium by treatment in a palladium sulphamate solution for 5–15 minutes prior to immersion gold plating for 15 minutes in the following solution at 90°C[10]:

Potassium gold cyanide	5 g/l
Ammonium citrate	20 g/l
Ammonium EDTA	25 g/l
pH (ammonia)	9–10

Duplex plates of this type are quite effective on copper. Good results have also been obtained using electroless nickel plating as a pretreatment, followed by immersion gold in a solution of the type just mentioned, and operated at a pH of 6–8. In contrast to the simple cyanide solutions, formulae of the above type sometimes give irregular results when applied directly to copper, in which context the type and metallurgical history of the basis metal appear to be the critical factors.

In the decorative field the best results on small items are obtained by barrel plating with bright nickel to increase lustre and to provide corrosion resistance, followed by barrel tumbling in the hot citrate gold described in the preceding paragraph. This will give a "24 Carat" colour. For lighter colours, controlled barrel gold electroplating is preferred. Larger items which cannot be treated in bulk can also be processed in electroless nickel followed by immersion gold.

Similar procedures are employed in the industrial field wherever wear resistance is not a critical requirement. Immersion coatings are often protected by an air drying or stoving lacquer to enhance their corrosion resistance.

SILVER

The solutions used for silver are generally similar to those described for copper.

The earliest types of solutions were those based on the neutralised gold chloride solution attributed to both Boyle[11] and later Elkington[12], large volumes of work being processed by this method. Graham[13] claimed an improvement by replacing the chloride by bromide or iodide. McNally[14] patented a mixture of gold chloride, citrates and free hydrochloric acid, and operated at a pH of 0.3–1 for the production of bright, adherent films of gold on silver plated copper coils. Porter[15] utilised an alkaline solution of gold chloride, an acetate and cyanide for a similar purpose. None of these solutions, however, are in active use today.

The hot gold cyanide solutions used for copper have probably had the major use over the longest period for the treatment of silver. In production, it is somewhat difficult to get such a bath "started". Increasing the temperature to 90–95°C is beneficial and the performance of the solution improves with use, probably due to reduction of the cyanide content and the introduction of foreign metal ions.

The acid or neutral citrate gold solutions[7,16,17] normally used for other basis metals rarely work well on silver. It has been suggested that a copper flash will produce good results. Improved results have also been achieved by raising the pH to 10 with ammonium hydroxide and adding a palladium complex salt[18,19]. Since quite heavy coatings (100 microinches) have been obtained in this way, the process is sometimes regarded as being in the "electroless" category, and it is possible that the palladium exerts an activating influence which makes the mechanism a borderline case with respect to a true immersion process. One undesirable feature is that as the gold is deposited the cyanide liberated complexes the palladium very strongly, rendering it ineffective for further participation in the reaction and leading to an increasing content of inactive palladium in the solution. Though with suitable recovery methods the process can be employed with advantage, the overall use is small.

NICKEL

The literature is almost devoid of references to immersion gold plating on nickel until 1950, when a patent[15] appeared covering a solution containing gold chloride (0.8 g/l Au), nickel chloride, phosphate, acetate and cyanide for the plating of radio tubes and similar items. No further reference to this process can be found.

At about the same time it was found possible to deposit up to 50 microinches of gold on to nickel electroplate and high nickel alloy (KOVAR) transistor headers using a standard high gold (8 g/l), medium cyanide plating solution at 90–98°C for 5–20 minutes[20]. However, this solution gave an irregular performance with respect to initial coverage of the basis metal and a variation in thickness of the immersion coating. If a gold colour did not appear on the parts within one minute of immersion, it was necessary to reprocess the items with additional acid activation. These difficulties led to its eventful displacement by the new acid gold barrel plating processes.

The acid citrate immersion golds described under copper,[7,10] work best on nickel and nickel alloy surfaces at a pH of 6–7 and a temperature of 70–90°C. The coating thickness approaches 7–8 microinches in 20 minutes, with some 80% of the deposition occurring during the first 5 minutes. This has proved a major use for this type of gold which shows very low porosity on nickel, while the nickel coating itself provides good corrosion resistance.

NICKEL ALLOYS

The methods described for nickel are usually effective on KOVAR and other high nickel alloys which do not contain chromium.

LEAD AND SOLDER

It is not normal practice to coat these metals with immersion gold. However, the acid and neutral citrate immersion types of solutions will deposit a coating of up to 100 microinches in 5–10 minutes at 70–80°C.

SILICON

In the development of silicon diodes another type of immersion gold had to be used to produce ohmic contacts. Pudvin[21] suggested the following strongly alkaline solution, operated at 40–70°C;

Potassium gold cyanide	10 g/l
Potassium hydroxide	200 g/l

Silverman[22] found that consistent results were obtained by this method only when the silicon surface was irradiated by infra-red light, when gold regularly plated first on the negative side of the junction. Turner[23] used hydrofluoric acid solutions of gold chloride to immersion plate silicon at room temperature and used the procedure to delineate the junction barriers between *p* and *n* type diodes.

Edson[24] also investigated acid solutions and obtained deposits of up to 20 microinches on germanium using a sulphuric acid bath at a pH of 2.5–3.0, but reported that addition of hydrofluoric acid was necessary in the plating of silicon. The operating temperature was high, 70–85°C, and bath life was short. Levi[25] used a similar acid gold solution with hydrofluoric acid at a pH of less than 3.0, but applied an electroless nickel deposit to the silicon prior to immersion gold. Acid versions of the standard citrate gold have also been used.

Mocanu[26] describes a process in which the silicon is first etched in a hot hydroxide solution and then transferred immediately, without rinsing, into a citrate type immersion gold solution at a pH of 7.

TUNGSTEN

The type of solution disclosed by Robinson[16], containing 5 g/l of potassium gold cyanide, 20 g/l of ammonium citrate, and 25 g/l of a chelating agent such as EDTA, will deposit gold on tungsten when used at a pH of 13, though undoubtedly a strong alkaline etch would improve results.

STAINLESS STEEL AND CHROMIUM

It is not difficult to produce immersion coatings of gold on these metals from acid solutions, but the adhesion is poor. Lyons[27] has claimed to have achieved an adherent coating on chromium by operating a bath of similar composition to that of Robinson[16], at 90°C and within the pH range of 2.5–4.5.

DEPOSIT CHARACTERISTICS

Since all immersion coatings are extremely thin, normally less than 10 microinches and often less than 3 microinches, the only tests that can be conducted are for porosity and lubricity. Porosity must be classified as poor since the replacement reaction is never complete. Also, in commercial practice the basis metal itself is rarely free of defects. It must be pointed out here that immersion coatings cannot be expected to resist the type of porosity or corrosion test applied to electroplated coatings. Short time tests such as immersion in nitric acid can be carried out, but these are to be regarded only as screening tests to eliminate coatings of very poor quality.

Coatings from freshly prepared solutions are extremely pure and hence very soft. In most cases it is preferable to use solutions on a batch basis without replenishment, since for every equivalent weight of gold plated out, an equivalent amount of the base metal goes into solution, and with this increase in the ratio of base metal to gold, there is an increasing tendency for deposition of an alloy rather than a pure gold coating. Indeed, in many cases, the addition of other metal ions may be made to produce alloy coatings of different colours, and possibly to improve the wear resistance of the deposits.

Immersion gold deposits on annealed polycrystalline copper substrates are epitaxial according to Sard[28]. It is quite possible that the same applies to coatings on nickel and other base metals, since the process involves atom for atom exchange.

APPLICATIONS

Immersion plating of gold probably experienced its greatest popularity just prior to the development of the cyanide electroplating processes in 1840, most of the applications for which the process was used being subsequently catered for by electroplating. The main applicational area is in the decorative and jewellery fields. In the past, decorative items of copper or brass were often given a gold "wash" followed by a chromate dip, with a final coating of lacquer to still further improve the protective value. As steel and zinc-base alloys became increasingly used it was necessary to apply copper and bright nickel undercoats prior to immersion gilding, and with the advent of the chelate immersion golds there has been a return to the use of the process on a bright nickel undercoat to impart a rich decorative appearance to many items. This finish has even been used to replace lacquered brass for trim on television sets and modern furniture, coffee pots and a variety of gift ware.

In the industrial field the process has clear limitations in respect of the limited thickness and corrosion resistance of immersion coatings. Nevertheless, it has found a number of useful applications in the coating of transistor headers and other electronic components, and the plating of printed wiring patterns which cannot be treated by conventional methods due to the lack of electrical continuity. Immersion gold plating has also been used to improve solderability (see Chapter 19) and to provide a dry lubricant film for the breaking in of hard mating contact surfaces.

Another useful characteristic of thin immersion gold films, especially over a bright plated or polished nickel surface, is that of good infra-red reflectance to reduce heat transfer (see Chapter 39).

Similar coatings have been utilised, in view of their strong de-wetting effect[29], to encourage dropwise condensation on condenser tubes used in de-salting equipment where, if the condensing surface floods, heat exchange is reduced and the columns are much less efficient.

REFERENCES

1 Latimer W. M., "Oxidation Potentials", Prentice-Hall, London (1952).
2 Sizelove O. J., *Mon. Rev. Am. Electropl. Soc.*, **18,** 4, 45 (1931).
3 Anon., *Mon. Rev. Am. Electropl. Soc.*, **12,** 3, 11 (1925).
4 "Metal Finishing Guidebook & Directory", Metal & Plastics Publications, Inc. Westwood, New Jersey (1937).

5 "Metal Finishing Guidebook & Directory", Metal & Plastics Publications, Inc., Westwood, New Jersey (1973).
6 Ehrhardt R., *Proc. Am. Electropl. Soc.*, **47,** 78 (1960).
7 Vrobel L., *Strojirenstvi*, **12,** 684 (1962).
8 Robinson H., *Engelhard Ind. tech. Bull.*, **33,** 86 (1962).
9 Vrobel L., *Korosea Ochrana Mater.*, 50 (1962).
10 Robinson H., U.S. Pat., 3,162,512 (1964).
11 Weisberg A., "History of Electroplating", unpublished work.
12 Elkington G. R., *London J. Arts*, May (1837); Dinglers polytech. J., 65, 42 (1837).
13 Graham A. K., Heiman S., Pinkerton H., *Plating*, **36,** 149 (1949).
14 McNally F., U.S. Pat., 2,836,515 (1958),
15 Porter R., Jones C., U.S. Pat., 2,501,737 (1950).
16 Robinson H., Brit. Pat., 872,785 (1961); U.S. Pat., 3,230,098 (1966).
17 Lareau E., U.S. Pat., 3,266,929 (1966).
18 Duva R., U.S. Pat., 3,396,042 (1968).
19 Moore T., Butler F. P., U.S. Pat., 3,458,542 (1969); U.S. Pat., 3,533,923 (1970).
20 Private communication.
21 Pudvin J. F., "Transistors Technology", Chap. 15 (1958).
22 Silverman S. J., Benn D. R., *J. electrochem. Soc.*, **105,** 170 (1958).
23 Turner D., *J. electrochem. Soc.*, **106,** 701 (1959).
24 Edson G., U.S. Pat., 3,214,292 (1965).
25 Levi C. A., U.S. Pat., 2,995,473 (1961).
26 Mocanu T., U.S. Pat., 3,099,576 (1963).
27 Lyons E., U.S. Pat., 3,502,548 (1970).
28 Sard R., *J. electrochem. Soc.*, **117,** 1157 (1970).
29 White M. L., *J. Phys. Chem.*, **68,** 3083 (1964); Erb R. A., Thelen E., *Ind. Engng. Chem.*, **57,** 10, 49 (1965).

Chapter 11

ELECTROLESS SOLUTIONS

Y. OKINAKA

The principle of electroless gold plating has recently attracted considerable attention from those concerned with the fabrication of electronic devices. The method is simple and economical as a means of depositing gold, for example, as a conductor or contact material in thin film circuits. It also has the unique capability of depositing the metal selectively, only on the areas of a substrate surface where a catalyst is present.

A survey of the literature reveals that a number of electroless gold bath formulations are available. One should be aware of the fact, however, that in both the patent and the open literature, confusion exists as to the usage of terminology. "Electroless plating" is actually a term coined by Brenner and Riddell[1] to describe *autocatalytic* processes of plating metals without the use of a source of external current. By definition, therefore, an electroless process should be capable of depositing a metal on a substrate of the same metal and continuing the deposition through the catalytic action of the deposit itself. Thus *immersion plating* which is based on the principle of galvanic displacement should be clearly distinguished from "electroless plating". In the literature, however, confusing phrases such as "electroless plating in solutions free of extrinsic reducing agent" are found. Impact (mechanical) plating has even been called "electroless", simply because the process does not require external current. Obviously, indiscriminate use of the word "electroless" to describe all plating processes which require no external current should be avoided. In the above examples, however, it is quite clear that the processes in question are not electroless in the sense that they are not autocatalytic. On the other hand, there are cases where the distinction is less straightforward. For example, some "electroless" gold baths deposit gold on basis metals such as copper or nickel to a considerable thickness at a relatively fast rate, but no plating occurs if a sheet of gold is used as the substrate. Because of the fact that the plating does not stop as soon as a thin flash of gold is deposited as it does in immersion (displacement) baths, and these baths contain a reducing agent, they would appear to be "electroless". Nevertheless, the fact that no plating occurs on a gold substrate would indicate that the plating reaction is not autocatalytic and hence not electroless.

There are therefore three possibilities which could explain the behaviour of such baths. (i) The reaction is a galvanic displacement and produces a porous deposit. (ii) The basis metal acts as a catalyst, but gold does not, and the deposit is porous. (iii) The reaction is indeed autocatalytic, but only the freshly deposited gold is active as a catalyst. Although this last possibility appears to be unlikely, it is not generally possible to characterise a specific bath without carrying out a detailed study.

This chapter reviews those baths which are, at least, claimed to be capable of continuously increasing the deposit thickness regardless of whether the deposition reaction is autocatalytic or not. For completeness, some of the early systems which have already been described in review articles[2,3] are also included. The development of baths capable of depositing gold on gold substrates is quite recent and in general, not much information is available, particularly on characteristics of the deposits plated from such baths. The only exception is the bath using an alkali metal borohydride as the reducing agent which was developed by the writer at Bell Telephone Laboratories.[4] A considerable portion of this chapter is therefore devoted to describing this bath and the resulting deposits.

LIMITATIONS

While the electroless gold plating process is attractive in principle because of its simplicity, economy, and ability to plate selectively, it has several limitations which must be clearly understood if one is to apply it successfully. Such limitations are more or less common to all electroless processes.

1 *Low plating rate.* The plating rate of electroless baths cannot be controlled over a wide range as in electrolytic methods. It is limited by the instability of the solution which is usually encountered if one tries to increase the rate beyond a certain value. The maximum plating rate of the electroless gold processes reviewed in this chapter appears to be generally around 5–6 μm/hr. An exception is the aldehyde-amine borane bath which was developed recently by Rich[5], who claims that 1 mil (25 μm)/hr can be achieved (page 100).

2 *Necessity for careful control of plating conditions.* Electroless plating systems are thermodynamically unstable; in other words, thermodynamics favour the reduction of metal ions by the reducing agent in the solution. However, the rate of this reaction is so slow that the solution can remain stable for a long time if the conditions are maintained properly.

Once the conditions are disturbed in such a way that the rate of the homogeneous reaction is increased, the bath decomposes quite rapidly, because as soon as some metal particles begin to form, they act as catalytic centres for the heterogeneous reduction. It is important that parameters such as temperature, concentration and pH are controlled more closely for electroless plating than one is normally accustomed to in electrolytic techniques. For example, since the deposition rate and stability of electroless baths are very temperature sensitive, a temperature control of $\pm 1°C$ may be required for critical applications. The use of a hot plate for direct heating of the bath may lead to decomposition due to local overheating. Because of the sensitivity of electroless systems to heterogeneous catalysis, contamination of the bath with particulate matter such as insoluble impurities in the chemicals or dust particles must be avoided.

3 *Necessity for thorough substrate cleaning.* This is especially important when an electroless bath is used for selective plating. One particular example is in the gold plating of fine line circuit patterns delineated using photolithography. If the areas where gold is not desired are contaminated with even a small amount of organics or metal particles (residues from etching), the metal may initially begin to deposit in the form of discrete particles in the contaminated areas. Such gold particles continue to grow in size and eventually lead to short circuits. Thus, for success in this type of application, it is essential to have available a carefully established specific cleaning procedure. It is also equally important that the catalytic regions are completely free from contamination which might poison the catalytic activity. Such precautions may not be necessary for other methods of gold deposition.

4 *Relatively short bath life.* The life of an electroless solution may be limited not only by spontaneous decomposition caused by extraneous factors such as contamination and so forth, but also by the accumulation of reaction products which may make the system unstable. Generally the life is considerably shorter than that of electrolytic plating baths. For economic reasons, therefore, the utilisation efficiency of gold becomes an important consideration in selecting the bath composition.

5 *Sensitivity of deposit thickness distribution to stirring conditions.* If the deposition rate is partially or totally controlled by mass transfer, a uniform distribution of deposit thickness can be obtained only under uniform mass transfer conditions. Such baths must be operated with uniform mass transfer conditions. A simple motorised stirring rod assembly or a magnetic stirrer may not be satisfactory and the use of a rotating substrate holder may be necessary. It has been stated that an advantage of electroless plating is that it has 100% throwing power. It is true that the deposit thickness distribution is unaffected by the iR drop in the plating solution because no external current is involved. However, it should be kept in mind that both macro- and microthrowing powers can become less than 100% if the deposition rate of a specific bath is sensitive to stirring and this condition is not uniform.

HYPOPHOSPHITE BATHS

In this and subsequent sections, various electroless gold baths are reviewed. As such, they are classified according to the reducing agent employed. The baths containing hypophosphite, the most commonly used reducing agent in electroless plating, are described in the first instance.

SWAN AND GOSTIN[6]

The solution formulation and operating conditions of the Swan-Gostin bath are as follows:

Potassium cyanoaurate	2 g/l
Ammonium chloride	75 g/l
Sodium citrate, dihydrate	50 g/l
Sodium hypophosphite, dihydrate	10 g/l
pH	7–7.5
Temperature, °C	93 ± 1

This bath is extremely temperature sensitive and, with mild agitation, deposits gold on electroless nickel at the following rates: 0.00001 in. (0.25 μm) at 91°C, 0.000092 in. (2.3 μm) at 93°C, and 0.000188 in. (4.8 μm) at 94.5°C in one hour, with a total thickness of 0.0009 in. (23 μm) in 15 hours. However, the deposition rate was found to decrease with time and a periodic activation (method not described) was necessary to build-up the 23 μm thickness. Swan and Gostin used this bath for basket or barrel plating. various electronic components. Jostan and Bogenschütz[7] also employed this process and obtained good results. On extremely smooth copper substrates which were specifically prepared for microelectronic applications, they successfully deposited 0.3–1 μm of gold.

Brenner[3] states that his experiments indicated that this solution deposited gold by immersion (galvanic displacement) rather than by electroless means. Analysis of a nickel substrate upon which gold had been deposited showed that a part of the nickel was lost into the solution. Furthermore, a piece of gold foil did not show any significant increase in weight.

Structure of Deposit

Tanabe and Matsubayashi[8] carried out an electron microscopic and diffraction study of gold deposits formed on single and polycrystals of iron in the Swan-Gostin bath. Their experiments showed that the growth mode in the initial stages of deposition is different from that observed in a galvanic displacement bath consisting of chloroauric acid and hydrochloric acid. Their results are summarised in the following paragraphs.

1 In the very early stages of gold deposition (<0.3 sec) gold particles with diameters less than 100 Å are formed. A number of gold nuclei, 20–30 Å in diameter, were observed with a maximum density of $1.4-2.2 \times 10^{12}/cm^2$. These particles have no preferred orientation with respect to the substrate.

2 These gold particles coalesce to form particles greater than about 130 Å in diameter. At this stage they begin to show epitaxy with respect to the orientation of the substrate, but misalignment still exists. The coalescence of gold particles does not accompany a change in their shape. With the displacement bath containing chloroauric acid, the initial particles grow to form a network structure.

3 The misalignment decreases as the gold deposit grows and it reaches a minimum when the substrate surface is just completely covered with gold crystals. The film was shown to be epitaxial at this stage and orientation relationships were established.

4 Upon further growth, the deposit becomes polycrystalline, because gold particles with random orientation grow at defect sites of the single crystals.

Although the purpose of Tanabe and Matsubayashi's investigation was to compare electroless and displacement deposits, it is doubtful whether their comparison can be generalised, especially when evidence is lacking to show that the Swan-Gostin bath is not a displacement bath. In a more detailed study of these deposits, Tanabe and Matsubayashi[9] observed two kinds of rotational Moiré patterns, stress due to lattice defects and the formation of microtwins.

BROOKSHIRE[10]

Brookshire also used sodium hypophosphite as the reducing agent. Two typical examples of solutions proposed by him are as follows:

	1	*2*
Gold cyanide, AuCN	2.0 g/l	20.0 g/l
Sodium hypophosphite, monohydrate	10.0 g/l	100.0 g/l
Potassium cyanide	0.2 g/l	80.0 g/l
pH	7.5	13.5
Temperature, °C	96	96

Since bath 1 does not contain excess KCN, only a part of AuCN dissolves in the solution in the form of $Au(CN)_2^-$. Initially, there are no free cyanide ions in the solution and the free cyanide ions generated as a result of plating will combine with the excess AuCN. The bath is therefore self-replenishing as long as solid AuCN remains in the bath. Bath 2 contains a large excess of KCN. The plating rates reported are 9.85 mg/cm^2/hr (5.1 μm/hr*) and 12.3 mg/cm^2/hr (6.4 μm/hr) for Baths 1 and 2, respectively. It is claimed that gold will plate on steel, iron, ferrous alloys, nickel, cobalt, gold, silver, platinum, copper, copper-base alloys, magnesium and aluminum. However,

* All conversions from weight to thickness were made assuming a bulk density of 19.3 g/cm^3 for the purpose of approximate comparison of the plating rates of various baths.

the experience of the writer with Bath 2 shows that it does not plate gold on gold and the deposition rate on a sheet nickel substrate is unaffected by the presence or absence of hypophosphite.

EZAWA AND ITO[11] (PURE GOLD)

Four examples of solutions proposed by these investigators are:

	1	*2*	*3*	*4*
Potassium cyanoaurate	2 g/l	1 g/l	1 g/l	3 g/l
Potassium cyanide	5 g/l	3 g/l	1 g/l	5 g/l
Sodium acetate	5 g/l	5 g/l	1 g/l	1 g/l
Sodium hypophosphite	15 g/l	12 g/l	9 g/l	15 g/l
Sulphamic acid	5 g/l	3 g/l	2 g/l	3 g/l
pH	3	4	3–4	3–4
Temperature, °C	75	70	70	80

Bath 1 was used to plate gold on a sheet (25 cm^2) of Fe-Ni (28 %)–Co(18 %) alloy. The deposition rate was 56.7 mg (1.18 μm) in 30 minutes. A 25 cm^2 aluminum sheet activated in a dilute palladium chloride solution was plated with gold in Bath 2. The amount of metal deposited in 30 minutes was 16.3 mg (0.34 μm). Bath 3 deposited 2.8 mg (0.058 μm) of gold in 30 minutes on a 25 cm^2 molybdenum sheet which had also been previously activated. In Bath 4, 33.8 mg (0.70 μm) of gold was deposited on a 25 cm^2 nickel sheet in 30 minutes, while only 9.6 mg (0.2 μm) was obtained in the same time on a copper sheet activated with palladium chloride. The advantages claimed are the lower temperature and higher plating rate as compared to the Swan-Gostin bath and Walton's diethylglycine bath.

EZAWA AND ITO[12] (GOLD-NICKEL ALLOY)

Bright gold deposits containing 0.1–1.2 % Ni were obtained on iron and Fe–Ni(28 %)-Co(18 %) sheets in the following baths:

	1	*2*	*3*
Potassium cyanoaurate	1 g/l	2 g/l	3 g/l
Glycine	10 g/l	5 g/l	20 g/l
Ethylenediamine tartrate*	50 ml/l	50 ml/l	50 ml/l
Nickel chloride	0.13 g/l	1.3 g/l	0.7 g/l
Sodium hypophosphite, monohydrate	5 g/l	5 g/l	3 g/l
pH(NaOH)	8.0	8.0	8.0
Temperature, °C	85	85	85

The nickel content of the gold deposit decreases with increasing plating time. The plating rate was found to decrease with increasing time and

* Saturated solution at 20°C.

ranged from 2.4–3.1 mg (0.25–0.32 μm) in 10 minutes, 4.5–6.5 mg (0.47–0.67 μm) in 30 minutes and 5.7–9.5 (0.59–0.98 μm) in 60 minutes on 5 cm^2 substrates.

HYDRAZINE BATHS

GOSTIN AND SWAN[13]

Gostin and Swan used hydrazine as the reducing agent in the following bath:

Potassium cyanoaurate	3 g/l
Ammonium citrate	90 g/l
Hydrazine hydrate	0.0002 g/l
pH(NH_4OH)	7–7.5
Temperature, °C	92–95

It is claimed that this bath deposits gold on nickel, copper and silver, and it is on these substrates that the greatest adhesion is obtained. It may also be plated directly onto iron, steel and ferrous alloys as well as brass, bronze and other non-ferrous alloys. Data presented by the inventors show that gold plating continues over many hours, is extremely temperature sensitive and yields a thickness of 0.001 in. (25 μm) in 20 hours at 92–94°C. The plating rate was fairly constant for many hours and ranged from 0.00005 in./hr (1.25 μm/hr) at 92–93°C to 0.00029 in./hr (7.25 μm/hr) at 95°C.

LUCE[14]

From a number of bath formulations given by Luce using hydrazine or its derivatives, the following three examples are cited:

	1	*2*	*3*
Chloroauric acid	0.025M(9.9 g/l)	0.013M(5.1 g/l)	0.05M(19.7 g/l)
Sodium sulphite	1.6M(202 g/l)	—	—
Sodium metabisulphite	—	1.0M(190 g/l)	—
Ammonium sulphite, monohydrate	—	—	1.3M(201 g/l)
Ammonium chloride	1.5M(80 g/l)	1.5M(80 g/l)	1.5M(80 g/l)
Cyclohexylamine*	1.3M(129 g/l)	1.3M(129 g/l)	—
Hydrazine	1.8M(58 g/l as N_2H_4)	1.8M(58 g/l) as N_2H_4)	—
Hydroxylamine	—	—	2.0M(66 g/l)
pH(NH_4OH)	>10	>10	>10
Temperature, °C	85	85	85

* Stabiliser.

Bath 1 was used on an electroless nickel plated steel sheet, the deposition rate being 0.00015 in./hr (3.75 μm/hr) with agitation. The solution is stable for only 1–6 hours. Bath 2 was likewise used for plating on a nickel sensitised steel surface, while Bath 3 was utilised when a silver sensitised polyester film substrate was involved. Nickel is said to be the most active catalyst, but the following metals also fulfill this purpose: Co, Cu, Ag, Rh, Au, Pt, Pd. Instability appears to be a problem of this system.

DIETHYLGLYCINE BATH

Walton[15], and Pokras, Sullens and Walton[16], prepared an electroless gold bath for plating on nickel and Fe-Ni-Co alloys such as KOVAR. Optimum solution compositions are as follows:

	1	*2*
Potassium cyanoaurate	28.0 g/l	28.0 g/l
Citric acid, monohydrate	60.0 g/l	—
Tartaric acid	—	60.0 g/l
Tungstic acid	45.0 g/l	45.0 g/l
Sodium hydroxide	16.0 g/l	16.0 g/l
N, N-Diethylglycine, sodium salt	3.75 g/l	11.25 g/l
Phthalic acid, monopotassium salt	25.0 g/l	25.0 g/l
pH	5.0–5.5	5.0–5.5
Temperature, °C	88–93	88–93
Agitation	mechanical, rapid	mechanical, rapid

In preparing this plating solution, it is necessary to follow the proper sequence in mixing the various constituents. The reader is therefore referred to the original paper for specific instructions.[15] It is reported that the most adherent and best appearing gold deposit on KOVAR is obtained from the citric acid solution (Bath 1) whereas the tartaric acid solution (Bath 2) results in the most desirable deposit on nickel base materials. The citric acid solution deposits 5.6 mg Au/in^2 (0.45 μm) in 1 hour and 20 mg/in^2 (1.6 μm) in 6 hours on KOVAR and yields similar results on nickel. The deposition rate on copper is considerably slower, 0.93 mg/in^2 (0.07 μm) in 1 hour and 1.5 mg/in^2 (0.12 μm) in 6 hours. The tartaric acid solution gives similar deposition rates.

It is of interest to note that the deposition rate is greatly dependent on substrate materials. Because of the observed continuous increase in deposit thickness with immersion time, Walton concluded that the process is truly electroless. However, the necessity of having a ferrous alloy or nickel as the substrate for obtaining a reasonable deposition rate appears to indicate that the process is not autocatalytic.

AMMONIA BATH

Schneble, McCormack, and Zeblinsky[17] suggested a rather unusual bath in which the conventional reducing agents are not required, and yet plating is said to continue up to a considerable thickness at a rate of 35–120 millionths of an inch (0.9–3 μm) per hour. The bath is characterised by the presence of a water soluble salt of transition metals or Group Ib metals such as $PdCl_2$, $CoCl_2$, or $CuCl_2$ and a rather large quantity of ammonia. The latter is claimed to help prevent formation of black specks or pits and to yield bright deposits. Examples of bath formulations are:

	1	*2*	*3*	*4*
Potassium cyanoaurate, (as Au)	6.5 g/l	6.5 g/l	6.5 g/l	6.5 g/l
Sodium citrate	—	30 g/l	30 g/l	30 g/l
Palladium chloride	5 g/l	2 g/l	2 g/l	2 g/l
Ammonium hydroxide (28 %)	500 ml	500 ml	500 ml	500 ml
Formaldehyde (37 %)	—	—	10 ml	—
Sodium hypophosphite, monohydrate	—	—	—	10 g/l

The following compounds are mentioned as optional bath stabilisers at concentrations less than 100 g/l, Rochelle salt, gluconic acid, gluconate, triethanolamine, glucono-δ-lactone, EDTA, and modified EDTA. Addition of surfactants such as organic phosphate esters and oxyethylated sodium salts are also reported (<5 g/l) to improve operating efficiency of the baths. An optimum temperature range is 70–80°C.

The reaction mechanism of these baths is not clear, but it is not of the displacement type according to the authors. It is stated that the process is capable of plating gold on Cu, Ni, Ag, Au (autocatalytic), Mo and Co, with particularly good results being obtained on Cu, Ni and Ag.

THIOUREA BATH

Oda and Hayashi[18] obtained a patent for an electroless gold plating solution containing cobalt chloride as a catalyst and thiourea as the complexing/reducing agent. Examples of bath formulations are as follows:

	1	*2*
Potassium cyanoaurate	5 g/l	5 g/l
Thiourea	25 g/l	20 g/l
Ammonium citrate	20 g/l	30 g/l
Cobalt chloride	30 g/l	15 g/l
pH (NH_4OH and citric acid)	6.5	7.0
Temperature, °C	83–87	85–90

A deposition rate of about 5 μm/hr is reported on nickel and KOVAR specimens with vigorous stirring. The bath can be used several times provided that the pH is adjusted prior to each plating process.

The writer has confirmed that this bath plates gold on gold substrates. In spite of the presence of a large quantity of $CoCl_2$ in the solution, a spectrographic analysis did not detect cobalt in the deposit.

BOROHYDRIDE BATH

BATH COMPOSITIONS

The writer[4] developed an autocatalytic gold plating bath using an alkali metal borohydride as the reducing agent. Three typical formulations are:

	1	*2*	*3**
Potassium cyanoaurate	5.8 g/l	1.45 g/l	0.86 g/l
Potassium cyanide	13.0 g/l	13.0 g/l	6.5 g/l
Potassium hydroxide	11.2 g/l	11.2 g/l	11.2 g/l
Potassium borohydride	21.6 g/l	21.6 g/l	10.8 g/l
Temperature, °C	75	75	75

This system has been studied quite extensively at Bell Telephone Laboratories.[4, 19–23, 25, 26] Some of the more important points are summarised in the following text.

SUBSTRATES

This bath deposits gold on metals such as Pd, Pt, Au, Ag and Cu in addition to Ni, Co, Fe and their alloys. However, since deposition on these three metals occurs initially by a displacement reaction, ions of these metals accumulate in the bath. This contamination leads to spontaneous decomposition of the bath and in the case of nickel, to a significant decrease in deposition rate.[23] Therefore, it is generally preferable that substrates of Ni, Co, Fe and their alloys be first coated with a thin layer of displacement gold before they are placed in the borohydride bath. The deposition on copper also begins with a displacement reaction, but the accumulation of dissolved copper in the solution does not lead to such undesirable side effects. The bath also deposits gold on non-conductors such as glass or plastics if they are properly sensitised and activated by, for example, the conventional $SnCl_2$-$PdCl_2$ method.[24] However, preliminary indications are that the adhesion of gold on such substrates is generally poor.

PURITY OF DEPOSIT

Unlike the nickel deposit formed in a borohydride bath for electroless nickel, the gold deposit is essentially pure and contains only 0.0001–0.0004% boron.

PLATING RATE

The plating rate of the borohydride bath decreases with increasing concentrations of KOH and KCN, and increases with KBH_4 concentration and

*A 5× concentrated stock solution can be prepared for this bath. It is stable for at least 2 months at room temperature.

temperature. The effect of $KAu(CN)_2$ concentration is unusual in that, as the $KAu(CN)_2$ concentration is increased, the rate first increases to a maximum and then decreases (Figure 11.1). The plating rate is also sensitive to external agitation. Bath 1, for example, gives a rate of 0.7 μm/hr on a gold sheet substrate without external agitation, while with vigorous agitation of the bath a maximum of about 3.5 μm/hr can be achieved. The variation of plating rate with agitation speed is illustrated in Figure 11.2. The data were obtained using

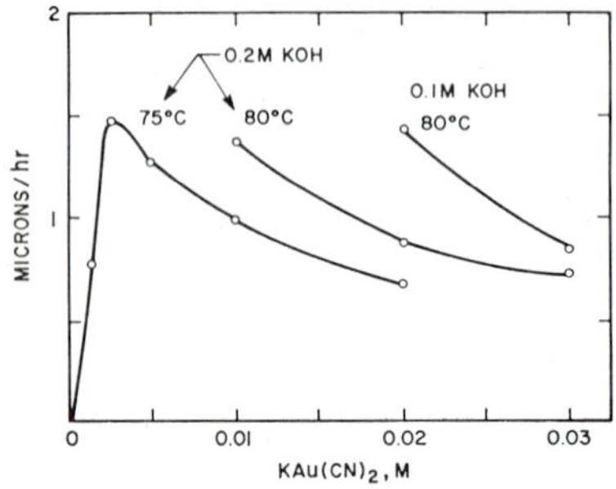

Fig. 11.1 Effect of $KAu(CN)_2$ concentration on deposition rate of the borohydride bath. (KCN 0.2M, KBH_4 0.4M); bath unagitated; gold sheet substrate.

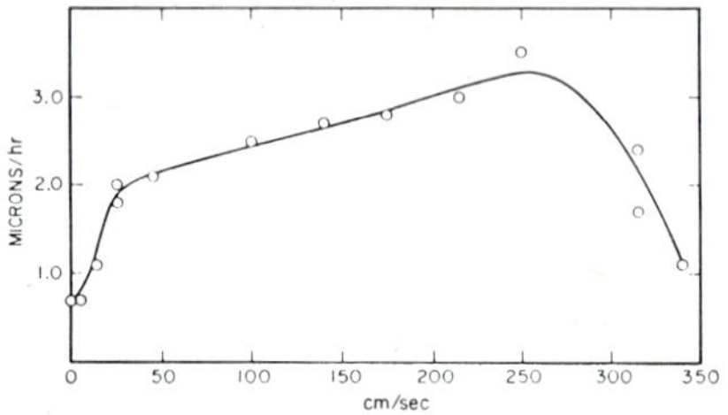

Fig. 11.2 Effect of rotation speed of substrate on deposition rate.

a rotating PTFE holder with substrates of evaporated Au/Ti/glass and plotted against the linear velocity of the substrate surface. It is seen that a maximum plating rate occurs at about 250 cm/sec. The decrease at higher speeds seems to be at least partly due to a compensating effect of the solution motion caused by the rotation of the holder. Bath 2 with vigorous agitation yields a rate of about 6 μm/hr. The plating rate of Bath 3 is about the same as that of Bath 2. Bath 1 yields deposits with satisfactory properties regardless of whether it is agitated or not, whereas Baths 2 and 3 must be used with agitation to obtain deposits of good quality. The reasons are discussed further on in this chapter under the headings "Structure of Deposit" and "Physical Properties". For general practical purposes, it is recommended that either the solution be stirred vigorously, or the substrate held in a suitable holder and rotated at a fast rate. More uniform deposits can be obtained by the latter method.

The deposition rate of the borohydride bath is dependent upon the crystalline orientation of the substrate surface. Using single crystals of copper as substrates, Sard and Wonsiewicz[25] discovered that with agitation, the deposition rate on [111] surfaces is about four times greater than that on [110], and the rate on [100] is intermediate. The magnitude of the orientation effect is reduced by a factor of two for unagitated solutions.

BATH STABILITY

A general problem common to all electroless plating systems is that of bath instability and the borohydride system is no exception. It is sensitive to soluble as well as insoluble impurities and since the material itself often contains insoluble particles, filtration is recommended after the bath is prepared. The plating vessel and substrate holder must be free from contamination. The temperature should not exceed 85°C even locally. A thermostatically controlled water bath regulated to ± 1°C is recommended for heating the plating bath in order to obtain reproducible plating rates. If such care is exercised, the borohydride bath can be operated for about 20 hours with periodic replenishment of constituents (see the next section). Alternatively, the bath can be operated more simply in a single batch operation with unused gold reclaimed by boiling. In this case, the bath can be used for 4–5 hours.

The borohydride bath does not decompose spontaneously at room temperature. However, the plating rate decreases slowly (about 1% per day) on storage because of the gradual hydrolysis of borohydride ions.

REPLENISHMENT OF BATH CONSTITUENTS

Both gold and KBH_4 can be added periodically to the bath for the purpose of replenishment.[20] In order to avoid decomposition due to local concentrations combined with a high temperature, the replenishment should be carried out at a temperature below 40°C. As the source of gold, AuCN rather than $KAu(CN)_2$ is recommended, because accumulation of excess CN^- tends to slow down the deposition reaction. The amount of AuCN to be added can be calculated from the initial $Au(CN)_2^-$ concentration and the weight of gold that has been deposited, or from a chemical analysis of the solution using, for example, an atomic absorption method (see Chapter 33). The amount of KBH_4 to be added can be found either by chemical analysis of the plating solution or by calculation using the rate constant of the first step of the hydrolysis reaction. The latter procedure is applicable provided that the total length of time for which the bath has been at the operating temperature is known. It has been established that the consumption of KBH_4 is mostly by hydrolysis and the quantity consumed for gold deposition is quite negligible. For Bath 1 at 75°C, the decrease in BH_4^- concentration with time has been determined and this is shown in Figure 11.3. Also shown in this figure are variations of the deposition rate (without agitation) and concentration of $Au(CN)_2^-$ with time during plating on an 80 cm^2 copper substrate in 500 ml solution. The decrease in plating rate with time shows that the bath should be replenished after 4–5 hours of operation. In one experiment, a bath was replenished four times with both AuCN and KBH_4 and operated for a total of 21 hours. The bath eventually became unstable because of the accumulation of $B(OH)_4^-$ (hydrated form of BO_2^-), a final product of oxidation and hydrolysis of borohydride.

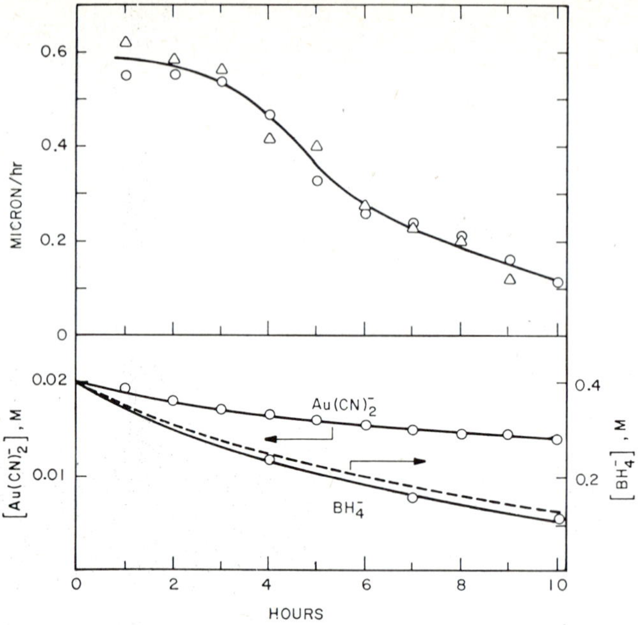

Fig. 11.3 Variation of deposition rate and concentration of $Au(CN)_2^-$ and BH_4^- with time for Bath 1 (unagitated). (Substrate, 80 cm^2 copper sheet; solution volume, 500 ml; 75°C; dashed line, calculated using rate constant).

REACTION MECHANISM

The writer has carried out an electrochemical and polarographic study[22] of this system. Results are briefly summarised in the following paragraphs:

The hydrolysis of BH_4^- proceeds in two steps.

$$BH_4^- + H_2O \xrightarrow{k_1} BH_3OH^- + H_2 \tag{1}$$

$$BH_3OH^- + H_2O \xrightarrow{k_2} BO_2^- + 3H_2 \tag{2}$$

The species that acts as the reducing agent for $Au(CN)_2^-$ is BH_3OH^-, the intermediate formed by reaction (1), and not BH_4^- itself. In other words, BH_3OH^- is a stronger reducing agent than BH_4^-. The pseudo first order rate constants for the above two reactions as determined by polarography are $k_1 = 3.29 \times 10^{-5}$ sec^{-1} and $k_2 = 1.43 \times 10^{-3}$ sec^{-1} in 0.2M KOH at 75°C.

The deposition reaction is a mixed potential reaction consisting of the following two partial reactions:

$$Au(CN)_2^- + e^- \rightarrow Au + 2CN^- \tag{3}$$

$$BH_3OH^- + 3OH^- \rightarrow BO_2^- + 3/2H_2 + 2H_2O + 3e^- \tag{4}$$

It appears that $Au(CN)_2^-$ and BH_3OH^- adsorb competitively on gold, the former being more strongly adsorbed and tending to inhibit the adsorption of the latter. This accounts for the peculiar dependence of the deposition rate on the $Au(CN)_2^-$ concentration shown in Figure 11.1.

STRUCTURE OF DEPOSIT

The growth and structure of gold deposits produced in the borohydride bath have been thoroughly studied.[19,21,23] In this section, a summary is presented of the surface morphology and the orientation of the electroless gold deposits. A brief description is also included concerning nodule formation and its prevention.

Morphology

The colour of the electroless gold coating formed in the borohydride bath varies, depending on the deposition conditions and the solution composition. For example, Bath 1 (p. 91) produces a typical yellowish gold colour. If the $KAu(CN)_2$ concentration of this bath is decreased to 0.005M (Bath 2) or less and the plating is carried out without bath agitation, the colour becomes orange to brown. This discoloration is also obtained at higher temperatures and KBH_4 concentrations. Since these two factors increase the speed of deposition, the discoloration would seem to be due to the high plating rate. However, if the rate is increased by stirring the solution, exactly the opposite trend is observed. Namely, the conditions which produce a dark deposit without agitation yield light yellow deposits with agitation.

It has been found[19] that the colour of electroless gold deposits is determined solely by the surface morphology. Three distinctly different morphologies of electroless gold deposits are illustrated in Figure 11.4 by scanning electron micrographs. Figure 11.4a shows the surface morphology of deposits about 2 μm thick on an evaporated Au/Ti/glass substrate plated in a solution with the composition of Bath 1 at 75°C, except the concentration of $KAu(CN)_2$ was decreased to 0.0025M (0.73 g/l). The plating was carried out without agitation. The deposit was brown in colour, consisting of particles smaller than about 1 μm and protruding outwards from the surface. This type of structure is referred to as "outward growth". Figure 11.4b illustrates the morphology of a deposit obtained from Bath 1 (0.02 M $KAu(CN)_2$) also with no external stirring. The deposit was smooth and yellow, with a fine structure consisting of a network of web-like features. This type of structure is called "lateral growth". Figure 11.4c shows the structure of a deposit obtained in Bath 1 with vigorous agitation. This deposit is characterised by well developed crystalline facets. The colour is also the typical light yellow of gold. This structure is believed to be a better developed form of the lateral growth structure. As described in a subsequent section, the porosity is directly related to the morphology.

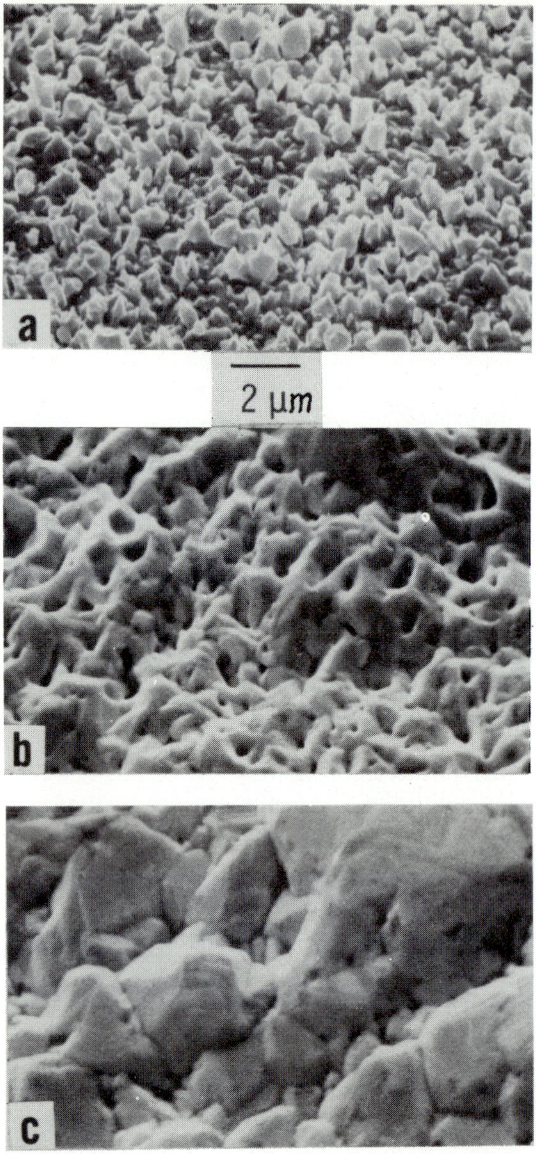

Fig. 11.4 Morphologies of electroless gold deposits obtained from the borohydride bath. Substrate, evaporated Au/Ti/glass (a) Bath 1 except [$KAu(CN)_2$] = 0.0025M, unagitated (b) Bath 1 [$KAu(CN)_2$] = 0.02M unagitated (c) Bath 1, strongly agitated.

Since space does not permit a detailed discussion of the different morphologies, interested readers are referred to the original papers.[19,21]

Orientation of Deposit

The crystalline orientation of electroless gold deposits is dependent on both substrate and deposition conditions, and it varies with the deposit thickness. For copper substrates, Sard[26] observed that the initial gold films of 1,000–1,500 Å are epitaxial, i.e., the gold films deposited onto [100], [110] and [111] copper surfaces yielded electron diffraction patterns of single crystal gold. This is not surprising because the initial deposition reaction on copper substrates is a galvanic displacement.

Sard[21] also carried out an extensive investigation of the relationship between deposit orientation and deposition conditions for different substrates. For electroless gold films (0.3 μm thick) on polycrystalline gold sheets, an X-ray diffraction study showed that decreasing $KAu(CN)_2$ and increasing KBH_4 concentrations favour increased [111] preferred orientation. However, for chemically activated roughened glass substrates (using a $SnCl_2$–$PdCl_2$ process), only randomly oriented deposits were obtained. Clearly, the initial stage of growth is greatly influenced by the nature of the substrate.

With copper and gold substrates, it was found that when conditions favour the lateral type of growth, the resulting deposit tends to remain in parallel orientation, while the outward type deposits form with [111] preferred orientation. This tendency becomes more pronounced as the deposition time is increased. It is an important feature of the deposition process that under conditions where lateral growth occurs, the substrate orientation continues into the deposit.

If the substrate has a [111] fibre texture such as for evaporated Au/Ti/glass, electroless gold deposits also show a very strong [111] fibre texture regardless of the deposition conditions (despite morphological variations). If the substrate does not have such a strong orientation (e.g., evaporated Pd/Ti/glass), deposit orientation depends on solution variables and deposition conditions.

Thus, there is no simple relationship between preferred orientation and surface morphology for electroless gold deposits because of the combined effects of substrate and deposition variables. To illustrate this complexity a schematic summary of X-ray data[21] is given in Figure 11.5 for various substrates. Arrow signs indicate the direction of variations that occur as deposit thickness increases. The horizontal axis is the ratio of diffracted intensities [111]/[200], which was used as an index of preferred orientation. More detailed experiments of deposit texture were carried out by Sard and Wonsiewicz[25] using single crystal substrates. As already mentioned, they found that the deposition rate is significantly affected by the crystalline orientation of the substrate.

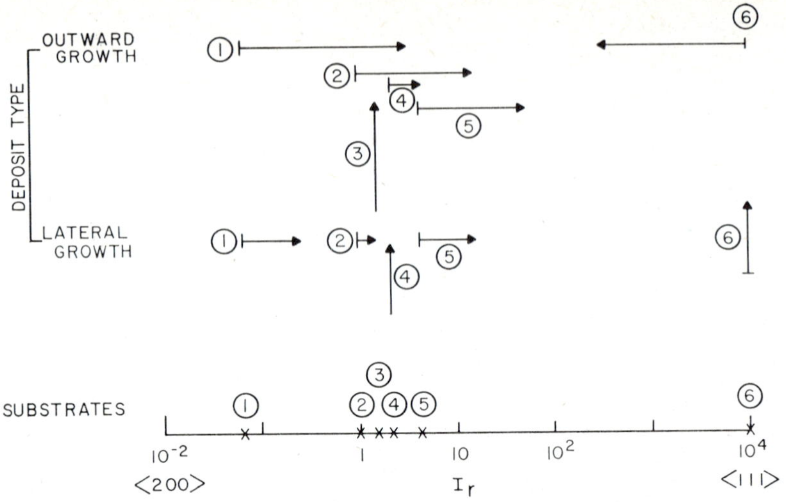

Fig. 11.5 Changes in degree of preferred orientation of electroless gold deposits on various substrates with increasing thickness. Substrates are (1) copper sheet (2) gold sheet (3) rough glass sensitised and activated by the $SnCl_2$–$PdCl_2$ method (4) evaporated Pd/NiCr/alumina (5) evaporated Pd/Ti/glass (6) evaporated Au/Ti/glass.

Nodule Formation and its Prevention

If the plating is carried out in an unstirred borohydride bath the resulting deposit may contain nodules, especially when the substrate surface is smooth as in the case of metallised glass or silicon slices.[4,23] These nodules protrude outward above the film surface and may grow to a size of 3–5 microns in diameter. Soluble and insoluble impurities in the plating solution and defects such as pinholes and scratches in the substrate surface contribute to the nodule formation. Among the impurities that promote nodule formation are certain organic surface active compounds and salts of Ni, Co and Fe. However, when vigorous agitation is used, nodule formation can be significantly reduced or eliminated. As already mentioned, bath agitation increases the plating rate. In view of these beneficial effects it is recommended for general purposes that plating should be performed with agitation.

PHYSICAL PROPERTIES

Although a number of bath formulations are reported in the literature and have been reviewed elsewhere in the text, extremely little data have been reported on physical properties of electroless gold deposits except for those deposited in the borohydride bath. For any practical applications of gold deposition, knowledge of physical properties of the deposit is essential. This is true, particularly for electronics applications, so it is highly desirable that at least some essential properties such as porosity, hardness and electrical resistivity be measured and reported for any new electroless gold processes that might be developed in the future.

The physical properties of deposits obtained from the borohydride bath have been extensively investigated.[4,19,23] The density of the deposit is essentially equal to that of bulk gold. Deposits obtained from Bath 1 have a Knoop hardness of 62–63 as measured with a 25 g load on a cross-section of a 15–25 μm thick deposit. Bath 2 gives deposits with a hardness of 82–84.* The corresponding hardness value of pure, annealed gold is 30–40. Measurements were also carried out on thin films and it was shown that under certain circumstances, it was possible to obtain reliable hardness values with a 1 g load for 4–5 μm films.[19] Bath agitation does not affect the deposit hardness.

The sheet resistance of electroless gold deposits is very close to that of pure, annealed gold.[19] The sheet resistance value decreases from 3×10^{-2} ohms/square at 0.15 μm to 1×10^{-2} ohms/square at 1.4 μm for deposits on substrates of unglazed alumina coated with 300 Å of evaporated Ti and followed by approximately 1 μm of evaporated copper.

The porosity of electroless gold deposits of the same nominal thickness depends greatly on the morphology. Deposits with a lateral structure are much less porous than those with an outward structure.[19] Also, deposits obtained with bath agitation and hence with a well developed faceted structure, are even less porous than those with "web-like" features obtained without bath agitation.[23] The least porous coatings are comparable to the least porous electrolytic gold. In general, it can be stated that non-yellow deposits with outward growth structure are quite porous and hence should be avoided.

For thin film circuit applications, thermocompression bonding capability is important. Tests carried out on thermocompression bonded wires on electroless gold deposits thicker than 0.5 μm on a variety of substrates, indicated satisfactory bondability and mechanical properties.

DIEMETHYLAMINE BORANE BATHS

OKINAKA[4]

The writer[4] prepared an autocatalytic bath using dimethylamine borane as the reducing agent. A typical bath has the following composition:

Potassium cyanoaurate	5.8 g/l
Potassium cyanide	1.3 g/l
Potassium hydroxide	45 g/l
Dimethylamine borane	23.6 g/l
Temperature, °C	85

* This is still considered to be soft gold relative to hard gold deposits produced electrolytically for contacts where Knoop hardness values are in the 140–250 range.

The high KOH concentration is necessary to obtain a reasonable deposition rate. It appears that, as in the case of the borohydride bath, the reducing species is the BH_3OH^- ion which may form as a result of the reaction of dimethylamine borane with OH^- ions:

$$(CH_3)_2NH \cdot BH_3 + OH^- \rightarrow (CH_3)_2NH + BH_3OH^- \quad (5)$$

The maximum deposition rate is only about half that of the borohydride system described in the preceding section. Because of the high alkalinity and low plating rate of the dimethylamine borane bath, the borohydride system is preferred for practical purposes.

McCORMACK[27]

More recently, McCormack obtained a patent for an autocatalytic system using dimethylamine borane as the reducing agent. The following example is cited in the patent:

Chloroauric acid	0.01 M (4.12 g/l)
Sodium potassium tartrate, tetrahydrate	0.014 M (3.95 g/l)
Dimethylamine borane	0.013 M (0.76 g/l)
Sodium cyanide	400 mg/l
pH (NaOH)	13
Temperature, °C	60

The addition of surfactants such as organic phosphate esters and oxyethylated sodium salts is recommended for best results. The amount of NaCN (stabiliser) is critical, the bath not functioning if the concentration exceeds 500 mg/l. It is claimed that the following metals are catalytic: Ni, Co, Fe, steel, Pd, Pt, Cu, brass, Mn, Cr, Mo, W, Ti, Sn and Ag. The deposit obtained from this bath is reported to be bright and ductile. The plating rate is not mentioned in the patent.

ALDEHYDE-AMINE BORANE BATH

Very recently, Rich[5] reported on an electroless gold plating solution with a synergistic reducing system consisting of an aldehyde and an amine borane. Neither of the reducing agents, if used alone yield a usable deposition rate, but in combination result in deposition rates up to 1 mil/hr. The following composition has been given as a typical example:

Chloroauric acid	0.0075–0.025 M (3.0–9.9 g/l)
Sodium sulphite	0.05–1.60 M (6.3–202 g/l)
Formaldehyde	0.61–1.23 M (18–37 g/l as HCHO)
Dimethylamine borane	0.05–0.17 M (2.9–10 g/l)
pH	10
Temperature, °C	85–90

Also reported are bath formulations containing $KAu(CN)_2$, sodium or ammonium citrate, sodium tartrate, acetaldehyde and trimethylamine borane. Rich believes that the aldehyde is the primary reducing agent, whereas the role of amine borane is to reduce the formate ion, the oxidation product of HCHO, back to the aldehyde. The process is autocatalytic and other metals such as Pd and Ni also serve as the catalyst. The deposit is 99.9% pure, contains 0.05% each of carbon and boron and yields Knoop hardness values of 60–66. The bath is stable for 3–5 hours. Because of the very high plating rate reported, this bath appears to be of practical interest. Further information, however, must await publication of the full paper.

APPLICATIONS

The literature does not contain any description of large scale practical applications of electroless gold. It is believed that this is due to the fact that truly electroless autocatalytic gold processes have been developed only very recently. The development of such baths has been specifically aimed at applications in the fabrication of electronic devices. Electroless gold may find use as contacts for passive components of integrated circuits, conductor paths, bonding pads for beam leads and plating beam leads on silicon integrated circuits.

The electroless gold process at its present stage of development is more complementary than it is competitive with the other methods of gold deposition, i.e., conventional electrolytic, immersion, aerosol and brush plating, vacuum evaporation and sputtering. The process is expected to be applied successfully where its unique catalytic feature is essential and other methods fail, and also where its inherent economical factors are properly realised. In many respects the current status of the electroless gold process is more or less comparable to that of the best electroless copper baths available at the present time. It is not yet comparable to electroless nickel systems, which are now quite highly developed from a technological aspect. In view of the fact that the electroless gold process is finding many important applications in highly sophisticated areas of technology such as the fabrication of integrated circuits, it is essential that the process should have a very high level of reliability. Efforts being made at the present time for further improvements are directed toward making the electroless gold process more compatible with the modern technology of electronic device fabrication. There is little doubt that this objective will soon be achieved.

Acknowledgments

The author is indebted to R. Sard and D. R. Turner who critically reviewed this manuscript and made valuable suggestions.

REFERENCES

1 Brenner A., Riddel G. E., *J. Res. Natn. Bur. Stand*, **37,** 31 (1946); *Proc. Am. Electropl. Soc.*, **33,** 23 (1946).
2 Rhoda R. N., *Plating*, **50,** 307 (1963).
3 Brenner A., "Modern Electroplating", 2nd Edn, F. A. Lowenheim (Ed.), Wiley, New York (1963).
4 Okinaka Y., *Plating*, **57,** 914 (1970).
5 Rich D. W., *Proc. Am. Electropl. Soc.*, **58,** (1971).
6 Swan S. D., Gostin E. L., *Metal Finish.*, **59,** 4, 52 (1961).
7 Jostan J. L., Bogenschütz A. F., *Plating*, **56,** 399 (1969).
8 Tanabe Y., Matsubayashi H., *J. Metal Finish. Soc. Japan*, **21,** 6, 335 (1970).
9 Tanabe Y., Matsubayashi H., *J. Metal Finish. Soc. Japan*, **21,** 8, 436 (1970).
10 Brookshire R. R., U.S. Pat., 2,976,181 (1961).
11 Ezawa T., Ito H., Jap. Pat., 40–1081 (1965).
12 Ezawa T., Ito H., Jap. Pat., 43–25881 (1968).
13 Gostin E. L., Swan S. D., U.S. Pat., 3,032, 436 (1962).
14 Luce B. M., U.S. Pat., 3,300,328 (1967).
15 Walton R. F., *J. electrochem. Soc.*, **108,** 8, 767 (1961).
16 Pokras D. S., Sullens T. L., Walton R. F., U.S. Pat., 3,123,484 (1964).
17 Schneble F. W., McCormack J. F., Zeblinsky R. J., U.S. Pat., 3,468,676 (1969).
18 Oda T., Hayashi K., U.S. Pat., 3,506,462 (1970).
19 Sard R., Okinaka Y., Rushton J. R., *Plating*, **58,** 893 (1971).
20 Okinaka Y., Wolowodiuk C., *Plating*, **58,** 1080 (1971).
21 Sard R., Paper presented at the Fall Meeting of the Electrochemical Society, Cleveland, Ohio (1971).
22 Okinaka Y., *J. electrochem. Soc.*, **120,** 739 (1973).
23 Okinaka Y., Sard R., Craft W. H., Wolowodiuk C., Paper presented at the Fall Meeting of the Electrochemical Society, Miami Beach, Florida (1972).
24 See, for example, Goldie W., "Metallic Coatings of Plastics", Vol. 1, Chap. 5, Electrochemical Publications, Ayr, Scotland (1968).
25 Sard R., Wonsiewicz B. C., to be published.
26 Sard R., *J. electrochem. Soc.*, **117,** 9, 1156 (1970).
27 McCormack J. F., U.S. Pat., 3,589,916 (1971).

PART 3

Techniques

Chapter 12

PRETREATMENT AND PLATING PROCEDURES

W. K. A. CONGREVE

Pretreatment of the basis metal is of fundamental importance in gold plating, as in all other branches of electrodeposition, since its effectiveness puts an upper limit on the quality of the final coating. In this chapter recommendations are given on pretreatment procedures for the gold plating of metals and semiconductors commonly involved in present day gold plating practices. A considerable amount of space is also devoted to processes for plating the less common metals, which today concern only a few specialist platers, but may be of more general interest in the future.

Pretreatment for gold plating is of course based on the same principles, uses the same techniques and has the same objective as for any other plating process. The only difference may be one of degree. Owing to the high intrinsic cost of gold and its widespread use in unusually exacting applications, which often involve very expensive components, greater emphasis is commonly placed on economy in the use of the metal and on the quality of the plating. Both these requirements demand careful planning and design, and rigorous control of the pretreatment processes.

Ideally the condition of the surface of a component immediately prior to entering the gold plating electrolyte should be as follows:

1 Completely free from grease and oil.
2 Free from significant oxide and other surface films.
3 Free from gross defects of surface finish such as scratches, tool marks, casting defects and pitting from over-pickling, corrosion, etc.
4 Of uniform appearance.
5 Smooth and preferably bright.
6 Free from severe work hardening.

To this list of surface requirements must be added certain conditions regarding the quality of the bulk metal, such as:

7 Freedom from excessive inclusions.
8 Freedom from porosity.
9 Freedom from excessive amounts of dissolved gas.

Of these requirements those relating to the bulk metal are outside the control of the plating shop and not only emphasise the need for close co-operation between plater and manufacturer, but also stress the importance of the thorough inspection of components before acceptance.

Within the plating shop itself some or all of the following steps are required to ensure that the article is in a suitable condition for immersion in the electrolyte:

1 *Inspection* The first, and essential, step in pretreatment of incoming components is a thorough inspection with the object of determining the nature and degree of pretreatment necessary. The efficiency and profitability of the jobbing plating shop will depend to a considerable extent on the effectiveness of this inspection.
2 *Mechanical Polishing and Buffing* Inefficient mechanical polishing can cause serious difficulties in plating, e.g., abrasive particles embedded in the softer metals, defects in the basis metal smeared over to reappear in subsequent processing, overheating of the surface and consequent charring of the buffing wax, etc. Platers having insufficient of this work to employ full time specialist operators usually prefer to use sub-contractors.
3 *Degreasing* Mineral oils and greases are best removed before the components enter the plating shop in order to avoid the danger of cross-contamination. The familiar methods of solvent vapour and emulsion cleaners must be carried out efficiently, since these compounds are not destroyed in the subsequent electrolytic cleaners and quite small amounts carried over can cause serious plating defects.
4 *Cleaning* In the writer's experience inefficient cleaning is the most common cause of plating defects. This is commonly attributable either to under-design, *i.e.*, an inadequate factor of safety to allow for unusually stubborn soils, or to a lack of proper maintenance and control.
5 *Descaling* Heat scale is often difficult and expensive to remove and often causes deterioration of the surface finish. It can usually be prevented by performing any necessary heat treatments including welding and brazing, in reducing or neutral atmospheres.
6 *Electrolytic and Chemical Polishing* These processes are widely used to improve the surface finish of articles before plating. Their action is one of brightening rather than smoothing, and they are frequently used as the final stage following mechanical polishing. In this context they are particularly useful as they remove the undesirable outermost layer of work hardened metal.
7 *Removal of Oxides* This step in the pretreatment of metals is usually called "etching". The word is an unfortunate carry over from the days when it was believed that roughening of the surface, and hence mechanical keying, was necessary in order to obtain good adhesion. In fact, with only a very few possible exceptions, the optimum surface for plating is one that is smooth, bright and free from work-hardening. The "etching" processes

are therefore designed either to remove a minimum of metal, or to combine oxide removal with chemical or electrolytic polishing.

8 *Strikes and Undercoats* Form an essential part of many plating processes and are discussed in general under their respective headings and are detailed in the preplating schedules for individual metals.

The field of pretreatment is clearly far too wide to be covered in a single chapter. It has therefore been necessary to assume that the reader is familiar with the principles and practice of mechanical polishing, degreasing and cleaning, and to refer to these only when for some reason it is necessary to depart from normal practice. Similarly, it has been necessary to severely limit or omit discussion of the principles underlying the processes recommended. In most cases, however, these can be found in the references cited.

It has been necessary to be highly selective between the very large numbers of published processes. The writer has been guided in this primarily by the requirements of the general plater who is interested in processes by which he can plate successfully onto a wide variety of substrates without having to maintain an unduly large number of reagents. The specialist plater, concerned with production line plating of a few articles, will be able, by referring to the literature and by experimentation, to find in many instances special purpose reagents and processes which will be more economical and which may give somewhat better surface finishes.

The majority of processes outlined have either been used successfully for a number of years by the writer in small scale plating or have gained general acceptance in the plating field.

Some of these processes for the common easy-to-plate metals are considerably more elaborate than those found in the literature. The reason for this difference is that the latter are often designed for the production of cheap decorative plate where adhesion, porosity, etc., are of less critical importance. At the other extreme these simpler processes may be suitable for high quality work in fully integrated mass production units where the quality of the basis metal is strictly controlled. The processes recommended in this chapter are considered to be the minimum necessary for a general plating shop to obtain consistently sound, adherent deposits with a reasonable factor of safety to allow for variations in the basis metal and for other contingencies.

STRIKE SOLUTIONS

In the processes which follow, frequent reference is made to the use of cyanide copper or acid nickel strikes. The function of these is to protect the surface from attack by the electrolyte, with possible production of non-adherent immersion deposits and/or contamination of the plating bath or, in the case of refractory metals, to activate the surface by electrolytic reduction of the oxide film and immediate replacement of this by a protective film of an easily plateable metal.

Where the quality of the basis metal is strictly controlled a gold strike may well be satisfactory as an activating treatment, as for example, in the plating of semiconductors (see Chapter 37) and of components used in space applications (see Chapter 39). Gold, however, is less effective than acid nickel or cyanide copper as a reducing activating strike and does not provide an adequate alternative when plating onto refractory metals, or even on the more easily plateable metals of the variable quality that may be encountered in the general plating shop.

Fully adherent deposits of gold can be deposited on copper or nickel without a gold strike.

GOLD

An effective cyanide strike formulation is as follows:

Gold potassium cyanide	1.5 g/l
Potassium cyanide (free)	10 g/l
Potassium hydrogen phosphate	15 g/l
pH	10.0–11.5
Temperature	50–60°C
Current density	1.0–1.5 A/dm^2

NICKEL

The Woods nickel strike has been used successfully for many years, typical composition and operating conditions of this type of solution being as follows:

Nickel chloride, hexahydrate	240 g/l
Hydrochloric acid*	125 g/l
Temperature	Room
Current density	2–4 A/dm^2
Polarity	Anodic-Cathodic or Cathodic only

A strike used successfully by the writer in the plating of nickel and nickel-containing alloys, stainless steels and many difficult-to-plate metals and alloys, is a low nickel/high sulphuric acid solution of the following formulation:

Nickel chloride, hexahydrate	20 g/l
Sulphuric acid	30% v/v

* When common mineral acids occur in formulations without further qualification, reference is to normal concentrated acids of the following specific gravities:

Sulphuric	1.84
Hydrochloric	1.16
Nitric	1.42
Phosphoric	1.75

Used cathodically at room temperature with an applied voltage of 8 volts for 20–30 seconds, this solution deposits a very thin layer of nickel, too thin in fact to do more than impart a greyish tinge to a brass surface, but which is nevertheless, an extremely effective basis for further plating.

COPPER

The use of copper as a strike has the special advantage of providing a visual indication of the effectiveness of the pretreatment. On a properly prepared surface the copper film will be completely uniform in colour and brightness. The presence of discoloured areas gives a valuable indication of faulty pretreatment at a stage when it is relatively simple to strip the very thin copper deposit and apply remedial measures. This is particularly useful in plating articles of complex shape where cleaning or etching processes may have failed to penetrate to the base of recesses, and also in the plating of bi-metallic components.

The conventional cyanide or Rochelle salt copper electrolyte operated at high current density with copious gassing makes an effective alkaline strike.

Dini and Helms[1] have recently described the following acid copper strike solution:

Copper sulphate, pentahydrate	0.375 g/l
Hydrochloric acid	37% v/v
Current density	4–6 A/dm^2

This was developed primarily for the pretreatment of certain stainless steels, but appears to have wider application. It is comparable to the Woods nickel solution in respect of low metal/high acid content, but perhaps with the added advantage of providing the diagnostic feature of copper just mentioned.

UNDERCOATS

Undercoats of substantial thickness, most frequently of copper or nickel, are often used in gold plating for the following reasons:

(*a*) As an alternative to bright dipping or electropolishing to improve surface finish.
(*b*) To reduce porosity and improve corrosion resistance.
(*c*) To act as a diffusion barrier and improve high temperature stability.

Many proprietary processes are available which are capable of producing bright deposits of these metals with good levelling properties when required. These, however, require a fair amount of control unless used more or less continuously, hence to meet occasional needs it is sometimes convenient to use dull or semi-bright processes and to improve the surface brightness as required by electropolishing or bright dipping.

Typical general purpose nickel plating baths are as follows:

Watts type (dull)	
Nickel sulphate, hexahydrate	300 g/l
Nickel chloride, hexahydrate	40 g/l
Boric acid	30 g/l
pH	3.5–4.0
Temperature	50°C
Current density	2–3 A/dm^2
Sulphamate (semi-bright)	
Nickel sulphamate	600 g/l
Nickel chloride, hexahydrate	5 g/l
Boric acid	40 g/l
pH	4.6
Temperature	60°C
Current density	2–5 A/dm^2

The following Rochelle salt copper bath produces semi-bright deposits:

Copper cyanide	50–55 g/l
Sodium cyanide	65 g/l
Sodium potassium tartrate	40 g/l
Sodium carbonate	40 g/l
Sodium cyanide (free)	6–7 g/l
pH	11.0–12.0
Temperature	65°C
Current density	2–3 A/dm^2

PRETREATMENT AND PLATING PROCEDURES

COPPER AND COPPER-BASE ALLOYS

Plating on copper and copper-base alloys presents no real problems. Effective processes for descaling, electropolishing and bright dipping are readily available.

Descaling

The following solution used at room temperature is quite satisfactory:

Sulphuric acid	4% v/v
Sodium dichromate	200 g/l

This may leave the surface in a passive condition, hence it should be followed by a cathodic activation treatment in an alkaline electrolyte, preferably containing cyanide, or in a non-chromate acid bright dip.

An alternative solution, which is particularly effective in removing heat scale from components with copper glass-to-metal seals, is as follows:

Ammonium chloride	260 g/1
Hydrochloric acid	5% v/v

Electropolishing

Copper and most copper alloys can be electropolished in phosphoric-chromic acid mixtures, the results usually being better with single-phase than with two-phase alloys. The composition and operating conditions[2] of a typical electrolyte are:

Orthophosphoric acid	74% v/v
Chromic acid	60 g/l
SG	1.60–1.62
Cathodes	Lead
Anode: cathodic ratio	1:2–1:3
Time	4–5 min

The specific gravity is controlled by adding water, or by evaporation at about 120°C. A prerequisite for satisfactory performance of the solution is pre-electrolysis to a total of 5 A-hr/l using a copper anode. For the first 50 A-hr/l of operation the optimum current density is 40–50 A/dm^2 at 35–45°C, which may later be reduced to 30–40 A/dm^2 at 20–30°C.

The electrolyte polishes over a wide range of current density (5–70 A/dm^2) and is therefore suitable for treating complex shapes. Other operating features of the process are discussed in detail by Fedot'ev and Grilikes[2].

Chemical Polishing and Bright Dipping

A number of very effective proprietary processes are available and these are probably best employed in regular production. However, the following formulations are useful in catering for infrequent requirements:

(*a*)	Hydrogen peroxide (100 vol)	150 ml/l
	Sulphuric acid	1.3 ml/l

The action of this solution is essentially smoothing rather than brightening. At room temperature metal is removed at the rate of about 12 microns/hour, the usual treatment time being 15–60 minutes. The process leaves a visible film of oxide on the surface, which may be removed by dipping in 10% sulphuric acid solution.

(*b*)	Nitric acid	20–25% v/v
	Glacial acetic acid	25–30% v/v
	Phosphoric acid	55% v/v
	Hydrochloric acid	0.3–0.5% v/v
	Temperature	90°C
	Time	2–5 min

This process is particularly suitable for copper-nickel alloys. Care must be taken to avoid drag-in of water.

In the chemical polishing of brass the relative rates of attack on copper and zinc may vary with the composition and structure of the alloy. The following solution provides some control in this respect:

(*c*)	Sulphuric acid	50% v/v
	Nitric acid	25% v/v
	Sodium chloride	5 g/l
	Temperature	Room
	Time	20–60 sec

An increase in nitric acid content will increase the attack on copper and an increase in sodium chloride concentration will favour attack on zinc.

Plating

COPPER

Following electropolishing or bright dipping in non-chromate solutions, copper may be gold plated immediately after a dip in 10% cyanide solution. However, since some passive areas may be present it is always preferable to follow the cyanide dip by a strike in cyanide copper before transferring to the gold plating bath.

If electropolishing or bright dipping has not been used, copper may be gold plated directly after light etching in one of the following solutions:

(*a*) Sulphuric acid, 10% v/v, for 1–3 min., followed by ammonium persulphate, 200 g/l, for 20 sec
(*b*) Nitric acid, 15% v/v
(*c*) Fluoroboric acid, 20% v/v

Etchant (*a*) produces a smooth matte surface.

BERYLLIUM-COPPER

Beryllium copper components may often carry a heavy uneven layer of scale as a result of improperly controlled heat treatment. This can be removed by the process described for copper, which should be followed by a dip in 30–50% v/v nitric acid solution or by chemical polishing, for which some particularly effective proprietary processes are available. Procedure is then as for copper, but in this case a high current density strike in cyanide copper must be regarded as essential.

BRASS

The etchants recommended for copper are also suitable for brass. Brass should always be given a cyanide copper strike prior to gold plating.

COPPER-NICKEL ALLOYS

Copper-nickel alloys which do not require bright dipping may be etched for 30–50 seconds in either of the following solutions:

(*a*)	Hydrochloric acid	50% v/v
	Methanol	50% v/v
	or	
(*b*)	Hydrochloric acid	30% v/v

In plating all alloys containing nickel, the writer prefers to follow polishing or acid etching by a cathodic activation treatment in the sulphuric acid nickel strike bath (page 108).

TELLURIUM-COPPER

Free machining tellurium-copper should be subjected to a bright acid dip, followed by a strike in a conventional acid copper sulphate bath prior to gold plating.

BRONZE

Alloys of copper containing tin can be etched in a 50% v/v solution of 48% fluoroboric acid. This should be followed by a cyanide dip and a high current density copper cyanide strike. These alloys should not be treated in etchants containing nitric acid since they lead to formation of a surface layer of insoluble tin oxide.

LEADED BRASS

Treatment is similar to that recommended for bronze.

CARBON AND LOW ALLOY STEELS

Due to the high electrochemical potential difference between the metals, ferrous alloys plated directly with gold show very poor corrosion resistance, hence it is more usual to employ thick undercoating deposits of copper or nickel. Undercoats of a bright levelling type serve additionally as substitutes for bright dipping or electropolishing. For optimum corrosion resistance a nickel undercoat is usually preferred.

Descaling

Where necessary, descaling can be carried out by pickling in hydrochloric, or more often sulphuric acid, the most economical process utilising 5% sulphuric acid at 60–65°C. Since mineral acids attack the basis metal where exposed by discontinuities in, or removal of, scale, frequent inspection is necessary to avoid over-pickling.

Electropolishing

A suitable electrolyte for carbon and most low and medium alloy steels is the following:

Orthophosphoric acid	70% w/v
Sulphuric acid	12% w/v
Chromic acid	6% w/v
Water	12% w/v
SG	1.73–1.74
Temperature	60–70°C
Current density	40–50 A/dm^2

Since the tolerance of this solution to variation in operating parameters is rather narrow, it is generally preferable to improve the surface finish when necessary by the use of a bright nickel or copper undercoat or by chemical polishing.

Chemical Polishing

Most mild and low alloy steels can be brightened in mixtures of oxalic acid and hydrogen peroxide. The degree of brightness obtained depends on the nature of the alloying elements and generally decreases with increasing carbon content. The following solutions[3] are effective:

(*a*)	Oxalic acid	2.5% w/v
	Hydrogen peroxide (30%)	1.3% v/v
	Sulphuric acid	0.007–0.010 v/v
	Temperature	16°C
	Time	30–60 min
(*b*)	Phosphoric acid	50% v/v
	Sulphuric acid	20% v/v
	Nitric acid	30% v/v
	Temperature	80°C
	Agitation	Moderate

Plating

In the plating of mild and low alloy steels, the choice of etchant depends on the surface condition of the component to be plated. For convenience, etchants may be classified as follows:

1 *Light etches* suitable for surfaces which have been electrolytically or chemically polished or heat treated in a reducing atmosphere only a few hours prior to plating, and carry no visible oxide film. These etches have no effect on surface finish.

2 *Moderate etches* for use when there has been a longer delay, from several hours to a few days, between polishing or heat treatment and plating. These etches may be suitable for well machined components when it is not considered necessary to remove a work-hardened surface before plating.

3 *Heavy etches* are required for most machined components and all those that have been subjected to severe cold deformation. These etches remove a considerable amount of metal and leave a matte surface.

A number of effective proprietary processes of types 1 and 2 are available, but the writer has encountered none suitable for heavy etching. Useful non-proprietary formulations of the three types are as follows:

1		Sulphuric acid	15% v/v
		Temperature	Room
		Time	30–60 sec
2		Hydrochloric acid	50% v/v
		Methanol	50% v/v
		Temperature	Room
		Time	20–40 sec
3	(*a*)	Hydrochloric acid	50% v/v
		Phosphoric acid	5% v/v
		Temperature	70°C
		Time	20–60 sec
	(*b*)	Sulphuric acid	60–70% v/v
		Voltage	12.5
		Polarity	Anodic
		Time	30–60 sec

After etching, gold plating may be carried out following a dip in cyanide solution, but more frequently a nickel or copper undercoat is applied.

STAINLESS AND HIGH ALLOY STEEL

Descaling

Posselt, Anderson and Shrode[4] have recently described a process for the descaling of ferrous alloys, stainless steels and a variety of special alloys. This comprises of alternate dips, ranging in duration from a few seconds to occasionally over ten minutes, in alkaline permanganate and inhibited hydrochloric or sulphuric acid solutions of the following compositions:

(*a*)	Potassium permanganate	50 g/l
	Sodium hydroxide	100 g/l
	Temperature	90°C

(*b*) Sulphuric acid, 10% w/w, inhibited, room temperature
or Hydrochloric acid, 20% w/w, inhibited, room temperature
or Hydrochloric acid, 30% w/w, inhibited, room temperature

This type of process was developed with the object of eliminating problems of basis metal attack and hydrogen embrittlement. The original paper presents procedures for descaling some twenty different alloys and should be consulted for further details.

Some acid mixtures containing hydrofluoric acid are also effective in scale removal, but are generally unpleasant to use and appear to have no advantage over the process just outlined.

Electropolishing

Many high alloy steels can be electropolished with solutions based on mixtures of sulphuric, phosphoric and chromic acids, but except in the case of stainless steel, there is little published information available, hence reliance must be placed on proprietary processes.

A published[2] formulation suitable for the electropolishing of stainless steels is the following:

Orthophosphoric acid	45% w/w
Sulphuric acid	34% w/w
Chromic acid	4% w/w
Water	17% w/w
SG	1.65
Current density	35–50 A/dm^2 reducing to 25–30 A/dm^2 when Fe content reaches 2.5–3%
Temperature	65–70°C, reducing to 60–70°C as Fe content rises
Time	5–7 min at 30 A/dm^2 1–3 min at 50 A/dm^2

Numerous proprietary electrolytes are also available, some of which are capable of producing equally good results at lower current densities and temperatures.

Chemical Polishing

There appears to be very little published information on chemical polishing of stainless steels, though some proprietary processes are available.

Plating

STAINLESS STEELS

This general description includes a large number of alloys, having in common a high chromium content (normally about 18%), and high resistance to corrosion due to the presence of a self-healing chromium oxide film on the surface, which also accounts for the difficulty in plating these alloys. The chromium oxide is reduced electrolytically by hydrogen at room temperature, but the process seems to be relatively slow and in practice effective only in a low efficiency acid strike solution.

After normal cleaning, or electropolishing, a stainless steel surface is in a more or less passive condition and must be activated before plating by electrolytic reduction of the oxide film. This is commonly carried out by anodic treatment for 30–60 seconds at 1 A/dm^2 in a Woods nickel bath (page 108), followed by rapid reversal of polarity and a cathodic strike in the same solution at 3A/dm^2 for 3–5 minutes. After rapid rinsing the surface may then be plated as required.

The writer, however, favours the following process:

(i) Etch/polish in 60–70% v/v sulphuric acid, the work being made anodic at a voltage of 12.5 for 30–60 sec.
(ii) Nickel strike in low nickel/high sulphuric acid solution (page 108) for 30 sec at 8 V.
(iii) Rinse rapidly and plate as required.

Good results have also been obtained in experimental plating using the copper strike recently described by Dini and Helms[1] (page 109) for 1–5 minutes, but the writer prefers to precede this by electropolishing or by the anodic etchpolish in sulphuric acid mentioned above in (i). As pointed out earlier, this process may well prove applicable to a wider range of steels than that specified by the original reference.

ALLOY STEELS

It is far beyond the scope of the present chapter to attempt a classification of the very large number of alloy steels commercially available, and recommendations for plating each of the groups of alloys. Very few alloy steels present any real difficulties but, nevertheless, an unfamiliar material may require some experimentation. The majority of high alloy steels may be plated by modification of the etch-polish/nickel strike process used for stainless steel, usually by interposing between the two treatments a chemical etch in one of the following solutions:

(*a*)	Hydrofluoric acid (48%)	50% v/v
(*b*)	Sodium hydroxide	100 g/l
(*c*)	Ferric ammonium sulphate	50 g/l
	Sulphuric acid	12.5% v/v
	Hydrochloric acid	15% v/v
(*d*)	Ferric chloride	100 g/l
	Hydrochloric acid	10% v/v
	Temperature	65°C

Solution (*a*) is particularly useful in plating titanium-containing steels and solution (*b*) in the treatment of high molybdenum steels.

When processing an unfamiliar alloy steel, the use of a copper undercoat as a check on effective pretreatment (page 110) is to be particularly recommended.

NICKEL AND NICKEL-IRON ALLOYS

Descaling

Heavily scaled nickel and nickel-iron alloys can be descaled by pickling in the following solutions:

(*a*)	Hydrofluoric acid (48 %)	12.5 % v/v
	Ferric sulphate	25 g/l
	Temperature	65–70°C

Care is necessary in handling this solution, though it is less hazardous than simple hydrofluoric acid solutions due to complex formation with ferric ions.

(*b*)	Phosphoric acid	40 % v/v
	Hydrochloric acid	5 % v/v
	Temperature	80°C

This solution is particularly useful for descaling nickel-iron-cobalt alloy (KOVAR or NILO K). The basis metal is attacked but not pitted, and glass-to-metal seals are not damaged by moderate exposure to the solution.

Electropolishing

Nickel can be electropolished quite satisfactorily in the phosphoric-sulphuric-chromic acid mixtures used for carbon steel, but rather better results are obtained with the following solution, which should be aged before use:

Sulphuric acid	50 % v/v
Phosphoric acid	41 % v/v
Citric acid	20 g/l
Water	9 % v/v
Temperature	20–25°C
Current density	35–40 A/dm^2

This process can be used to electropolish electrodeposited nickel. In polishing dull deposits, up to 10 microns of nickel may be removed. A reasonably good polish can be obtained, even on dull nickel plate, by electropolishing in the 60–70 % sulphuric acid bath used for anodic etching of stainless steel and many other metals.

Nickel-iron alloys are not readily electropolished.

Chemical Polishing

Nickel may be brightened and smoothed by treatment in the following:

Phosphoric acid	55% v/v
Sulphuric acid	25% v/v
Nitric acid	20% v/v
Temperature	80°C
Time	2–3 min

Up to 10–15% of water may be added.

The NILO and KOVAR iron-nickel and iron-nickel-cobalt alloys may be polished with the following acid mixtures. All these reagents have short lives.

(*a*)	Glacial acetic acid	600 ml
	Nitric acid	200 ml
	Hydrochloric acid	2–5 ml
	Temperature	70–75°C
	Time	5–10 sec

The article must be rinsed immediately and plunged into a mild alkali solution, ammonium hydroxide or carbonate, to prevent staining.

(*b*)	Glacial acetic acid	75 ml
	Nitric acid	27 ml
	Hydrochloric acid	1 ml
	Time	3–5 min at room temperature 10–15 sec at 60°C
(*c*)	Hydrogen peroxide (100 vol)	35% by vol
	Sulphuric acid	20% by vol
	Temperature	room
	Time	30 secs

These three processes are suitable for glass-to-metal seals.

Plating

Nickel and its alloys are notorious for their tendency to become passive. Passivation can develop even during transfer from an acid etch to the plating bath or from one plating bath to another. It should therefore be regarded as mandatory to activate nickel and all nickel-containing alloys immediately prior to plating.

NICKEL

Nickel which has been electrolytically or chemically polished may be activated by cathodic treatment in 10% sodium cyanide solution for 30 seconds only at 2.5 A/dm^2, or in the low nickel/high sulphuric acid strike solution for 30 seconds only with an applied voltage of 8 volts. When no polishing has been carried out the following etches are suitable:

(*a*) *Moderate etch*	Hydrochloric acid	50% v/v
	Methanol	50% v/v
(*b*) *Heavy etch*	Ferric chloride	150 g/l
	Hydrochloric acid	10% v/v

NICKEL-IRON ALLOYS

The surface layers of nickel-iron alloys, particularly those which contain cobalt, e.g., NILO K, often show oxide contamination in depth in the as-received condition and if this contamination is over-plated, adhesion is likely to be poor as indicated by typical blister formation on heating.

A number of alternative pretreatments may be used for these alloys, including the following:

(*a*) Heat treat in hydrogen* at 800°C, followed by a swill† in 25% hydrochloric acid solution.
(*b*) Chemical polish (as above)
(*c*) Etch

Ferric ammonium sulphate	5%
Sulphuric acid	12.5 v/v
Hydrochloric acid	15% v/v
Temperature	Room
Time	1–2 min

(*d*) Electrolytic etch

Sulphuric acid	60% v/v
Voltage	12.5
Time	30 sec

The writer prefers to follow any of these treatments by a low nickel/high sulphuric acid strike activation as for nickel.

MOLYBDENUM AND TUNGSTEN

The plating of molybdenum and tungsten has been the subject of a number of investigations over the past twenty years, reflecting a growing demand, especially in the electrical and semiconductor industries. As a result, it is now possible to obtain fully adherent deposits on both metals by a variety of methods, the ultimate choice depending primarily on the geometry and size of the components, and on any special restrictions which may be imposed by requirements for intermediate layers, dimensional tolerances and so forth.

* Richards[5] has shown that heat treatment of NILO 50 (50% Fe–50% Ni) in wet hydrogen can result in formation of numerous blisters. To avoid this, the moisture content of the hydrogen must be controlled at below about 100 ppm.

† The acid swill may be omitted if components are transferred from the furnace to the nickel strike within a few minutes.

Both metals are difficult to plate because secondary oxide films form rapidly on rinsing and are only slowly removed by electrolytic reduction in a strike bath.

Descaling

Heat scale can be removed from both metals by soaking in the following solution:

Potassium permanganate	40 g/l
Sodium hydroxide	80 g/l
Temperature	90–95°C
Time	30 min

Electropolishing

Molybdenum can be electrolytically polished in aqueous sulphuric acid solutions over a wide range of compositions. Tegart[6] recommends a 20% v/v solution used at room temperature with an applied potential of 12 volts. An alternative solution is the following:

Sulphuric acid	12.5% v/v
Methanol	37.5% v/v
Time	1 min

The blue film of molybdenum oxide formed during these processes is removed by swilling in sodium hydroxide solution.

In the writer's experience, the sulphuric acid etch/polish used in the pretreatment of stainless steel (page 116) produces a satisfactory finish on molybdenum, but pitting may occur if the sulphuric acid concentration is allowed to fall below about 50%.

Tungsten can be electropolished in a 10% potassium hydroxide solution at room temperature, with a current density of 3–6 A/dm^2.

Plating

MOLYBDENUM

The writer has found the following processes to give sound deposits on molybdenum:

1 (i) Electropolish—66–70 v/v sulphuric acid, 12 V, 0.5–2 min.
(ii) Rinse in (*a*) water (*b*) 10% sodium hydroxide (*c*) water.
(iii) Strike* in low nickel/high sulphuric acid solution.
(iv) Gold strike†—30 sec in acid gold solution at 1.0 A/dm^2.
(v) Heat treat in hydrogen for 10 min at 800°C.
(vi) Plate as required.

---

* The nickel strike may be omitted if transfer times are kept to a minimum. The initial gold film may then be somewhat lacking in cohesion and must be handled with care.

† The initial gold deposit should be bright, and sufficiently coherent and adherent to permit handling.

2 In this procedure the initial electropolish of schedule 1 may be replaced by heat treatment in hydrogen as above. This modification is useful when dimensional tolerances prohibit removal of metal, and also in the plating of bimetallic components.

3 Electropolishing may also be replaced by dry blasting with a fine abrasive (200 mesh), a process which is again useful in treating bimetallic components.

4 Fine and Bracht[7] recommend the following process, based on the deposition of a thin layer of gold on to a porous molybdenum oxide film and subsequent reduction of the latter in hydrogen.

(i) Heat treat in dry hydrogen (<2 ppm water vapour), 10 min, 1000°C.

(ii) Oxidise by immersion for 8 sec at room temperature in the following solution:

Ammonium hydroxide, SG 0.880	80% v/v
Hydrogen peroxide (30 vol)	20% v/v

(iii) Gold strike to deposit 1–2 μm from an acid gold bath.

(iv) Heat treat in dry hydrogen, 10 min, 900°C to reduce oxide.

(v) Gold plate as required.

In the writer's view the conditions specified for the heat treatment appear to be a little over rigorous. Experience suggests that a temperature of about 850°C would be suitable for both heat treatments, and that a considerably higher moisture content could be tolerated. In practice, it is extremely difficult to maintain the very low level specified in the exit gas from a tube furnace.

5 Seegmiller, Gore and Calkin[8] have developed the following procedure, based on the use of a new high acid cobalt strike:

(i) Electropolish—70–80% v/v sulphuric acid, 3–6 A/dm^2, 45°C, 1–3 min.

(ii)

Cobalt sulphate, pentahydrate	200 g/l
Sulphuric acid	9% v/v
Temperature	Room
Current density	30 A/dm^2
Time	1–6 min

(iii) Plate as required.

The present writer has found this process to give good results, provided that it is followed by heat treatment in hydrogen. A practical limitation, however, is the use of comparatively concentrated sulphuric acid for electropolishing, which rapidly absorbs water from the air and requires frequent adjustment. If the concentration falls significantly below the specified 70%, a swill in 10% sodium hydroxide solution should be incorporated following water rinsing after electropolishing.

6 In many general plating shops facilities for controlled heat treatment are not available, hence there is considerable interest in procedures which do not require this step. A process of this type, communicated to the present writer a number of years ago, but of unknown origin, involves an AC etch in a gold cyanide solution followed immediately by a DC strike in the same solution. Adhesion is adequate for many purposes, though the deposit may blister if heated. Solution composition and current densities are as follows:

Potassium gold cyanide	2 g/l
Sodium cyanide	2 g/l
Sodium hydroxide	15 g/l
AC etch, current density	15 A/dm^2
DC strike, current density	60 A/dm^2
Counter electrode	Platinum

The strike is continued until the surface is visibly covered by a thin continuous gold deposit. Plating may then be continued in a normal gold plating bath.

As an alternative, a copper strike may be used in the same technique, to provide a basis for subsequent gold plating. Conditions are then as follows:

Cuprous cyanide	1.5 g/l
Sodium cyanide	6.7 g/l
Sodium hydroxide	53 g/l
AC etch	100 A/dm^2
DC strike	300 A/dm^2

In passing, it is of interest to record the observations by the writer that molybdenum layers on ceramic, applied in the form of a suspension and fired at 1600°C in hydrogen, can be plated with adherent coatings of gold or copper without additional pretreatment, even after exposure to normal laboratory atmospheres for at least one or two days. This result appears to be associated with the finely divided state, or "microgeometry", of the layer, since metal in normal sheet form must be plated within about one hour of heat treatment if further surface cleaning is not to be required.

TUNGSTEN

Tungsten is considerably easier to plate than molybdenum. An intermediate heat treatment is not essential, though desirable as a precautionary measure. In the writer's experience the most reliable process involves as the first essential stage, an AC etch for 1–2 minutes in a strongly ammoniacal nitrate solution of the following composition. This is carried out at room temperature using a voltage sufficient to produce moderate gassing at the metal surface.

Langmuir's solution*	600 ml
Ammonium hydroxide (0.880)	400 ml

As for molybdenum, the etch may be replaced by heat treatment in hydrogen at 800°C or by abrasive blasting. The reject rate for these procedures is very low, but it may be still further reduced by plating an initial layer of 1–2 microns of gold, heat treating in hydrogen at 700°C, then increasing the thickness of the gold deposit as required.

When removal of metal is not permissible and heat treatment not practicable, moderate adhesion of gold to tungsten may be obtained by immersing the metal in a boiling solution of 10% sodium hydroxide as a replacement for the initial etch.

Seegmiller, Gore and Calkin[8] recommend the use of a cobalt strike as in their procedure for molybdenum, but following an initial anodic etch for 2 minutes in 10% potassium hydroxide solution at 3 A/dm^2. The latter may be omitted when plating tungsten sheet.

CHROMIUM

Freshly deposited chromium can be plated by transferring, via a good rinse, directly to a high acid nickel strike bath. If an interval of only a few minutes occurs, satisfactory results may still be obtained by soaking in concentrated hydrochloric acid for about 1 minute before nickel striking.

Chromium which has been exposed to the atmosphere for longer periods is extremely difficult to prepare for plating. The following process due to Reid and Ogburn[9] gives excellent results, but is restricted to thick chromium deposits since several microns may be removed in an electrolytic etch. The first step is an anodic etch in the following solution:

Glacial acetic acid	85–90% v/v
Sulphuric acid	10–15% v/v

This is operated at room temperature with a current density of 3–7 A/dm^2 and the process is continued until a uniform dark brown film is produced, usually about 5 minutes. The surface is then rinsed quickly and transferred immediately to a Woods nickel strike for 2–4 minutes at a current density of 20–25 A/dm^2, following which it may be gold plated as required.

---

* Prepared by boiling together the following solutions:

(*a*)	Sodium nitrate	76 g
	Potassium nitrate	76 g
	Sodium carbonate	1.9 g
	Water	to 1 litre
(*b*)	Sodium hydroxide	11 g
	Water	27 ml

The life of the solution is limited to about 1 week at room temperature.

ALUMINIUM

Aluminium is a highly reactive metal, oxidising virtually instantaneously on exposure to air or water. The oxide film is coherent and protective, and is not reduced by heat treatment in hydrogen or *in vacuo*, nor electrolytically. Plating methods involve the replacement of the natural oxide film by a protective film of zinc or by a porous anodic oxide layer. Electroless nickel may also be deposited directly on aluminium.

Electrolytic and Chemical Polishing

Owing to its widespread use for reflectors and decorative purposes a very large number of processes have been developed for the polishing of aluminium[10]. However, these processes are not employed to any extent in electroplating because an initially bright surface is inevitably degraded to some extent by subsequent pretreatment operations.

A process developed by the Batelle Memorial Institute[11] is as follows:

Sulphuric acid	14% w/w
Phosphoric acid	57% w/w
Chromic acid	9% w/w
Water	20% w/w
Temperature	65°C
Current density	16 A/dm^2
Agitation	Moderate mechanical

A widely used chemical polishing process developed by the General Motors Corporation[12] is the following:

Nitric acid	3.75% w/w
Ammonium bifluoride	0.65% w/w
Chromic acid	0.65% w/w
Ethylene glycol	0.6% w/w
Copper nitrate	0.0025% w/w
Temperature	95°C
Time	0.5–5 min

Plating

Aluminium and its alloys may be etched in either acid or alkaline solutions, the former usually being preferred as giving a better surface finish. After etching, certain alloying elements may be unattacked and remain on the surface as a smut, which can generally be removed by dipping in 50% nitric acid solution.

Cleaning and etching of aluminium has been extensively reviewed by Spring[13]. In practice these processes are based largely on proprietary formulations, but some useful general purpose solutions are as follows:

(*a*)	Hydrofluoric acid (48 %)	100 ml/l
	Nitric acid	100 ml/l
	Temperature	40°C
(*b*)	Sulphuric acid	150 ml/l
	Chromic acid	50 g/l
	Temperature	50–60°C
	Time	30–60 sec
(*c*)	Sodium hydroxide	40 g/l
	Temperature	70°C
	Time	1–2 min

Some of the more widely used plating processes are as follows:

1 Modified Zincate Process[14]

(i)	Clean and etch	
(ii)	Immersion zincate	
	Zinc oxide	100 g/l
	Sodium hydroxide	525 g/l
	Ferric chloride	1 g/l
	Sodium potassium tartrate	10 g/l
(iii)	Strip in 50 % v/v nitric acid	
(iv)	Second zincate	As above
(v)	Strike	Low pH copper cyanide solution, 2.5 A/dm^2, 2 min, room temperature
(vi)	Gold plate	As required

A modification[15] of this process employs a brass strike in place of the copper strike. When good corrosion protection is required, an undercoat of 5-10 microns of nickel is advisable.

A number of proprietary zincate immersion processes are available to provide thinner and sounder deposits from alkaline solutions with minor amounts of other metals[16]. The writer has found one process of this improved type to be very satisfactory for the treatment of aluminium/steel bimetallic components.

2 Travers Process[17]

In this process the basis for plating with an initial nickel deposit is provided by a highly porous oxide film produced by anodising in phosphoric acid or a phosphoric-sulphuric acid mixture, preferably containing nickel. The optimum anodising conditions vary for different alloys and details should be sought from the reference cited. The process is therefore best suited to production line plating of components of uniform composition. The essential steps are as follows:

(i) Clean

(ii) Anodise

Phosphoric acid	15% v/v
Sulphuric acid	15% v/v
Temperature	Room
Current density	3 A/dm^2
Time	5 min

(iii) Nickel plate in Watts or sulphamate bath at low current density for 5–10 min., then gold plate as required.

MAGNESIUM

The following pretreatment[18] is generally satisfactory for electroplating magnesium and its alloys:

(i) Pickle

Phosphoric acid (conc.)	
Temperature	room
Time	30 sec

(ii) Activate

Phosphoric acid	13.5% v/v
Potassium fluoride	70 g/l
Temperature	room
Time	for 30 sec after gassing ceases

(iii) Zincate dip

Zinc sulphate	50 g/l
Sodium hydroxide	180 g/l
Potassium fluoride	6.0 g/l
Sodium carbonate	4.0 g/l
pH	10.6
Temperature	90–95°C
Time	5 min

(iv) Copper plate

Conventional copper cyanide electrolyte. Immerse "live" into electrolyte and deposit at least 5 μm

(v) Gold plate — As required

The plating of magnesium and its alloys has been reviewed by Spencer[19,20] who recommends a chromic acid pickle for minimum attack on the basis metal. He also specifies a copper plating bath at pH 9.6–10.4 with the free cyanide strictly controlled at 5.6 g/l and an initial current density not greater than 1.0 A/dm^2 to minimise hydrogen absorption.

ZINC-BASE DIE CASTINGS

Zinc-base die castings normally consist of about 95% zinc, 3–5% aluminium, with minor amounts of copper and magnesium. Since a high degree of skill and experience is required for the production of good quality castings, sub-standard components are often encountered.

Poor design or casting techniques can cause a large number of different defects such as shrinkage holes, cracks, gas voids, foliations, microporosity, etc., any of which may render the component unplateable.

Good die castings normally have a chilled surface layer of dense, non-porous (and hence plateable) metal a few mils thick. This should not be removed by excessive buffing or over-etching.

Degreasing and preliminary soak cleaning should be particularly thorough because the components often carry buffing wax residues. Silicone oils may have been used as die lubricants and traces of these may well remain on the components.

Hydrogen pick up may cause blistering of a porous casting, so cathodic cleaning or strong acid etchants should be avoided.

The following pretreatment cycle is suitable for zinc and most zinc-base alloys including die castings.

(i) Electrolytic clean	Anodically, in a mild silicate containing alkali cleaner
(ii) Acid etch	
Sulphuric acid	0.5–1.0% v/v
Temperature	room
Time	30–60 sec
(iii) Copper plate	Cyanide, preferably the Rochelle copper bath
pH	11.5–12.5
Temperature	55°C
Current density	2–3 A/dm^2
Time	6–8 min

It is good practice to inspect the components after the first 30–60 seconds, at which stage they should be covered with a uniform, bright copper plate. The presence of darker spots suggests porosity in the casting or inclusions. The presence of larger, dark or unplated areas indicates incorrect pretreatment.

Safranek, Miller and Faust[21] report that the quality of plating on zinc-base die castings is markedly improved by using ultrasonic agitation to remove smut during etching in dilute acid and also during the copper cyanide strike.

TITANIUM

Titanium is undoubtedly one of the most difficult metals to plate. It is rapidly attacked by oxygen and once the oxides are removed, it re-oxidises immediately during rinsing in water or on exposure to air. The oxides are soluble only in strong acids and are not reduced by heat treatment in a reducing atmosphere or electrolytically in aqueous solutions. Conventional methods of acid etching followed by water rinsing cannot therefore be applied.

Owing to its importance in industry, a great deal of research work has been carried out on the plating of titanium and several successful processes have been developed.

Descaling

The descaling of titanium and its alloys has been comprehensively reviewed by Spencer[22]. For thick scale the author recommends the use of an oxidising molten salt bath followed by a pickle in a nitric-hydrofluoric acid mixture.

For moderate and light scale, Spencer recommends a four stage process involving pickling and electrolysis in nitric-hydrofluoric acid mixtures and oxidation in a caustic-permanganate solution.

Plating

Cramer[23] and his associates at the US Bureau of Mines succeeded in depositing thick films of palladium on titanium by depositing a thin preliminary film of platinum from a fused cyanide bath followed by palladium from an aqueous electrolyte. The same fused bath platinum plating pretreatment is a suitable basis for gold plating.

The same authors have obtained adherent electrodeposits on titanium by an immersion treatment for 10–15 minutes in a mixture of glacial acetic acid (85% v/v) and hydrofluoric acid (12% v/v), followed by an AC etch for 10–15 minutes at 40–60 volts in the same solution as recommended by other workers[24,25].

This treatment leaves the substrate with a matte grey appearance and the authors report that it can be dried and weighed without adversely affecting the adhesion of subsequent electrodeposits. It has been suggested[26] that a protective film of TiF_4 is formed on the surface which prevents oxidation but dissolves in the plating bath to expose clean metal.

A satisfactory process based on a method developed by Friedman[27,28] is as follows:

Scour with pumice moistened with 5% sodium hydroxide and 5% sodium carbonate solution. Soak for 5 minutes at room temperature in:

Nitric acid	20 g/l
Hydrofluoric acid	50 ml/l

Immerse for 10 minutes in boiling concentrated hydrochloric acid and without rinsing, iron plate in the following solution:

Ferrous chloride	375 g/l
Calcium chloride	15 g/l
Hydrochloric acid	to pH 0.6–1.0
Temperature	88°C
Time	not specified

Heat treat in neutral atmosphere or *in vacuo* for 1 hour at 540°C and plate as required.

In a process described by Marshall[29] a prolonged etch in concentrated hydrochloric acid is followed by rinsing directly in a Rochelle salt solution. From this the part is transferred, again without intermediate rinsing, to a Rochelle salt copper plating bath. Titanium forms soluble tartrate complexes and by this process the surface appears to be protected from oxidation. Details are as follows:

(i) Scour with pumice as in the previous process, wash and drain.

(ii) Soak in concentrated hydrochloric acid (SG 1.18). The addition of about 1 g/l of chloroplatinic acid to this solution is beneficial in treating certain alloys. Treatment time is from 10 min to 2 hours at 30°C or 5 min at 90°C.

(iii) Rinse thoroughly with good agitation in 5% Rochelle salt solution.

(iv) Without rinsing, transfer to the following copper bath and plate for 5 min (maximum) at 0.4 A/dm^2.

Copper sulphate, pentahydrate	60 g/l
Rochelle salt	160 g/l
Sodium hydroxide	50 g/l
Temperature	Room

(v) Plate as required.

(vi) Heat treat in neutral atmosphere or *in vacuo* for 1 hour at 450°C.

This process appears to be suitable for nearly all commercially used titanium alloys. It is reported, however, to be unsatisfactory for the 4% Mo, 4% Al, 2% Sn alloy, for which the following procedure is recommended by the British Standards Institution[18].

(i) Soak for 1 min at 95°C in concentrated hydrochloric acid, SG 1.18.

(ii) Silver strike

Sodium cyanide	62 g/l
Silver cyanide	3 g/l
Temperature	Room
Time	15 sec

(iii) Silver plate

Potassium cyanide	37 g/l
Silver cyanide	25 g/l
Potassium carbonate	25 g/l
Temperature	Room
Current density	0.3–0.4 A/dm^2
Time	10 min

(iv) Plate
Nickel recommended, but may be replaced by gold
(v) Heat treat *in vacuo* better than 10^{-6} mm Hg at 555–650°C, raising temperature gradually and maintaining vacuum at all stages.

Huddle and Flint[30] have patented a method in which the metal is subjected to a dry blast with 120 grit iron powder, followed by immersion deposition of copper. The present writer has found that equally good results can be achieved by abrasive blasting with 200 mesh aluminium oxide, followed directly by a cyanide copper strike and subsequent gold plate. The final coating appears to be stable and is suitable for soldering and thermal compression bonding.

BERYLLIUM

Beryllium is a highly reactive metal, chemically similar to magnesium and aluminium. The metal and its compounds are highly toxic, the maximum permissible concentration in air being only about 2 micrograms/m^3. Hence it should be handled only under supervision by personnel fully conversant with the hazards and the necessary safety precautions.

The surface of the metal in the as-received condition is commonly oxidised and crazed to a depth of several microns, and this surface layer must be removed before plating, for which purpose the following chemical polishing process due to Beach and Faust[31] is suitable:

Sulphuric acid	5% v/v
Phosphoric acid	95% v/v
Chromic acid	12% w/w
Temperature	Room
Time	5 min

About 7 microns are removed by this process, which has the advantage of clearly showing up surface cracks, etc.

Missel and Titus[32] have described a double zincate process for plating beryllium similar to those employed for aluminium. The stages are:

(i) Descale by wet abrasive blasting (80 mesh grit, 80 psi)
(ii) Immerse for 10 min in 10% v/v sulphuric acid
(iii) Zincate immersion

Sodium hydroxide	500 g/l
Zinc oxide	100 g/l
Time	3 min

(iv) Strip zinc coating in 33% v/v nitric acid
(v) Second zincate treatment as above
(vi) Copper strike for 1 min at 2.5 A/dm^2, followed by 10 min at 1.2 A/dm^2 in the following solution:

Sodium copper cyanide	37.5 g/l
Sodium cyanide (free)	4.0 g/l
Sodium carbonate	7.5 g/l
Temperature	Room

In connection with this procedure the use of grit blasting, either wet or dry, seems questionable to the writer in view of the toxicity hazards.

PRECIOUS METALS

Platinum Group Metals

The platinum metals, unlike gold, develop passive films of oxide or absorbed oxygen on exposure to the atmosphere, but simple activation treatments such as the following are satisfactory:

Platinum	(i) Cathodic alkaline clean
	(ii) Immersion for 2–5 min in concentrated hydrochloric acid
Rhodium	(i) Cathodic alkaline clean
	(ii) Immersion for 5–10 min in 20% v/v sulphuric acid
Palladium	(i) Anodic alkaline clean
	(ii) Immersion for 1–2 min in concentrated hydrochloric acid

Gold

It is possible to electroplate pure gold following a simple cathodic alkaline clean. Gold alloys may well be covered with thin oxide films derived from the alloying elements. If the latter are known, it is good practice to apply the pretreatment appropriate to these metals, otherwise a swill in concentrated hydrochloric acid or cathodic activation in dilute sulphuric acid solution or cyanide solution is generally adequate.

Silver

Light tarnish films may be removed before plating by cathodic cleaning, preferably in a cyanide-containing cleaner, followed, if necessary, by alternate immersion in potassium cyanide solution and dilute fluoroboric acid. For more resistant tarnish films the cyanide immersion may be replaced by short anodic treatment in 10–20% potassium cyanide solution at about 1–2 A/dm^2.

The general subject of plating onto precious metals has been reviewed by Foulke[33].

TANTALUM, NIOBUM, ZIRCONIUM

These metals have many similarities from the plating viewpoint. They are prone to hydrogen embrittlement, but hydrogen absorbed during electrolytic treatments is readily dispelled by heating *in vacuo* at 800–850°C.

They form exceptionally refractory oxide coatings which are extremely resistant to attack by acids and alkalies and are not reduced by heating in hydrogen or *in vacuo*. These are usually removed prior to plating by hydrofluoric acid based etches or by abrasive blasting. The secondary oxide films on these metals are thin and are absorbed into the body of the metal on heating to 800–900°C *in vacuo*.

Many of the published plating procedures are based on the following general scheme:

(i) Etch in hydrofluoric acid mixtures to remove natural oxide films
(ii) Deposit a thin film of a metal readily permeable to hydrogen such as iron or nickel
(iii) Heat treat *in vacuo* to 780–850°C to expel absorbed hydrogen and to disperse the thin film of secondary oxide formed during rinsing throughout the basis metal. The plating should then be firmly bonded to the substrate.

Descaling

Descaling is normally carried out as part of the fabrication process. For the methods used reference should be made to standard works on the metallurgy of these metals[34,35,36].

Polishing

ZIRCONIUM

Zirconium, hafnium and their alloys can be electropolished in a neutral or alkaline solution of ammonium fluoride and nitrate.[37] The recommended electrolyte composition and operating conditions are:

Ammonium fluoride	20–40%
Ammonium nitrate	20–40%
pH	6–8.4
Temperature	30°C
Current density	0.5–3 A/dm^2
Time	4–10 min

Gentle agitation is recommended, and the pH is maintained by purging the electrolyte with a stream of inert gas to remove excess ammonia.

McGraw and Jakobson[38] have claimed a method for chemical polishing of zirconium and its alloys in a sulphamate-bifluoride solution of the following composition, followed by de-smutting in 50% nitric acid.

Potassium sulphamate	9.5%
Ammonium bifluoride	4.0%
Sulphamic acid	3.5%

At pH values of 2.7–3.7 the rate of metal removal is 5–10 μm/min.

TANTALUM AND NIOBIUM

Miller[34] recommends the following processes for the chemical and electrolytic polishing of tantalum:

Chemical polishing	
Nitric acid	40% v/v
Sulphuric acid	40% v/v
Hydrofluoric acid (48%)	20% v/v
Electropolishing	
Sulphuric acid	90% v/v
Hydrofluoric acid (48%)	10% v/v
Temperature	35–45°C
Current density	10 A/dm^2
Time	9 min

Both solutions are also suitable for niobium, in this case electropolishing is carried out at room temperature with gentle agitation.

Plating

The fused salt bath platinum strike pretreatment[23] referred to earlier in connection with titanium can also be applied to niobium. Although deposits produced on tantalum and zirconium by the process were not satisfactory, indications are that the use of a fused salt bath in the pretreatment of these very difficult metals appears very promising and may well form the basis of commercial procedures for all three.

The process developed by Huddle and Flint[30] also noted earlier (page 131) is claimed to be generally applicable to the plating of refractory metals, including tantalum, zirconium and niobium.

ZIRCONIUM

Beach[36] recommends the following process:

Iron plate	
Ferrous chloride	300 g/l
Calcium chloride	150 g/l
pH	1.5–2.0
Temperature	93°C
Current density	4 A/dm^2

Pre-bake at 200°C for 2–4 hours in air, followed by heat treatment at 850°C for 45 minutes *in vacuo* and plate as required.

Nickel may be used instead of iron as the initial deposit. The writer has obtained sound deposits of gold on zirconium by the following process:

(i) Etch in 4% hydrofluoric acid for 1 min (ii) Without rinsing, transfer to low nickel/high sulphuric acid strike (page 108) (iii) Gold strike in acid gold solution (1 μm) (iv) Nickel plate in Watts bath (2.5 μm) (v) Bake at 200°C in air for 1–2 hours (vi) Heat treat at 850°C *in vacuo* for 30 min (vii) Plate in acid gold solution.

The interface is somewhat brittle.

NIOBIUM

Beach[36] reports the following method:

(i) AC etch in 48% hydrofluoric acid, 20–100 A/dm^2 at 20–30°C

(ii) Desmut

Hydrofluoric acid (48%)	2% v/v
Nitric acid	50% v/v

(iii) Iron plate

Ferrous sulphate	300 g/l
Ferrous chloride	42 g/l
Ammonium sulphate	15 g/l
Sodium hydroxide	15 g/l
Boric acid	30 g/l
Temperature	60°C
Current density	3 A/dm^2

(iv) Pre-bake at 200°C for 2 hours in air, followed by heat treatment at 700°C for 1 hour *in vacuo*.

TANTALUM

Saubestre[39] has proposed a method based on the production of a hydride film by cathodic treatment in a dilute acid or alkali solution. This film is over-plated with a thin coating of nickel from a Watts solution and the hydrogen then removed by heat treatment at 450–750°C *in vacuo*. Following this treatment the surface can be plated after a Woods nickel strike.

A process patented by Van Gilder[40] involves the deposition of a preliminary gold coating from a chloride solution, thus:

Heat treat *in vacuo* (10^{-16}mm Hg) at 1995°C

Gold Plate

Gold chloride	3.75 g/l
Sodium cyanide	15 g/l
Temperature	70°C
Current density	1 A/dm^2

Heat treat *in vacuo* at 1100°C.

The writer has obtained reasonably satisfactory plating on tantalum by the procedure described for zirconium, but with the preliminary etch in hydrofluoric acid replaced by an abrasive blast with 200 mesh alumina at 60 psi for $\frac{1}{2}$–1 minute.

Gold may be deposited directly onto tantalum, but in the absence of a nickel interlayer excessive Kirkendall porosity develops in the gold coating due to the much more rapid diffusion of gold into tantalum than of the substrate into gold.

MULTI-METAL COMPONENTS

The plating of components comprising of two or more metals may require considerable ingenuity in the selection of a pretreatment procedure which will be effective for all the metals concerned without degrading the surface finish of any individual one. A particular danger is that of preferential attack by etchants at the junction between two metals due to difference in electrochemical potential. This is most likely when the potential difference exceeds about 200 mV. For practical purposes this can be measured with sufficient accuracy by immersing samples of the metals concerned in the etchant and connecting them through a high impedance millivoltmeter.

As in the plating of any unfamiliar metal it is advisable to incorporate a thin cyanide copper plate as a check on the effectiveness of the pretreatment employed, as mentioned on (page 110).

Heat treatment in hydrogen, followed by a high acid nickel strike offers an effective pretreatment for many multi-metal components. This procedure is of course not permissible when the component includes metals of low melting point, or the oxides of which are not reducible by hydrogen. A minimum temperature of 700°C is necessary to reduce oxides on ferrous alloys, and about 800°C for those on molybdenum and tungsten. Limitations to heat treatment temperature may also be imposed by the presence of brazed joints in the assembly, when it is inadvisable to heat at a temperature higher than 50°C below the melting point of the braze material.

Where heat treatment is not practicable, wet or dry blasting of the surface with fine abrasive, e.g., 200 mesh alumina powder is a generally useful alternative which is effective for all the commonly plated metals except aluminium and magnesium, and often permits the plating of otherwise unplateable multi-metal components. The limitations of this type of process are:

(*a*) surface is roughened and work hardened
(*b*) grit may become embedded in soft metals
(*c*) not applicable to beryllium, due to health hazard.

Following vapour blasting, a high acid nickel strike is often desirable, and plating should in any case be carried out without undue delay.

The writer has found the following etches particularly useful in the pretreatment of multi-metal components:

(*a*) Sulphuric acid, 60% v/v (anodic etch-polish)
(*b*) Fluoroboric acid (48%), 5% v/v (anodic)
(*c*) 50% hydrochloric acid (v/v) in methanol (immersion)
(*d*) Fluoroboric acid (48%), 20–25% v/v (immersion)
(*e*) 15–20% ammonium persulphate solution with excess ammonia (immersion)
(*f*)

Ferric ammonium sulphate	5%
Sulphuric acid	12.5% v/v
Hydrochloric acid	15% v/v

(*g*) Langmuir's solution (for tungsten and tungsten-rich alloys (page 124).

When ferrous or nickel alloys, molybdenum or tungsten are involved, etching should be followed by a high acid nickel strike.

When only a few components are to be plated, older techniques such as hand scouring with pumice or fine alumina paste may often be effective in supplementing a somewhat inadequate pretreatment, though not of course applicable to regular production runs.

The general principles just outlined are illustrated by the two following examples of procedures that have been successfully used by the writer for plating specific multi-metal components.

1 Molybdenum with iron alloy (not stainless steel)

(i) cathodic alkaline clean (ii) anodic etch in 60–65% v/v sulphuric acid, 12.5 V, 30–60 sec (iii) rinse in 10% sodium hydroxide solution (iv) low nickel/high sulphuric acid strike (v) gold plate (2 μm) (vi) heat treat in hydrogen at 850°C for 10–15 min (vii) plate as required.

For combinations of stainless steel with molybdenum the same procedure is satisfactory, but the initial gold coating should be 4–5 microns and heat treatment at 750°C in very dry hydrogen (less than 5 ppm water vapour).

2 Component comprising of NILO K and copper (both with heavy heat scale), a tungsten base alloy, nickel, and a silver-containing braze metal.

The following combination of descaling and etching treatments was used:

(*a*) NILO K descale	
Phosphoric acid	40% v/v
Hydrochloric acid	5% v/v
Temperature	80°C
(*b*) Copper descale	
Nickel chloride	20% w/v
Hydrochloric acid	5% v/v
Temperature	Room
(*c*) Tungsten alloy etch	
Sodium hydroxide	10%
Time	5–10 min
Temperature	Boiling

(*d*) Nickel and Nilo K etch

Ferric ammonium sulphate	5%
Hydrochloric acid	12.5% v/v
Sulphuric acid	12.5% v/v
Temperature	Room
Time	1 min

(*e*) Copper and silver braze metal etch

Fluoroboric acid (48%)	20% v/v
Time	1 min
Temperature	Room

This treatment was followed by a low nickel/high sulphuric acid strike, and finally a cyanide copper strike at 2 A/dm^2 for 2 minutes as a base for final gold plating.

SOLDERED COMPONENTS

Soldered brass or copper components can be plated successfully after etching in a 25% solution of fluoroboric acid (48%), followed by a cyanide dip and copper plating to about 2.5 microns from a cyanide bath.

In the case of copper components preferential attack on the solder occurs in fluorobroic acid of this strength due to the potential difference between copper and solder (approximately 450 mV). Proprietary etches are available to overcome this difficulty, and a 50% solution of hydrochloric acid in methanol may also be used. Alternatively a nitric acid based bright dip may be employed, followed by ultrasonic agitation in 5% fluoroboric acid or hydrochloric acid solution.

Special difficulties arise when the solder is porous or joints are not completely filled, a condition more commonly encountered with higher melting point solders than with the conventional tin-lead alloys. In such cases black staining may appear along the soldered joint on transfer to the gold bath, even after applying a relatively thick coating of copper. Satisfactory coverage can often be achieved however by following the 2.5 microns of copper with about 5 microns of silver.

THIN FILMS

Thin gold films of less than about 1 micron deposited on non-conducting surfaces are frequently thickened by electrodeposition. The pretreatment used depends primarily on the type and extent of adhesion between the initial film and the substrate, and the thickness of the former.

When a strong chemical bond has been established over the whole of the interface, as for example, by firing a metal film on ceramic, normal cleaning, striking and etching processes can be used, provided that the latter are not sufficiently vigorous to dissolve a significant amount of the film.

If, however, the bond is either weak or incomplete, as is sometimes the case with chromium or titanium on glass, or if no significant bonding at all occurs, as for evaporated films of gold on glass and ceramics, any pretreatment which involves evolution of hydrogen, e.g., cathodic degreasing, is ruled out by the danger of lifting or exfoliation of the film. Anodic cleaning at low current densities, up to 1.4–2 A/dm^2, is usually permissible. Soak cleaning may also be used, but ultrasonic agitation should be employed only with caution. Where possible, as for example when the initial very thin film has been produced by electroless deposition, problems of pretreatment should be avoided by transferring immediately to the electroplating bath.

Films deposited by physical vapour deposition are invariably covered with a film of oil (usually silicone based) derived from the vacuum pump. This may be removed by wiping with hot xylene or toluene or preferably by soaking in a suitable detergent cleaner. Anodic alkali cleaning will sometimes remove such films, but the oils are not destroyed and tend to build up in the cleaning solution. Anodic alkali cleaning or prolonged soaking in alkaline solutions cannot be used on very thin films of gold with a chromium underlayer, since the latter may be dissolved through pores in the gold film.

SEMICONDUCTORS

Electroless and electrolytic plating as well as physical and chemical methods such as physcial and chemical vapour deposition have all been extensively used in the metallising of semiconductors for the production of ohmic or rectifying contacts, but it is not clear from published details which of these techniques is favoured in commercial production. Processes were reviewed by Duffek[41] in 1964, and the same author reviews present practices in Chapter 37 of the present volume.

A comprehensive description of processes is also to be found in the report of a Conference organised by the Electrochemical Society[42].

Jet etching and plating techniques for selective plating of germanium and silicon using very high current densities have been described by Schnable and Lilker.[43]

The present section is therefore limited to brief descriptions of a few unsophisticated processes which have been successfully used by the writer for the gold plating of certain common semiconductor materials.

Germanium

Chemically polish at room temperature in the following solution:

Nitric acid	45.6% v/v
Hydrofluoric acid (48%)	27.2% v/v
Glacial acetic acid	27.2% v/v
Bromine	10 drops/50 ml

Remove residual oxide film by cathodic treatment in 5% v/v sulphuric acid solution at 2–20 A/dm^2 for $\frac{1}{2}$–1 minute and electroplate as required.

Alternatively the oxide film may be removed by a quick dip in hydrofluoric acid which, however, degrades the surface polish to some extent.

Silicon

Owing to the greater stability of the oxide it is much more difficult to plate onto silicon than germanium. As with other semiconductors, common practice is to deposit an initial film by physical, or occasionally chemical vapour deposition methods and to increase the thickness if necessary by conventional electroless or electrolytic plating. However, the following methods are reasonably satisfactory for gold plating by electroless or electrolytic plating processes only:

1 (i) Remove oxide by immersion in 20–50% v/v 48% hydrofluoric acid solution for 10–30 sec at room temperature, followed by 2–3 min treatment in a boiling solution of the following composition:

Sodium carbonate (anhydrous)	4.25 g/l
Methanol	40% v/v
Water	Balance

(ii) Transfer immediately, without rinsing, to a cyanide plating bath.

2 (i) Remove oxide as in 1

(ii) Transfer immediately without rinsing to electroless nickel plating solution:

Nickel chloride, hexahydrate	30 g/l
Di-ammonium hydrogen citrate	65 g/l
Ammonium chloride	50 g/l
Sodium hypophosphite	10 g/l
pH, (adjusted with ammonia)	8–10
Temperature	92±1°C
Time	10 min

(iii) Heat treat in hydrogen at 650°C for 5 min

(iv) Rinse in hydrofluoric acid to remove brown stain

(v) Repeat electroless nickel plating step

(vi) Electroplate as required.

Turner[44] has reported that silicon can be nickel plated directly in a bath containing hydrofluoric acid, with a 1 minute soak to remove the oxide film before applying the plating current. Adhesion is reported to be good though there is some evidence of a fluoride film between nickel and silicon.

Since silicon is a highly resistive material (10^{-3}–10^{-5} ohm-cm), use of a well insulated jig is necessary in electroplating to avoid short circuiting the substrate. To achieve uniform plating, connection should be made across the whole of the back of the slice. It should also be noted that although the surface of oxidised silicon is hydrophilic, the clean surface exposed after boiling in sodium carbonate solution is hydrophobic, hence the well known "water break" test cannot be applied in this case as a criterion of cleanliness.

Gallium Arsenide

Gallium arsenide is commonly metallised by physical vapour deposition of two or more metals successively, followed by heat treatment to form a eutectic alloy, which may be electroplated with a few microns of gold for thermo-compression bonding. Metallising techniques have been described by Cox[45].

Peterson[46] has described the following procedure for deposition of electroless nickel, which may be followed by immersion gold plating.

(i) Lap with 12.5 micron alumina
(ii) Soak-clean in alkaline cleaning solution containing sodium meta-silicate, sodium carbonate and wetting agent
(iii) Rinse in deionised water for at least 5 min
(iv) Immerse for 2 min at room temperature in stannous fluoroborate solution of the following composition:

Stannous fluoroborate (47%)	15–20 g/l
Fluoroboric acid (42%)	50 ml/l
Hydrofluoric acid (48%)	10 ml/l
Wetting agent (10% solution)	10 ml/l

(v) Rinse thoroughly for 2 min in cold water, followed by hot water
(vi) Immerse for 2 min in palladium chloride solution:

Palladium chloride	0.1–0.2 g/l
Hydrochloric acid	15 ml/l

(vii) Electroless nickel plate

Silicon carbide

Ohmic contacts to silicon carbide have generally required metallisation by physical methods, followed by alloying heat treatment. However, Brander and Lewis[47] have described a process involving successive deposition of titanium and gold to give a film which can be thickened by electroplating if necessary, without heat treatment.

REFERENCES

1 Dini J. W., Helms J. R., *Plating*, **57,** 906 (1970).
2 Fedot'ev N. P., Grilikhes S. Ya., "Electropolishing Anodizing and Electrolytic Pickling of Metals", Draper, Teddington, England (1959).
3 Marshall W. A. J., *Proc. Electropl. Depos. tech. Soc.*, **50,** 78 (1953).
4 Posselt H. S., Anderson F. J., Shrode, L. D., *Plating*, **55,** 883 (1968).
5 Richards B. P., *Metal Finish J.*, **17,** 208 (1971).
6 Tegart W. J. McG., "The Electrolytic and Chemical Polishing of Metals", Pergamon Press, London (1956).

7 Fine R. M., Bracht W. R., *J. electrochem. Soc.*, **112**, 47(c) (1956).
8 Seegmiller R. Gore J. K., Calkin B., *Proc. Am. Electropl. Soc.*, **49**, 67 (1962).
9 Reid W. E., Ogburn F. J., *J. electrochem. Soc.*, **117**, 93 (1964).
10 Wernick S., Pinner R., "The Surface Treatment and Finishing of Aluminium and its Alloys", Draper, Teddington, England (1964).
11 Faust C. L., U.S. Pat., 2,550,544 (1951).
12 General Motors Corporation, U.S. Pat., 2,625,453 (1951).
13 Spring S., *Metal Finish.*, **56**, 8, 66 (1968).
14 U.S. Pat., 2,676,916 (1960).
15 Berg R. V. V., *Trans. Inst. Metal Finish.*, **45**, 4, 161 (1967).
16 Wyszynski A. E., *Trans. Inst. Metal Finish.*, **54**, 4, 147 (1967).
17 Wittrock J. J., *Proc. Am. Electropl. Soc.*, **48**, 52 (1961).
18 British Standard Code of Practice for Cleaning and Preparation of Metal Surfaces, CP3012 (1972).
19 Spencer L. F., *Metal Finish.*, **69**, 12, 43 (1970).
20 Spencer L. F., *Metal Finish.*, **70**, 2, 43 (1971).
21 Safranek W. H., Miller R. H., Faust C. L., I.L.Z.R.O. Project No. 2, f-12, Progress Dept. No. 12, Batelle Memorial Inst., Columbus, Ohio.
22 Spencer L. F., *Metal Finish.*, **56**, 8, 52 (1968).
23 Cramer S. D., Kenahan C. B., Andrews R. L., Schlain D., R. I. 7016, U.S. Dept. of the Interior, Bureau of Mines, Washington, D.C. (1967).
24 Domnikov L., *Metal Finish.*, **60**, 3, 59 (1962).
25 Tripler A. B., Beach J. G., Faust C. L., B.M.I. 1097, Batelle Memorial Inst., Columbus, Ohio (1956).
26 Bartlett E. S., Ogden H. R., Jaffee R. I., "Coatings for Protection of Molybdenum from Oxidation at Elevated Temperatures", A. D. 210, 978 (U.S. Government R & D Rep.), March 6 (1959).
27 Friedman I.. *Plating*, **54**, 1035 (1967).
28 Friedman I., U.S. Pat., 2,834,101 (1958).
29 Marshall W. A., *Trans. Inst. Metal Finish.*, **44**, 3, 111 (1966).
30 Huddle A. U. H., Flint O., Brit. Pat., 788,721 (1952).
31 Beach J. G., Faust C. L., "Modern Electroplating", F. A. Lowenheim (Ed.), Wiley, New York (1963).
32 Missel L., Titus R. K., *Metal Finish.*, **65**, 10, 59 (1967).
33 Foulke D. G., *Plating*, **51**, 685 (1964).
34 Miller G. L., "Tantalum and Niobium", pp. 328, Butterworths, London (1959).
35 Beach J. G., "Technology of Columbium" (Niobium), B. W. Garson & E. M. Sherwood (Eds), pp. 58, Wiley, New York (1958).
36 Beach J. G., "The Metallurgy of Zirconium", B. Luetman & F. Kerze (Eds), MacGraw Hill, New York (1955).
37 Sietnicks J. A., U.S. Pat., 3,234,111 (1966).
38 McGraw J. W., Jakobson K., U.S. Pat., 3,264,219 (1966).
39 Saubestre E. B., *J. electrochem. Soc.*, **106**, 305 (1959).
40 Van Gilder R. D., U.S. Pat., 2,492,204 (1945).
41 Duffek E. F., *Plating*, **51**, 877 (1964).
42 Conf. on Ohmic Contacts to Semiconductors, Montreal, Canada, Oct. (1968); The Electrochemical Soc. (1969).
43 Schnable G. L., Lilker W. M., *Electrochem. Technol.*, **1**, 203 (1963).
44 Turner R. T., *J. electrochem. Soc.*, **106**, 786 (1959).
45 Cox R. H., *Solid St. Electron.*, **10**, 1213 (1967).
46 Peterson M., *Metal Finish.*, **61**, 81 (1963).
47 Brander R. W., Lewis D. T., Brit. Pat., 1,234,976 (1969).

Chapter 13

VAT PLATING

B. D. Ostrow & F. I. Nobel

Although the choice of equipment for gold plating may on the surface appear to require only the same consideration as for non-precious metals, greater care must be taken because of the exceptionally high cost of the metal being plated. The proper selection of equipment may result in substantial economies in gold plating that are not significant in base metal plating.

In addition to using the minimum volume of gold electrolyte consistent with technical quality, other substantial economic savings can be effected by the proper selection of ancilliary equipment and in the judicious choice of operating conditions. Since gold is an expensive metal, its efficient distribution on components is essential, otherwise the marginal excess of gold over that required to achieve the minimum thickness specified becomes unreasonable and therefore represents wastage. Any factors which affect distribution, other than the inherent characteristics of the solution, are therefore a prime consideration. As such, many features of vat plating must be taken into account, not the least of them being the geometric configuration, fabrication material, spacing of anodes, rack design, characteristics of agitation and so forth.

TANKS

The function of a plating tank is to contain the electrolyte without contributing any contamination and without deteriorating through normal use in the plating room. Since the solutions concerned have such high intrinsic value, the utmost care must be taken to ascertain that there is no leakage or any likelihood of this occurring. As a safety precaution, many plants utilise an auxiliary tank to house the plating tank; alternatively, this may be placed under the latter to catch any possible spillage or leakage.

MATERIALS

Plastic

The most commonly used plating tank material for gold is welded polypropylene because of its good resistance to the electrolytes, high strength, high temperature characteristics and durability. Large tanks fabricated from this material generally have steel or other reinforcement on the outside for increased strength, and the tanks are generally set on steel legs held within a framework. Tanks fabricated from PVDC (polyvinyl dichloride) and high temperature polymethylmethacrylate are also satisfactory. Moulded, seamless PVC and polyethlyene tanks of fairly heavy construction are available and can be satisfactorily employed by inserting them into a fibre glass, steel, or stainless steel tank for support. However, these materials will soften and lose strength at temperatures exceeding 60°C.

All plastic tanks must be thoroughly washed, then leached for a few hours to remove plasticiser prior to use. This may be accomplished by using a warm solution of 5% sulphuric acid followed by further leaching with a 5% solution of sodium or potassium hydroxide. The tank should then be rinsed clean.

Steel

Stainless steel tanks are satisfactory for alkaline cyanide solutions used for "flash" deposits. However, they should not be considered for the more modern neutral or acid gold solutions. These tanks have the disadvantage of "stray currents" which are detrimental to good metal distribution. Steel and stainless steel are quite suitable if they are lined with PVC or other suitable plastic which is impervious to the solution. Rubber linings are *not* satisfactory.

HEATERS

Since many gold plating solutions operate at elevated temperatures some source of heating is required. This can be supplied by electric power, steam, hot water, or gas. Heat transfer may be by direct immersion of the heating unit in the tank, by circulation of electrolyte through an external heat exchanger, or by the use of a jacket around the tank. For the heat exchanger, steam is the usual heat source, whereas for a jacketed tank either hot water or steam may be employed.

HEATERS FOR STAINLESS STEEL TANKS

Alkaline cyanide solutions in stainless steel tanks may be heated by a gas flame underneath the tank with thermostatic control; however, this method is not very satisfactory since temperature control is poor. Stainless steel coils heated with steam have been employed with satisfactory results, but in most cases, these solutions are heated with thermostatically controlled stainless steel electric immersion heaters.

HEATERS FOR PLASTIC OR PLASTIC LINED STEEL TANKS

Electric immersion heaters with a quartz sheath and thermostatically controlled, are satisfactory for all solutions except those that are highly alkaline. Since quartz is breakable and very easily cracked, the heater is generally protected with a plastisol coated framework. The plastisol coating must be inspected periodically for breaks or pinholes which might lead to contamination of the gold plating solution by the exposed metal. The plastisol employed should be checked for compatibility with the plating solution and should be leached prior to use.

The most satisfactory heater for all types of gold solutions is an electric immersion heater with a titanium sheath and thermostatically controlled. Titanium is inert in these solutions and unlike quartz, is not subject to breakage. Furthermore, no plastisol protection is required. These heaters have been known to give satisfactory service for long periods of time.

FILTERS

Most gold plating solutions require either periodic or continuous filtration. Since the electrolyte is so expensive, the prime requirement of a filter system must be the ability to operate without leaks. Furthermore, the materials used in the construction of the filter must be inert to the electrolyte otherwise contamination of the solution and deterioration of the filter will occur.

IN-TANK FILTERS

These are generally preferred to external filters since they may be installed in the bath without any external piping which could become subject to leaks and siphoning action. These filters must be fabricated from inert plastics and all parts coming in contact with the solution must likewise be inert. Any leaks or seepage from such a pump must flow back into the solution.

EXTERNAL FILTERS

If an external filter and pump is used, it should be checked for leaks not only when operational, but also when idle. As a safety measure, the filter is sometimes placed in a polyethylene lined steel drum or tank to catch any possible leaks.

FILTER TUBES

Most filters used with gold solutions contain cylindrical tubes wound with polypropylene or cotton fibre. Tubes can be obtained in various mesh sizes, the finest being capable of removing particles down to 1–3 microns. In most cases, a 10 micron filter tube is preferred since it offers less resistance to solution flow.

After use, when the filter tube has become dirty or clogged, it should be saved for recovery of any gold that may be present.

Since these filter tubes contain soluble materials that can contaminate some gold solutions, they must be thoroughly leached with hot 5% sulphuric acid prior to use in acid baths, or hot 5% sodium or potassium hydroxide before using in neutral or alkaline baths. After leaching, the tube must be thoroughly rinsed before use by flushing with clean water.

Porous carbon or porous stone tubes are available for highly sensitive solutions but they are more expensive than cotton or polypropylene tubes.

AGITATION

"Flash" deposits of 2–3 microinches in thickness do not require agitation. When plating heavy deposits, however, agitation is required to prevent streaking and to assist in obtaining uniform deposits.

SOLUTION AGITATION

In many cases, agitation of the electrolyte is provided by the pumping action of the filter. Discharge from the latter can be directed at, or near, the work being plated or else it can be channelled into piping along the bottom of the tank with the flow directed upwards through holes in the pipe. Installations of this type provide a satisfactory method of obtaining both filtration and agitation from the same unit.

In barrel plating this is the only type of agitation required.

WORK AGITATION

In addition to the solution movement created by filtration, agitation may be supplemented by moving the work with a cathode rod agitator or similar device to move the plating rack. When plating through-hole printed circuit boards, plating racks should be arranged so that the solution flows through the holes as the board is moved back and forth by the cathode rod.

AIR AGITATION

Air agitation may be used with some types of gold plating solutions, but its suitability for any particular process should be checked with the supplier. Generally it is not desirable with alkaline cyanide gold baths.

ANODES

STAINLESS STEEL

Use in Alkaline Cyanide Baths

For many years, 18–8 stainless steel anodes have been successfully employed in alkaline cyanide gold solutions used for decorative thin or "flash" deposits. In many such installations, especially those used for pure gold "flash" deposits, the tank is fabricated from stainless steel which is made the anode. Many of these gold solutions are used for obtaining an alloy of a specific colour and for these applications, better colour control is possible when anode rods are used, insulated from the tank and connected to separate stainless steel anodes.

Stainless steel tends to be inert in pure alkaline cyanide gold electrolytes and will not dissolve in the bath. However, if the solution should build-up in chloride ions from water additions, or if complexing agents are used in the bath formulations, there will be a slight attack on the stainless steel, the rate of which will depend on the amount of chloride ion, and the amount and type of complexing agent. The result of this attack is a gradual build-up of nickel, iron, and chromium impurities in the plating bath.

Use in Acid and Neutral Baths

Stainless steel anodes should not be used in acid or neutral gold baths where high purity, soft deposits are required. Some proprietary electrolytes have been developed for depositing hard gold containing small quantities of codeposited nickel or cobalt and a few of these have been formulated for use with stainless steel anodes. Stainless steel should not be used in any acid or neutral gold solution unless specifically approved by the supplier as a suitable anode material. With the proper solution formulation, the rate of attack on the stainless steel that would otherwise take place is significantly retarded.

GOLD

Use in Alkaline Cyanide Baths

Soluble gold anodes have been used successfully in alkaline cyanide solutions containing free cyanide. For effective operation both anode and cathode efficiencies should be equal and close to 100%. When operating with gold anodes in such a balanced system, significant economies are realised, since pure gold metal is the cheapest form of replenishing gold in the bath during operation. The anodes do not dissolve cleanly, however, and tiny particles are formed which can cause roughness and a loss of gold. In such cases, the anodes are usually bagged with cotton bags that have been properly leached prior to use. A serious disadvantage in the use of soluble gold anodes is the considerable monetary investment involved and the danger of possible pilfering.

Use in Acid and Neutral Solutions

In some types of neutral and acid gold solutions, it is permissible to use titanium, electroplated with gold or wrapped with pure gold foil as an "insoluble" anode material. Although such anodes operate quite satisfactorily, there is a disadvantage in that the gold does in fact dissolve from the anode at a slow rate (approximately 1g/100g deposited) which necessitates periodic replating or replacement of the foil. When using gold plated titanium, it is common practice to replate the anodes at least once a week, this of course depending on the amount of use.

GRAPHITE

Graphite or carbon anodes may be used for most gold solutions in any pH range. These anodes are readily available and are relatively inexpensive. They have the disadvantage of forming a carbon powder during use since the graphite slowly disintegrates, hence many platers utilising graphite anodes prefer to "bag" them in cotton or polypropylene bags as a preventative measure.

Graphite anodes are available as extruded 1 in. round rods which can be drilled and tapped, and fitted with a Monel or stainless steel hook, with the solution level below the hook. Compressed graphite sheet is also available and these sheets can easily be sawn into square or rectangular anodes which can be fitted with hooks as for extruded rod.

PLATINUM COATED TITANIUM OR TANTALUM

Platinum is an ideal anode material, but can seldom be employed as the pure metal because of its high cost. It is widely used, however, in the form of a plating or cladding on "valve" metals, principally titanium and tantalum. These metals cannot be used in the bare form as anodes since they are readily passivated under anodic conditions with the formation of insulating films. However, this property constitutes a prime advantage when the metals are employed as supports for a conducting layer such as platinum or gold, since any pores or other discontinuities that may be present or may develop in the substrate are automatically plugged by the passive anodic film.

Since titanium and tantalum are poor conductors (about the same as stainless steel), an anode installation of this type requires sufficient numbers of "straps" with the proper cross-section connecting the anode with the anode bar to provide a cross-section capable of carrying the anode current without overheating.

EFFECT OF ANODES ON GOLD SOLUTIONS[1]

Since most anodes used in gold plating are insoluble, consideration must be given to the possible anode reactions taking place and how they affect the operation of the particular process being used. The following anode reactions are possible during electrolysis:

Anode Reactions

1 $4OH^- - 4e \rightarrow 2H_2O + O_2$
2 $Au + 2CN^- - e \rightarrow Au(CN)_2^-$
3 $2CN^- + 8OH^- - 10e \rightarrow 2CO_2 + N_2 + 4H_2O$
4 Organic material $-$ e $\rightarrow$ Oxidation products
5 Stainless steel $-$ xe $\rightarrow$ Fe^{2+}, Cr^{3+}, Ni^{2+}

Reaction 1: Takes place with the formation of oxygen at the anode. This reaction reduces the hydroxyl ion content of the bath, thereby tending to lower the pH of the solution. This tendency may be offset by reactions taking place at the cathode.

Reaction 2: Occurs only if a gold or gold plated anode is used. The degree to which this reaction takes place depends on the chemistry of the electrolyte.

Reaction 3: Illustrates the oxidation of cyanide at the anode. The overall reaction shown, is the sum of the oxidation of cyanide to cyanate and the further oxidation of the cyanate ion to carbon dioxide and nitrogen.

Reaction 4: Proprietary gold baths may contain organic materials in the form of organic acids, brighteners and other addition agents. These materials may participate in anodic oxidation reactions with the formation of polymeric or degradation products which, in turn, may have significant effects on the operation of the bath as well as on the physical characteristics of deposits.

The extent to which such reactions may occur depends on the nature of the anode material and the organic materials concerned, and also on the bath operating conditions.

Reaction 5: May occur when stainless steel anodes are used. Again it must be emphasised that the chemical composition of the bath may greatly influence the extent of this reaction. The chromium in the stainless steel dissolves first as Cr^{3+}, and may be further oxidised to Cr^{6+}. Similarly, the Fe^{2+} may be oxidised to Fe^{3+}. There is also the possibility of the Fe^{2+} and Fe^{3+} forming cyanide complexes. The rate at which solution of the stainless steel anode occurs will be greatly accelerated by the presence of chloride or fluoride ion in the bath.

CHOICE OF ANODE MATERIALS

It may be seen from the previous text that anode reactions may produce materials that act as contaminants in the gold plating bath. In order to keep these contaminants at a minimum, it is important to choose the anode material carefully.

The choice of the proper anode material is dictated primarily by the chemistry of the solution and secondly by cost. Since many gold plating solutions used today are proprietary, the suppliers recommendation should be sought concerning the type of anode material best suited for a particular solution. When plating thick gold deposits, platinum is generally preferred. When cost considerations become important, graphite is the second choice. When gold "flashing" thin deposits for decorative work, stainless steel is the most economical.

EFFECT OF ANODES AND RACK DESIGN ON METAL DISTRIBUTION

In order to improve metal distribution throughout any plating rack or any large part, the low current density areas should be closer to the anode than high current density areas. It may become necessary to make conforming anodes, or to use short anodes located closer to low current density areas. In some cases, metal distribution may be improved by using fewer anodes and, generally speaking, large numbers of anodes that "cover" the entire work area should be avoided, since it is common to find with such an arrangement that the thickness of gold deposit on the top and bottom of the rack is about twice that at the centre.

In this chapter it is only possible to give the general principles involved in racking since the exact rack design for any given installation will depend on the parts being plated, spacing of the work, distance from the anodes, type of anodes and the geometry of the set-up. Generally speaking, the parts should be racked fairly close to one another on the same rack so that the high current density area of one piece "robs" current from the high current density area of the adjacent part. If the parts are too close to each other, however, then these high current density areas may become low current density areas and the coating may be too thin. This problem of spacing can readily be resolved by making a few runs and measuring gold thickness on various sections of the piece being plated, then making the necessary correction in spacing until the best deposit uniformity is achieved.

In addition to achieving the most uniform deposit over the surface of any single part, it is also important to achieve uniformity of thickness with relation to disposition of parts on the plating rack. To assess this, thickness is commonly measured on parts from the top, centre, and bottom of the rack. A variation of 10% is considered good. In order to achieve this kind of uniformity, it may be necessary to build the rack so that the parts at the centre of the rack are considerably closer to the anodes than those at the top or the bottom. If it is difficult to design the rack in this manner, the same result can be achieved by bending the anodes so that the distance from the anode to the parts in the centre of the rack is far less than the distance of the anode to the bottom or the top pieces on the rack. Considerable savings in gold are realised in production plating shops carrying out heavy gold plating when the rack design and anode spacing have been properly adjusted.

These principles can also be applied to gold plating of large printed circuit boards where the centre or low current density area of the board tends to receive less gold than the outside edges. In these cases the anode can be shorter and narrower than the board and bent in toward the centre of the board so that the distance from the centre of the board to the anode is considerably less than the distance from the edges of the board to the anode. Other methods for achieving maximum uniformity of coatings on printed circuit boards are dealt with in Chapter 35.

Good rack maintenance is very important and the area exposed at the tips should be kept to a minimum in order to minimise the amount of gold being deposited on the rack tips. These tips should not be permitted to build-up in metal and they should be periodically cleaned or stripped, and the strippings should be sent to the refiner. Rack coatings must be inert to the gold plating solution and the coatings must be maintained to prevent any exposure of rack metal to the plating solution.

RINSING

To preserve the life of the electrolyte, adequate rinsing prior to the gold plating operation is extremely important. A drag-in of wetting agents from a previous cleaning operation could be detrimental to the gold plating bath in reducing the brightness range and cathode efficiency. A drag-in of any base metal can impair the effectiveness of the bath in its ability to produce the correct purity of deposit required, the proper colour, cathode efficiency and bright plating range.

In some plating installations, especially those that perform heavy production barrel plating, a chemically compatible gold strike bath is utilised prior to the plating stage, with drag-out recovery and rinse tanks used prior to the gold-plating bath in order to keep the latter clean. In these installations the strike bath builds-up in impurities and is replaced as frequently as necessary. When plating semiconductor or high reliability parts for electronic applications, it is important to use deionised water rinses prior to the gold plating bath and the last rinse prior to drying the work after plating should also be deionised water. In these cases the plating solution make-up water must also be deionised.

When barrel plating, sufficient time must be allowed in the rinses with the barrel rotating continuously in the rinse tank to permit sufficient rinsing after the gold plating operation. Generally two water rinses followed by a deionised water rinse are sufficient and the rinsing time in each operation depends upon the load and can be from 1–5 minutes duration. Drag-out recovery tanks are always used immediately following the gold tank and prior to the rinses in order to save the gold. The level of water in the rinse tanks should be at least 1 in. higher than the level of any other solution in the plating line to make certain that the barrel is completely rinsed after each step.

Rinse tanks can be constructed from stainless steel or preferably polyethylene. Counterflow rinsing can be utilised where two rinse tanks are used together. The principles of rinsing are further discussed in Chapter 18.

DRYING

When drying work that has been gold plated for jewellery or industrial use, it is very important to obtain a surface that is spot free and without any residual material present which would cause a stain or adversely influence soldering and/or any other type of bonding.

In the jewellery industry, the old fashioned method of tumbling the parts in warm sawdust or corncob dust followed by an air blast, is still in use today for some types of work. In most cases, however, rack parts are dried in an oven or a tank which is equipped with a strong hot air blast blowing directly on to the work. In some cases, spot-free drying has been achieved by using a specially adapted degreaser, operating with perchloroethylene or with a FREON type of solvent.

Barrel plated electrical contacts are most commonly dried by using a final hot water rinse, followed by a hot centrifugal spindryer. Care must be taken when plating electrical contacts having blind holes. Parts of this nature must be left spinning in the hot dryer long enough to permit all of the water to evaporate and the vapour to be completely driven off.

Some semiconductor operations have achieved successful drying by rinsing in fresh alcohol or acetone after the deionised water stage. The final operation consists of a blast of clean hot air. In this case the solvents are used only once and it must be understood that a fire hazard has been introduced. A vapour degreaser utilising FREON solvent is also a highly successful method for drying critical semiconductor parts in order to obtain chemically clean surfaces after gold plating.

PURIFICATION OF GOLD SOLUTIONS[2,3]

ORGANIC IMPURITIES

It has been shown by many investigators that organic polymers are generated at both the anode and the cathode in certain gold plating solutions.[4,5] These polymers and organic breakdown products, together with drag-in of wetting agents from previous cleaning or acid dipping tanks, are the organic contaminants the gold plater must contend with. In some gold plating processes these contaminants can cause dullness, brittleness, poor solderability and low cathode efficiency. They will also discolour the solutions, turning them dark brown or tan in colour, with the intensity of the colour becoming deeper as plating is continued over a period of time.

Most of such impurities can be readily removed by treatment of the solution with activated carbon. In a typical batch treatment 1–2 lbs of activated carbon is added per hundred gallons of plating solution, the solution agitated for about 1 hour, then returned to the tank through a filter. Powdered carbon is preferred to granular since it offers a higher surface area for absorption. Some gold will also be absorbed, the amount increasing with higher treatment temperatures, hence the treatment should not in general be carried out at a temperature greater than about 38°C, and the gold content of the electrolyte should be checked on completion. Overnight treatment is not recommended since with prolonged contact some absorbed impurities may tend to go back into solution.

METALLIC IMPURITIES

Where high purity deposits are required or when specific physical and metallurgical properties are sought, metallic impurities must be closely controlled. Since many gold plating solutions are proprietary, specific recommendations as to which metals are to be regarded as harmful impurities can only be provided by the supplier. It should also be noted that the effect of base metal contamination varies with the nature of the bath. For example, in an alkaline cyanide solution iron is complexed to such a degree that it does not codeposit with the gold, whereas in an acid type citrate or phosphate bath a few parts per million can codeposit and adversely affect the metallurgical characteristics of the deposit, such as heat resistance, etc.

To minimise the effect of metallic impurities many proprietary baths employ suitable complexing and chelating agents, the action of which is, however, effective only to a limited extent and within a limited concentration range. Methods may also be recommended by the supplier for the removal of specific contaminants, such as lead, copper, silver, iron and nickel.

Nevertheless, the best approach to the problem of contamination is to prevent it as far as possible by the judicious application of good basic plating practice, such as the use of a gold strike prior to gold plating, thorough rinsing before and after striking, the use of adequate thicknesses of undercoating deposit where necessary, prompt removal from the strike and plating bath of parts which may fall from jigs, and so on.

Lead

This is one of the worst metallic contaminants since only a few parts per million may have severe deleterious effects, particularly in low current density areas. Many solutions contain specific chelating agents to render lead inactive. If the solution is such that chelating agents for lead are not compatible, then a contaminated bath is usually best discarded and replaced, though for some proprietary processes the supplier may recommend a purification procedure based on precipitation and filtration.

Copper

Copper contamination can cause dullness in low current density areas and, in some solutions, will codeposit with the gold, resulting in off-colour deposits and a decrease in purity. Similar comments concerning the use of chelating agents and purification procedures apply as in the case of lead. In alkaline cyanide baths codeposition of copper can be reduced or prevented by increasing the gold concentration, free cyanide and solution temperature, and increasing the degree of agitation, all of which factors will tend to decrease the deposition potential of copper in relation to that of gold.

Iron, Nickel and Cobalt

These metals have a greater tendency to codeposit in acid baths or alkaline sulphite baths than in alkaline cyanide solutions, with deleterious effects on deposit purity and colour. Coatings may also show poor solderability and brittleness. As before, codeposition may be reduced by the use of proprietary chelating agents, techniques of purification being available from suppliers. In alkaline cyanide solutions the adverse effects of contamination may be reduced by changes in electrolyte and deposition parameters similar to those indicated in the case of copper.

Silver

Silver is the most common contaminant in gold salts and it tends to plate out very rapidly with gold. When plating high purity, soft gold deposits, it is important to make up and replenish the bath with high purity gold salts since traces of silver in the deposit tend to harden the gold and change its metallurgical properties.

CONTAMINATED SOLUTIONS

Since there is very little metal loss when the gold solution is refined, the majority of plating personnel refine their own contaminated solutions once the metallic contamination has built-up to such a degree that it has become troublesome. In such cases a new solution should be prepared and recovered gold from the old one sent to the refiner for credit.

GOLD CONSERVATION[5]

When plating with base metals such as zinc, copper or nickel, it is common for the prudent plater to be aware of the cost of electroplated metal even though this may not be very significant in relation to the total plating cost. When heavy gold plating, however, the value of the metal deposited is so great that it is the most important factor in determining the total plating cost. Because of this, it is mandatory that adequate measures should be taken to minimise gold losses and prevent any loss for unaccountable reasons.

SOURCES OF GOLD LOSSES[6]

THEFT

Theft cannot be ignored as a source of gold loss and for this reason, an elaborate control system must be set up for the handling and distribution of gold salts to personnel. Such a system should attempt to establish and control a balance of the weight of gold introduced into the plant with the weight of gold electrodeposited. Although the best system cannot prevent theft, it can establish whether or not an unaccountable loss is taking place.

DRAG-OUT LOSS

Drag-out loss represents the amount of gold that is removed from the solution by the parts, racks and barrels as they are removed from the plating bath at the completion of the plating cycle. This loss can be substantial and the proper measures must be taken to recover this gold before it is lost in the rinse tank effluent.

THE DRAG-OUT TANK

In the sequence of operations, the gold plating tank should be immediately followed by a still rinse which has no overflow, the drag-out tank. The gold content in this tank should not be permitted to rise to any appreciable level since the next step in the plating sequence is a running water rinse which overflows to waste. Any drag-out of gold from the drag-out tank will then tend to be lost.

The most efficient installations remove all or most of the gold from the drag-out rinse water before the next load of work enters this tank. For this purpose, ion exchange resins are used.

ION EXCHANGE RESINS[7,8]

Ion exchange resins can be successfully employed to salvage gold from solutions that would otherwise be lost in waste water. Commercially available materials will absorb the gold complex ion from solution on contact with excellent efficiency, one cubic foot of resin being capable of retaining as much as 120 Troy ounces of metal.

In use, the resin is contained in units of proprietary design through which the drag-out solution is circulated, and by this means the gold content of the drag-out is reduced almost to zero while the resin remains active. The effluent from the equipment is checked periodically for gold content in order to make certain that the resin is not saturated. When the resin is saturated and is no longer capable of removing gold from the drag-out solution, it is removed and sent for recovery.

In some installations, all waste water from the plating line may be treated in a similar way to absolutely ensure the complete recovery of gold.

OTHER LOSSES

Solution Leakage

This subject has been discussed in previous sections on tanks and filters. All pipe lines carrying gold solution to and from the tanks should be frequently examined. Plumbing of this type should be kept to a minimum or preferably eliminated completely.

When removing work from the plating solution, sufficient dwell time over the tank should be allowed to permit the solution to drain as completely as possible into the tank. This is especially true for barrels, and if the parts tend to "cup" and trap solution, the barrel should rotate over the tank until dripping stops.

Samples

Frequent chemical analysis requires the removal of solution and wherever practicable, the unused sample should be returned to the plating tank. A record should be kept of gold solution removed since, in some cases, this may be significant.

Similarly, gold plated parts that are sampled for test purposes may in some cases represent a significant gold loss and should be accounted for.

THICKNESS MEASUREMENTS AND OVERPLATING[9]

ESTABLISHING LOCATION OF THICKNESS MEASUREMENT

Since the thickness of gold will vary over the surface of a plated part, it is extremely important to establish the critical location where the measurement is to be made. If this happens to be in a low current density area, more gold will be required due to overplating on high current density areas of the part. It can be readily seen, therefore, that establishing the location for making thickness measurements has a far reaching effect in determining the overall gold requirement.

THICKNESS MEASUREMENT PROCEDURE

Although thickness measurement procedures are discussed in considerable detail in Chapter 28, it cannot be over-emphasised that the method used must be reliable and reproducible in order to prevent overplating with the resultant loss of gold, or underplating with resultant rejects.

STATISTICS

Once the reliability of thickness measurements and the critical location have been established, a statistical approach should be applied to the plating process in order to optimise all the process variables with the resultant savings in gold usage. The use of statistics is discussed more completely in Chapter 14.

ELECTRICAL CONTROLS

AMPERE-HOUR METERS

The ampere-hour meter measures ampere-hours that pass during the plating cycle and can be used with good advantage in gold plating in two ways:

1 To regulate the replenishing additions of gold and addition agents to the solution. When the number of ampere-hours reaches a predetermined value, a signal is activated which alerts the plater to make an addition to the solution.

2 Since the amount of gold deposited is proportional to the number of ampere-hours passed, the meter can be used to automatically turn off the power, thereby discontinuing plating and/or to activate a signal.

CONSTANT CURRENT

Automatic current control built into the rectifier is especially useful in gold plating since the area of work is always predetermined. A pre-set current with the desired current density may be set by the plater at the start of the cycle and this current will be maintained regardless of line voltage changes, solution temperature variations, anode polarisation changes, etc. The advantage of this is to obtain uniform plating from one load to the next. The constant current control can operate in conjunction with an ampere-hour meter and both units can be incorporated into the same rectifier.

REFERENCES

1 Nobel F. I., Kessler R. B., Thomson D. W., Ostrow R. F., *Plating*, **54,** 926 (1967).
2 Foulke D. G., *Metal Finish.*, **67,** 6, 99 (1967).
3 Silverman L., Bernauer B., Pettinger F., *Metal Finish.*, **68,** 8, 48 (1970).
4 Munier G. B., *Plating* **56,** 151 (1967).
5 Silver H. G., *J. electrochem. Soc.*, **116,** 5, 591 (1969).
6 Donaldson J. G., *Plating*, **56,** 719 (1967).
7 Pritchard E. J., Poihoda W. W., *Plating*, **56,** 1044 (1967).
8 Szkudlapaki A. H., Poihoda W. W., *Metal Finish.*, **67,** 9, 54 (1967).
9 Wilson R., *Trans. Inst. Metal Finish.*, **47,** 42 (1967).

Chapter 14

BARREL PLATING

F. I. Nobel & B. D. Ostrow

Barrel plating is the only method suitable for the treatment of large quantities of small components. The technique is successfully applied to a large variety of parts, ranging from small contacts which form tightly packed loads that roll freely, to large transistor bases which form loosely packed loads tumbling only with difficulty.

In general, barrel techniques are utilised to plate:

(*a*) parts which are too small or too difficult to rack
(*b*) large volumes of parts economically
(*c*) parts which cannot be satisfactorily plated by any other technique
(*d*) a wide range of parts using one piece of equipment.

Since some of the equipment and methods used in vat plating are also applicable to barrel plating, some of which have already been discussed in the previous chapter (rinsing, drying, anodes, etc.), they will not be enlarged upon any further.

TYPES OF BARREL[1,2]

HORIZONTAL BARRELS

Horizontal barrels are most commonly used in gold plating. These consist of a horizontal cylinder, either round, hexagonal, or corrugated, and are equipped with cathode contacts inside the barrel for making electrical contact with the work. The barrels are generally fabricated from polypropylene, high temperature polymethacrylates, or other suitable plastic that is inert to the cleaning and plating solutions employed. For gold plating, the most popular material is polypropylene.

The cylinder is rotated by plastic drive gears that are impervious to all of the solutions used in the plating sequence. The speed of rotation is generally fixed and can vary from 6–12 rpm as required. For optimum mixing and tumbling of the work load, a rotational speed of 10–12 rpm is normally recommended.

The cylinder walls contain holes for solution and current transfer and should be as large as possible, but small enough to contain the work inside the barrel. To prevent small parts from sticking or being trapped in the holes of the cylinder walls, an internal lining of SARAN (PVC) cloth may be used. This loosely woven lining permits the enlargement of the hole dimensions beyond the size of the parts and consequently promotes better solution movement through the cylinder. A further improvement in solution transfer is claimed with a barrel only partially immersed in the bath, using a cylinder wall with a criss-cross, waffle shape that tends to trap solution, which then flows by gravity through the barrel wall when at the top of the rotation cycle.

Cathode Contacts in Horizontal Barrels

When plating large quantities of small parts that tumble freely and form a tightly packed load with good electrical continuity, the most common type of cathode contacts are flexible danglers. These danglers are loose and tend to remain surrounded by the work load throughout the cycle, which considerably decreases the amount of gold deposited on the dangler contact.

Another type of contact used for these freely tumbling loads consists of small buttons placed on the inside walls of the cylinder. Small horizontal strips have also been used in place of buttons. Both tend to break up the load and help move the parts about, but their main disadvantage is that a greater build-up of gold occurs since the contacts are not surrounded by parts when on top of the rotation cycle. When the contacts are "exposed" in this way they tend to plate, thereby causing a loss of gold which would otherwise be deposited on the parts. Attempts have been made by some barrel manufacturers to overcome this difficulty by using a commutator, which disconnects the contacts from the cathode connection when they are at the top of the cylinder and out of contact with the load being plated.

For the barrel plating of loads that are loosely packed and tumble very poorly, if at all, an exposed centre rod is most commonly used as the cathode contact. The rod tends to break up the load during rotation since the work is carried up by the cylinder rotation and tumbles over the rod. Some manufacturers prefer to hang loose chains or wires from the rod while others prefer to insulate it, with 4 or 5 exposed discs making the cathode contact. In the majority of cases, the centre rod approach is the most efficient for these loads.

OBLIQUE BARRELS

Oblique barrels differ from horizontal ones, the cylinder being open at the top and at an angle of about 45°C to the horizontal. These barrels can be constructed in two ways:

(*a*) with solid walls that contain both the solution and the work being plated. There is no tank, the anode being suspended in the solution from the open end of the cylinder.

(*b*) with perforated walls similar to the horizontal barrel, in which case the barrel is wholly or partially immersed in the plating solution and external anodes are used.

Type (*a*) can be used for depositing thin "flash" coatings on small parts, but is not usually recommended for heavy gold plating since, without an adequate solution reservoir, the metal concentration changes too rapidly. Type (*b*) has been used for plating small parts that form freely tumbling loads. The loads are smaller in oblique barrels compared to the equivalent horizontal ones and the barrel sizes used in production are small. Large oblique types are not generally used for gold plating. Parts forming loosely packed loads or those that tumble poorly, are generally not plated in oblique barrels.

The cathode contact in the oblique barrel is on the bottom and generally consists of buttons or a central disc. Since these contacts are partially outside the body of work at the top of their rotation cycle, they tend to build-up in gold. For this reason, they are kept as small as possible and closer to the centre of the cylinder base.

USE OF TUMBLING MEDIA OR BALLAST

NON-CONDUCTIVE MEDIA

Many parts which do not move freely in the barrel require the addition of non-conductive media to the load to improve tumbling as the barrel rotates. Good tumbling action is important in order to achieve a more uniform distribution.

The most common type of non-conductive media is glass beads, generally with a diameter of 0.125–0.25 in. If the density of the ballast must be higher than that of glass, ceramic balls can be used.

The quantity of balls required with the load depends on the nature of the parts and a few experiments can readily determine the minimum amount which gives the best tumbling action without adversely affecting electrical continuity.

CONDUCTIVE MEDIA

When barrel plating parts having one or more electrically isolated areas, better results can be achieved if the load also contains a quantity of conductive media to improve the electrical continuity of the load. In some cases, the use of both conductive and non-conductive media is beneficial for both tumbling action and electrical contact.

Conductive shot can be selected from the following:

(*a*) steel balls, cylinders or cones
(*b*) metallised plastic balls
(*c*) metallised ceramic balls.

The uniform mixing of ballast with loads is highly desirable and is somewhat dependent on the density of the parts and the shot. If the density of the shot is too low compared with that of the parts, the shot tends to concentrate at the top of the load and mixing is poor. Similarly, if the density of the shot is much greater than that of the parts, the shot will be concentrated at the bottom of the load and again mixing will be poor. The following relationship can be used as a guide to determine the proper density of both conductive and non-conductive ballast.

$$\frac{Ds - Bs}{D_p - B_p} = 1$$

Ds = density of the shot

Bs = buoyancy effect of the solution on the shot

D_p = density of the parts

B_p = buoyancy effect of the solution on the parts.

CHOICE OF CONDUCTIVE OR NON-CONDUCTIVE BALLAST

When the use of ballast is indicated, a non-conductive material should be selected whenever possible, since gold plated on the conductive shot represents a loss of metal that would otherwise be deposited on the components. Conductive shot should only be employed to carry current to electrically isolated areas and the amount used should be the minimum required to produce satisfactory work. This minimum is dependent not only on the type of parts in the load, but also on the type of gold plating solution.

BIPOLARITY[3]

If a part to be plated is suspended in the plating bath without making contact to the cathode, the area of the part nearest the anode becomes cathodic while that nearest the cathode becomes anodic.

Bipolarity plays a very important role in the plating of barrel loads having poor electrical continuity, particularly when the parts also form loosely packed loads with poor tumbling characteristics. With parts of this nature, there will be many times during plating when certain areas are not making contact with the cathode and hence will exhibit bipolarity. When this occurs, the part that becomes anodic, as well as any gold deposited thereon, may tend to dissolve in the plating bath, and if the bipolarity condition is maintained then the part may emerge from the bath with a bare spot, devoid of gold, and may even be etched.

The effects of bipolarity can be overcome by the following methods:

1 Incorporate large quantities of conductive shot with the barrel load. This improves electrical continuity throughout the load thereby preventing or sharply reducing, bipolarity. The obvious disadvantages in the use of conductive shot to overcome the "bare spot" problem is the loss of gold that deposits on the shot.

2 Either avoid the use of conductive shot or use only very small amounts in extreme cases, and employ a gold plating solution whose chemistry is such that there is a sharply reduced tendency to etch the basis metal or dissolve gold when the part becomes anodic due to bipolarity. This method is preferred for economic reasons since there is little, if any, loss of gold.

In neither method should the barrel be overloaded as tumbling action is important. Bright or smooth deposits are preferred since they promote better tumbling action whereas dull and coarse deposits are often instrumental in causing tumbling action to cease completely.

APPLICATION OF STATISTICS TO THICKNESS CONTROL

Although a knowledge of statistics is important in all types of plating, it is of far greater importance in gold plating due to the intrinsic value of the metal being deposited. It has been shown by many investigators that statistics can readily be applied to barrel plating and can be a valuable tool in determining the minimum amount of metal required for a barrel load. Statistics can also be used to study the overall process with a view to making improvements and to obtain optimum results with the greatest economy.

Thickness measurements of gold deposits on groups of parts in various barrel loads indicate that there is always a spread of thickness from one part to the next throughout the load and variation of thickness over the surface of any one part. It is obvious that considerable savings can be realised if barrel plating of gold is performed with the least amount of thickness spread. Costly overplating to meet thickness requirements would then be kept to a minimum.

FACTORS INFLUENCING METAL DISTRIBUTION[4,5]

There are many variables in barrel plating influencing the spread of thickness throughout the load. These may be conveniently classified under two headings:

(*a*) *Equipment Factors*

1. Type of barrel—horizontal, oblique, etc.
2. Shape of barrel—hexagonal, round, etc.
3. Type, size, number, and location of cathode contacts.
4. Type, size, and number of holes in the barrel.
5. Speed of rotation.
6. Load size.
7. Type of part.
8. Type and quantity of ballast used, if any.
9. Type, surface area, number and location of anodes.
10. Number of barrel rotations throughout the run.

(*b*) *Process Factors*

1 Type of gold plating process.
2 Solution composition.
3 Throwing power.
4 Cathode efficiency.
5 pH.
6 Thickness of deposit.
7 Current density.
8 Time.

It can be readily seen that many of these factors are interdependent. For example, when plating to a specified average thickness, a change in current density would also necessitate a change in plating time, total number of barrel rotations, cathode efficiency and throwing power.

The best way to evaluate and adjust these factors in order to achieve the desired results, is to apply statistical techniques.

STANDARD DEVIATION

The average deposit thickness on the parts in any barrel load is the arithmetic mean of all the measurements made. This figure is important since it determines the total amount of gold used and also the plating time required at a given barrel efficiency. It gives no indication at all, however, concerning the spread of thickness through the load. For example, consider the following two hypothetical series of measurements on sets of five samples taken from different barrels:

Barrel A	3, 5, 1, 7, 9
Barrel B	5.5, 4.0, 6.0, 5.0, 4.5

In each case the arithmetic mean is 5, but it is clear that parts from B are far superior in respect of range or spread of thickness.

It is also obvious that a better indication of the degree of spread would be obtained by averaging the deviations of individual measurements from the mean value. Since, however, deviations may be either positive or negative their average must always tend to zero unless the signs are ignored. Hence, it is customary to use the root mean square of individual deviations to express the spread of thickness about the mean. This quantity, of basic importance in statistical method is usually symbolised by the Greek letter sigma (σ), and may be expressed as:

$$\sigma = \sqrt{\frac{(\bar{t}-t_1)^2+(\bar{t}-t_2)^2+\ .\ .\ .\ (\bar{t}-t_n)^2}{n-1}}$$

$$= \sqrt{\sum\frac{(\bar{t}-t)^2}{n-1}} \qquad (1)$$

where t = average thickness

t_1, t_2, etc. = thicknesses on individual parts

n = number of parts.

DETERMINATION OF STANDARD DEVIATION[6,7]

Since it is not practical to measure the thickness on every piece from a load, a random sample representative of the entire load must be taken and sigma calculated. Two methods for making this calculation will be presented. Method A is based on a fairly complicated formula, representing an alternative form of equation (1), and requires a great deal of arithmetic involving the square root of the sums of squares. Method B is simple to use and is based on groups of measurements and their ranges.

Method A

$$\sigma = \sqrt{\frac{\Sigma t^2 - (\Sigma t)^2/n}{n-1}}$$

where t = measured thickness on a sampled part

n = total number of measurements

Σt^2 = sum of the squares of each thickness measurement

$(\Sigma t)^2$ = the square of the sum of all the thickness measurements.

A more detailed discussion of the standard deviation and the application of this equation has been presented in reference 6.

Method B

Random samples for measurement are taken in groups of five. Then sigma is given to a good approximation by:

$$\sigma = \frac{\bar{R}}{2.326}$$

where R_5 = range within a group of 5 samples (i.e., highest thickness—lowest thickness)

$$\bar{R} = \frac{\Sigma R_5}{n}$$

n = number of groups of five.

2.326 is a numerical factor applying when each group contains five samples. This is derived from statistical calculations beyond the scope of the present discussion.

A typical example of the application of the method is as follows:
A random sample of 5 groups, consisting of 5 measurements from each group is taken, totalling 25 measurements. They are arranged in a box thus:

Group	Measurements					R_5
1	50	45	41	52	49	11
2	47	51	50	39	52	13
3	51	53	42	53	49	11
4	52	50	48	40	45	12
5	48	52	51	54	43	11

$$R_5 = 58$$

$$\bar{R} = 58/5 = 11.6$$

$$\sigma = \frac{11.6}{2.326} = 5.0$$

For comparison the same group of measurements were used in method A and the formula yielded $\sigma = 4.9$.

CALCULATION OF SAMPLE SIZE[8,9]

The sample size must give an acceptable approximation of the whole load and yet must not require an excessive number of measurements. Obviously, there will be some uncertainty about the average thickness calculated from a sample since it cannot be guaranteed that it is identical to the entire load. The statistician speaks of a "degree of confidence" as a measure of this uncertainty. It can be readily seen that the degree of confidence increases as the sample size increases and reaches a maximum when the sample is equal to the entire load. The maximum allowable difference between the average calculated from the measurements of the random sample and the average which would be obtained if the entire load were measured, must be arbitrarily fixed. An estimate of the standard deviation must also be made and the equation presented in the ASTM Manual[6] applied as follows:

$$n = \left(\frac{3\sigma}{\Sigma}\right)^2$$

n = sample size

σ = estimate of the standard deviation

Σ = the maximum allowable difference between the estimate of the average thickness obtained from the sample, and the result if all units were tested.

A simplified method for obtaining the proper sample size is described in references 8 and 9, which adequately present charts and all the necessary procedures.

NORMAL DISTRIBUTION

Since the thickness of deposit on any individual part in a barrel load depends on a chance combination of circumstances, the thickness distribution among all the parts will conform to a normal distribution curve (Figure 14.1). The greatest number of parts will show the average thickness and only a very small number will be outside the thickness range of $\bar{t} \pm 3\sigma$, where $\bar{t}$ is the average thickness. In fact, 99.7 % of the parts will lie within this range, 95.4 % will lie within the range $\bar{t} \pm 2\sigma$, and 68.3 % in the range $\bar{t} \pm \sigma$.

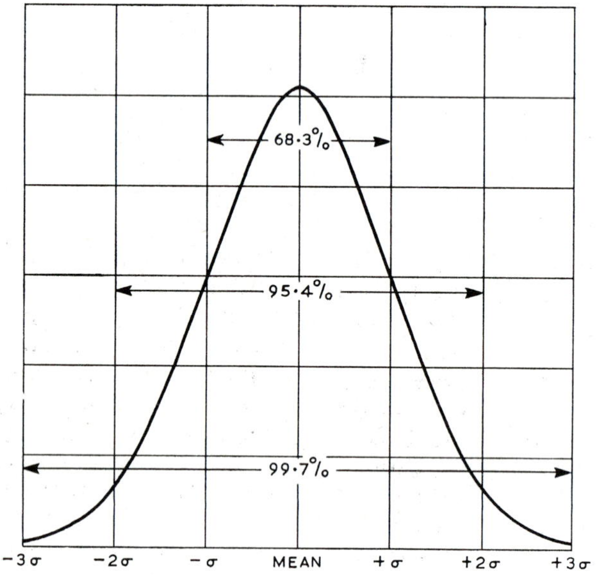

Fig. 14.1 Normal distribution curve.

CALCULATION OF REQUIRED DEPOSIT THICKNESS

In order to calculate the required mean deposit thickness for a given load, a minimum deposit thickness must be specified and an acceptable quality level (AQL) must also be agreed. For average work, this level is generally about 2–2½ % and for high reliability, it is usually 0.1 %. The normal distribution curve should be referred to for the following discussion:

(*a*) *Acceptable Quality Level is* 50 %

If the mean or average thickness of the load is equal to the minimum thickness required, then the entire area under the curve to the left of the mean would be below specification and 50 % of the parts would fail.

(*b*) *Acceptable Quality Level is* 16 %

If the mean thickness is equal to the minimum requirement plus σ, then no more than 16 % of the parts would fail. This percentage represents the area of the curve to the left of $-\sigma$.

(*c*) *Acceptable Quality Level is* 2.3 %

If the mean thickness is equal to the minimum requirement plus 2σ, then no more than 2.3 % of the barrel load would fail. This represents the area of the curve left of -2σ.

(*d*) *Acceptable Quality Level is* 0.15 %

For this AQL, the mean thickness must be equal to the minimum required plus 3σ. This 0.15 % level represents the very small area to the left of -3σ.

GOLD SAVINGS

A study of the preceding discussion indicates the saving of gold that can be achieved by improving the factors governing metal distribution, thereby reducing sigma, particularly for high quality work. The following is an example of how much gold may be saved by taking the necessary steps to reduce standard deviation:

Minimum thickness requirement	=	100 microinches
Acceptable Quality Level (AQL)	=	0.15 %
Standard Deviation	=	16 microinches.

In this example the necessary mean thickness is the minimum thickness requirement plus 3σ or $100+(3)(16)$ or 148 microinches. If the plating process is improved to reduce the standard deviation to 8 microinches, then the mean should be $100+(3)(8) = 124$ microinches. This improvement has saved 24 microinches which represents a savings of 24/124 or 19 %. A 19 % saving of highly priced gold metal is considerable, thus underlining the importance of working on the plating process to reduce the standard deviation. This saving indicates the importance and the necessity to become familiar with the statistical method and to utilise it to maximum advantage.

BARREL EFFICIENCY

The overall barrel efficiency is a measure of the amount of gold plated on the parts compared with the total amount of gold deposited. The cathode efficiency of the electrolyte, as measured by normal methods, is the maximum that the barrel efficiency can attain. Any value less than this maximum represents a loss of gold that would otherwise be plated on parts. This loss is due to gold being deposited on to cathode contacts, conductive ballast, exposed cathode connections under the solution level due to insulation breakdown, parts stuck in the corners or holes in the barrel, treeing and so forth. Heavily plated cathode contacts should be removed, sent to the gold refiner, and replaced by new ones. Good barrel maintenance is essential for high barrel efficiencies.

Barrel efficiency can be determined as follows:

$$BE = SE \times \frac{Wp}{(7.36)\,(AH)}$$

BE = Barrel efficiency, expressed as percent.

SE = Efficiency of the gold plating solution measured at the average current density in the barrel, expressed as percent.

Wp = Weight of deposit on parts, expressed in grams.

7.36 = The electrochemical equivalent for monovalent gold, g/A-hr.

AH = Current in amperes multiplied by plating time in hours.

When plating tightly packed loads that tumble freely, the barrel efficiency should be close to the solution efficiency since very little gold need be lost to cathode contacts and conductive shot is not required.

Parts that tumble poorly and have electrically isolated areas present a more complicated situation. If the electrical continuity of the load is poor, the cathode contacts as well as those areas on parts electrically continuous with the cathode contact, will take the bulk of the current and plate at very high current densities. At these high current densities, the solution efficiency may be considerably lower than that measured at the average current density for the load. Furthermore, the contact will tend to plate rapidly, causing the barrel efficiency to be reduced even further.

Loads of this type require conductive ballast to produce better electrical continuity, thereby spreading the current over a greater area and reducing the current density on those areas that are actually plating at any given time. In this way, the use of some conductive shot can raise barrel efficiency. If the amount is greater than that required, the barrel efficiency again will be reduced because of the gold that is plated on the shot. A few experiments can determine the optimum amount required.

For the most complicated parts of this kind, it is difficult to predict barrel efficiency beforehand. This will necessitate keeping records for each type of part to facilitate setting current and plating times to achieve the required deposit thickness.

AVERAGE CURRENT DENSITY

In barrel gold plating, the average current density is the total current divided by the total surface area of metal plated. The total surface area includes the respective area of cathode contacts, conductive shot and parts. The average current density assumes that all areas in the load are plating at the same time. This is not a correct assumption in barrel plating and the actual current density is difficult to determine and may vary throughout the cycle. The average current density as just defined is always used.

The proper average current density for any given load depends on the type of gold solution, type of parts, load and barrel size. Generally speaking, larger loads and larger barrels require lower current densities. A given part plated in a 4 in. × 6 in. barrel can plate at 2–3 times the average current density used for the same parts in a 10 in. × 18 in. barrel. A larger load size within a given barrel requires a lower average current density than a small load.

In order to achieve better metal distribution, the average current density should be as low as is practical. Lower current densities generally yield lower standard deviations. On the other hand, average current densities that are exceedingly low should be avoided since plating time will be prolonged, production rate will drop and the operation can become uneconomical. For any given production run, the best average current density to use with respect to both good production and metal distribution, can be advantageously determined by a statistical evaluation of a few experimental runs.

SIZE OF BARREL, WORK LOAD AND TANK

BARREL SIZE—TANK SIZE

The best barrel size to use depends upon production requirements, the type of parts to be plated and the size of the plating tank. The cylinder must be large enough to permit parts to tumble without jamming between cylinder walls and cathode contacts, yet small enough to fit comfortably in the plating tank.

The tendency is to choose a barrel size that will give the highest production output in the shortest time. To achieve the best metal distribution, however, it should not be too large since lower standard deviations for any given part are obtained from smaller barrels. In production gold plating, barrel sizes rarely exceed 10 in. × 18 in. and usually vary from 6 in. × 12 in. to 8 in. × 18 in.

The size of the tank should be large enough to hold a sufficient volume of solution to provide a satisfactory reservoir of gold ions so that replenishment additions of gold can be made at reasonable intervals.

WORK LOAD

In any given barrel, the size of the work load should be adjusted to give good tumbling action and be large enough to satisfy production requirements.

When plating small parts that tumble freely and form tightly packed loads, the barrel is usually loaded from one quarter to two thirds its capacity.

Transistor parts and others that form poorly tumbling, loosely packed loads, should generally fill about one third to two thirds the capacity of the barrel when tumbling in water or solution. The barrel is filled to just above or below the centre rod in this case, since the load volume increases markedly when tumbling in any solution or rinse tank.

In order to adjust the load size of any part to improve metal distribution, a few runs can be made and the results evaluated using the statistical method. Generally speaking, load sizes for any given part should be adjusted to give the best tumbling action so that the most uniform metal distribution can be achieved.

ADVANTAGES AND DISADVANTAGES OF BARREL PLATING

Advantages of barrel plating have already been discussed in the introduction to this chapter. For many parts that can be either rack or barrel plated, the latter is preferred since there is a saving in labour for the same production output, large rack inventories are eliminated and metal distribution can often be superior to rack plating.

Some production parts do not require a uniform gold deposit over the entire surface area, the gold only being required in one specific area, with little or no gold on the rest of the part. With parts such as these, barrel plating is wasteful of gold since selective plating of this type cannot be achieved at all or only when using costly masking procedures. Selective plating is therefore limited to plating on racks, strips or continuous strip.

Highly polished decorative work susceptible to surface scratching, and delicate parts subject to distortion, are not processed in barrels.

REFERENCES

1 Alina W., *Products Finish*, **35**, 5, 90 (1970).
2 Alina W., *Products Finish*, **35**, 8, 96 (1970).
3 Nobel F. I., Thomson D. W., *Plating*, **57**, 469 (1970).
4 Nobel F. I., Ostrow B. D., Kessler R. B., Thomson D. W., *Plating*, **53**, 1099 (1966).
5 Nobel F. I., Kessler R. B., Thomson D. W., *Plating*, **58**, 1198 (1971).
6 A.S.T.M. Manual on Quality Control of Materials, Special tech. Publ. No. 15C.
7 Statistical Quality Control Handbook, Western Electric Co. (1956).
8 U.S. Dept. of Defence, Mil. Std. 105D, April (1963).
9 U.S. Dept. of Defence, Mil. Std. 414, June (1957).

Chapter 15

AEROSOL DEPOSITION

W. GOLDIE

An alternative method to depositing metals by chemical reduction using immersion processes is to spray the reactants so that when the solutions impinge on the surface a metallic film is formed by their interaction. The reducing agent and the metal ions are kept separate until metallisation occurs so that decomposition is not possible when not in use. The apparatus normally used for carrying out this process consists of a gun, although air "brushes" have been employed and aerosol cans containing the fluorohalogen refrigerants for small scale and laboratory work have been reported[1,2].

The conditions which favour spraying should, in effect, be the opposite of that required for immersion processes, that is instability. The more unstable the solutions are on interaction, the easier deposition should be. Weak complexing agents, powerful reducing agents and high absolute concentrations are advantageous. Other factors which facilitate high plating rates in immersion solutions and hence decomposition, such as increased temperatures, also favour spraying techniques. A compromise is normally made. The choice of specific reactants and concentrations is selected to obtain maximum efficiency and a deposit which is qualitatively sound.

ADVANTAGES AND DISADVANTAGES OF AEROSOL DEPOSITION

ADVANTAGES

Speed

A conducting film may be formed much more rapidly by aerosol deposition. The time taken is dependent on the area to be metallised, whereas with immersion techniques it is independent of area. As a rough guide, using aerosol techniques, an average sized printed circuit board measuring 5 in. × 4 in. will take about 20–30 seconds to form a satisfactory conducting film,

---

Note—Gold deposited by aerosol techniques is applicable, on an almost exclusive basis, to non-conducting substrates and as such, Chapter 15 is written with this in view. The immersion processes referred to throughout are those covered by Chapter 11 and not those described in Chapter 10.

but will take anything upwards of 2 minutes using the immersion processes currently available which are capable of depositing gold directly onto polymeric materials (see Chapter 11).

Efficiency

Under properly controlled conditions and choice of reactant concentrations, 100% efficiency can be obtained in specific cases resulting in no wastage of expensive metallic salts or an additional cost of reclamation. In order to achieve adequate deposition rates, some efficiency may have to be sacrificed.

Stability

Since the reducing solution is separated from the metal ions until metallisation occurs, there is no danger of decomposition taking place.

Adaptability

The technique is portable, allowing for the metallising of very large objects such as reflectors which could not be carried out satisfactorily by other means.

Contamination

Since a number of plastics are susceptible to the ingress of moisture, the speedy deposition of a metallic film will act as a barrier to prevent this. As the substrate is only in contact with the solution for a short period, the risk of contamination is reduced. Furthermore, any leaching of impurities to the solution can be an embarrassment with immersion solutions.

DISADVANTAGES

Economics

When aerosol deposition was introduced into the mirror industry, a significant reduction in processing costs was claimed[1]. This amounted to about 20%. As this estimate was compared against the "homogeneous" deposition of silver, the comparison is not valid when using autocatalytic plating techniques. In general, aerosol plating is more expensive than the equivalent electroless technique.

Configuration

Irregular shaped objects, particularly fine bore tubes, are much better processed by immersion techniques. Although the use of aerosol techniques is limited to geometric configurations which are not too complex, it is, nevertheless, a more versatile process than, for example, vacuum evaporation.

ATOMISERS

Atomisers are primarily designed to cause a liquid to accelerate and disintegrate as quickly as possible in a controlled manner and to direct the resultant particles in a preferred direction. They are normally classified according to the source of energy employed to effect atomisation. Various sub-headings within each classification exist but these are not discussed here since they are beyond the scope of the present work.

(*a*) Centrifugal
(*b*) Pressure (otherwise known as hydraulic, airless, mechanical)
(*c*) Gaseous (pneumatic, air, steam).

Fundamentally there is no difference in the basic mechanism of atomisation by the above methods. Pneumatic (compressed air) is the technique usually adopted because of its overall effectiveness, although liquid pressure (hydraulic) is employed in particular cases.

HYDRAULIC

The main disadvantage that this technique suffers from is that the flow rate is low since it varies with the square root of the pressure. Higher flow rates can be achieved by increasing the size of the orifice but this results in an increased droplet size as well. Droplet size is relatively large and as plating efficiency is improved by a decrease in the droplet size, the overall result is that plating efficiency is not particularly good. If a small droplet size is to be obtained by this process, very fine orifices are essential and there are innumerable complications in their production. A further disadvantage is that erosion of the orifices presents a severe engineering problem. This technique is therefore only resorted to when pneumatic means cannot be employed, such as the spraying of hot solutions which cannot be satisfactorily carried out due to the cooling effect of the compressed air, or reducing solutions are used which are very easily oxidised by the flow of air.

PNEUMATIC

This is the most satisfactory method in existence at present for producing fine sprays and is particularly attractive because fairly large orifices can be used. The mean diameter of the particles may be as small as 10–20 microns but they lack uniformity, a characteristic which is not so evident in the two techniques previously described. If uniformity could be improved, plating efficiency would increase. Although the overall plating efficiency is very high due to the relatively small droplet size, there is a finite optimum drop size for maximum efficiency. When this is reached, efficiency decreases with decreasing droplet size.

The guns used in pneumatic atomisation of liquids are available with various modifications. For example, mixing of the solutions may take place internally or externally. The former has certain disadvantages in that precipitation of the metal may occur inside the gun, since the reaction takes place on mixing. Consequently the orifices become blocked and have to be cleaned.

Single, double and treble nozzle guns may be used. When three solutions have to be sprayed, Y-tube connections may be utilised on single or dual nozzle guns and be reasonably effective in segregating solutions until commencement of spraying.

Stainless steel is normally used in the construction of the gun although plastic parts may be incorporated where necessary.

LIQUIFIED PROPELLANTS

An aerosol pack contains the solution together with a liquified propellant gas under pressure; the latter is normally of the fluorinated hydrocarbon family, such as Freon or Arcton. The free space is saturated with the vapour of the propellant, which on release of the valve ejects a stream of liquid from the orifice. On leaving the orifice, the reduction in pressure obtained on reaching the atmosphere causes the gas to vaporise. This is so rapid that the liquid stream is broken up, or atomised, resulting in fine droplets of solution being propelled towards the substrate in a stream of vapour.

When heated slightly, aerosols become more efficient since a rise in temperature increases the kinetic energy of the molecules in the enclosed space, which results in a greater pressure being exerted when equilibrium is attained between the vapour and liquid. Overheating should be avoided as the consequences could be dangerous.

Compatibility of the solution and propellant, and the solution and container should be investigated prior to their use.

ADVANTAGES AND DISADVANTAGES OF LIQUID PROPELLANTS

ADVANTAGES

1 Portability makes it extremely useful for small scale laboratory work.
2 Cleaning of equipment is unnecessary.

DISADVANTAGES

1 The use of aerosols is not a cheap method of producing a metallic film.
2 Since a maximum flow rate of only about 30 ml/min is obtainable, this imposes a restriction on the technique for production purposes.
3 Container is not refillable.
4 Mechanical failure can render the contents useless.
5 Intermittent spraying can only be carried out. Continuous use may result in a momentary reduction in pressure, since the gas inside the container must vaporise during spraying to replace the volume vacated by the liquid and so continue to exert a pressure on the surface of the remaining liquid.

FACTORS INFLUENCING ATOMISATION

The general purpose of spraying is to increase the surface area of a mass of liquid and direct it towards the substrate and to speed up a chemical reaction. In order to design suitable equipment to administer this technique the fundamentals of fluid kinetics of rapidly moving liquid jets and their subsequent disintegration should be understood by the designer. They are discussed briefly in the following paragraphs.

ATOMISERS

The source of energy used for atomisation, which has already been discussed, and the design of the atomiser contribute directly to the efficiency of the process. As the droplet size decreases the greater the area that the same volume of liquid will cover, hence the greater the efficiency because of the better chemical interaction of the particles. Smaller droplet size is not only accomplished by selecting the appropriate energy source but in the choice of orifice diameter, the smaller the diameter the smaller the droplet size.

EFFECT OF LIQUID PROPERTIES (PHYSICAL)

The properties which exert most influence in atomisation processes are (1) surface tension and (2) viscosity. Surface tension tends to resist the formation of new surfaces as does viscosity. During the final stages of disintegration, viscosity will oppose surface tension unless the viscosity is very low. As aqueous solutions are normally used, this point is not pertinent. It is difficult to achieve a small droplet size with low viscosity liquids when hydraulic methods are employed.

Although low concentration of surfactants usually improve the wetting of substrates, it is inadvisable to use excessive amounts of wetting agents to lower the surface tension of the liquid since the physical properties of the deposited film are frequently affected. Interference with the optical properties may also be encountered.

EFFECT OF FLOW RATE

Liquid flow rate affects the process of atomisation considerably. High flow rates increase the rate of deposition, but only at the expense of plating efficiency. This is due to the formation of a large droplet size, resulting in ineffective mixing and retardation of the chemical reaction.

In pneumatic atomisation the ratio of rate of air flow to rate of liquid flow is important and should be high to achieve a low particle size.

The principles of atomisation are exceedingly complex but the more pertinent points raised above will give the reader a general insight into what is involved.

TYPICAL SOLUTIONS

Prior to the published work by Levy on aerosol gold solutions, only two references have been uncovered pertaining to the deposition of this metal. Andres[3] suggests the following formulation:

Solution "A"

Gold chloride	25 gm/l
Sodium carbonate	25 gm/l

Solution "B"

Formaldehyde	40 gm/l
Sodium carbonate	40 gm/l

Later the following solutions were proposed by Schneider[4]:

Solution "A"

Gold chloride	75 gm/l

Solution "B"

Glycerine	500 ml/l
Sodium hydroxide	50 gm/l
Mannitol	trace.

Nothing further materialised for the next 25 years until the published work of Levy.[1,2,5,6,7] As a result solutions are now available commercially under the trade name "Lockspray-Gold".

The deposited metal has a purity of 99.99% and a resistivity of 2.4 μ ohm-cm, similar to that of bulk gold. Grain size is between 20 and 50 Å. The process is carried out at room temperature and, as the efficiency is about 60%, it is desirable to collect and reclaim the gold from the spent solution.

A patent[8] assigned to Lockheed Aircraft Corporation describes the use of hydrazine as a reducing agent in aerosol gold plating. The source of metal ions is a soluble gold salt such as the chloride or bromide, and this is complexed with a suitable complexing agent like ammonium carbonate, ammonium hydroxide, or an aliphatic amine. Ethylenediamine is preferred.

A typical solution is:

Solution "A"

Gold chloride (0.2 M)	50 ml/l
Ethylenediamine	1.5 ml/l

Solution "B"

Hydrazine (20 M)	100 ml/l
Sodium hydroxide (1 M)	50 ml/l

The solutions are sprayed simultaneously at room temperature, the pH of the resultant solution being between 6 and 8.

PREPARATION OF SOLUTIONS

Solution "A"

1 50 ml of 0.2 molar gold chloride are added to 700 ml of water and neutralised with a solution of sodium hydroxide.

2 1.5 ml of ethylenediamine are added to 50 ml of water and neutralised with hydrochloric acid. Solution (1) is added to solution (2) and thoroughly mixed after dilution to 1 litre with water.

Solution "B"

The reducing solution is prepared by mixing 50 ml of 1 molar sodium hydroxide with 100 ml of 20 molar hydrazine and then diluting to 1 litre with water.

FACTORS INFLUENCING AEROSOL DEPOSITION

A considerable amount of work has been carried out on the spray parameters of this process. It is interesting to note that the variables recorded earlier by Upton[9] in his investigation on silver have been verified by Levy[1,2] for gold.

RATE OF FILM FORMATION

Following an initial rapid rate of formation, it drops off somewhat and proceeds to increase linearly with time (Figure 15.1). This initial rapid reaction is attributed to the reducing action of the stannous ion prior to metallising.

EFFECT OF SPRAYING DISTANCE

Increasing the distance between the nozzle and substrate being metallised using a single nozzle gun, results in a decrease in deposition rate. A reduction in efficiency is also obtained (Table 15.1). This is probably caused by a homo-

Table 15.1
The Effect of Spraying Distance on Efficiency

Spraying Distance (in.)	*Efficiency %*
3	32.8
6	30.7
12	28.7
15	27.7
18	23.6
21	19.7
24	18.3

geneous action taking place before the metal reaches the substrate and is due to the design of the gun.

Where dual nozzle spraying apparatus is used with intersecting sprays, the deposition rate and efficiency are at a maximum when the substrate is located in the zone of mixing of the two reactants. At greater distances efficiency decreases, homogeneous reaction taking place prior to the metal reaching the substrate. At shorter distances the two sprays do not mix before reaching the substrate, hence there is a sharp decrease in the amount of metal deposited.

EFFECT OF LIQUID FLOW RATE

When the liquid flow rate increases, the plating rate increases. In fact it increases proportionately according to the square root of the liquid flow rate. Efficiency declines as flow rate is increased. This is because the liquid droplet size becomes larger and results in poorer mixing.

EFFECT OF CONCENTRATION

As would be expected, the rate of deposition increases with increased concentrations. However, efficiency decreases (Figure 15.2). At low concentrations, 100 % efficiency can be achieved.

AIR PRESSURE

An increase in the air pressure means that air velocity and air flow also increase which results in an improvement in the rate of deposition (Figure 15.3), assuming liquid flow rate remaining constant. This is because droplet size decreases and better mixing of the solution takes place. The mass ratio, air flow/liquid flow, should be high, the higher the ratio the greater the efficiency, this being due to decreasing droplet size. Droplet size plays an important role in aerosol deposition. When hydraulic means are used, a smaller droplet size is obtained by using higher liquid pressures.

It has already been mentioned that orifice diameter influences atomisation considerably. The relationship between this and air pressure, flow and liquid flow is important.

In Figure 15.4 the effect of air pressure on plating efficiency is shown together with the effect of decreasing the orifice diameter. It can be seen that a minimum air pressure of about 20 psi is required to effect good atomisation with the type of nozzle employed.

APPLICATIONS*

Gold has many outstanding properties, the most important of which are its

---

* The reader is referred to Chapter 39 where further information may be obtained regarding the applications cited in this section.

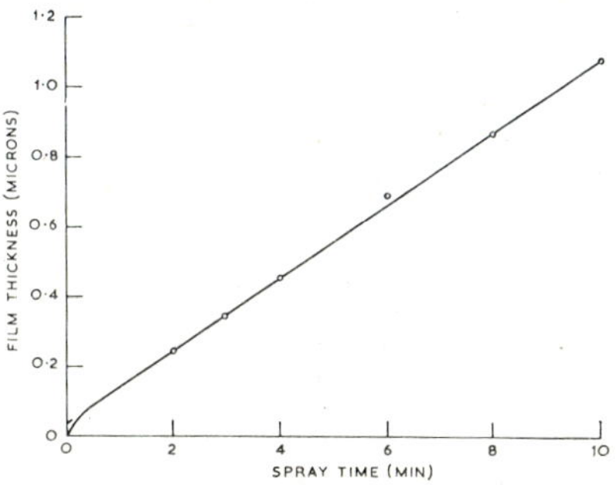

Fig. 15.1 Effect of plating time on deposition rate of gold. A glass plate, 5 cm^2 was sensitised in stannous chloride prior to spraying with a 0.0187 molar gold solution and a reducing agent using a dual nozzle sprayer. (Binks 181 SS) Liquid flow rate 22 ml/min; spraying distance 10 in. [Levy & Delgado].

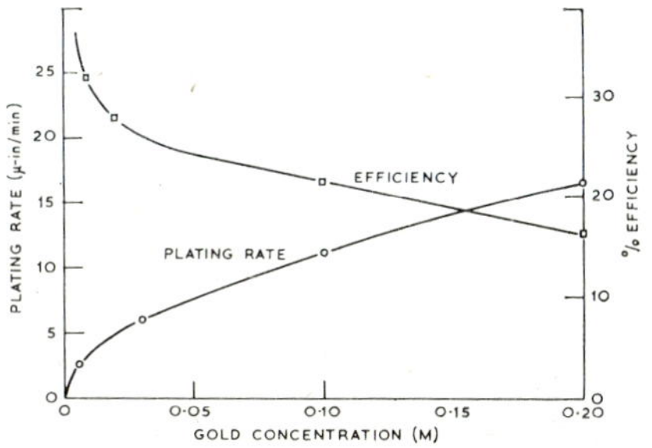

Fig. 15.2 Effect of gold ion concentration on plating rate. Conditions of spraying similar to that of Fig. 15.1. The plating rate increases with the concentration of gold in the plating solution. Non-linearity indicates a decrease in plating efficiency which is highest in very low metal concentrations. [Levy & Delgado].

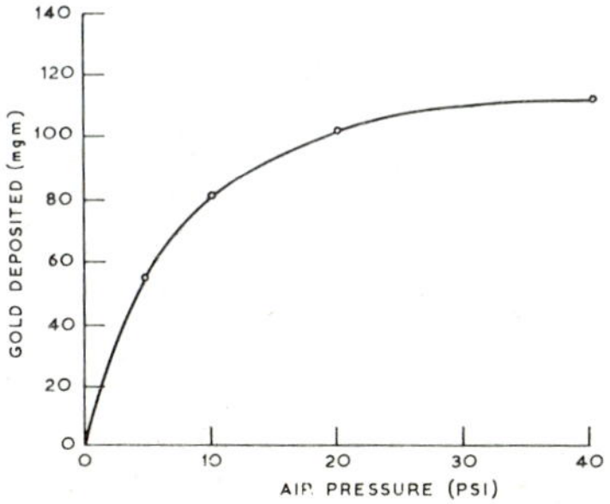

Fig. 15.3 Effect of air pressure on gold deposition. Increasing air pressure breaks the liquid stream into smaller droplets. The resultant improved mixing of reactants increases efficiency of deposition. Spraying conditions were 70 ml/min at 10 in. distance using sensitised plates. Air flow is 9 scfm at 50 psi. [Levy].

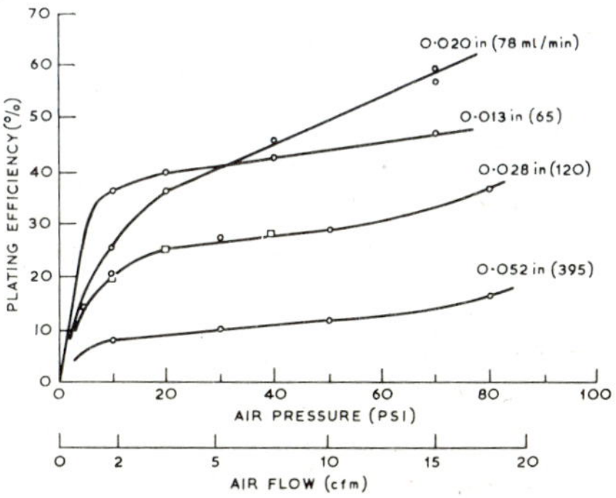

Fig. 15.4 The highest plating efficiency is achieved with a high air flow rate and a small liquid orifice. Liquid flow rate and orifice diameter are indicated. [Levy & Delgado].

(1) high electrical conductivity, (2) high infra-red reflectance and ultra-violet absorption which make it useful for thermal control applications since it is classified as a solar absorber. It also finds use in optical control applications, particularly as visible light may be transmitted through thin transparent films. The remaining property (3) is that of its high degree of chemical inertness which supports the application in which it is used.

Because of the characteristics exhibited by gold, it is a considerable asset in the field of space research. As many of the applications involved are large, it is more suitable to deposit the metal by spraying techniques rather than by immersion processes since objects up to 300 ft^2. (assembled from panels of up to 50 ft^2.) have been plated[1,10].

OPTICAL CONTROL

Thin gold films are used on "optical" plastics since they reflect eye damaging infra-red radiation, absorb ultra-violet and allow good transmission in the visible spectrum. A typical application is a sunshade visor worn by astronauts. By codepositing other metals with gold, transparent optical films may be selected on the basis of their transmission spectra. Applications envisaged[11] include eye protective filters with normal colour response and solar filters for optimising plant growth systems.

THERMAL CONTROL

One of the outstanding properties possessed by gold is that of its excellent heat reflecting power. Polished gold will reflect 97% of the incident infra-red radiation which is the highest reflecting surface of any metal. It is therefore used in thermal control applications. One such application is a thin gold film deposited on both sides of a lightweight plastic heat shield that insulates a subliming solid attitude control rocket motor in satellites.

ANTENNAS

Synthetic fibres have been plated with thin gold films to make lightweight high reliability unfurlable antennas.

REFERENCES

1 Levy D. J., *Proc. Am. Electropl. Soc.*, **51,** 139 (1964).
2 Levy D. J., Delgado E. F., *Plating*, **52,** 1127 (1965).
3 Andres F. O., U.S. Pat., 1,953,330 (1934).
4 Schneider H., U.S. Pat., 2,136, 024 (1938).
5 Gomes G. S., Levy D. J., *Products Finish.*, **27,** 10, 36 (1963).
6 Levy D. J., "Materials for Space Vehicle Use", Vol. III, Soc. of Mater. Process Engrs. Symp., Seattle (1963).
7 Levy D. J., *Metal Finish. J.*, **10,** 115, 157 (1964).
8 Brit. Pat., 1,027,652 to Lockheed Aircraft Corp. (1966).
9 Upton P. B., Soundy G. W., Busby G. E., *J. Electropl. Depos. tech. Soc.*, **28,** 103 (1952).
10 Lockspray-Gold Process, Technical Data Bull., 6-76-66-9, Lockheed Missiles & Space Co., Palo Alto, California (1966).
11 Levy D. J., Shellito K. K., "Gold Spray Process Solves Spacecraft Thermal Control Problems" (unpublished work).

Chapter 16

HIGH SPEED SELECTIVE (BRUSH) PLATING

M. RUBINSTEIN

Most platers are familiar with brush plating, a technique which is known synonymously as selective, swab, touch-up, stylus and tampon plating. It is also sometimes referred to as doctoring. Since the most advanced versions are proprietary, it is also popularly known as SELECTRON* and DALIC† plating.

As a generic name, the writer prefers "Selective Plating", though there is grave danger of confusion with bath selective plating processes involving specialised masking or controlled immersion techniques (see Chapter 17). Brush plating, the most popular term, is not truly descriptive, since it conjures up in the minds of most people the more primitive variation of the technique used primarily for retouching or doctoring defective electroplate, while over 90% of present day production involves original applications. In the remainder of this chapter, therefore, the term "Selective Plating" will be used to emphasise this difference.

Selective plating is a true electrodeposition technique, though in appearance it more closely resembles arc welding. A power supply provides DC current through two flexible cables, one of which is clamped to the work. The anodic cable is plugged into a special working tool, known as a stylus, brush, or spatula. The most up-to-date tool, the stylus, is provided with an anode wrapped with an absorbent cotton swab. This is immersed in a special high speed selective plating solution, or the electrolyte is flowed or pumped to the swab. The wrapped anode, saturated with solution, is rapidly moved over the surface of the area to be plated, thus completing the circuit (see Figure 16.1). Gold or other metals are deposited from the swab directly on to the area of contact.

HISTORICAL BACKGROUND

Historically, selective plating is as old as electroplating itself. It is possible that an old time plater removed an article, such as a teapot for example, from

* SELECTRON—Registered Trade Mark of Selectrons, Ltd., New York.
† DALIC—Registered Trade Mark of Laboratories Dalic, Paris.

a plating bath, only to find that one small area remained unplated. Instead of having to strip the deposit and replate, he connected a wire to the anode bus bar and attached the other end to an anode wrapped with a rag. The teapot was then held against the cathode bus bar, the wrapped anode dipped into the bath and swabbed over the unplated area. This procedure left much to be desired, since the deposit was thin, adhesion only fair, and the operation physically messy. Furthermore, the article would start to corrode only after it had been purchased. In short, the technique was used primarily to deceive the customer, and as such, fell into disrepute. Nevertheless, this form of touch-up selective plating was, and still is, used by many plating shops throughout the world for simple repairs.

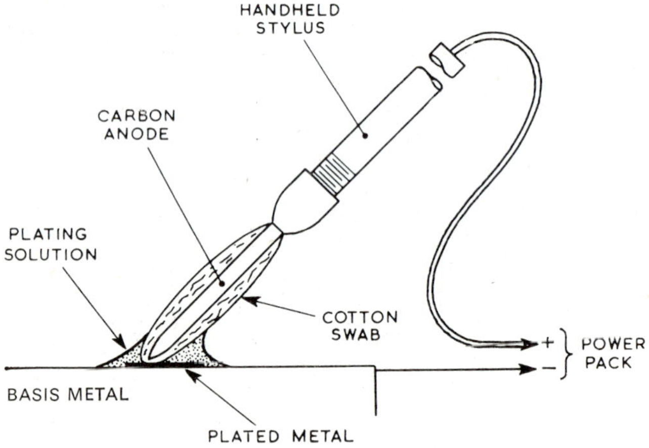

Fig. 16.1 Schematic diagram illustrating the principle of selective (brush) plating.

Modern high speed processes came into being only within the last 25 years. The first plating school instruction manual on this technique was published in 1950[1], while the first detailed article in a US technical journal only appeared in 1954.[2] The writer delivered a paper before the American Electroplaters Society in 1956,[3] the first extensive report on this subject ever presented to that professional body, although similar papers had previously been delivered to British and French electroplating associations between 1950 and 1955. Several patents on the latest selective plating tools and solutions were granted in France, Britain and the United States between 1948 and 1960.[4] These represented the first fresh approach to the subject in many years, though significant prior art was cited dating back to 1899. Mechanised or automated selective plating, sometimes known as "flow plating" was described in US Patents granted in 1965 and 1967.[5]

MODERN DEVELOPMENTS

Three basic improvements were necessary before this technique could find extensive use other than simple doctoring applications. First of all, special electrolytes were developed, these being aqueous solutions of organometallic chelates, with metal contents appreciably higher than those of standard plating solutions. Cyanides were eliminated or kept to a minimum. The second new development was an assortment of working tools, called tampons in France and stylii in the US and Britain, incorporating cooling fins for the dissipation of heat. Attached to the stylus handle was an anode made of a special grade of heat resistant graphite, which was highly purified to prevent contamination of the electrolytes. The third development was a group of specially designed power packs which incorporated very precise ampere-hour meters for close thickness control and other devices to simplify operations and insure safety.

These three material improvements created the breakthrough to modern selective plating. Deposition could now take place at much higher rates and result in controlled thicknesses of metallurgically sound metal deposits with excellent adhesion and desirable mechanical properties. This in turn enabled this technique to be widely accepted where conventional processes proved impossible, complicated or expensive.

Another interesting result of these improvements was that this method came to be used more frequently by non-platers than by platers. Most modern proprietary selective plating units are not located in plating shops, but are used as in-process tools in machine shops, maintenance departments, research laboratories, electronic assembly areas, heat treatment departments and similar places. Personnel are normally not platers but shop workers trained for special metal deposition operations.

ADVANTAGES AND LIMITATIONS

What are the specific advantages of selective plating over more conventional plating methods? Some are obvious. When only a small area of plate has been damaged, it is much simpler to retouch that than to strip and replate. Components must often be disassembled before plating, while others are too large for existing facilities. Even if large tanks are available elsewhere, extensive packing, shipping and paper work (as well as lost time) may be required. Many parts cannot be immersed, because solutions contaminate existing components. Typical examples are assembled electronic components like printed circuits, wired electrical assemblies such as motors and commutators, rubber-metal and plastic-metal combinations, and seam welded or riveted assemblies. In all the above cases, selective plating may provide a simple alternative.

On many occasions, it would be more convenient to bring plating facilities to the component rather than take the component to the plating shop. This is particularly relevant when items are very large, very heavy, fastened down or located in a remote area. Responding to such a need, selective plating has been used on church domes, religious objects in churches, statuary in parks, electrical power lines in remote areas, oil drilling rigs in the field, ships' engines in the engine room, aircraft on the maintenance line, and for repairing damaged areas on rotogravure cylinders and chemical mixing vats in the respective departments where the equipment is located.

There are also less obvious areas where selective deposition can prove invaluable. For instance, greater adhesion may be achieved to refractory metals, aluminum, stainless steel and carbon than is possible with conventional electroplating approaches. It can also be used to avoid elaborate masking on complicated components where only limited areas need plating. Furthermore, selective plating can replace bath processes, or else may be used in conjunction with them, for plating into difficult recesses or blind holes where inside anodes would normally be needed. Finally, it is an ideal method for prototype and experimental work. For large scale bulk production, conventional electroplating is cheaper and more convenient, since bath hours are normally used rather than man-hours of labour. In carrying out a cost analysis, however, the plater should take all production factors into account. Since proprietary selective plating solutions are invariably more expensive than equivalent bath electrolytes, a true comparison can only be made if plating costs such as disassembly, masking, jigging, packing and paper work are also considered.

Even today, with greatly improved techniques of mechanised selective plating, by far the vast majority of general electroplating will continue to be carried out by conventional means. A good general rule, therefore, is that bath plating should always be used, unless some special advantage is either quality, time, convenience or cost can be achieved through the use of selective plating. The recommended areas where these special advantages are applicable are as follows:

1 Repair of defective electroplate.
2 Partnership applications (e.g., conventional plus selective plating).
3 Components which become contaminated if immersed in an electrolyte.
4 Parts too large for existing facilities.
5 Parts requiring costly disassembly and/or handling.
6 Parts needing considerable masking before bath plating.
7 Improving adhesion to aluminium, stainless steel, carbon and refractory metals.
8 Plating ultra-high strength steels with minimal hydrogen embrittlement.
9 On-site plating in the field.

DEPOSIT CHARACTERISTICS

Since the use of this technique is growing in critical areas such as the aerospace industry, it is pertinent that the characteristics of the deposits should be discussed. Unfortunately, not all of the desired tests have as yet been documented, but the following information is currently available:

1 Corrosion Resistance—Approximately on a par with conventional electroplated coatings, thickness for thickness.
2 Hardness—In general, hardness is somewhat greater than that of bath deposits. The hardest selective plating deposits obtainable are a cobalt-tungsten alloy of 810 Vickers and a nickel-tungsten alloy of 750 Vickers. Gold has a hardness of 100–110 Knoop, the hardest 24 carat gold available. A 23+ carat, hard gold alloy is also obtainable with a Knoop hardness of 140–160.
3 Porosity—75% less than an equivalent thickness of metal deposited by normal electroplating procedures.
4 Structure—Random or amorphous structure for gold with no clearly defined crystalline boundaries.
5 Adhesion—As good or better than conventional plating, assuming proper preparation in both cases. On some base metals, adhesion is far superior.
6 Fatigue Loss—Generally less than that of bath plating.
7 Hydrogen Embrittlement—Considerably less than that of bath plating.

With regard to specifications, the attitude of the US Government is that their Mil Specifications require certain performance tests, such as salt spray for corrosion, tape and bend tests for adhesion, etc. If these are passed, it is of no consequence how the metals are deposited.[6] Selective plating deposits, properly applied, meet almost all Mil Specifications, including Mil-G-45204A, Types I and II for gold.

MATERIALS AND EQUIPMENT

ELECTROLYTES

Many improvements in modern selective plating are due to better materials and equipment. Of primary importance are the electrolytes. While it is sometimes possible to utilise standard bath plating solutions, this is inadvisable, since specialised proprietary electrolytes plate at rates from 5–50 times faster than conventional ones. More is lost in man-hours than can be saved by using less expensive solutions. Furthermore, proprietary electrolytes are designed to give deposits of desirable characteristics at any temperature from room to boiling, while operating at wide current density fluctuations.

Selective plating techniques normally utilise aqueous solutions of organo-metallic chelates, specifically developed for this process. These are much more concentrated than conventional electrolytes, two of the better known proprietary systems, for example, using gold plating solutions with a metal concentration of 100 g/l, 25 times more concentrated than an average standard gold cyanide bath. Electrolytes are available for any metal which can be deposited from an electroplating bath. This chapter, however, deals only with those solutions used for gold plating or in conjunction with gold plating in specialised applications. Table 16.1 lists solutions for these specific metals, together with data on the speed of deposition, etc.

Table 16.1
Typical Selective Plating Solutions used in Gold Plating Applications

Solutions	*Current Density (A/ft^2)*		*Approx. Metal Conc. (g/l)*		*Avg Time to Deposit 0.001 in. (sec)*	
	C	*S*	*C*	*S*	*C*	*S*
Copper (Alkaline)	10–75	1500	20–50	60	120	12
Gold	5–50	400	4–12	100	480	15
Nickel (Acid)	10–100	1000	35–80	100	150	40
Nickel (Low Stress)	10–100	1000	60–100	70	120	60
Nickel (semi-bright)	10–100	1500	60–80	75	150	20
Rhodium	10–80	1500	1.3–2	20	1000	60
Silver	5–100	1250	30–95	190	240	10

C = Conventional.
S = Selective.
* Deposition rates are given for a total area three times that contacted by the anode pad. Figures are for air-cooled stylii.

ANODES

The earliest selective plating working tool was a metal anode wrapped with a rag. As the technique became more sophisticated, a variety of working tools and connecting anodes became available. An early forerunner, still used in proprietary hobby kits, is a stainless steel spatula covered with a canvas sock. Some kit manufacturers later switched to nylon bristle brushes with soluble anode inserts, but these provide severe limitations in deposition rates due to intense anode polarisation at higher current densities. The introduction of inert anodes however, overcame this difficulty. Their adoption also simplified solution control, since it could be arranged that metal would deposit from the solution at approximately the same rate that solvents evaporated.

Perhaps the most sophisticated working tool for selective plating is the "stylus". This consists of a conductive core, a plastic handle cover to protect the operator, aluminum cooling fins to dissipate heat, and a device for fastening an anode. Most anodes are made of a special grade of graphite, with metallic impurities kept to a bare minimum to avoid solution contamination, and are resistant to heat and chemical activity, so that breakdown is slow. They can be manufactured or machined to various sizes or shapes, providing the anode is not too small. Graphite anodes with small cross sections are brittle, so that an alternative is required. The only other inert materials which will withstand the proprietary solutions are platinum and its alloys, platinised titanium and tantalum. The recommended material is an alloy of 90%

platinum and 10% iridium, the presence of the latter metal conferring additional rigidity to the anode.

Typical shapes are round, flat, half round (for building up inside diameters while a bearing or circular part rotates in a lathe) and concave (for building up outside diameters of shafts as they rotate). Flat anodes can have a slight curvature worked into them mechanically, whereas small ones can be sharpened to a point in a pencil sharpener or ground flat to a screwdriver edge.

In use, the anodes are wrapped with an absorbent material the most popular of which is cotton batting. A sterile, long fibre variety is necessary, since most commercial non-sterile types have a starch sizing which contaminates solutions. An alternative absorbent wrap is DACRON or polypropylene wool. The swab is often covered with a sleeve to provide extra resistance to wear and tear. Surgical tubing provides a convenient type of sleeve and special materials such as DACRON or 60% DACRON—40% cotton are also used.

Stylii and stylus handles of different capacities are available. A popular general purpose stylus for small and medium sized anodes should have a current carrying capacity of approximately 25 amps. Larger ones for use with anodes making contact with, for example, 12 in.^{2} of surface, require a current carrying capacity of up to 125 amps. These handles are naturally made with larger cooling fins, and the flexible cable is provided with a twist-lock device for locking into the stylus handle, since with larger tools, heavier cable must be used and there is a tendency for the cable to detach itself unless securely connected.

POWER SUPPLIES

Standard bath plating power packs can be used and as such, may be quite acceptable. However, specially designed units do provide additional advantages, such as low amperage and relatively high voltage capacities. A popular standard proprietary selective plating power pack, for instance, has a current carrying capacity of 30 amps to an output of 0–30 volts. It is perhaps pertinent at this stage to point out that a total of 30 amps working on a small area might very well provide a current density close to 5000 A/ft^2 of contact area.

Certain auxiliary controls are necessary and useful. Infinitely variable voltage is preferred. A reversing switch for stripping and etching operations is essential, and a safety cut-out device to protect both work and operator is very important. One proprietary system uses a cut-out switch which actuates in $\frac{1}{2}$ cycle. When plating through a cotton swab at 15–20 volts, any break through could result in an intense arc, which might do more damage to a valuable part than caused by the original defect. Furthermore, in the rare event that the rectification itself breaks down, it is inadvisable to have an operator standing with wet hands (and possibly a weak heart) getting the full 110 or 220 line voltage. The cut-out switch must be grounded in such a way as to shut off the equipment when this occurs. Finally, every good selective plating power pack has a precise ampere-hour meter for thickness control. Up to date units have digital solid state ampere-hour meters with six position readouts, readable to three places after the decimal point. These are accurate to within 0.5% over 99% of the range. Since every bottle of proprietary solution is marked with an ampere-hour factor, this meter provides a precise method of regulating thickness. A trained operator should be able to maintain control to within 10% of the thickness desired.

In addition to the solutions, working tools and power supplies just described, certain other accessories are necessary and valuable. A selective plating kit normally includes cotton batting, sleeves, small porcelain evaporating dishes (for holding solution), rubber trays, a test tube holder (for unused stylii), a chest of small plastic drawers (for separating and storing used anodes), an assortment of wire brushes, emery paper, abrasive stones, razor blades or small trimming knives, burnishing tools, plater's tape and various other accessories.

TECHNIQUES AND PROCEDURES

The selective plating operator requires information on specialised procedures. For example, both the speed of motion and the pattern followed influence final results. If an operator moves the swab back and forth over a rectangular shaped area, an extra heavy build-up occurs at the two ends. Consequently, trained personnel employ a combination circular and back-and-forth motion, sometimes described as "orbital". Wherever possible, most effective results can be obtained by rotating round parts in a lathe. The wrapped anode can therefore make contact with the rotating surface, the tool itself remaining stationary. Small reversible turning heads, with variable speeds from 0–600 rpm, are available for this type of operation.

While in most cases the solution is applied by dipping the anode into a small dish, in some applications the solution can be pumped to the anode swab or drip fed from a funnel. Various mechanical tricks must be acquired in order to obtain the best results. Excessive hand pressure should be avoided, as it dries out the swab too quickly and causes burning. Too slow a motion will give rough crystalline deposits. Too rapid a motion cuts down on current efficiency and results in slower deposition rates. For optimum results in gold plating, a relative anode-cathode motion of 30–60 ft/min is recommended. Temperatures must be monitored, particularly on large components which provide a substantial heat sink, since high speeds normally obtained are in part determined by heat developed through solution resistance. Proper preliminary cleaning steps and, in some cases, intermediate activation steps, are necessary.

While a knowledge of conventional plating can be very helpful to the selective plater, the two techniques do not exactly follow parallel lines. Certain procedures familiar to the electroplater should be ignored when attempting to gain a thorough understanding of this subject.

PREPARATION OF BASE MATERIALS

Cleaning methods prior to selective plating are similar to those for conventional plating, but with some significant modifications. Grease and oil are best removed with a solvent. Heavy corrosion, rust or tarnish should be removed by mechanical techniques, using wire scratch brushes, steel wool, emery paper or abrasive blasting. Parts are then ready for electrocleaning, etching and activation in accordance with accepted selective plating practice.

On mild steel, copper and brass, the only preliminary preparation normally required is a quick electroclean with a specialised selective electrocleaning solution. Unlike normal electroplating, different alkaline cleaning solutions are not required for different base materials. A single formulation can be used by varying voltage and amperage output and by changing cleaning times. For example, electrocleaning copper usually requires 10–15 seconds at 8–12 volts (depending on the contact area), whereas in the case of mild steel, approximately 30 seconds would normally be necessary at 10–15 volts. White metals and die castings need only a few seconds at 6–8 volts.

Stainless and high alloy steels require direct current activation* after preliminary electrocleaning and this is normally followed by a nickel flash without an intermediate rinse. A special nickel solution* provides the best adhesion to stainless and high alloy steels, nickel and chromium, and their respective alloys. After the initial flash, the surface is rinsed, and any subsequent desired metal is deposited.

High carbon steels normally require electrocleaning followed by two reverse current etching operations. The first uses a strong acidic etch* which removes some surface metal, leaving carbon exposed as a grey to black smut. This smut is then removed with a neutral etchant* and immediately plated, preferably with an initial deposit of nickel.

Preparation techniques are available for other basis materials, such as cast irons, refractory metals, carbon and semiconductors. Table 16.2 outlines the preliminary preparations of some of the more common ones. The time required for each activation step varies with the size of the stylus being used and with the area being prepared.

Most selectively plated metals can be deposited on to copper, brass, or steel. There are, however, a few exceptions. Silver requires a preliminary deposit of gold, palladium, or a silver strike, whereas acid copper requires an undercoat of alkaline copper, nickel, or both. When plating over soft white metals, solder alloys, zinc and cadmium, a deposit from an alkaline or neutral electrolyte is necessary before acid solutions are used. On castings, alkaline solutions should be employed prior to plating with acid electrolytes.

---

* Proprietary solution available from Selectrons, Ltd.

Table 16.2
Preparation of Base Metals Prior to Selective Gold Plating

Base Metal	*Precleaning & Electrocleaning*	*Activating*	*Preplate*
Brass, Copper, Copper Alloys, Gold, Silver, Mild Steels	Electroclean, 8–12 V, normal current; rinse.		If gold is desired metal, a preplate of nickel is often recommended to prevent diffusion. If silver is desired metal, use a preplate of gold or palladium.
Aluminum	Electroclean, 6–14 V, normal current; rinse.	Acid activator,† 6–14 V, reverse current; rinse.	Nickel (0.0003 in.) normal current, 6–14 V; rinse.
High Silicon Aluminum Alloy	Electroclean, 6–14 V, normal current; rinse.	Acid activator,† 6–14 V, reverse current; rinse; electroclean, 6–14 V, reverse current; rinse.	Nickel (0.0003 in.) 6–14 V, normal current; rinse.
Nickel, Chromium, Stainless Steel, Nickel and Chromium Alloys	Electroclean, 10–15 V, normal current; rinse.	Acid activator*, 10–15 V, normal current; NO RINSE.	Special nickel, 8–14 V, normal current (0.00005 in.); rinse.
High Carbon Steels	Electroclean. 10–15 V, normal current; rinse.	Acid activator†, 8–14 V, reverse current; rinse; neutral activator‡, 10–20 V, reverse current; rinse.	Nickel, 6–14 V, (0.0003 in.), normal current; rinse.
High Alloy Steels with Carbon and Nickel and/or Chromium	Electroclean, 10–15 V, normal current; rinse.	Acid activator†, 8–14 V, reverse current; rinse; neutral activator‡, 10–12 V, reverse current; rinse; acid activator*, 10–15 V, normal current; NO RINSE.	Special nickel, 8–14 V, normal current; rinse.
Ultra-High Strength Steels	Al_2O_3 dry abrasive blast; NO RINSE.		Immediately plate with gold.
Carbon	Do *not* electroclean; solvent clean only.		Plate with alkaline or neutral solutions.
Refractory Metals	Al_2O_3 dry abrasive blast using bottled Argon gas as a propellant.		Within 30 seconds, preplate with a special nickel at 8–14 V rinse.
Solder, Lead, Tin, Babbitt Alloys, Cadmium, Zinc Die Castings	Abrade area lightly; electroclean, 6–8 V, normal current; rinse.		Plate with desired white soft metal; if a hard metal or copper is desired, use a neutral nickel as a preplate.

Proprietary solutions are available from Selectrons, Ltd., with the following designations *Activating Solutions #1; †Activating Solution #2; ‡Activating Solution #3.

APPLICATIONS

Repair of worn or defectively plated areas with gold is historically the oldest application. For decorative purposes, selective plating has been used on areas of cigarette cases and lighters, where the buffing wheel has been applied a little too vigorously. Assembled jewellery has been retouched with gold where disassembly is expensive or where stones could possibly crack if immersed in a heated electrolyte. For conspicuous consumption, bathroom hardware and fancy gold plated automotive and pleasure boat hardware have been replated *in situ* where disassembly would obviously be very expensive. As already mentioned, statuary and ecclesiastic articles have also been catered for where the part could not easily be moved from its location.

For the most part, however, this process has been of greatest value in mechanical, electrical and electronic engineering applications. Worn areas on printed circuit edge connector contacts have been replated as well as worn or chipped areas of electrical contacts. Electrical contacts on computers have been replated by a mechanised selective plating technique. Dies for PVC moulding have been repaired, when excessive pressure at gate inlet points resulted in eroding the previous hard gold deposit. Inside surfaces of cryogenic valves have also been retouched with gold.

When repairing defective plating, it is essential that all loosely adherent earlier deposits first be removed. Plating over old gold which is poorly bonded to the base metal, is essentially like painting over flaking paint. It is bound to fail. The selective plating will adhere to the original deposit, but this will peel and flake from its base removing the selective gold with it in the process. Defective gold can easily be removed with a rubber (electrotyper's) eraser containing an abrasive material.

Exact colour matching is difficult in repair work. This is particularly true for alloy plating, but even with a 24 carat deposit, the colour of selectively plated gold may vary from that of the conventionally electroplated metal due to differences in grain structure. In industrial applications the slight mismatch of colours may be of no consequence, but in those applications where it assumes some significance, the situation can frequently be corrected by finishing with a thin flash of gold over the entire component, including the area which has been retouched. Still, one can often repair defective areas so that the naked eye finds it difficult to distinguish old plate from new. This is carried out by blending with a little fine rouge powder or, for matte deposits, a soft nickel wire scratch brush.

PRINTED CIRCUITS

Probably the most widely used application for selective gold plating is in the manufacture and repair of printed circuits and related printed wiring.[7] The General Electric NASA Division in Mississippi, USA, claimed savings upwards of $344,000 in 1969[8] by using this technique for circuit maintenance.

Similar savings are documented by such companies as IBM,[9] Litton Industries[10] and Collins Radio.[11] Some printed circuit applications are as follows:

1 Repair of gold plating on edge contacts of assembled circuit boards without disassembly or elaborate masking or jigging.
2 Economical small scale production plating of specialised printed circuits in lots up to about 300 per day.
3 Selective plating of flexible circuitry and tape cable. Conventional electroplating often creates problems of solution contamination in chinks or microscopic cracks.
4 Prototype manufacture.

Edge Connector Contact Tabs

Printed circuits are frequently manufactured in large numbers, with many components soldered or connected to the board. When, at some stage in production, it is discovered that the gold on edge contacts has flaked or peeled, it is too late to replate these areas by conventional means. Immersion in an electrolyte would contaminate and ruin many of the assembled components. By short circuiting the edge contacts, they can be readily selectively plated without endangering the assembly.[12] Figure 16.2 illustrates the repair of gold plated contacts by selective plating, the first application of this technique in the printed circuit field. In a similar case, the original plating may be excellent, but after a number of boards have been manufactured and assembled, a change in specifications is introduced. The extra thickness of gold may be deposited by this means.

The procedure for this repair is relatively simple. All loose or defective electroplating must first be removed. The edge contacts must then be short circuited, the area selectively electrocleaned and plated. To avoid endangering neighbouring components, rinsing is sometimes carried out with a small sponge or swab of wet cotton rather than with a stream of water.

The only real problem encountered is in obtaining a dependable connection across all of the contacts. The circuit in Figure 16.2 has been short circuited with a new copper tape, conductive on both sides, which gives excellent results. Another method involves the use of a polyester type of tape with a $\frac{1}{4}''$ or $\frac{1}{8}''$ wide strip of copper running along the length. A different approach involves the use of a harp-like holder, where contact is made by means of two tightly stretched fine stainless steel wires. Finally, a method which is useful in special cases is to "paint" connecting lines using a conductive silver paste, air dried for 2–3 minutes prior to the selective plating operation, and subsequently removed with a solvent. This works extremely well on contacts or on fine points or dots of circuits in disconnected areas.

For double sided circuits incorporating through-hole plated connections, contact can be made from the back. On these boards, through-hole selective plating can only be achieved by using a stainless steel backing plate, the solution collecting in the holes and acting as miniature plating baths whenever the swab passes over a given area.

Using one or more of the preceding methods, a quick efficient plating technique is possible for single, small groups or entire lengths of edge contacts. It is also useful if, for any reason, the entire circuit board has to be plated.

Small Scale Circuit Production

Most printed circuit production of any volume is processed most economically by normal electroplating procedures. However, specialised production applications do arise where selective plating can prove valuable. A typical case involves the manufacturer who only has small production runs of up to 300 a day. A second possible user is the manufacturer who bath plates large numbers of circuit boards, but periodically has small, important batches requiring special handling or jigging.

For the small scale manufacturer, selective plating provides an immediate economic incentive, since a proprietary installation, suitable for plating gold on 300 circuit boards a day can be purchased for as little as £400–600 ($1,000–1,500). In addition, conventional plating normally requires a trained electroplater, while selective plating can be carried out by an intelligent bench operator after a brief specialised course. With a little training and a proper bench set-up, such an operator can easily process 300 boards a day.

In production plating applications, it is necessary to devise a simple means of mounting the circuit board and establishing electrical connection quickly across the contacts. This is frequently accomplished by using a foot controlled air clamp, which brings a shorting bar down across the board. Since the level of edge contacts may vary, it is important to line the bottom of the shorting bar with aluminium or lead tape to allow a little "give".

Flexible Circuits

Flexible circuits and tape cable are normally plated by conventional means. There are, however, special problems. Flexing of the circuit often imparts microscopic cracks or openings between the conductive copper and the underlying plastic. In the case of tape cable, sections are often cut to size and some of the plastic cover removed at both ends to expose contact areas. This may well result in some weakening of the bond between the copper and underlying plastic. For both flexible circuitry and tape cable, the presence of microscopic cracks or gaps creates difficulties. Solutions are "sucked up" into these openings and become trapped. No amount of rinsing completely removes them, and the residual plating solutions later cause "sweating", corrosion and changes in conductivity and resistivity.

Selective plating avoids this type of contamination. Without immersion, there is no hydrostatic pressure to "push" the solution into the openings. The time required for plating a given thickness is appreciably less than that for bath plating and consequently there is less time for the solution to be entrapped. Finally, the very low free cyanide content (less than 0.5 g/l) of selective gold plating solutions creates fewer corrosion problems.

A number of flexible circuit and tape manufacturers, with previous reject rates up to 30%, have reduced this figure to zero by introducing this procedure into their establishment.[13]

ELECTRICAL CONTACTS

Replating gold on worn or damaged areas of electrical contacts provides another excellent application, since such contacts can no longer be easily removed. An attempt to replate them in a bath without disassembly normally results in contamination of neighbouring components as well as wastage of gold on non-essential areas.

A typical contact repair was carried out in the United States by Remington Rand Univac. In some of their earlier computer models, they experienced difficulties with changes in electrical conductivity caused by corrosion after gold plating had worn off pyramid shaped electrical contacts. In the newer models, this problem has been overcome by using a gold or silver layer of considerable thickness brazed over the contact before assembly. However, by the time the need for this thick layer was realised, connections on the older models had already been assembled into vast batteries of electrical contacts on computer units in the field.

These studs were selectively plated using a small hand held electrical drill, with a shaped carbon anode substituted for the drill bit. A felt liner covered the end of the carbon anode. The anode tip was dipped into the gold plating solution and contact was established with the male stud, while the drill rotated at a predetermined speed. A small foot pedal controlled electrical timer signalled the length of time required. A skilled operator could do thousands of contacts per day, depositing approximately 0.000050 in. of gold on each stud over a six second period. When a bell rang indicating the time cycle was completed, the operator would merely re-dip the tip of his anode into the solution, re-establish contact at a new point and start the drill so as to insure constant movement across the surface.

Some of the added advantages of gold plating connector pins are indicated by the history of the altimeter servo unit illustrated in Figure 16.3, manufactured by Litton Industries. Since this device is to be used as part of the Dew Line System*, reliability is of the utmost importance. Consequently the US Government has a 23 hour test specification before the manufacturer can ship this product. During the test cycle, the male connector plug at the right is subjected to numerous insertions and withdrawals. Frequently, some of the gold plating fails on the connector pins. Previous repair involved removal of

---

* Dew Line System—A radar intercept system being used by the United States in Alaska and Northern Canada to warn of unauthorised transpolar flights.

the plug and replacement with a completely new assembly. However, specifications state that any time any single component is replaced in the assembly, the 23 hour test cycle must be repeated. Now that selective plating is used, the defective plate is removed from the pins using a wire brush and fresh gold is replated without removing the component. Since, strictly speaking, no single component has been replaced in the assembly, there is no need to repeat the lengthy test cycle. This may sound like sophistry, but it does indicate some of the logistic benefits of selective plating as a repair technique, certainly when compared to disassembly and reassembly.

IMPROVING SOLDERABILITY

Certain metals, such as aluminum and stainless steel, provide soldering difficulties. While it is true that one can solder to these metals using an acid type flux, such fluxes are normally forbidden in the electronics field since acid residues contaminate neighbouring components. Selective plating avoids this difficulty and is therefore widely used to improve solderability of aluminum and stainless steel, as well as refractory metals such as molybdenum, titanium and tungsten. It is also employed to obtain a solderable surface on carbon and semiconductors.

The technique is relatively simple. The particular base metal is cleaned, activated and a thin film of nickel or cobalt plated on the surface to a thickness ranging from 0.0003 in.–0.001 in. This is followed by rinsing and a flash deposit of approximately 0.000030 in. of gold. Since adhesion of all these materials to the base metal is excellent, a gold surface is obtained which is intimately bonded to the underlying metal. Gold can easily be soldered using a neutral type flux. In fact, using a controlled atmosphere oven, it is possible to solder to it with no flux whatsoever.

In the USA, innumerable soldering departments use selective plating as a preparatory technique, the equipment being set up in the soldering section so that components do not have to be dispatched to the plating shop.[14]

MOULDS

With the increasing use of polyvinyl chloride (PVC) and polyvinylidene chloride (PVDC) for containers and bottles, production problems have arisen that are peculiar to these two materials. A major problem with these polymers is that the moulding process liberates minute quantities of hydrochloric acid, which tends to etch or pit the die cavities. This causes drag marks and other flaws in finished pieces which affect their cosmetic value and results in rejects. It also shortens the life of the moulds and dies. The same corrosion problem occurs when seams have to be fused or when other production steps bring hot PVC into contact with metal.

A simple answer to this problem, currently practiced in a number of moulding shops throughout the world,[15] is to selectively plate these mould surfaces with a thin 0.0001 in. deposit of a hard gold alloy (140–160 Knoop) which is highly resistant to hydrochloric acid attack, as seen in Figure 16.4. This prevents penetration of the hydrochloric acid released by the PVC and considerably increases the life of the mould. Gold can be deposited directly on steel or other mould metals. In the case of chromium surfaces, a simple prior activation technique is necessary.

An additional advantage is that repairs to the gold plate can readily be made while the mould is still in the machine. Since certain areas of the surface experience greater wear, the coating wears off in a reasonably short time. A selective plating unit can then be wheeled up to the machine, the area quickly cleaned and additional gold deposited. The new deposit is normally blended into the old with a little rouge.

AIRCRAFT AND MARINE

Specific uses are found in the maintenance depots of almost every commercial airline[16] and Air Force[17,18] station in the USA and Europe.[19] Numerous marine maintenance depots, both commercial and Navy,[20] are also starting to find this technique valuable, particularly for saving down-time.

Most aircraft and marine applications involve the use of cadmium, nickel, copper and tin. For certain specialised work, however, selective gold plating has also proved valuable. For example, contact areas on the slip ring of the ADF Loop Collector on Douglas DC-8[16] and Boeing B-720[16] aircraft have been replated with gold, as have numerous similar components.

Gears, bearings, bushings and cylinder liners of aircraft engine components are often selectively silver plated to prevent galling and fretting or to provide an interference fit. Selective gold is always used as a flash preplate on these components. On the Douglas DC-8,[16] gold plus silver has been used to repair piston head grooves on the main gear door activator piston rod and the internal diameters of worn gland piston rod bores on aileron hydraulic power unit glands. On Pratt & Whitney JT3D, JT4 and JT8D aircraft engines,[16] gold plus silver has selectively built up internal diameters on worn cylinder bushings, cover bushings, corroded and undersized main accessory drive gear box liner bores and worn seal lands of bearing nozzle assemblies. Similar deposits on outside diameters of bearing oil transfer tubes, worn journals where adapter mates with gear drive, and on pressure ratio control yoke shafts have brought these parts back to original dimensions or have enabled interference fits.

Other important aircraft work for gold plating involves improving wear on copper commutators for the large number of high speed servo motors used throughout the aircraft.[21] A flash of gold over a highly finished commutator surface followed by rhodium and a final flash of gold not only lengthens life, but radically reduces operating temperature and radio interference caused by arcing. A typical commutator where performance improved with this type of selective plating is on the booster pump on Sud Aircraft CVL-VI-R,[16] where

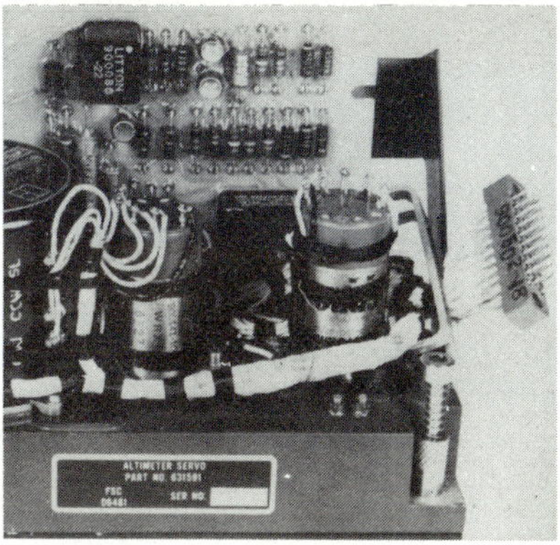

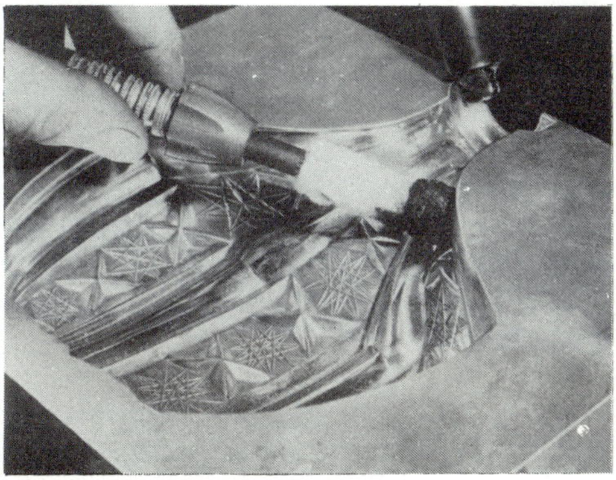

Fig. 16.2 *Top left*: Gold plating of printed circuit edge connector contacts.

Fig. 16.3 *Top right*: Instead of having to dismantle this harness assembly of an altimeter servo in order to service multiple connector, far right, faulty pins are now replated in place. [Courtesy of Litton Industries].

Fig. 16.4 *Bottom left*: Split die for blow moulding PVC containers receives final deposition of hard gold by means of selective plating. By guarding against hydrochloric acid released during moulding, mould life is considerably extended, downtime is reduced, and a better product at lower cost is achieved. [Courtesy of Diemould, Inc.].

Fig. 16.5 *Bottom right*: One of four bronze equestrian statues recently restored and selectively gold plated on site. These statues are located at the Memorial Bridge Plaza in Washington D.C. [Courtesy of Roloc].

Various Applications of Selective (Brush) Plating.

life of the part was doubled as compared to operating with an unplated copper commutator.

In the marine field, similar components have been restored to an active life and their performance improved by selective gold plating, gold plus silver, or rhodium plus gold. An excellent example of a marine application involves replating of cylinder liner adapter faces and seatings on Napier Deltic Engines.[22] In order to combat corrosion of the spot faces on the liner, the area is electroplated with tin during production by conventional means. After assembly or during repair, it was found that the addition of 0.0002 in. of gold plate to the tin layer enhances corrosion resistance. In some cases, gold followed by an additional 0.0003 in. deposit of tin has improved performance even further. Similar operations have been carried out on other parts of the cylinder liners, such as in the threads and on the seatings for the nozzle air start and blanking plug adapters. Selective plating has been approved by Lloyd's Register of Shipping[23] for prevention of frettage, and for resizing and corrosion protection.

ON-SITE APPLICATIONS

Selective plating provides an ideal technique for components which cannot be moved or are too large for existing plating facilities. Typical examples are gilded statues in parks as illustrated in Figure 16.5. When the gold layer slowly disappears due to diffusion or wind and rain erosion, selective retouching is often the only answer. (In the United States, flame gilding cannot be used legally.)

Similarly, it provides a method for gold plating portions of ecclesiastic items in churches and museums. These are extremely valuable and institutions are somewhat reluctant to allow them to leave the premises. The Smithsonian Institute, for example, replates components on antique aircraft *in situ*. Several American hotels have chandeliers and other components gold plated without dismantling. Probably the most impressive application ever, was the regilding of two church domes in California.[24] The entire domes, (7,000 ft^2 in area) were first nickel and then gold plated to a thickness of 0.001 in. prior to final polishing and lacquering. Similar renovations have been carried out on gilded columns previously coated with gold leaf.

IMPROVING WEAR RESISTANCE

A number of articles which are gold plated have certain areas which experience extra hard wear. Typical examples are watchcases, the backs of expansion bracelets, and the bottoms of the bowls of teaspoons and tablespoons. Parts such as these, having been electroplated by conventional means, can subsequently be selectively plated with additional gold only on those areas where major wear occurs.

AS AN UNDERPLATE

Selectively plated gold is frequently used as an underplate for silver, rhodium and other metals. Since silver deposits non-electrolytically from plating solutions, a strike of some description is invariably required to prevent the formation of poorly adherent immersion deposits. Selective silver plating does not normally require a silver strike. Instead, it has been found that a thin colour flash of gold over copper, brass or underlying steel will provide a surface more noble than silver. Silver can then be applied directly to the gold with excellent adhesion and with no fear of an immersion deposit being formed initially. This technique, incidentally, has proved so successful in selective plating that the writer knows several US companies who now specify a dilute cyanide gold bath as a preplate prior to conventional silver plating.

Gold also provides an excellent underplate for rhodium plating. Most rhodium solutions are highly acidic and tend to attack underlying metals such as brass or copper. Such attack not only destroys lustre, but contaminates the rhodium solution, with the result that subsequent deposits are dark in colour. This problem is normally overcome by a nickel underlayer, but nickel in thin deposits tends to be porous, so a thin flash of gold serves as a superior preplate. Gold has also been used as a preplate under nickel to reduce porosity on engineering applications where porosity is detrimental.

JEWELLERY

Selective plating is often used to repair or replate worn or defective gold plating on jewellery. This is particularly valuable on antique assemblies where precious stones may be mounted in such a manner as to make removal difficult. Since immersing the entire piece of jewellery in a hot gold plating bath might tend to crack stones, selective plating can be used to refinish either sections or complete pieces. This technique is also being employed by several companies for obtaining two-tone effects in production plating. A Canadian jewellery manufacturer,[25] for instance, has made two-tone tie clips, cuff links and belt buckles by first of all rhodium plating them in a bath and then having limited areas selectively plated with gold. The alternative procedure is also possible. This technique eliminated substantial masking and resulted in lower costs.

MISCELLANEOUS

There exist many other applications for selective gold plating. For example, gold plated bus bar connections are not uncommon for certain specialised types of work. Normally, the connections are tin or silver plated, so as to provide positive contact. However, the writer knows of at least one installation in a chlorine plant, where chlorine gas tended to attack the silver. A deposit of 0.0001 in. of gold on the copper bus bar prior to making connections provided a superior contact area, uneffected by chlorine attack.

Gold plating has also been used selectively to fill pinholes and reduce porosity. A particular example is in a very sensitive low pressure safety valve used in a satellite system. The entire device, fabricated from 0.004 in. thick aluminum stock, was made non-operative by a pinhole in the centre. An attempt was made to weld an aluminum shim over it and although this succeeded in sealing the hole, it destroyed the accuracy of the device. Selective gold plating over the pinhole successfully closed the hole so that the part could be returned to service.

In a sealing operation of a different nature, low temperature diffusion seals at a gold-indium plated interface (equivalent in bond strength to a gold-copper weld) have been obtained at 370–400°C with a 30 minute time cycle.

MECHANISED SELECTIVE PLATING

Most selective plating is manual and as such, requires man-hours of labour and the continuous attention of an operator. Certain mechanised techniques, however, can reduce the labour involved. For example, parts are often rotated in a lathe or turning head while the operator holds the tool in steady contact with the moving surface. As a variation, the tool may be mounted in position. Where the part itself cannot be turned, a rotary tool is available commercially. This works on a flexible shaft similar to a dentist's drill and spins the anodes at speeds from 0–900 rpm reversible. Such a device is absolutely necessary for plating into small diameter holes, where hand motion is insufficient. It is also useful for plating larger surfaces, both flat and round, in that rotation of the stylus and anode provides the intense agitation necessary. The operator can move his hand much more slowly, thus reducing fatigue. A reciprocating machine is also available for back-and-forth motion to cover rectangular areas or rotating parts where the area to be plated is too wide for a stationary held tool.

In addition to the mechanical approaches to rotation and mounting, solution replenishment can be handled by means of a small pump feeding the solution through modified anodes. This saves considerable time, since dipping is time consuming and provides a hiatus to the plating operation. Furthermore, the wetter the surface, the higher the current density and consequently the greater the deposition rate. Although several types of pumps are satisfactory, the writer prefers a small peristaltic pump. Since the solution never touches any metal components, it can be used for a variety of solutions without contamination. Other types of mechanisation include turning the anode in a drill press so that it can be repeatedly lowered into small recessed areas or holes, either for plating a round spot on a flat surface or for plating the inside walls and bottom of a blind hole.

At one IBM installation,[26] small transistor tabs were plated on a production basis. Jigs holding 81 transistor parts were placed on an oscillating platen mounted under the stylus. The anode, held in contact with the tabs by a spring clamp, was periodically given a quarter of a turn mechanically so as to obtain even wear around the wrap. The solution was fed to the top surface of the jig from a separator funnel.

A more sophisticated form of mechanised selective plating is known as "flow plating".[27] This is a combination of selective plating and bath electroforming techniques, in that a wrapped anode is no longer used. A hollow inert anode is provided, which is mounted in relation to the work so as to provide a gap of approximately 0.060 in. from the surface where plating is desired. In plating shaft or housing surfaces, the anode is rotated within or around the component, or alternatively the part is turned while the anode is held stationary. Solution is pumped through holes in the walls of the anode, so that it is constantly changing in the 0.060 in. gap. This method provides an excellent approach for continuous production.

SUMMARY

High speed selective gold plating provides one more additional tool for the plating fraternity. While not an economical procedure for most high production applications, it provides a simple ancillary technique for applications which give rise to special problems when using more conventional means.

In evaluating this technique, platers should strive not to restrict their thinking only to retouching and doctoring. Nor should they conclude *a priori* that, since selective plating normally requires manual manipulation, it does not lend itself to some types of production.

If these two prejudices are overcome, intelligent platers can find selective plating an excellent money and time saving device for a wide variety of applications in electronics, plastic moulding, aircraft and marine maintenance, jewellery manufacture and numerous other industries. All that is required is an awareness of possible applications and a willingness to learn the procedures needed for a somewhat different approach to electroplating.

REFERENCES

1 Kusher J. B., "A Simplified Manual of Modern Brush Plating", Joseph B. Kushner Electroplating School, Evansville, Indiana (1950).
2 Rubinstein M., *Mater. Meth.*, **40,** 6, 98 (1954).
3 Rubinstein M., *Proc. Am. Electropl. Soc.*, **43,** 246 (1956).
4 Icxi J. J. G., U.S. Pat., 2,748,069 (1956); U.S. Pat., 2,961,395 (1960).
5 Schwartz B. A., U.S. Pat., 3,183,176 (1965); U.S. Pat., 3,313,715 (1967).
6 Wright Air Development Center, Ohio, private communication (1959).
7 Rubinstein M., *Electron. Mfr.*, **15,** 2, 41 (1971).
8 *Metalwkg Prod.*, **114,** 17, 37 (1970).
9 Moxley A., Mutnick S., *Electron. Packag. Prod.*, **6,** 11, 22 (1966).
10 Litton Industries, Guidance and Control Systems Division (USA), Technical Manual SRM/E-38 (1964); *Electron. Prod.*, **4,** 5, 33 (1964).
11 Collins Radio Co., Dallas, Texas, private communication (1964).
12 *Missiles & Space*, **11,** 3, 20 (1963); *Products Finish.*, **17,** 6, 82 (1964).
13 Machine & Tool Blue Book (U.S.A.), 61, 4, 55 (1966).
14 *Product Engng*, **38,** 15, 102 (1967); Ryder, H. W., *Assembly & Fastener Methods*, **4,** 5, 45 (1966); *Des. Electron.*, **5,** 4, 2 (1968).

15 *Rubb. Plast. Age*, **48,** 2, 125 (1967).
16 Vanguard Pacific Inc., Santa Monica, California, Index for Selective Plated Components by Airframe and Engine Manufacturers (1970).
17 U.S. Air Force, Hill Air Force Base, Utah, MIL-STD-865 Para. 5.5.4.6, Note 1, (1969).
18 Selectrons Ltd., New York, Military Aircraft Applications (1969).
19 Rubinstein M., *Soc. Aerospace Mater. Process Engrs*, **13,** 145 (1968).
20 *Corrosion*, **17,** 17, 25 (1961); Long Beach Naval Shipyard (U.S.A.), *The Digest*, 23, 45, 1 (1966).
21 *Flight*, **73,** 2557, 108 (1958); *Power Engng*, **62,** 10, 84 (1958).
22 D. Napier & Son (London, U.K.), Napier Service Bull. No. 137 and SK No. 12067 (1962).
23 Lloyd's Register of Shipping (London, U.K.) private communication (1958).
24 *Products Finish.*, **30,** 8, 77 (1966).
25 E. S. Currie, Ltd. (Toronto Ontario, Canada) private communication (1959).
26 *Electronics*, **33,** 10, 70 (1960).
27 *Metal Finish. J.*, **10,** 111, 116 (1964); *Product Engng*, **40,** 13, 55 (1969).

Chapter 17

MISCELLANEOUS TECHNIQUES

T. WARLOW

1—THE TURBOJET

This machine was developed originally to cater for the specialist problem of the continental watch and decorative industries, where barrel plating of small, delicate parts, resulted in a high incidence of component damage. Since then it has seen limited use in the gold plating of industrial components such as delicate electronic connectors and flatpacks.

ADVANTAGES AND DISADVANTAGES

ADVANTAGES

1 A lower incidence of damage to delicate components is afforded by this process compared to more conventional methods of electroplating.

2 In barrel plating, flat components may tend to adhere to each other thereby leading to deposits of varying thickness. In such cases the turbojet can minimise these tendencies and achieve much more uniform deposits.

3 Barrel plating of small intricate components has always been a problem because of the tendency to lodge themselves inside barrel recesses. This leads to a wide variation in plating quality, burning of some components and often an unacceptable standard in the quality of the electrodeposit or its degree of adhesion to the substrate. These problems are considerably reduced by the turbojet process.

4 Higher operating current densities than those employed for barrel plating can be used. This tends to offset the smaller batch quantities necessitated by this process.

5 Scratching or other surface finish defects caused by barrel plating damage can be drastically reduced. This is particularly applicable in the decorative finishing of jewellery and watch components.

DISADVANTAGES

1 Due to the large exposed area of solution, atmospheric oxidation can be a problem with some electrolytes.

2 Build-up of the gold deposit on the cathode mesh results in excessive screening characteristics and hence a low plating rate of the components concerned. This also leads to a high drag-out of the solution due to poor draining.

3 The quantity of components capable of being processed by this technique at any one time is somewhat restricted.

EQUIPMENT

The construction of the machine is illustrated diagramatically in Figure 17.1. A rotating electrolyte injector is placed in an offset position underneath an open cylindrical basket with a mesh base. This basket holds the parts to be plated and rotates slowly on a vertical axis with periodic reversal of direction.

The unit is constructed within a standard plastic coated rectangular hollow section steel framework and usually employs a circular polypropylene plating tank with integral lip level fume extraction. The climatiser, agitation motors and filtration equipment are also housed with the tank inside this console. It is also possible to accommodate one or more units in the same solution, using a rectangular tank.

For gold plating, platinised titanium anodes are used in the form of a large cylindrical shell of expanded mesh around the tank walls, with a circular disc anode of the same material below the process basket. All internal anode couplings are of titanium and joining is by nuts and bolts of the same metal. The titanium contact strip has a thin coating of platinum metal deposited on the contact surface to improve the electrical connection.

The climatiser unit is fitted with an electric heater and also a water cooling attachment to allow for automatic temperature control, which may be necessary when the unit is used for other processes such as bright tin plating, in which high current densities are necessary for satisfactory operation and control of temperature is required to avoid degradation of any brighteners present.

The wire mesh in the base of the process basket provides cathodic contact to the components and is available in different mesh sizes to offer optimum electrolyte interchange, bearing in mind component dimensions. The standard brass wire meshes are plastic coated and are rubbed down with fine emery paper before use so as to expose the high spots on one side only. This system offers adequate cathodic contact to the components, while at the same time minimising build-up of gold on the mesh itself. Needless to say, the deposit eventually builds-up on these high spots and a tendency to "bridging" does occur. Unless the mesh is changed at this stage, plating will occur preferentially at these sites resulting in a marked reduction in deposition rate on the components and costly wastage of precious metal on the mesh.

The double jetted solution injector is fed from a standard climatiser/ filtration unit and is self-rotating due to the pressure of the solution being discharged through the angled jets. These two rotating jets of solution

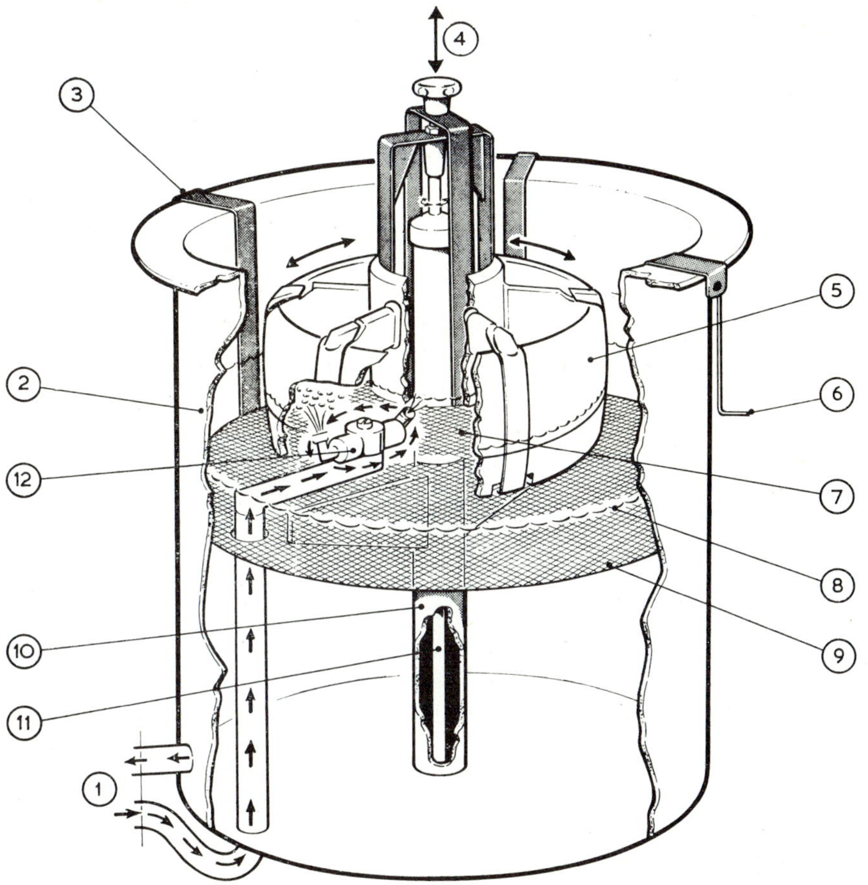

1. CLIMATISER ELECTROLYTE FEED
2. PLATING TANK WALL
3. PLATINISED TITANIUM ANODE STRAP
4. COMBINED CATHODE PICK UP/CARRIER HANDLE
5. TURBOJET BASKET
6. ANODIC CURRENT FEED
7. TURBOJET BASKET MESH
8. ELECTROLYTE LEVEL
9. BOTTOM ANODE MESH (PLATINISED TITANIUM)
10. PLASTIC SHROUD FOR CATHODE SUPPLY/ROTATORY SPINDLE
11. COMBINED CATHODE SUPPLY/ROTATORY SPINDLE
12. ROTATING ELECTROLYTE IMPELLER UNDER TURBOJET BASKET

Fig. 17.1 The Turbojet [Courtesy of Sel-Rex Company].

directed into the basket provide good solution interchange, while at the same time tending to reduce the effective weight of components so that they may change position without damage. Simultaneous rotation of the basket in alternating directions further improves the uniformity of plating conditions for all parts.

The flow of solution to the injector unit is controlled by a valve so that the degree of agitation can be adjusted to suit the particular component being plated. For example, very gentle agitation is required to avoid scratching of very delicate components requiring a decorative finish.

Apart from variation in flow rate, different jet sizes may be employed which are readily interchangeable. The angle of the jet may also be adjusted from near horizontal to near vertical positions and the combination of these factors is important when dealing with solutions of different densities.

The choice of cathodic mesh will, in the first instance, be dictated by component size and geometry. Due to difficulties in production, mesh sizes below 0.5 mm are at present available only in uncoated form. This leads to relatively short lives since a greater mesh area is offered to the electrolyte and bridging will occur at an earlier stage. Apart from wastage of gold in deposited form, excessive reduction in mesh dimensions will reduce the efficiency of solution replacement within the process basket and possibly increase drag-out losses due to slow drainage on removal from the plating tank.

It is possible to obtain baskets with a number of segments, each with its own separate mesh base. Apart from the obvious advantage of being able to process small quantities of different components at the same time, a saving of gold is achieved when only small quantities are being processed since segments not in use may be removed, thereby avoiding deposition on the mesh.

OPERATING CONDITIONS

The number of components capable of being plated by this process at any one time is dependent upon their surface area, geometry and weight. It is recommended that the surface area should be approximately 15 dm^2 and the maximum load about 300 grams. It is not possible to plate efficiently more than two superimposed layers of parts.

Pretreatment of the components is usually carried out in a fine mesh stainless steel dipping basket. Conventional trichloroethylene degreasing is followed by bright acid dips and alkaline cleaners in a sequence pertinent to the particular base metal in question. Care must be taken during the loading operation to avoid contamination of the clean components. If required, undercoats of copper, nickel or silver can be deposited in separate turbojet units depending upon the particular specifications in operation.

After plating, the components are carefully transferred to a suitable basket for rinsing. Drying of delicate components is best carried out using established solvent drying techniques based on halogenated hydrocarbon solvents coupled with suitable surfactants.

DEPOSIT CHARACTERISTICS

The appearance, metallurgical characteristics and electrical performance of components plated by this process are very similar to those obtained by more conventional techniques. However, due to the fact that the unsatisfactory current density distribution encountered in barrel plating is improved by this process, a much more uniform thickness of gold is achieved. This, of course, is a pertinent point to be considered when assessing processes for the deposition of precious metals since thickness distribution is of paramount importance.

SUMMARY

This hybrid system of electrodeposition intermediate between that of vat and barrel plating processes offers some of the advantages of both systems. It should, however, be remembered that it does not replace these techniques, but merely supplements them. It is a system capable of plating small and delicate components which can be handled only with great difficulty, or with a high incidence of damage, by more conventional electroplating techniques.

2—SELECTIVE PLATING

In the highly competitive electronics industry there is a continuing urge to reduce component production costs. When gold plating is involved as part of the manufacturing process an obvious way to contribute towards this end is to restrict to a minimum the amount of precious metal used, with the essential proviso that operational reliability will not be impaired.

In the case of an electronic connector, in particular, the intrinsic value of the gold required for a specified coating of 200 microinches can account for a very high proportion of the total manufacturing cost. This type of component therefore readily lends itself to cost reduction studies along these lines.

Various approaches to the problem of minimising gold consumption are possible. These methods, which may be used either alone or in conjunction with each other, are as follows:

1 Reduction in coating thickness. To be effective this approach must include close attention to the surface finish of the basis metal and to the selection of a suitable undercoat, with regard to both nature and thickness. It may be combined with post-plating treatments designed to chemically plug pore sites.[1,2,3]

2 Use of lower carat gold alloy deposits. This approach is limited by the specific technical requirements of the coating.

3 Use of gold dotting or inlaying techniques. In principle this is an attractive alternative to selective electroplating, but presents a number of problems which are discussed later.

4 Selective plating to confine the functional coating to operational areas only, possibly with a thinner overall deposit of gold for the sake of uniform appearance.

Selective plating (method 4) with which the present chapter is primarily concerned, although shrouded in a great deal of secrecy, is now well established in the electronics industry, the competitive nature of which has provided the stimulus for the development of even newer and more effective techniques.

ADVANTAGES AND LIMITATIONS

The obvious advantage of selective plating is the cost reduction possible by the restriction of heavy gold deposits to operational areas only. However, the application of selective plating techniques to established piece part components designed to be handled by conventional electroplating methods, often involves a high labour investment in jigging and other operations. Hence, if all the possible benefits are to be reaped from the selective approach, a high degree of design collaboration is necessary, from the manufacturing press operation through to the final electroplating stage.

Reductions in costly labour content have also been achieved by handling components on continuous reels or in very long strips. In this connection the greater uniformity in component operating characteristics obtained has provided further inducement for the introduction of selective plating on a continuous flow basis. Even so, a high labour content may be involved in the production of selectively plated items, hence the method is usually reserved for the application of relatively heavy precious metal deposits, where the reduction in metal consumed more than offsets the extra labour involved.

The production output of selectively plated components will often fall below that of work obtained by barrel plating unless full use is made of high speed plating and pretreatment processes. A working compromise must be established based on production output, labour involvement, capital investment and the expected savings in precious metal consumption.

Definition of the plated area is important in relation to wear performance, associated with which may be problems arising from transfer of base metal onto the contact surface, or of the migration of base metal corrosion products from those areas adjacent to the gold plated contact zone (see Chapter 21).

Problems may arise due to the high plating speeds often necessitated in selective plating. For example, physical properties of the deposit, such as ductility and hardness, can be very different from those obtained by more conventional methods, this often being due to dramatically different secondary metal codeposition rates.

High electrode polarisation may also be encountered, leading to increased voltage, gas evolution and reduced cathode efficiency. Deviations from the deposition rate calculated from an ideal primary current distribution, so necessary for super-selective, ultra high speed plating, are often met, and can only be resolved by experimentation with equipment design, electrolyte selection and operating conditions.

TECHNIQUES

The methods most generally used in selective gold plating are based on the following techniques:

(*a*) Selective immersion, (*b*) Masking, (*c*) Brush Plating, (*d*) Jet Plating, (*e*) Selective sensitisation, followed by electroless plating (restricted to the plating of non-conductors).

As to be expected, there is considerable secrecy concerning practical details of these processes. In the general discussion which follows, brush plating techniques are not included, since they are dealt with separately, in considerable detail, by Rubinstein in Chapter 16.

SELECTIVE IMMERSION

Vat Plating

Jigged components are selectively immersed in the electrolyte which is contained in a standard open top plating tank. Either solution agitation, or moving cathode rail are used to improve plating speeds and deposit characteristics. This sytem is often used for the gold plating of printed circuit board edge connector contacts, where strips of masking tape are used to accurately control the plating depth.

This system of batch rack plating gives intermittent production output and in many cases poor thickness distribution characteristics, but offers the advantages of cheap simple equipment along with an inherent relative freedom from mechanical failures.

Continuous Plating by Selective Immersion

This system is used for the selective plating of strips or jigs of components on a unit which gives straight through, in-line processing using a system in which the electrolyte is pumped over variable height weirs, thereby producing an adjustable head of solution through which the components can pass.[4,5]

Components or strips are passed horizontally through the weirs at each end of the process sequence troughs, e.g. electrocleaners, acid dips, copper or nickel undercoats, gold plate, drag-out, rinse and drying stages.

Plating solutions are stored below the plating troughs in heated sump tanks, each solution being pumped up to the process trough and then returning by gravity back to the sump tank. Rinses are usually fed by mains water supply and take the form of a spray rinse with the used water returning by drain to waste.

Plating height control in the gold bath can be obtained by altering the height of the monorail guide over the bath or cog system for strip plating, throttling the pumps, using a by-pass drain to the sump with a variable outlet, or by altering the location or design of the weir gates.

The process sequence or choice of undercoat can be quickly changed by the use of alternative sump tanks. Relatively high plating speeds can be achieved using this system, i.e., 200 microinches (5 microns) in four minutes, with deposits capable of meeting the forty eight-hour, or similar, gas tests. The choice of electrolyte must be carefully considered and for these plating speeds it is usual to use gold solutions containing 15–32 g/l of gold metal.

This self-contained system offers high output speeds, uniformity of process, and continuous production flow together with the flexibility to handle different components and plating specifications. Automatic bowl feeding of loose components into the plating jigs can often be accomplished and offers great reductions in labour involvement.

Disadvantages of the system lie in the capital cost, the relative complexity of the equipment, and associated possibilities of intermittent mechanical failure.

Selective strip plating can be accomplished on similar equipment and, with suitable modification, one side, or even a selected strip width of the raw material, can be gold plated on a continuous flow basis.

Anodic solution flow is sometimes used with these systems to improve plating speeds and deposit characteristics. In this system the anode is constructed from a hollow stainless steel tube through which the solution is pumped. Along the length of the tube is a line of small holes positioned so as to direct a stream of electrolyte towards the cathode face.

MASKING

In this process the non-functional parts of the component are lacquered by hand, coated by photoresist techniques or screened by the use of plastic inserts, masking formers or tape prior to conventional, or near conventional, vat or barrel plating.

This system can quite easily be introduced, but is often rendered relatively inefficient due to time consuming masking and stripping operations which frequently require a high labour involvement.

JET PLATING[6,7,8,9,10,11,12,13,14]

Many forms of jet plating are at present being developed, often under high security conditions. Essentially the system involves the impingement of a high velocity jet of electrolyte onto a very small cathode target area. Under these conditions of extremely high solution interchange, cathodic polarisation can be virtually eliminated. Plating by primary current distribution tends towards unity and under carefully controlled conditions electrodeposition by secondary current distribution can be eliminated, or at least reduced to an acceptable low level.

Discrete dots or stripes of gold can be obtained in extremely localised target areas at extremely high plating speeds (200 microinches in about 10–15 seconds). Dots of 0.040–0.100 in. can be built-up on any part of the component by this technique.

Sometimes the design of this equipment incorporates a rubber mask or screen which shields off that part of the component not requiring plating.

One such machine known to the writer was designed to selectively plate the small functional areas of integrated circuit leadframes. This machine used an integral mask and solution jetting system. Leadframe strip was fed through the machine in pulses of 15 seconds, during which the lengths of strip passed through built-in high speed cleaning operations followed by the gold plating stage. Here the mask clamped onto several frames simultaneously and the plating operation took place. The built-in jetting units utilised integral anodes in their design and permitted the deposition of 100 microinches (2.5 μm) of gold in 15 seconds. The utilisation of this machine by the intended customer was stated to be capable of offering gold savings as high as 95%, with a production rate of 24 ft/min of IC strip, using a triple strip mask and feed system.

PHOTOSELECTIVE DEPOSITION[15]

This form of selective plating is applicable only to the finishing of non-conducting materials such as plastics, where it is possible by the use of photoresists, screens or other techniques to selectively sensitise and electroless plate chosen areas.

Photo-oxidation by ultra-violet light of 2537 Å wavelength of the hydrated stannic oxide layer (nominal formula $Sn(OH)_2$) after the stannous chloride sensitisation stage in plating on plastics is a more recent development.

$$SnCl_2 + 2H_2O \rightarrow Sn(OH)_2 + 2HCl$$

$$Sn(OH)_2 + \tfrac{1}{2}O_2 \xrightarrow{h\nu} SnO_2 + H_2O$$

Normal palladium chloride activation of the unexposed areas then takes place.

$$Sn(OH)_2 + PdCl_2 \rightarrow Pd^0 + SnO_2 + 2HCl$$

It is reported that resolution of one micron can be obtained by these methods.

MISCELLANEOUS TECHNIQUES

Passive Deposit Method[16,17,18,19]

Here a "non-plateable" metal such as titanium is selectively plated in areas of the substrate where no subsequent deposit is required. This passive metal thus acts as a mask so that subsequent electrodeposition of gold occurs only in the required areas, after which the "non-plateable" metal is removed.

Current Shaping[20,21]

The system of screening areas of the cathode with perspex is often used to shape the primary current distribution patterns so as to enhance plating rates in preferential areas. This is particularly effective in cases where electrolytes with very poor throwing power are employed.

Sometimes one surface can be significantly arranged opposite the electrode, and by the use of a second electrode placed at the back with a potential relatively cathodic compared to the part that is desired to be plated, enhanced, preferential plating rates can be achieved.

Gold Dots and Inlays

Since the use of gold inlays or dots offers similar cost reductions to those obtained by some selective plating methods, it may be appropriate here to refer to some of the disadvantages of the former techniques which may restrict their use.

Materials can be purchased in strip form containing inlays and these can be fed directly into production press equipment, often eliminating the need for further electroplating. Typical inlay materials are alloyed with 25% silver, or smaller amounts of nickel, copper, or even platinum group metals.

High alloy inlays often fail in corrosion resistance when gas tested. Inclusions of base metal may also show up unfavourably. Forming operations sometimes damage the inlay or its hard undercoat, thus precluding the use of an inlay material for some applications. Proprietary costs for the material cropped off as scrap are lost when this material is sent to a refiner for gold reclamation, along with a further charge for refining. The cost of a narrow inlay is often much more expensive than the cost of selective plating. Furthermore, the cropped off raw edges, an unavoidable part of the process, invariably suffer extensive corrosion during gas test operations with associated creep of migrating base metal corrosion products on to the functional contact faces, causing a subsequent increase in contact resistance. The choice of substrate material is often limited, especially if heat treatment operations follow the press forming stage. For this reason beryllium-copper, a common material in the electronics industry, cannot usually be used in inlaid forms.

Gold dotting can be achieved by resistance welding or ultrasonic techniques. Gold alloy materials used take the form of thin metal wire or tape fed into the welding head. The cost of welding equipment and material often prohibits the use of this system when compared to some selective plating methods and there is often a problem of weld strength. Quality assurance usually means a built-in system of 100% testing of the welds *in situ*. Again, some base materials are more susceptible to giving poor welds. As in the case of inlay materials, corrosion products from surrounding base metal can creep over the contact area and lead to higher contact resistance.

Ultrasonic welding techniques are often hampered by an inability to make formed spring components "acoustically dead", so that much of the ultrasonic energy is dissipated during the welding operation, thus giving low interfacial energy concentration and hence poor welds.

Pastes and Paints

Metallic paints or pastes can be applied to selected areas of components, and after air drying or stoving operations produce acceptable contact sites.

Graphite or silver-based paint is sometimes used to conduct the cathodic current to selected metallic areas embedded in non-conductors, after which conventional electroplating can take place.

APPLICATIONS

Probably the most familiar example of selective plating is in the edge connector plating of printed circuit boards which involves fairly simple masking by tape or photoresist and is often carried out in conjunction with partial immersion. A similar technique is used in the plating of switch patterns and other printed components. For applications within the microelectronics industry, the reader is referred to Chapter 37.

The advantages of selective plating in the production of electronic components have been reviewed by Cooley[11], with emphasis on the flexibility of processes and the cost reductions which can result from the plating of components in continuous strip form. The following examples are taken from this source.

IC Leadframes

In the case of integrated circuit leadframes, lead strip length may vary depending on the packaging technique used by the manufacturer, which is often a major factor in the selection of methods for producing the strip leads. For leadframes in the form of continuous stamped strip, selective plating techniques available offer alternative possibilities for plating, e.g.,

(*a*) Spot plating in die attach/lead bond area only
(*b*) Stripe plating down the centre
(*c*) Heavy gold plating in the centre, with a flash on the lead ends.

The most suitable of these methods must be decided by the process production engineer in conjunction with the plater.

Transistor Headers

One advantage of the plating of components in strip form is that it can allow subsequent processing operations to be conveniently carried out before separating individual parts. Thus one approach to the plating of transistor headers involves the selective plating of KOVAR strip with a central stripe after which the device is assembled in a continuous operation through chip mounting, lead bonding and tab and post-welding before separation of components.

Diode Frames

Diode frames may require gold only on the contact area. Originally the whole of the frame was completely plated with the required thickness of gold at a plating cost of 15¢ per unit. The next stage of development was to electroplate the functional areas only with the required thickness and gold flash the remainder at a cost of 9.5¢ per unit. The final step was to selectively stripe the prestamped stock, resulting in gold only where it is actually needed, the final total cost of plating being 4.7¢ per unit.

MOS and LSI

One of the most exciting of the new selective plating techniques is with individual units such as MOS and LSI flat packages. In these packages, the die pad and sealing ring are plated with the required heavy gold, usually 100 microinches. The leads, which need only a flash for solderability, are plated with 15–20 microinches of gold. The selvedge receives little or no deposit. Since the lead and selvedge area is several times that of the centre pad and ring, the savings in gold costs are dramatic.

Connector Terminals

The technique of selectively plating components in strip form is particularly advantageous when internal surfaces are to be plated. Cooley cites the case of female connector terminals which would normally be barrel plated after forming, when the achievement of a specified thickness of gold on the interior entails the deposition of a much thicker coating on exterior non-functional surfaces. By selectively plating one side only of strip prior to stamping and forming, the required thickness of gold is produced evenly over the internal surface without wastage. Cost reduction in a particular case is quoted as being from \$3.78 to \$0.83 per thousand.

Economics

The greatest motivation for the introduction of selective plating techniques is that of reduced costs. Electronic components requiring heavy gold electrodeposits for wear resistance are wide open for cost reduction exercises based upon selective electrodeposition.

Frequently the gold deposit represents the highest cost factor in the manufacture of an electronic component and greater cost reductions can be achieved by selective plating studies than by any other manufacturing cost reduction exercise.

The following examples illustrate further specific instances of cost saving resulting from selective plating. The first is from the writer's* own experience while the others are quoted by Cooley†.

1 Connector Pin

Barrel electroplate with 5 μm gold. Total surface area 0.305 in^2. Selectively plate area 0.0682 in^2. Production forecast 2.35 million contacts per annum.

Barrel plating:

$0.305 \times 2.35 \times 10^6$ = 717,200 in^2

Weight of gold at 63 mg/in^2 for a 5 μm deposit = 44,400 grams

Proprietary cost of metal deposited per gram = 96p

Total material expenditure = £42,624 per annum

Selective Plating: $0.0682 \times 2.35 \times 10^6$ = 160,200 in^2 (equivalent to 9,930 g gold)

Material cost = £9,533

Total material savings using selective plating = £33,091

2 Etched IC Lead Frame

		Cost per unit
Traditional method:		
Raw material (KOVAR) and etching		4.0 ¢
Plating (50 microinches all over)		7.2 ¢
	Total	11.2 ¢
Selective Plating:		
Raw material (KOVAR) and etching		4.0 ¢
Plating (80 microinches in 1 in. dia)		3.0 ¢
	Total	7.0 ¢

* Price of gold as at March, 1973.
† Price of gold as at September, 1970.

Note—With the rapid rise in the price of gold in the latter half of 1972 and the early part of 1973, the case for selective plating is of course further strengthened—Editors.

3 Stamped IC Leadframe (typical)

Traditional method:

Raw material (KOVAR)		1.0 ¢
Stamping (18 in. strips)		1.0 ¢
Plating (50 microinches all over)		4.2 ¢
	Total	6.2 ¢

Selectively plated (*stripe*)

Raw material (KOVAR)		1.0 ¢
Stamping		1.0 ¢
Plating (50 microinches in centre)		2.0 ¢
	Total	4.0 ¢

4 Strip Connector (per contact)

Traditional method:

Raw material (nickel silver)		0.20 ¢
Stamping and Forming		0.38 ¢
Plating (50–100 microinches all over)		0.78 ¢
	Total	1.36 ¢

Selective Plating:

Raw material (nickel silver)		0.20 ¢
Stamping and Forming		0.40 ¢
Plating (50–60 microinches in contact area)		0.27 ¢
	Total	0.87 ¢

SUMMARY

Despite the technical problems frequently associated with the selective plating of precious metals, the realistic and attractive cost savings potential of this approach will ensure continuing development. Many processes are being studied at present, often under the strictest security restrictions, to make a breakthrough in some form of selective deposition, in an attempt to reduce manufacturing costs.

The electronics industry is an expanding and competitive one, and quite often the highest cost centre in a production chain is that of the plating operation. Under such conditions, the future use and development of selective plating techniques is assured.

REFERENCES

1 Antler M., *Plating*, **54,** 915 (1967).
2 Krumbein S. J., Antler M., IEEE Trans. on Parts, Materials & Packaging, PMP-4, 1, 3 (1968).
3 Garte S. M., *Plating*, **53,** 1335 (1966).
4 Warlow T., *Metal Finish. J.*, **15,** 174, 204 (1969).
5 Baldock J., Miles J. J., U.S. Pat., 3,657,097 (1972).
6 Strelchun F., Burndy Corp., private communication.
7 Nobel F., Lea-Ronal Corp. private communication.
8 Vaughan R. T., *et al.*, U.S. Pat., 2,854,387 (1955).
9 Tiley J. W., *et al.*, U.S. Pat., 3,067,114 (1962).
10 Zimmerman E. M., U.S. Pat., 2,854,387 (1955).
11 Cooley R., Proc., Inter-Nepcon, Brighton (1970).
12 Warlow T., Oxy Metal Finishing Congr., Amsterdam (1971).
13 Zarb G., Interfinish, Basle (1972).
14 *Electroplg Metal Finish.*, **25,** 10, 30 (1972).
15 Sharp D. J., *Plating*, **58,** 786 (1971).
16 U.S. Pat., 3,514,379 (1970).
17 German Pat., 1,621,123 (1966).
18 German Pat., 2,059,333 (1966).
19 Brit. Pat., 1,175,667 (1967).
20 Rousselot R. H., *Metal Finish.*, **59,** 3, 57 (1961).
21 Christie J. J., Thomas J. D., *Plating*, **52,** 855 (1965).

Chapter 18

POST-TREATMENT

W. K. A. Congreve

In many cases a gold coating of specified thickness which has been properly applied, requires only thorough rinsing and drying to meet minimal requirements of surface appearance and adhesion, and hence can be put into service without further treatment. In some cases, however, further treatment may be applied with the specific object of improving such factors as adhesion, structure, surface condition, mechanical properties and so forth, either to meet immediate service requirements or to facilitate subsequent processing, possibly after periods of storage. Such processes are described in more detail in the applicational section of the present work, but for the sake of completeness it seems appropriate to conclude the present section with a brief general review, under the following headings:

1 Rinsing
2 Drying
3 Cleaning
4 Heat treatment
5 Polishing
6 Lubrication
7 Stripping

RINSING

The theory and practice of rinsing has been reviewed by Pinkerton[1], and, in the context of high reliability precious metal plating, by Tweed[2]. Some practical aspects are discussed elsewhere in the present volume (see Chapters 13, 14, 34, 37).

On withdrawal from the plating bath a metal surface is covered with a film of electrolyte and the function of rinsing is to dilute this film to the point where the concentration of plating bath residues is not significant.

This process occurs by diffusion and the rate of, as well as the maximum, dilution obtainable is therefore dependent on the cleanliness of the bulk of the rinse water. Agitation also speeds up the process by reducing the thickness of the contaminant film.

The degree of agitation required depends on the geometry of the plated articles. Small components of simple shape may be rinsed adequately by immersion for a few minutes in a slowly moving stream of water, while complex configurations require more vigorous agitation and/or spray rinsing.

For rinsing gold plated components it is advisable to use three separate tanks, the first for drag-out and the second and third for running water rinsing, with clean water entering the third and overflowing into the second. This counter-current system provides the most efficient rinsing with minimum water consumption. Once the rinsing parameters, such as agitation, flow rate and temperature, have been adjusted for optimum results in terms of efficiency and water consumption, control can be maintained by monitoring the resistivity of the overflow from the final tank. (A resistivity of one megohm is equivalent to a concentration of approximately 120–180 ppm of ionised inorganic compound).

It is sometimes economical to recover rinse water by ion exchange processes, but it should be noted that these alone will not generally extract organic materials and a complementary activated carbon treatment will be necessary when the latter are present.

In electronic applications demineralised water is commonly used for rinsing. It is perhaps not generally appreciated that both demineralised and distilled water are more corrosive than normal tap water towards certain base metals. For this reason gold coatings on ferrous alloys should be rinsed and dried rapidly to avoid possible corrosion of the basis metal through pores which may be present in the coating.

DRYING

Drying should be rapid, preferably in a hot air oven. Centrifugal drying is also widely employed in the treatment of barrel plated components. Alternatively, drying may be assisted by dipping in a simple solvent such as methyl alcohol or acetone. However, this process is expensive, due to rapid contamination of the solvent. A number of proprietary de-watering agents are available for different applications; these replace the water by a film of an organic compound, usually designed to provide some protection against tarnishing and corrosion, and some are claimed to have no deleterious effect on solderability. However, it is wise to carry out extensive evaluation of such treatments before adoption in any specific case.

CLEANING

Articles which have been gold plated, rinsed and dried may require additional cleaning, either immediately or after storage in order to reduce surface resistivity, to improve solderability or ease of alloying in semiconductor bonding, or to remove visible stains. The more common contaminants present on gold plating are:

1 Plating residues left by inadequate rinsing. Such contaminants may be detectable only by performance tests or by electron diffraction, etc., but may nevertheless have significant effects in service.
2 Corrosion products derived from reaction between relatively high concentrations of plating bath residues with the substrate, particularly in recesses where both initial gold coverage and subsequent rinsing are likely to be unsatisfactory.
3 Films derived from atmospheric corrosion of the substrate through a porous coating.
4 Oxidation of metals codeposited with the gold. This can occur during the rinsing and drying of deposits from certain acid gold baths contaminated by iron.

Plating bath residues which have been allowed to dry cannot be removed by simply passing the component again through the rinsing cycle. The best method is to use a cathodic alkali clean followed by an efficient rinse.

Corrosion products can usually be removed by prolonged soaking in hot water, possibly with ultrasonic agitation, followed again by cathodic alkali cleaning and rinsing.

A suitable treatment for improving contact resistance, solderability and bondability of gold coatings is immersion in the following solution, with agitation:

Sodium hydroxide	200 g/l
Sodium cyanide	50 g/l
Sodium heptonate	30 g/l
Temperature	55–60°C
Time	3–5 min

This treatment does not significantly attack gold alone, but attack may occur rapidly if the gold coating is in electrical contact with exposed base metal, as for example, in the case of partially plated nickel-iron alloy.

Alternatively, immersion in 10% acetic acid or sodium gluconate solution is often useful for the above purpose.

HEAT TREATMENT

Gold plated components may be heat treated for a number of reasons, of which the following are the more important:

(*a*) For alloying purposes
(*b*) To relieve internal stress
(*c*) To promote adhesion to certain substrates.

A general point to note is that diffusion of certain base metals with gold is very rapid even at only moderately elevated temperatures, and may in some cases degrade the surface without noticeable change in appearance. Such a case, from the writer's experience, occurred in the heat treatment of a test piece of NILO K alloy (Fe-Ni-Co) plated with 10 microns gold. After 10 minutes in hydrogen at 450°C the surface appearance was quite unchanged, but examination by electron diffraction showed that the surface was covered by a film of an unidentified spinel type oxide. The maximum temperatures and times which can safely be used in heat treating plated articles depend on a large number of variables, hence precise data are seldom available, and conditions must be determined by practical testing in particular cases.

The use of heat treatment to modify the structure of electrodeposited gold alloys, as in certain watchcase plating processes, is referred to in Chapter 42, which also covers the aspect of stress relief. Diffusion heat treatment of separately applied coatings may also be a more economic and practicable method than alloy plating for the application of thin alloy coatings for special purposes, as in the plating of reed switch blades (see Chapter 38).

Examples of the use of heat treatment to promote adhesion of thin gold coatings on molybdenum, tungsten and other refractory metals are given in the relevant sections on Pretreatment (see Chapter 12).

POLISHING

Since polishing inevitably involves the removal of metal, it is used only very sparingly on gold coatings. Brightness requirements are usually met by polishing of the substrate, either mechanically or electrolytically, in conjunction with the use of bright plating processes, hence post-plating polishing is used only to rectify minor blemishes. This may be carried out by very light buffing or by hand, using very fine abrasives such as jewellers rouge (red ferric oxide). Electropolishing processes for gold have been described by Tegart[3], and one[4] which has long been used in the jewellery trade is the following:

Potassium cyanide	65 g/l
Sodium potassium tartrate	15 g/l
Phosphoric acid	18 ml/l
Ammonium hydroxide	2.5 ml/l
Temperature	50–60°C
Voltage	8–10.

Such processes should, however, be used with great caution on gold plate due to the relatively rapid rate of metal removal.

As for other metals, gold plating in the matte condition readily picks up stains as a result of handling, which are extremely difficult to remove. This can be prevented by a light burnish with a soft brass scratch brush. In trials made by the writer this treatment removed only about 0.1 mg/cm^2 from the surface of a pure acid gold deposit and, contrary to expectation, zinc contamination was not detectable on the surface when examined by X-ray fluorescence spectrometry.

LUBRICATION

Lubricants may be applied to gold plated surfaces to improve lubricity and are sometimes employed in conjunction with corrosion inhibitors to simultaneously improve the protective value of thin coatings by plugging of pores. These aspects are fully discussed by Antler in Chapter 21.

STRIPPING

Gold can be rapidly dissolved in a solution of iodine in potassium iodide or in an alkaline solution of sodium cyanide with an appropriate oxidising agent. The latter is used primarily for etching of gold films on non-metallic substrates to form thin film circuits.

For stripping gold electrodeposits, a variety of cyanide-based proprietary mixtures are available, some of which are claimed to be without significant attack on copper, silver and aluminium. Anodic dissolution in alkaline cyanide solution can be used to remove gold from ferrous metal substrates and a similar procedure is also often effective in the stripping of gold from aluminium.

In a recent paper[5] the theoretical aspects of chemical and electrolytic stripping solutions for the removal of gold from a number of base metals and from plastics are reviewed.

REFERENCES

1 Pinkerton H. L., "Electroplating Engineering Handbook" 2nd Edn, A. K. Graham (Ed.), pp. 589, Reinhold, New York (1971).
2 Tweed R. E., *Metal Finish.*, **67,** 8, 48 (1968).
3 Tegart W. J. McG., "The Electrolytic and Chemical Polishing of Metals", pp. 62, Pergamon Press, London (1959).
4 *Electroplg Metal Finish.*, **19,** 7, 230 (1966).
5 Ludwig R., *Oberfläche* (*Berlin*), 2, 84, (1972).

PART 4

Properties

Chapter 19

SOFT SOLDERING

C. J. THWAITES

The ability of gold to retain its original appearance and not to produce any visible tarnish films in most environments, are obvious features rendering this metal of interest as a solderable protective finish.

Although it is well known that soldered connections made to gold plated surfaces can lead to unreliable or mechanically weak joints, due to the rapid formation of intermetallic compounds during soldering, there are many instances where careful control of the gold thickness and the soldering conditions have resulted in a satisfactorily low incidence of failure. For example, Whitfield and Cubbin[1], reporting on their service experience found that soldered gold plated parts in telephone equipment gave failure rates below 0.0001 % joint hours, while in the guidance system of the early Minuteman I American space vehicles, gold plating was extensively utilised with no reported joint failures in 9×10^{11} joint hours[2].

However, numerous organisations have carried out studies on the soldering of gold because of problems of intermittent and unexplained unreliable joints that had come to light. The studies were mostly concerned with the apparent low mechanical strength of soldered connections on gold plating, but solderability has also been examined by some investigators and it is this latter subject which is dealt with first.

SOLDERABILITY

COMPARISON WITH OTHER COATINGS

Costello[3] of the General Electric Company in the USA was one of the first investigators to publish data comparing the solderability of gold and other coatings. He used an area-of-spread solder test, which is generally accepted to provide a measure of solderability of a surface[4]. This involved using a standard volume of solder or rosin-cored solder wire (in Costello's work activated rosin flux) which is melted for a fixed time on the various samples of electroplated coatings on copper. After cooling, the maximum total thickness of specimen plus solder is measured, these readings being related to the areas to which the solder has spread over the surface. For convenience, the solder-spreads were converted by Costello to "spread factors"[3] and some of his results are seen in Table 19.1.

Table 19.1
Solder-Spread Tests on Gold and Other Coatings[3]

Coating	*Spread Factor*
5 μm Au	88.6
5 μm Ni+1.2 μm Au	86.0
5 μm Sn–Ni*+1.2 μm Au	88.6
5 μm Ag+1.2 μm Au	86.2
5 μm Ni	83.5
5 μm Ag	85.1
5 μm Sn–Ni*	93.6
5 μm Ni+0.3 μm Rh	79.7

Copper substrates used throughout. Samples stored 4–6 weeks in laboratory prior to testing.
*65% Sn – 35% Ni

Later work by the writer[5,6] does not confirm the superiority of tin-nickel alloy plating as suggested by Table 19.1, even with more highly active fluxes. It is also somewhat difficult to assess these results because the paper gives no indication of the significance of the difference in the spread factor, but in another paper, Pessel[7], who proposed this factor for quantitatively indicating solderability, considered that factors of 70–80 were fair to good, 80–90 good to very good and 90 was excellent.

In an investigation concerned with trying to avoid the use of gold plating in printed circuits, Keller[8] found a mean area of spread of solder on gold plating of 2.9 cm^2, which compares unfavourably with a spread of 6.9 cm^2 on tin coated copper. It was observed that the spread of solder on gold was uneven, tended to "skip" small areas and eventually gave rise to de-wetting, but it is probable that the latter was due to incorrect preparation of the copper prior to plating.

Hyde and Goodwin[9] studied further the substitution of gold plating on circuit boards by tin-nickel. After one month of "natural" storage the gold plated boards could easily be dip soldered or soldered over a wave at a conveyor speed of about 2 m/min, whereas the tin-nickel required an acid dip to render it suitable for wave soldering even at 0.7 m/min. Following this, Castillero[10] suggested that an economic replacement for all gold plating consisted of 0.25 μm (or even as little as 0.03 μm) gold over a 5 μm thick coating of tin-nickel.

According to Korbelak and Duva[11] and a reputable manual on soldering[12], gold is unquestionably the most readily soldered metal, the former authors giving equal merit to gold, platinum and palladium. Silver and rhodium are stated[11] to lose solderability due to the formation of surface films. However, these authors suggest that the type and amount of alloying additions in the gold coating will influence solderability as well as change the appearance of the solder when the coating dissolves in it. In the soldering of semiconductor devices, Korbelak and Duva also stated that a tenfold increase in reliability was obtained by using a 1 micron undercoat of nickel beneath the final 1–1.5 microns of gold plating. In the solder dipping of component leads a 0.5 micron thick coating on copper always gave good solderability, but it was recommended that 1.2 microns was required on a nickel substrate.

In another semi-quantitative study of solderability, Thompson and Bjelland[13] placed thick gold coatings second in order of merit after hot dipped or electroplated tin and solder, while thin immersion coatings of gold and tin were shown to give a very inferior performance after one year of storage.

DIFFERENT TYPES OF GOLD

The area of spread of a given volume of solder was found to be highest on a bright gold deposit produced from a cyanide electrolyte according to Weil, Diehl and Rinker[14]. These investigators performed solderability tests at four temperatures, but only the tests at the higher temperatures showed significant differences in solderability. Those values are given in Table 19.2. Nickel-containing gold coatings were second in order of merit. Two substrates, silver and nickel, were employed and it was acknowledged that some solution of the former could occur if all the gold plating dissolved. It was observed that the bright gold deposits tended to give de-wetting of the solder, especially on a nickel intermediate layer. This may well have been a result of incorrect preparation prior to gold plating or perhaps to organic material in the bright nickel plating.

Table 19.2

Solder-Spread at Two Temperatures on Gold-Plated Nickel[14]

Electrolyte Constituents (g/l)		*Solder-spread (cm^2)*	
$KAu(CN)_2$	*Others*	*260°C*	*274°C*
15	100 Organic acid + Na salt 50	0.80	1.33
8	Co < 1; Organic acid + K salt 50	0.40	0.53
8	Co 1; Organic acid + K salt 50	1.02	0.75
8	Ni 1; Organic acid + K salt 80	1.13	1.30
8	Co 1; In 0.5 + K salt 100	0.41	1.13
12	KCN 90; $K_2Ni(CN)_4$5; Brightener 0.2	0.35	1.23
6	KCN 15	0.78	0.94
12	KCN 90; Sb complex; Brightener 0.2	1.20	1.32

It was claimed that the coatings of pure gold or antimony-gold alloy had capillary channels in the surface when examined by electron microscopy, which assisted with the flow of solder on the surface. These authors proposed that acid and nickel-alloyed types of gold plating were superior in solderability to the more commonly used cobalt-gold deposits. Parker[15], considered that the presence of 0.5% of nickel, cobalt or indium in gold electrodeposits decreased the level of solderability.

Organic Additions

Munier[16] at Bell Telephone Laboratories showed that substantial amounts of polymer were codeposited in hard gold plating, but not in soft cyanide deposits. Elements such as cobalt, in particular, increased the amount of polymer as measured by the carbon present which may reach as much as 0.7%, and the amount was shown to increase as the cathode efficiency dropped. Microscopic examination showed that the polymer was dispersed or layered throughout the deposit. In discussing this work, Davis[17] proposed that poor solderability of thick gold deposits was due to the presence of codeposited polymer which prevented wetting and was worse with thick coatings. Coatings less than 1.25 microns thick were therefore specified by this author's company to ensure satisfactory solderability.

INFLUENCE OF AGEING

The effect on solderability of diffusion occurring between a gold coating and the underlying copper was examined by Modjeska and Kann[18], who concluded that an undercoat layer is necessary, especially with thin gold deposits, to prevent this phenomenon. Solderability was tested by an undisclosed procedure after one hour of ageing at 237° in air, and those coatings showing best retention of solderability were 1.2 μm cyanide gold over a 0.5 μm minimum of nickel plating. An additional layer of silver, 0.5–5 μm thick, between the nickel layer and the copper substrate, appeared to confer no advantage. These results appear to be somewhat ambiguous since some other duplex coatings of much lower overall thickness showed equally good solderability. Rothschild and Kilgore[19] also demonstrated that diffusion of gold coatings on copper under dry heat conditions gave a substantial fall in solderability.

A detailed quantitative comparison of the solderability of different types of electrodeposited gold was reported by Harding and Pressly[20]. The results of their solderability tests, expressed once again as spread factors, are given in Table 19.3. It can be seen that a pure, matte deposit from a citric acid solution gave the largest area of solder-spread and was approximately equal in merit to conventional cyanide gold plating. The comments relating to the sensitivity of spread-factors earlier in this chapter should again be borne in mind. They also investigated the effect of the type of gold plating on the strength of solder joints, an aspect which is discussed later on in this chapter.

The behaviour of gold electrodeposits or duplex gold coatings after ageing was reported by Bester[2]. Figure 19.1 shows how the wetting time of copper, plated with 1.2 μm of gold, increases with the time of ageing at 77°C. The author also found that the same thickness of gold over nickel exhibited immediate wetting after 24 hours at 150°C, but the wetting time increased to 2.2 and 4.9 seconds after 24 hours of ageing at 205 and 260°C, respectively.

Table 19.3

Solder-Spread at 204°C on Gold-Plated Copper[20]

Electrolyte Constituents	*Solder-Spread Factor*
Conventional cyanide	90
Citric acid (no additive)	91
Citric acid + traces Co	87
Citric acid + Ni	87
Phosphoric acid + 1 % Co	83–88
Alkali CN + Ni + 0.5–1 % Ag	87
Blank—unplated Cu	81

Note—Results given only for thick (3 μm) coatings to obviate effect of substrate.

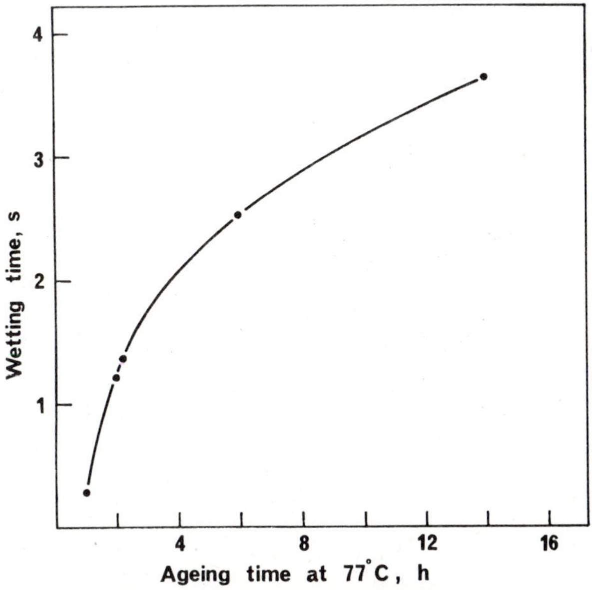

Fig. 19.1 Effect of ageing time at 77°C on wetting-time in molten solder of gold plated copper. [Bester[2]].

The writer has carried out both wetting-time and area-of-spread solderability tests on a variety of surface finishes on copper including thick gold coatings, duplex gold deposits and immersion coatings,[6] after various types of normal and artificial storage. The results of the wetting-time tests are shown in Figure 19.2. It will be noted that 5–8 μm of tin or tin-lead plating gave the best overall performance and that short wetting-times were retained generally for cobalt-or silver-alloyed gold. The pure (cyanide) gold deteriorated after ageing for 24 hours in steam. The thin immersion deposits of gold and of tin showed long wetting-times after 21 days long term damp heat (40°C, 90–95% RH) but both the duplex coatings of thin gold over thick coatings of silver or tin-nickel behaved well. The gold containing silver, gave the highest solder-spread of the unaged gold coatings, but the area was much less than on tin or tin-lead by a factor of 3–5. Both the duplex gold coatings showed relatively poor solderability in the solder-spread tests, but contrary to the wetting-times, the area of spread did not alter substantially during storage.

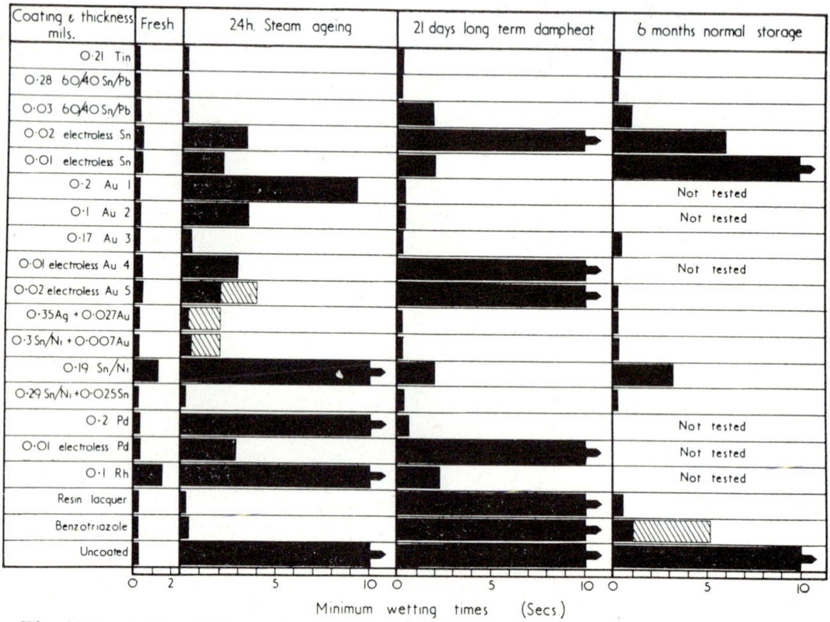

Fig. 19.2 Solderability, as indicated by minimum wetting-time, of various platings on copper after several different types of storage. [Thwaites[9]].
Notes:

(a) Values in 1st column are coating thickness in 0.001 in. units.
(b) 0.03 60% tin–40% lead was a hot roller solder coating.
(c) Au1 = pure cyanide deposit.
(d) Au2 = cobalt-alloyed deposit.
(e) Au3 = nickel-alloyed deposit.
(f) Sn/Ni deposits contained 65% tin.
(g) All tests made with non-activated rosin–alcohol flux and 60% tin–40% lead solder at 250°C.

METALLURGICAL REACTIONS DURING SOLDERING

When solder spreads over the surface of gold it appears that some diffusion of tin along the surface ahead of the advancing solder may occur. This has previously been illustrated for solder spreading on copper[21]. Heat treatment at 160°C of a solidified drop of solder on gold plating produces a halo of composition approximating to $AuSn_4$ beyond the edge of the solder drop, according to the experiments of Goldfarb[22]. It is well established that during the soldering of gold the latter is rapidly dissolved in the molten solder. Wild[23] found no influence of the amount of dissolved gold either on the wetting-time of clean copper or on the contact angle of the solder when measured during spreading drop tests. The results of Wild's tests are shown in Figure 19.3. Similar results with solder-spread tests were reported by Bester[2] who also showed clearly how the time of soldering using an iron influenced

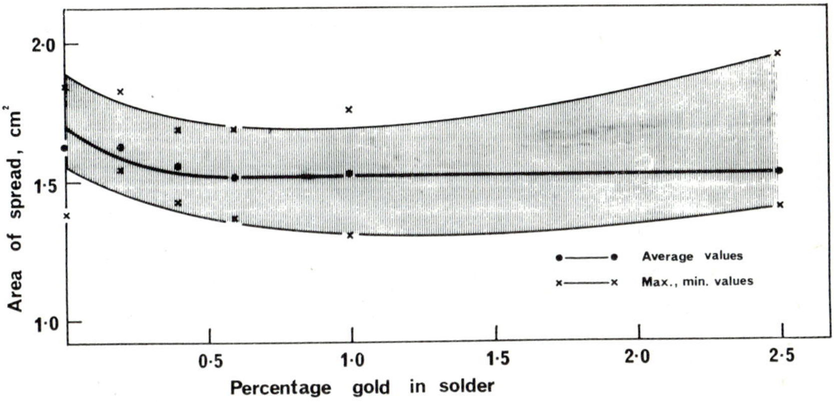

Fig. 19.3 Effect of gold content of solder on its wetting power. [Wild[23]].

the amount of gold dissolved in the solder. This in turn altered the appearance of the solder droplet. For example, 2 seconds soldering time resulted in a bright, smooth solder surface while 4 seconds produced a dull, gritty appearance on the solder. The present writer also found that in solder-spread tests, a cobalt-alloyed gold electroplate gave an irregular spread of solder while the pure (cyanide) gold produced a virtually circular drop (Figure 19.4). A dull, rough surface on the solder was observed on the alloyed gold (Figure 19.5) when it was dipped for a few seconds in a bath of 60% tin–40% lead solder at 250°C. This is due to the formation of intermetallic compounds (see following text).

THE GOLD-SOLDER SYSTEM

A study of the equilibrium diagrams for the binary systems gold-lead and gold-tin show close similarities. There are eutectic points at 90–90.5% lead and 214–217°C in the former system while in the latter, there are eutectics formed at approximately 85% tin, 215°C and at 20% tin, 280°C[24]. With tin, gold is known to form three distinct intermetallic compounds, AuSn, $AuSn_2$, $AuSn_4$ while Au_2Pb and $AuPb_2$ form in the lead-gold binary system[24]. The hardness of gold-tin compounds has been reported by McNeil[25] and Prince[26]. The solid solubility of gold in tin at 200°C is small (0.3%) and is negligible in lead (0.08%). In gold-tin and gold-lead alloys the change in liquidus tempera-

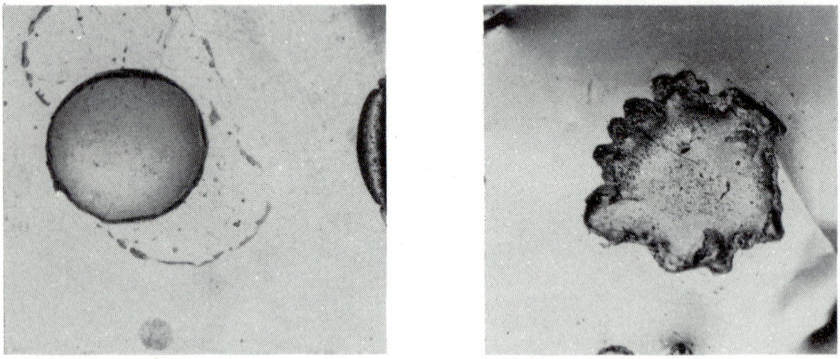

Fig. 19.4 Difference in uniformity of spread of solder on pure gold (left) and cobalt-gold deposits (right).

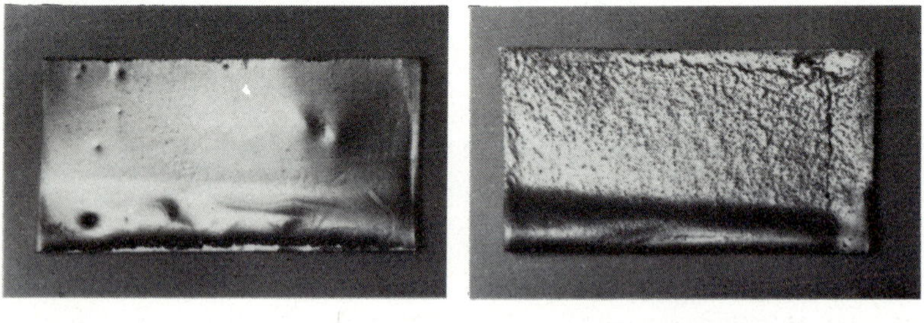

Fig. 19.5 Comparison of smooth and "gritty" surface of solder on pure and cobalt-gold deposits, after a short immersion in a solder bath at 250°C.

ture with increasing concentration of gold is relatively small. One would therefore not anticipate noticeable physical changes in the properties of a molten tin-lead solder bath operating at normal temperatures as it became contaminated with gold. The work of Karnowsky and Rosenweig,[27] and Prince[26] indicated the presence of a ternary eutectic point at 175–177°C and they illustrated a section of the ternary Au-Sn-Pb equilibrium diagram showing the phases present, when gold is added to a commonly used solder alloy containing 60 to 63% tin, balance lead (Figure 19.6). Wild[23] also published, at about the same time, the details of the liquidus and solidus boundaries for

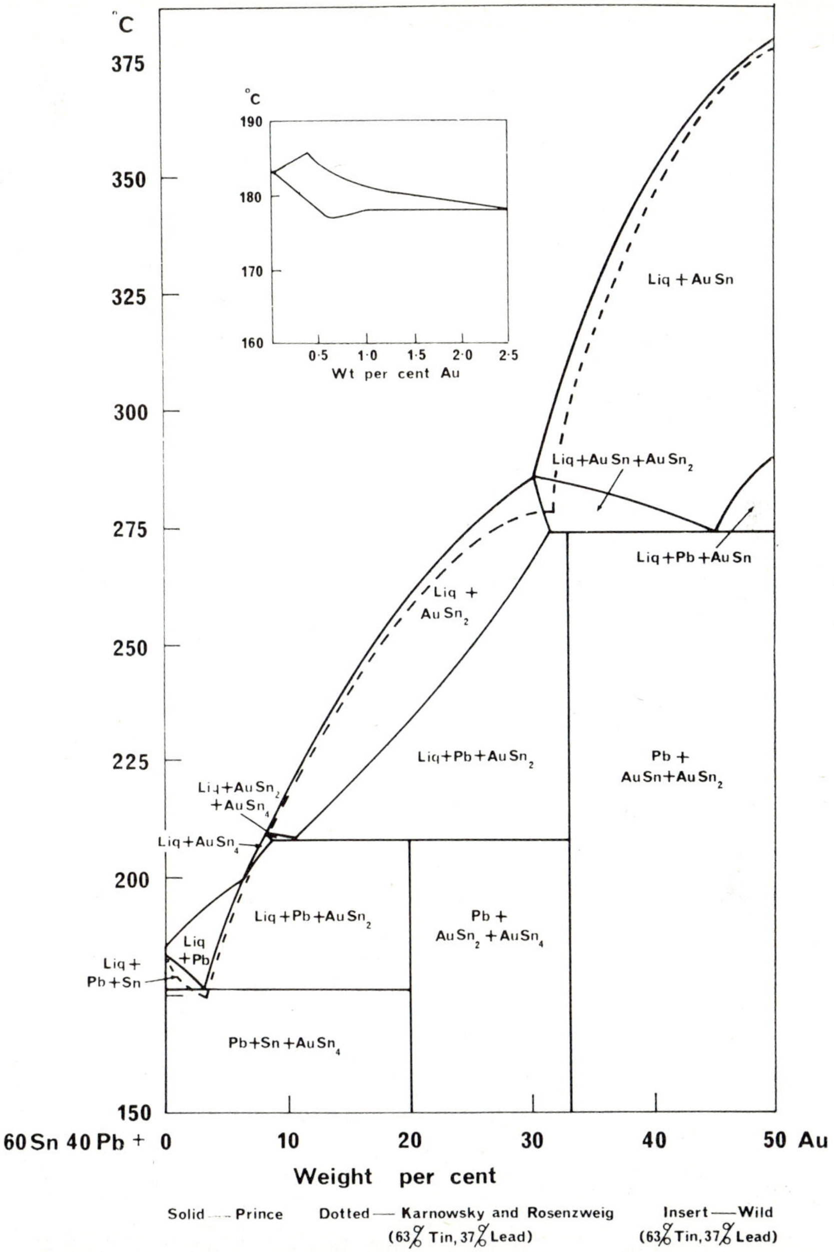

Fig. 19.6 The equilibrium diagram for gold in approximately eutectic tin–lead solder as determined in recent investigations. [Wild[23], Prince[26], Karnowsky & Rosenweig[27]].

63% tin–37% lead and 62% tin–36% lead-2% silver alloys containing concentrations of gold up to 2.5%, obtained with a differential scanning calorimeter. The results with the former alloy have been included in Figure 19.6.

Bader[28] experimentally compared the rate of solution of thin wires of differing metals (Figure 19.7) and confirmed the high reaction rate of gold that had been found in practice early on. From metallurgical considerations, one would expect that some of the compounds mentioned above would be observed in soldered joints. In hand soldered joints particularly, crystals of intermetallic may, in fact, be seen and do affect the resultant joint-strength. The conditions during making a joint, such as the rate of cooling, also affect the microstructure, which in turn can influence joint-strength.

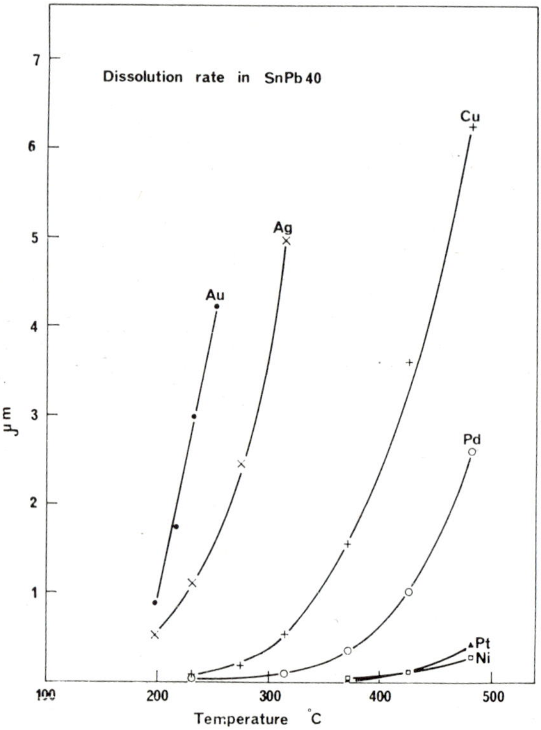

Fig. 19.7 Dissolution rates at various temperatures for a number of metals in 60% tin–40% lead solder. [Bader[28]].

SOLDERED JOINT STRUCTURES

A considerable number of excellent photomicrographs have been published, for example, by Bester[2], Foster[29] and many others which illustrate the rapid build-up of thick layers of intermetallic compounds when gold is brought into contact with molten solder. According to the electron micro-probe analyses reported by Bester[2], the predominating layer comprises $AuSn_4$, which normally exhibits a strong acicular growth. Where dissolution of a thick (e.g., 5 μm) coating of gold occurs within a limited joint space, a high concentration of gold in the solder can be reached and acicular $AuSn_4$ crystals are then evident in the bulk of the solder (Figure 19.8). These are strikingly white in colour when seen in a polished microsection and on breaking a joint containing quantities of these compound crystals, bright, white facets are seen in the fracture surface. Figure 19.9 is a photomicrograph of a section of a soldered joint made to a thick gold coating, showing that slight stressing which results during cooling, may cause rupture of the bond between the layer of $AuSn_4$ and the remaining gold beneath[6]. Fracture will also occur under stress between the $AuSn_4$ crystals in the solder and the solder itself. If copper

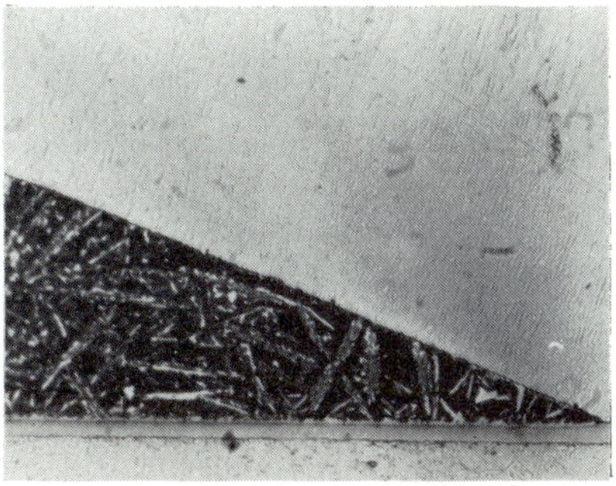

Fig. 19.8 Grey-white acicular crystals of $AuSn_4$ suspended within solder resulting from dissolution of a thick gold coating. (500×). [Tin Research Institute].

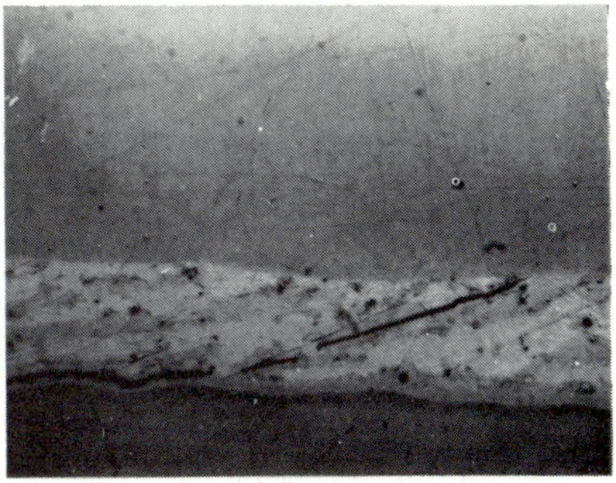

Fig. 19.9 Fracture between intermetallic compounds layer and the remaining gold plating due to stresses caused by cooling after soldering. (2000×) [Tin Research Institute].

wire soldered to a gold coating is removed in a peel test, separation occurs at the $AuSn_4$—gold interface and a surface is revealed which may become a characteristic brown-blue colour after a relatively short exposure to the atmosphere. Berry and Johnson[30] and others also illustrated fractures of this type along the $AuSn_4$ surface. This inherent weakness naturally may result in joints having low mechanical strength, an aspect which is discussed in some detail in the following pages.

A further feature of soldered joints on gold coatings is that abnormally high porosity is sometimes observed in soldered thick gold coatings, associated with the entrained crystals of intermetallic compound. Harding and Pressly[20] felt that this was more noticeable when soldering alloyed gold coatings, but the reasons for this phenomenon have not yet been found.

Because of the high rate of reaction between solder and gold, which leads to the risk of embrittled joints, efforts have been made to develop soldering alloys which have a lower reaction rate (see under "Solders").

MECHANICAL PROPERTIES OF SOLDERED JOINTS ON GOLD

BULK SOLDER PROPERTIES

In view of the wide usage of gold plating in electrónic equipments where the operating conditions are sometimes relatively severe, there have been a considerable number of investigations on the effect of gold on the bulk strength of solder. This work, together with actual measurements of the strength of joints made to gold, resulted from numerous service problems involving brittle type failures in soldered joints.

Bester[2] and Wild[23] showed that there was no significant change in the tensile strength of solders containing 62–63% tin (balance lead) for concentrations of gold up to about 12–14%. Wild also found no significant change in Young's modulus up to the highest gold content he examined (2.5%) whereas the yield strength showed a consistent and slight downward trend over this range. The elongation and reduction of area at failure increased significantly with up to 2.5% gold in a 63% tin–37% lead alloy, but not with a similar alloy containing 2% silver, a result which is open to question.

Bester[2] added increasing amounts of gold to a 63% tin–37% lead solder and found that the ductility was close to zero when 8–10% gold was present (Figure 19.10). He also found a simultaneous increase in hardness with gold content (Figure 19.11). This would normally be expected to lead to increased tensile strength, but the strengthening was counteracted by the simultaneous loss in ductility.

In 1963, Foster[29] published the results of an investigation on the ductility of solders containing gold, determined by bend tests on cast sticks of 4.5 mm^2 cross-section, bent over a 1.5 mm radius. It was found that a 5% gold-tin alloy could be bent through an angle of 90° without cracking, whereas a similar concentration of gold in lead could only be bent slightly before cracking occurred. 10% of gold in either metal produced extremely brittle alloys with no ductility. On adding various amounts of gold to a 60% tin-40% lead alloy, complete brittleness was found with 10% of gold, 5% produced failure after a 60° bend, but with additions of up to 2.5% gold the solder could be bent through 90°.

Walker and Waldie[31] carried out a somewhat similar test for ductility in which a fixed volume of solder was melted on to various thicknesses of gold plated copper. On bending the strips round a mandrel, with the solder drop on the outside of the bend, there was a distinct "end-point" at a gold plating thickness of 100 microns, at which the solder blob developed cracks.

A more sensitive mechanical test for detecting and measuring brittleness is one involving impact loading. Figure 19.12 records the results of Izod impact tests[2] showing that whereas 4% of gold in a 63% tin–37% lead solder causes

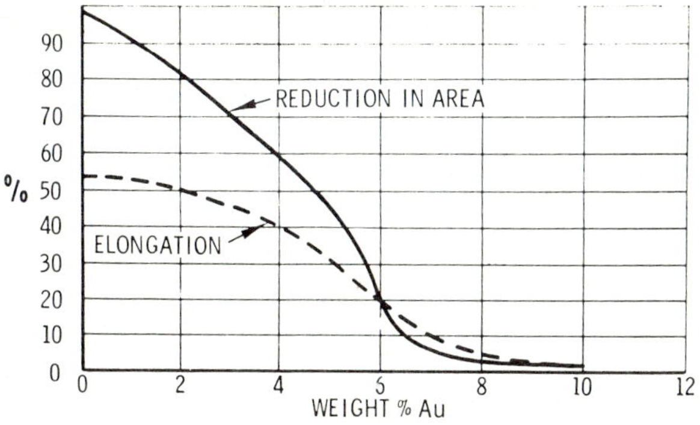

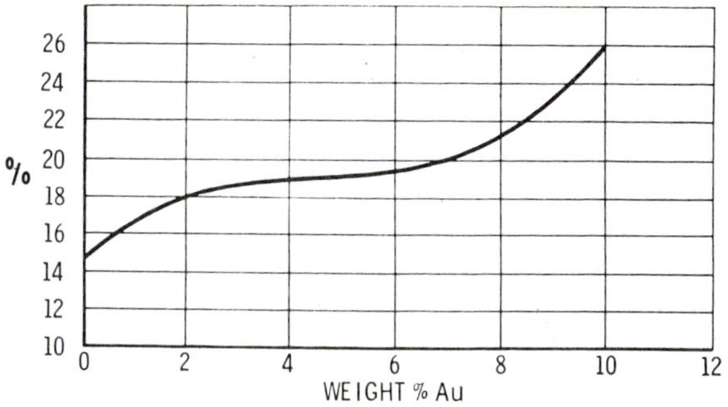

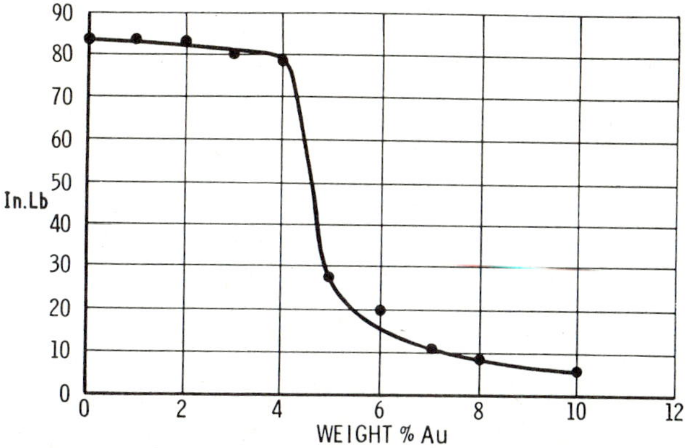

Fig. 19.10 *Top*: Influence of gold content in 63% tin–37% lead solder on tensile ductility. [Bester[2]].
Fig. 19.11 *Middle*: Diamond hardness of 63% tin–37% lead solder for various gold contents. [Bester[2]].
Fig. 19.12 *Bottom*: Effect on impact strength of 63% tin–37% lead solder of increasing concentration of gold. [Bester[2]].

no deleterious effects, a 5% addition results in a drop in impact strength from 80 to about 25 in.-lb.

STRENGTH OF JOINTS ON GOLD PLATING

An indication of the problems that may arise in the use of solder joints on electroplated gold is obtained from a paper by Bester[2] who reported the case of a lap joint made to 3.25 microns of gold with an insufficient volume of solder, which fractured when subjected to vibrations producing 20,000 G.

A very practical method of checking the strength of solder joints on printed circuit boards is to measure the force required to peel a wire which has been soldered along the surface of a conductor. Keller[8] gave the average force required for separation of twenty wires on a gold plated circuit as 1.8 lb and 4.2 lb for joints made to clean unplated copper. The range of forces was 0.5–3.0 lb for the gold coatings and 3.2–5.9 lb for joints to copper. Goldfarb[32] however, found no significant difference in the peel strength of thirty joints made to KOVAR leads with solders of high and low gold content. It is perhaps fair to comment that soldered joints are particularly susceptible to tearing so that a peel test may be considered to be an unsatisfactory method of estimating joint strength. Nevertheless it is true that a tearing component often exists in practice.

Thompson and Bjelland[13] were apparently the first authors (1961) to report on a detailed investigation on the strength of soldered joints made to a variety of coatings, including gold. Pairs of terminal posts were butt-soldered end-to-end using solder preforms. The tensile strength of the joints was determined. The theoretical tensile strength of the cross-section of solder in the joints was 104 lb, but many of the joints examined gave values well above this level. Their main purpose was to examine the effects of storage on the solderability of the coatings and on the resultant joint strength. Freshly made joints on coatings of gold in the region of 2.5 microns thick, plated over undercoat layers of copper, silver and nickel, gave a joint strength of 105, 107 and 117 lb, respectively, which appears to compare reasonably well with that of 125 lb obtained for tin-lead plating over nickel. The coefficient of variance for gold over copper was 5.8, whereas the gold plating over nickel and over silver showed high variance values of 21 and 11 respectively. From this it would seem that there might well be a risk of a larger scatter in the strength of joints made to some gold plating combinations. These results did not give a particularly clear picture of the effects of gold on joint strength, partly because of the extensive use of electroless nickel undercoats which caused de-wetting of the solder.

Together with his study of the microstructural effects, Foster[29] measured the shear strength of copper-copper joints made with 60% tin–40% lead solder containing various amounts of gold and found an initial 6% increase in strength with 2.5% gold, followed by a decrease in strength of 15% when 10% gold was present.

The investigations of Harding and Pressly[20] may be considered as the classical work on the subject. Not only did they examine gold coating thickness and type of gold used, but also the influence of soldering time and temperature. Some of their results for the latter aspect are illustrated in Figure 19.13 for a pure gold coating 7.5 microns thick, electrodeposited from a citric acid solution without additives. They noted that the soldering temperature had a far greater influence on the shear strength of lap joints than did the soldering time.

These investigators made the point that the alloyed gold deposits, particularly those containing cobalt, produced significantly weaker joints. Their results are best summarised by the curves shown in Figure 19.14, indicating the effect of gold plating thickness on joint shear strength for pure and cobalt-alloyed gold. Their paper included excellent photographs illustrating the influence of gold on the structure and on the fracture surfaces of failed joints. Harding and Pressly concluded that thin gold plating (up to 1.2 μm) will

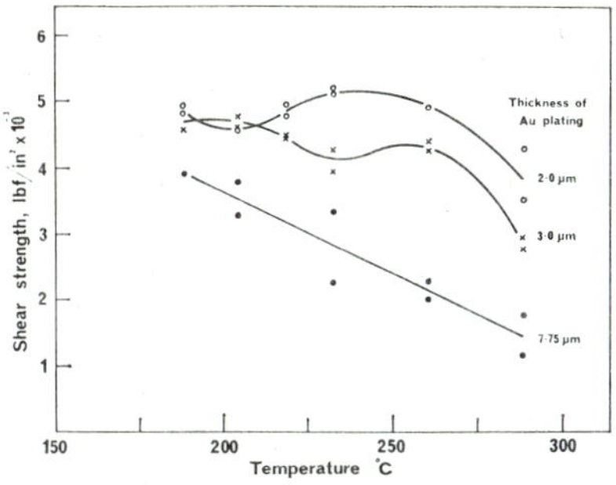

Fig. 19.13 Relationship between shear strength of soldered joints on various pure gold plating thicknesses and soldering temperature. [Harding & Pressly[20]].

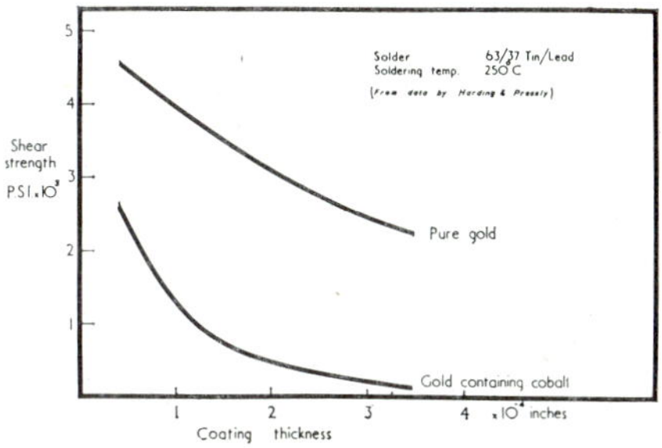

Fig. 19.14 Effect of type of gold plating thickness on shear strength of soldered joints. [Harding & Pressly[20]].

dissolve completely in molten solder but will have no measurable detrimental affect on joint strength, whereas the thick coating necessary for good protection of the substrate metal is likely to produce brittle soldered joints, particularly with alloyed gold plating. They summarised by saying that gold coatings above 1.2 microns should be avoided when strength is important and that the soldering operation should be closely controlled to minimise the time and temperature.

Weil, Diehl and Rinker[14] also determined the strength of butt joints made to a variety of types of gold plating of thickness 2.5 microns using several different solders. They did not, however, compare these results with other coatings or examine the effect of thickness of the plating. Table 19.4 shows some of their results, indicating that maximum joint strength was obtained with a 95% tin–5% antimony solder, some of the strengths being twice that for 60% tin–40% lead alloy. The strength of joints made at lower than normal temperatures by utilising alloys containing indium[33], was much lower for all types of gold plating. Only the results of tests with an underlayer of silver are reported here since they found that with gold direct on the nickel substrate the solder tended to de-wet, which would produce variable and low strengths.

Table 19.4
Tensile Strength of Butt-Joints on Gold Coatings with Three Different Solders[14]

Electrolyte Constituents (g/l)		*Tensile Pull (lb) to Break Joint Made at*		
		260°C		*204°C*
$KAu(CN)_2$	*Others*	*60 Sn–40 Pb*	*95 Sn–5 Sb*	*Pb–Sn–In Alloy*[31]
15	Organic acid + Na salt 100	42.5	65.0	28.5
8	Co < 1; Organic acid + K salt 50	30.8	48.5	6.9
8	Ni < 1; Organic acid + K salt 80	38.8	47.0	21.5
8	Ni 1; Organic acid + K salt 80	29.8	53.0	23.3

The shear strength of lap joints between copper plated steel soldered with tin-lead containing 63% tin, to which additions of up to 2.5% gold had been made, are shown in Table 19.5[23]. For test temperatures of 25 and 115°C, the shear strength of the joints tended to increase up to 1% of gold and then decrease, so that the strength with 2.5% gold was virtually the same as the basic solder. Tests at −60°C, however, showed a continuous increase in joint strength as the gold content became greater. Measurements of the strain at failure showed in fact a greater ductility when gold was present than the control material. The author concluded that there was no evidence from the mechanical tests that 2.5% of gold in solder, solidified in a reasonable time, would lead to embrittlement of soldered joints.

Table 19.5
Shear Strength of Soldered Joints Made with 63% Tin–37% Lead Solder Containing up to 2.5% Gold[23]

% Au	*Shear Strength (lbf/in²) at*		
	−60°C	*25°C*	*115°C*
0	10,160	6,120	2,900
0.4	10,340	6,560	3,155
1.0	10,860	7,020	2,985
2.5	11,150	6,310	2,510

The test on lap joints of Whitfield and Cubbin[1], Berry and Johnson[30], and Walker and Waldie[31] confirmed the effect of gold on the microstructure of solders, but could not detect any deleterious influence on the shear strength of soldered joints. However, Berry and Johnson showed that the formation of an excessive thickness of intermetallic compounds in joints to gold and to palladium surfaces could give rise to premature brittle fracture at the solder-compound interface.

SUMMARY

Summarising the influence of gold on the strength of soldered joints, the evidence indicates that an excessive amount of gold intermetallic compounds in the solder may cause a brittle type of fracture, but this will be evident mostly with tearing stress conditions. There is a much smaller influence on the tensile strength of the alloy or on the strength of joints tested at low strain rates. Cracking also readily occurs between the intermetallic compound layer and the remaining gold plating when the soldering conditions are such that not all of the gold is dissolved in the solder. In the majority of soldering processes even thick plating (e.g., 5 μm) is likely to be completely dissolved. The embrittlement of tin-lead solders appears to be worse with some alloyed gold platings. Solders with compositions giving low melting points, and hence lower rates of reaction with the gold, appear from the published data to give much lower joint strength.

SOLDERING METHODS FOR GOLD COATINGS

FLUX

All types of gold plating appear to be solderable with rosin-type non-corrosive fluxes whether of the activated type or not and often fluxless soldering is possible because of the absence of oxide films. The ease of soldering depends to some extent on the type and amount of addition agent present in the deposit. In general, however, it has already been indicated that gold has reasonably good solderability, but does not compare with the very large solder-spread obtained on tin and solder coatings of similar thickness under identical test conditions. A change in the type of flux does not achieve any substantial improvement in solderability.

One phenomenon that has come to light when soldering gold coatings is the occasional occurrence of a green coloration of the rosin flux residues. Tests have shown that the colour is probably due to the presence of copper abietate resulting from reaction of abietic acid, a major constituent of rosin[34]. This coloration should not be confused with corrosion products arising from the presence of excessive halides in the flux. The problem has arisen in hand soldering and it is the flux residues that remain at the periphery of the spreading pool of solder that become green. Activated rosin flux heated on gold alone and not in contact with solder does not usually change colour. No explanation of this phenomenon has been given, but it is tentatively proposed by the writer that it arises from the extraction of copper from the substrate through pores in the gold plating. The presence of gold perhaps promotes a chemical reaction, giving rise to green salts, which does not occur when soldering on uncoated copper or in the presence of other metals less noble than gold.

Gold plating is often used on KOVAR terminations for components having glass-to-metal seals and Reich[35] has demonstrated that such leads are subject to stress-corrosion cracking if acid chloride flux residues are incompletely removed.

SOLDERS

Since the major problem in the soldering of gold coatings is the rapid formation of intermetallic compounds, the use of lower temperatures and alloys of lower melting point than normal tin-lead solders is a point to be considered. Experience indicates, however, that the use of low melting point tin-lead-bismuth and tin-lead-cadmium alloys does not entirely eliminate the problem of brittleness due to excessively thick intermetallic compound layers.

In making soldered connections to thin or thick film circuits where extremely thin layers of gold are present, the rapid dissolution of the gold by tin-lead solders could lead to poor bonding between the solder and the underlying metal which may be, for example, a nickel-chromium layer. A means of avoiding this has been proposed which utilises a 20% tin–80% gold alloy (eutectic composition) in powder form, mixed with flux and screen-printed on the circuit in the required location[36]. For the mass soldering techniques used for printed circuit boards, normal tin-lead alloys near to eutectic composition are used and such alloys are included in BS 219 as Grades A and K containing 65 and 60% tin, respectively. In most cases gold coatings are completely dissolved in the solder bath during mass soldering.

Braun[37] investigated a range of alloys based on tin, lead and indium and showed by solid state diffusion experiments that the latter element drastically reduced the rate of compound formation. This was shown to arise from the formation of $AuIn_2$ in preference to, and much slower than, $AuSn_4$. Zinc also had a beneficial effect on the compound formation and as a result of this work an alloy containing 53% tin, 29% lead, 17% indium and 0.5% zinc was suggested and patented by Braun[33]. This alloy also has a low liquidus temperature (156°C) which allows soldering to be carried out at temperatures below those required with the normal tin-lead alloys. However, indium is an expensive metal and at the time of writing, this alloy does not enjoy much popularity as a solder. Other alloys have been examined by Brewer[38] but it was difficult to find a solder having all the desired properties of good wetting, high joint strength and a low rate of attack on gold. The effect of gold as an impurity in the solder bath on the wetting properties is not detectable[29] although at small percentages the solidified solder at the joints may have a dull and "frosted" appearance (Figure 19.5). It is likely, however, that a solder bath would be replaced for economic reasons before a significant build-up of gold had occurred. For example, an average wave soldering bath containing 100 lb of solder would contain 2 lb of gold if the concentration of this element reached 1 %, thereby representing a large security risk.

PROCESSES

The dip or wave soldering of components and printed circuits that have been gold plated presents no difficulties provided that the substrate material has been correctly prepared. Even thick deposits (5 μm) of gold will dissolve completely during the normal soldering process so that the solder is then required to wet the underlying metal. Thus, copper cleaned by a dry abrasive method may suffer from de-wetting of the solder[39] once the gold has been dissolved. Similarly, bright electrodeposited undercoats of copper and nickel or coatings of electroless nickel may be reluctant to wet easily due to organic inclusions.

Hand soldering with an iron is readily carried out, but it is in such circumstances that embrittled joints can be obtained, either when not all of the gold plating is dissolved or alternatively if the volume of solder is so small as to result in a concentration of gold exceeding about 5%. One recommended procedure, therefore, is to use a "wicking-off" technique[8] in which the first pool of molten solder is removed on the bit with a suction device or with a fine stranded wire. This is subsequently replaced by a further quantity of clean solder. Sometimes it may be necessary to deposit a barrier coating of nickel over the gold, followed by tin to provide solderability.

Flatpacks frequently have gold plated terminations which may cause difficulties in attaching them to circuit boards by reflow soldering. For effective joints to be produced by this method, a total coating thickness of 50 microns of solder is required, distributed almost equally between the surfaces to be mated. The use of gold on the terminations together with, for example, a roller solder coating of 1 micron thickness on the circuit board, provides an inadequate amount of solder. This would also give rise to an excessive concentration of gold in the solder joint, thereby leading to brittleness.

The best solution, where practicable, for both bit and reflow soldering, is to remove the gold plating on the relevant terminations in a bath of molten solder and re-dip in an uncontaminated solder bath so as to leave a thick solder coating of high solderability.

SERVICE RELIABILITY OF GOLD-SOLDER JOINTS

One characteristic of soldered joints made to gold plating is that there is an appreciable rate of solid-state diffusion which becomes highly deleterious at elevated temperatures. Braun and Rhinehammer[40] illustrated the rate of reaction by the thickness of intermetallic compound formed on solid gold wires at various temperatures. Rothschild and Kilgore[19] pictorially showed that heating of gold plating 1.2 and 2.5 microns thick for times in excess of 1 hour at 135°C, gave a drastic reduction in area of spread of a solder drop, while at 66°C, 96 hours treatment produced no detectable change in solderability. Thus, go!d plated components which have been in service for some time at the level of elevated temperatures that is usual for some electronic assemblies, might be difficult or even impossible to replace or re-solder due to diffusion of the gold coating into the substrate. Compound growth would also lead to increased joint brittleness.

The diffusion process, apart from destroying solderability, may cause loss of adhesion of the plated coating due to porosity at the interface forming by the Kirkendall effect. This physical phenomenon is a result of there being different diffusion rates in either direction between the gold and the underlying copper. It may be prevented by the use of a diffusion barrier of nickel or tin-nickel alloy. The use of such undercoats must also be considered in relation to the corrosion protection afforded by the complete coating system.

REFERENCES

1 Whitfield J., Cubbin J. A., *Automatic Telephone & Electric Co.* **21,** 1, 2 (1965).
2 Bester M. H., Proc., Inter-Nepcon, Brighton (1968).
3 Costello D. J., General Electric tech. Rep. No. R58EMH45, Aug. (1958).
4 Thwaites C. J., *Brit. Weld. J.*, **12,** 11, 543 (1965).
5 Thwaites C. J., *Trans. Inst. M :al Finish.*, **36,** 203 (1959).
6 Thwaites C. J., *Trans. Inst. Metal Finish.*, **43,** 143 (1965).
7 Pessel L., Proc. 59th Ann. Meeting A.S.T.M., Atlantic City, N.J., 1956, 1159 (1963).
8 Keller J. D., A.S.T.M., Special tech. Publ. No. 319, pp. 3 (1963).
9 Hyde W. B., Goodwin D. A., General Electric tech. Rep. No. R60EMH-39, June (1960).
10 Castillero A. W., *Metal Finish.*, **62,** 2, 61 (1964).
11 Korbelak A., Duva R., *Proc. Am. Electropl. Soc.*, **48,** 142 (1961).
12 "Soldering Manual", Am. Weld. Soc., New York (1959).
13 Thompson J. M., Bjelland L. K., *Proc. Am. Electropl. Soc.*, **48,** 182 (1961).
14 Weil R., Diehl R., Rinker E., *Plating*, **52,** 1142 (1965).
15 Parker E. A., *Electron. Des.*, 15th Feb. (1962).
16 Munier G. B., *Plating*, **56,** 1151 (1969).
17 Davis M. V., *Plating*, **57,** 262 (1970).
18 Modjeska R. S., Kann S. N., *Proc. Am. Electropl. Soc.*, **50,** 117 (1963).
19 Rothschild B. F., Kilgore L. C., West Reg. Tech. Session Am. Electropl. Soc., San Fransisco, March (1966). (North American Aviation/Autonetics Rep. X6-362/3111).
20 Harding W. B., Pressly H. B., *Proc. Am. Electropl. Soc.*, **50,** 90 (1963).
21 Bailey G. L. J., Watkins H. C., *J. Inst. Metals*, **80,** 57 (1951).
22 Goldfarb H., (North American Aviation/Autonetics, Rep. No. TR66-502) Nov. (1966).
23 Wild R. N., I.B.M. Federal Systems Div., Owego, Rep. No. 67-825-2157, Jan. (1968).
24 Hansen M., "Constitution of Binary Alloys", McGraw Hill, New York (1958).
25 McNeil M. B., *J. electrochem. Soc.*, **110,** 1169 (1963).
26 Prince A., *J. less-common Metals*, **12,** 107 (1967).
27 Karnowsky M. M., Rosenweig A., *Trans. Met. Soc. Am. Inst. Min. Engrs*, **242,** 2257 (1968).
28 Bader W. G., *Weld. J.*, **48,** 551s (1969).
29 Foster F. G., A.S.T.M., Special tech. Publ. No. 319, pp. 13 (1962).
30 Berry R. D., Johnson R. W., Paper Inst. of Welding, Autumn Meeting, London (1964)

31 Walker E. V., Waldie F. A., *P.O. elect. Eng. J.*, **58,** 268 (1966).
32 Goldfarb H. (North American Aviation/Autonetics, Rep. No. TR63-372) Aug. (1963).
33 U.S. Pat., 3,226,226 (1965).
34 Goldie W., private communication.
35 Reich B., *Solid St. Technol.*, pp. 36, April (1969).
36 Anon., *American Metal Market*, pp. 14, 27th Jan. (1970).
37 Braun J. D., *Trans. Am. Soc. Metals Quarterly*, **57,** 568 (1964).
38 Brewer D. H., *Weld, J.*, *Res. Suppl.*, pp. 465-s, Oct. (1970).
39 Thwaites C. J., Mackay C. A., *Metal Finish. J.*, **14,** 165 (1968).
40 Braun J. D., Rhinehammer T. B., *Trans. Am. Soc. Metals*, **56,** 870 (1965).

Chapter 20

WELDABILITY

M. H. Scott & K. J. Clews

The term weldability has been defined in various ways, but probably the most descriptive is that a metal has good weldability if sound welds can be produced in it by any required method and if the properties of the welds are such that full use can be made of the parent metal properties. The property that is usually most important is strength, but in the applications where electroplated gold is normally welded (electronics and more particularly microelectronics, in solid state devices such as transistors and integrated circuits), electrical conductivity is the primary requirement. However, this does not mean that strength is a property that can be ignored. In aircraft and missile applications, it is still very important, since the joints have to withstand the forces developed under rapid acceleration.

POSSIBLE JOINING PROCESSES

It is appropriate to begin by first considering the possible ways in which a joint can be made, and then the suitable processes.

Joints can be divided into two broad categories

(1) those which have been made without melting any of the components of the joint and
(2) those made by melting at least one component.

The former type depends upon intimate (i.e., atomic) metallic contact being made between the components. Force is therefore required, both to produce the required contact and if necessary, to break up the oxide or other films at the interface. Bonding by this method is possible at room temperature, but can be assisted by increasing the temperature to increase atomic mobility. In the second type of joint, where melting is involved, this can be confined to specially added metal of lower melting point or the main components of the joint can themselves be melted locally. Soft soldering (see Chapter 19) is an extreme case of the former, with the filler being supplied separately.

An allied process is where the main components of the joint do not themselves melt, but react together to give a lower melting point material at the interface. This process has been defined as liquid-phase diffusion bonding, but such joints are often referred to as brazed joints. Similar effects can also be obtained when the filler is provided as an electrodeposit and, as such, gold is frequently added in this way.

PRESSURE WELDING

The simplest joining process is pressure welding, where a high pressure is applied to the specimen to promote sufficient deformation at the interface to break up any surface films and result in intimate metal-to-metal contact. However, at room temperature the deformation required is high and for this reason, in the electronics field, it is normal to assist bonding by raising the temperature. The process is then usually referred to as thermocompression bonding[1] and is further sub-classified, according to the actual method employed, into (*a*) knife edge, chisel or wedge bonding, (*b*) ball or nail-head bonding and (*c*) eyelet bonding[2]. In the first variation, a heated wedge is pressed into the top component. In the second, a wire passes through a heated capillary tube, a ball is formed on the end of the wire by melting and the ball then pressed onto the surface to which it is to be welded. In the third instance, a heated rod having a suitably shaped tip is used as an indentor.

ULTRASONIC WELDING

Ultrasonic welding is, essentially, also a variant of pressure welding, the films at the interface being disrupted by the ultrasonic vibration. There is also a temperature rise due to the vibration, but this is of less importance.

RESISTANCE WELDING

Of the joining techniques that involve melting, the one of widest application in the electronics field is resistance welding, where the interfacial electrical resistance is used to produce sufficient local heat (at the faying surfaces) to cause melting. (In fact, joining can also be achieved without melting when it becomes a variant of thermocompression bonding, but the degree of control required is such that the process is not often practiced). The normal resistance welding configuration is with opposed electrodes but the electrodes can both be upon the same side of the joint[3]. The process is then known as series welding if the electrodes are spaced far enough apart to give two welds, or parallel gap welding if the electrodes are close together and only one weld is obtained[1,3,5]. The latter process is extremely important in the electronics field.

PERCUSSION WELDING

Percussion (capacitor discharge) welding in which the energy stored in a capacitor is used to produce a spark between the components to be joined, is best suited to joining a wire end-on to some other component. The spark causes local melting and when the components are pushed together, a joint is formed.

LASER AND ELECTRON BEAM WELDING

Some slight use is made of the laser for welds in electronic assemblies, but it is limited to joints where the joint face is accessible to the light beam. The electron beam process can also be used, but rather less conveniently and with the same restriction of accessibility for the beam.

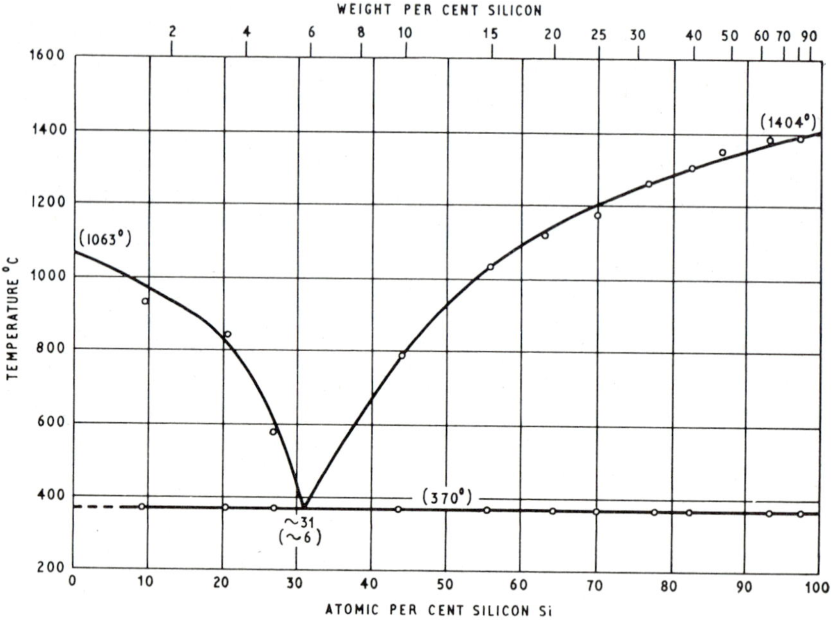

Fig. 20.1 Gold–silicon phase diagram.

LIQUID PHASE DIFFUSION BONDING

Liquid phase diffusion bonding is used especially for joining silicon chips to gold plated KOVAR. The two are heated in contact to a temperature of 370°C, at which temperature they form a eutectic (Figure 20.1)[6].

MATERIALS

There are three classes of material which can be encountered in the welding of electroplated gold. These are as follows:

(i) gold of such thickness that the underlying material on to which it is plated plays no part in the joining process.

(ii) gold thinner than (i), so that the underlying material does affect joining.

(iii) other metals which are to be joined to electroplated gold or to which electroplated gold is to be joined, e.g., aluminium, copper, nickel, silicon and also gold itself.

Clearly, there is no sharp distinction between the first two classifications. It depends not only on the process used, but on actual joining parameters for a given process. However, in general terms, where melting is not involved, as in thermocompression or ultrasonic welding, materials will more often fall into Class (i) than with processes where melting occurs.

An important use of electroplated gold is on wire for interconnection purposes. The wire itself is frequently an iron-nickel-cobalt low expansion alloy (KOVAR) where glass to metal seals are required. However, such components as the headers of transistors (on which the active devices are mounted and which usually form one of the contacts) are also gold plated. Depending upon the application and joining process, these materials may fall into Class (i) or Class (ii) as previously defined. A less common example is in the direct joining of an active device to a printed circuit, without the use of interconnecting wires. In such cases, the active device, or chip, has pillars formed on the face at appropriate positions and is bonded face down on to the printed circuit. (Because of this inversion, the process is frequently called flip-chip bonding). Alternatively, the pillars may be formed on the printed circuit. One way in which the pillars can be produced is by electrodeposition of gold, in which case the chips can be ultrasonically welded to the circuit. This is an example of a Class (i) material.

Of Class (iii) materials, aluminium is usually encountered as wire for interconnection purposes such as between an active device and a gold plated lead-out, although gold wire itself may be used. As described in the previous paragraph, silicon chips are bonded to gold plating while nickel and copper are frequently used for printed circuit tracks and as such, joints between these and gold plated connectors are required.

METALLURGICAL ASPECTS

When considering the joining of components, it is essential to take account of the metallurgy involved, both with respect to the time when the joint is made and to what may happen subsequently. The simplest case is that of identical metals being joined, which in the present context means gold to gold, a typical example being a gold wire used to connect a chip to a gold plated lead-out. Here, there are no metallurgical problems.

The next simplest case is where two metals have complete solid solubility, as with silver and gold. Again there are no metallurgical problems. Another simple system is one that shows little or no solid solubility, but does show a eutectic with a melting point lower than either of the pure metals. This happens in the gold-silicon system where the eutectic, which contains 6% silicon, has a melting point of 370°C as compared with 1063°C for gold and 1412°C for silicon (Figure 20.1). This lowering of the melting point is utilised as previously explained, for the joining of silicon to gold. The presence of a eutectic is not in itself an undesirable feature, even when brittle, if the layer is thin. However, it is frequently associated with the presence of intermetallic compounds, these having an atomic structure of their own, usually more complicated than that of the parent metals and differing in mechanical properties. Such compounds exist in the gold-aluminium and the gold-copper systems and undoubtedly the best known of these is the rather dramatically named "purple plague"[7]. This compound, which has the formula $AuAl_2$, is unusual in that it is highly coloured (purple, as the name suggests) and its presence can often be detected upon visual examination. It was noted that the presence of this phase frequently coincided with a marked brittleness in the aluminium-gold joint and hence the sobriquet "plague" was given to it. In fact, more recent work suggests that the brittleness is due to another compound Au_2Al, which is white and therefore not apparent on visual inspection, but only on metallographic sectioning. The formation of this phase is apparently aided by the presence of silicon[8].

Compounds such as these may be formed in two ways; either during making the joint, if this involves elevated temperatures (e.g., thermo-compression bonding) or after the joint has been made, if the temperature for subsequent manufacturing techniques and testing or during service is high enough for diffusion to occur. The latter is unlikely to occur with gold-copper joints, but is a very real risk with gold-aluminium.

Thus, in joining different metals, it is essential that not only the electrical, but the metallurgical characteristics of the system be considered. At the very least, the phase diagrams for the metal combinations concerned should be examined.

Another metallurgical aspect to which some attention should be paid is that of corrosion. Dissimilar metal combinations are always suspect in terms of their corrosion behaviour and with gold present (one of the most noble metals known) the risk exists, even in the absence of an external EMF. The risk is reduced to the extent that, in normal use, films of water should not form on electronic components. (Work at The Welding Institute showed that no corrosion occurred at 95% RH). However, if by accident or bad usage they do form, the gold-KOVAR system will be affected, the KOVAR containing sufficient iron to rust where it is exposed.

QUALITY OF JOINTS

It is perhaps pertinent at this stage that some examples of joints between gold and other metals, in various configurations and made by different processes, should be considered.

As already discussed, the gold-gold joint is free from metallurgical complications and the most likely cause of failure is mechanical damage. In making a pressure or an ultrasonic weld between a wire and a gold plated header, considerable indentation of the wire occurs: this is necessary to make the joint, but can be carried too far. Savage indentation of the type shown in Figure 20.2 is likely to lead to failure at the edge of the joint.

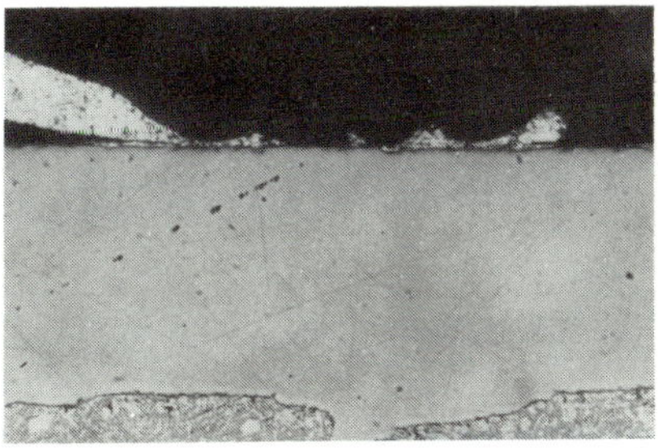

Fig. 20.2 Pressure or ultrasonic weld between wire and pad on silicon chip, showing excessive indentation. (250×).

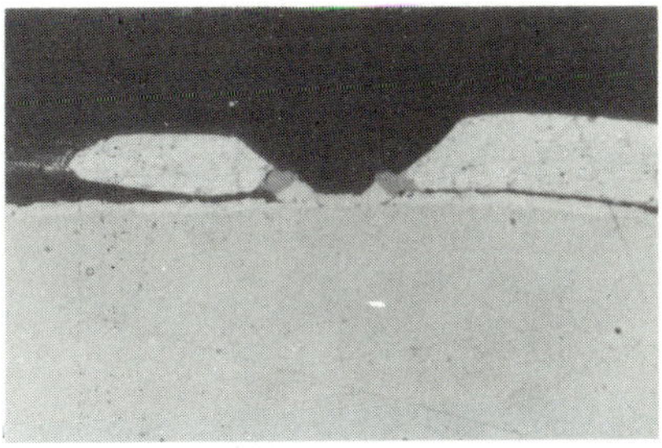

Fig. 20.3 Pressure or ultrasonic weld between aluminium wire and gold plated KOVAR post, showing diffusion between wire and gold plating. (350×).

When aluminium wire is used instead of gold, the metallurgical aspects must also be considered. Diffusion between the two will not occur during cold pressure welding, but it can in thermocompression bonding. Moreover, it is (or at any rate was) the practice on certain types of transistor to hold them at 300°C for up to one week for purposes of quality control. At this temperature, appreciable diffusion occurs. Figure 20.3 shows a section through a joint between an aluminium wire and a gold plated KOVAR post. This joint requires more deformation and therefore more indentation than gold to gold, hence tends to be mechanically weaker. The joint is further weakened by the diffusion that has occurred. In the figure, there are two distinct diffused areas visible, dark and light. The two dark areas adjacent to the aluminium wire (a red-brown colour under the microscope) were found by electron probe X-ray microanalysis to have a composition corresponding to $AuAl_2$, "purple plague". Cracking has occurred between the aluminium and the $AuAl_2$ on one side. Between the two dark areas, there is a paler region which completely penetrates the gold plate and is in contact with the KOVAR. Electron probe microanalysis showed this to contain not only gold and aluminium, but iron, nickel and cobalt. Thus, in this case, diffusion has also occurred from the KOVAR into the joint, a class (ii) situation.

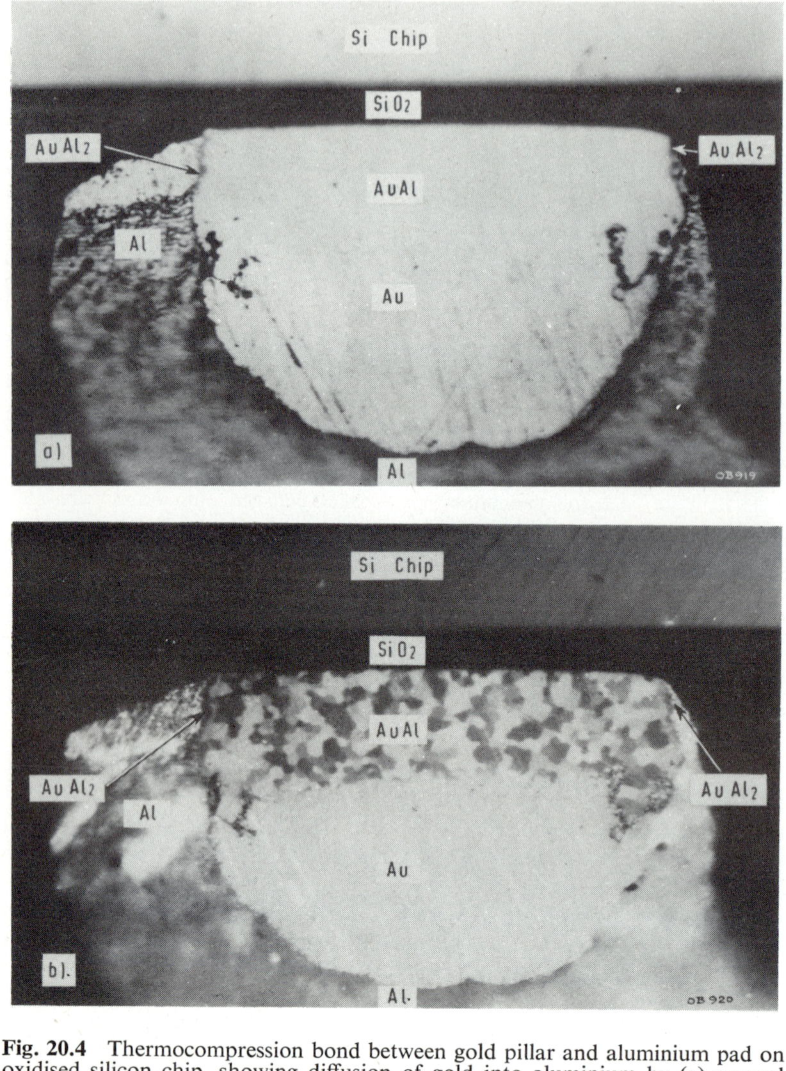

Fig. 20.4 Thermocompression bond between gold pillar and aluminium pad on oxidised silicon chip, showing diffusion of gold into aluminium by (a) normal light (b) polarised light. Taper section at 10°. Magnification-horizontal (660×); vertical (3,300×).

Diffusion reactions also occur in flip-chip bonding. Figure 20.4 shows a taper section (i.e., sectioned obliquely to increase the magnification in one direction) through an electroplated gold pillar bonded at 350°C to an aluminium pad on a silicon chip. Despite the time being short, gold has diffused through the depth of the aluminium pad. (It is worth noting the benefit obtained by examination under polarised light by comparing Figure 20.4(a) with 20.4(b).) Because of this diffusion, the joints were brittle.

Figure 20.5 shows a joint between a silicon chip and a gold plated KOVAR header. In this case, heating was prolonged so that a very thick band of gold-silicon eutectic has formed and there is no evidence of any gold plate remaining. This joint is defective, a crack having formed between the silicon chip and the eutectic.

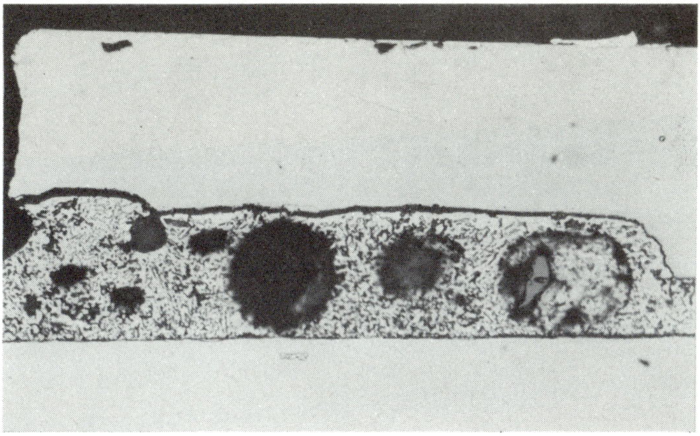

Fig. 20.5 Joint between silicon chip and gold plated KOVAR header, showing excessive alloying, porosity and cracking between chip and braze metal. (120×).

During the parallel gap resistance welding of lead-out to the tracks on printed circuit boards, a variety of joint structures can be obtained depending on the materials and the actual welding conditions used (heat input, electrode force, time). The type of joint will also be affected by the spacing between the electrodes. If they are set too closely together, the current path is mainly through the lead-out which can be melted without producing a joint. Conversely, if the electrodes are set too far apart, the maximum current flows through the circuit track (Figure 20.6 a) and melts the latter (Figure 20.6 b). The correct electrode spacing gives a balanced current flow through both the lead-out and the circuit track (Figure 20.6 c) which gives the most desirable type of joint (Figure 20.6 d), a brazed joint.

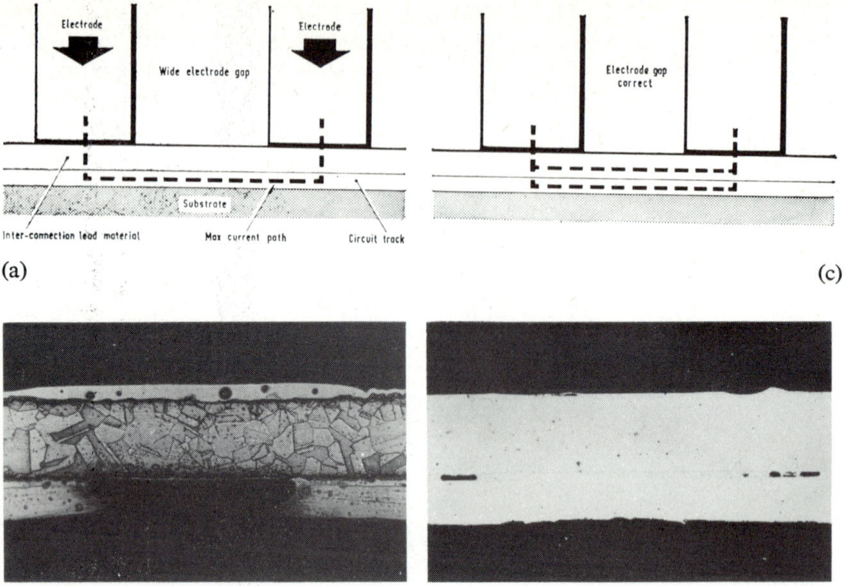

(a) (c)

(b) (d)

Fig. 20.6 (i) Parallel gap welding with electrode gap too wide. (a) main current path (b) consequent melting of nickel circuit track. (100×).
(ii) Parallel gap welding with correct electrode gap. (c) main current path (d) brazed joint produced between gold plating on KOVAR and nickel circuit track. (85×).

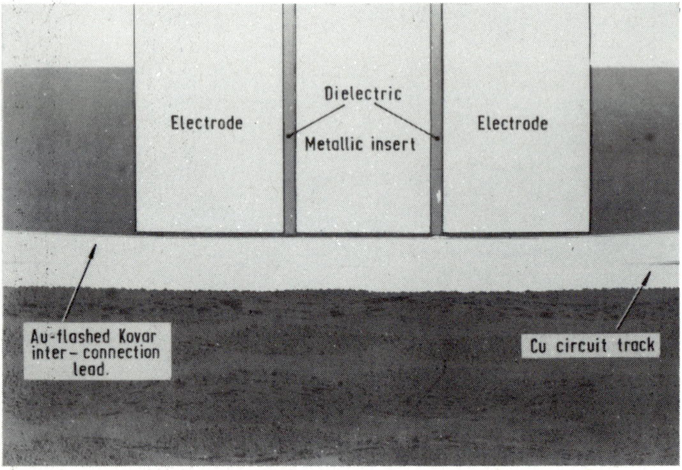

Fig. 20.7 Schematic representation of Welding Institute parallel gap welding electrode.

It is difficult to arrive at the welding conditions and electrode gap when the lead-out and circuit track have the same order of resistance such as KOVAR and nickel. When the circuit track is copper, having very high conductivity compared to gold plated KOVAR, it is almost impossible to arrive at satisfactory conditions. To overcome this difficulty, a special electrode has been developed at The Welding Institute, with a metallic insert in the gap and insulated from the electrodes (Figure 20.7). This insert acts as a heat sink and allows joints to be made between gold plated KOVAR and copper circuit tracks.

INSPECTION

It is appropriate at this stage to introduce the question of inspection of joints. Since normal inspection techniques used for welds (dye penetrant, radiography and ultrasonics) are, with very minor exceptions, unsuited to microwelds, inspection is therefore confined to visual and low power microscopic examination and electrical checks. While these will pick out unwelded joints they are not usually able to discriminate between other joints. However, in the case of gold plated leads parallel gap welded to copper printed circuit board, a convenient visual method of assessment is possible. If the heat input is low, a form of thermocompression weld is obtained. This is clearly visible in a transverse section (Figure 20.8 a) which shows that no melting of the gold has occurred, but this can also be detected from above the joint by the absence of fillets either side of the joint (Figure 20.8 b). With a higher heat input, melting of the gold occurs giving a brazed joint, which is the most desirable type, and the formation of small fillets on either side of the joint (Figure 20.8 c & 20.8 d). With a further increase of energy, the fillets are larger and there is some risk of reducing the strength of the bond between the copper track and the circuit board (Figures 20.8 e & 20.8 f) as the result of degradation of the polymer. A further increase of heat leads to melting of the copper as well as the gold, producing large fillets and certain loss of track to circuit board bond strength (Figures 20.8 g & 20.8 h).

The relationship between fillet length and heat input is typified by Figure 20.9 a, which also shows the relationship between peel strength and fillet length (Figure 20.9 b). Thus, by measuring the latter an indirect measure of peel strength can be made for quality control purposes.

Notwithstanding the difficulties encountered in making joints in and between electronic components, it must be emphasised that the presence of gold makes possible the production of joints which are difficult, if not impossible, to make in its absence.

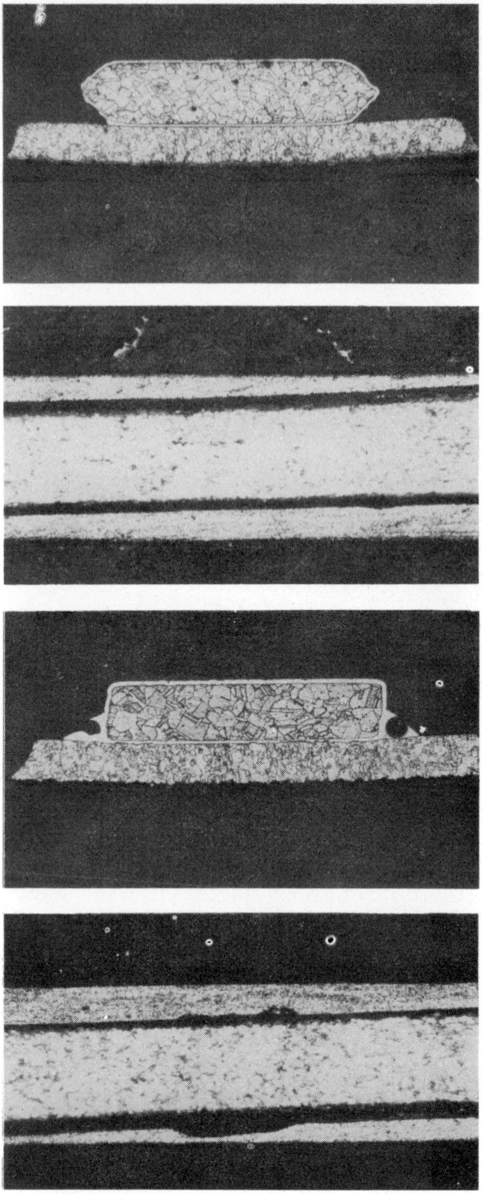

(a) (b) (c) (d)

Fig. 20.8

(a) Parallel gap weld with too little heat, showing no melting. Transverse section. (80 ×).

(b) As (a), no fillets visible. (30 ×).

(c) Parallel gap weld with correct heat, showing melting of gold plating and slight alloying with copper circuit track. Transverse section. (80 ×).

(d) As (c), small fillets visible. (30 ×).

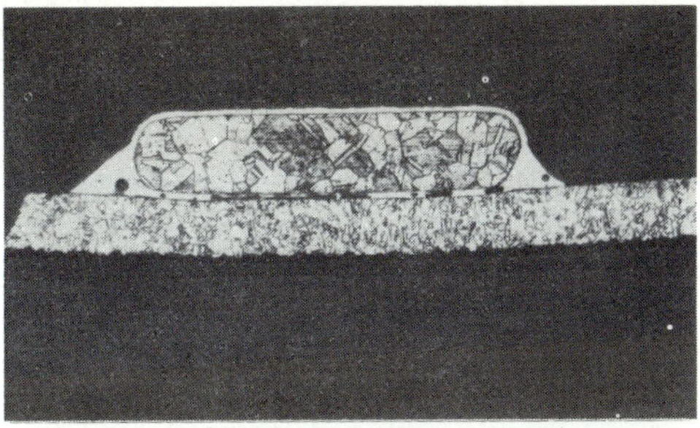
(e)

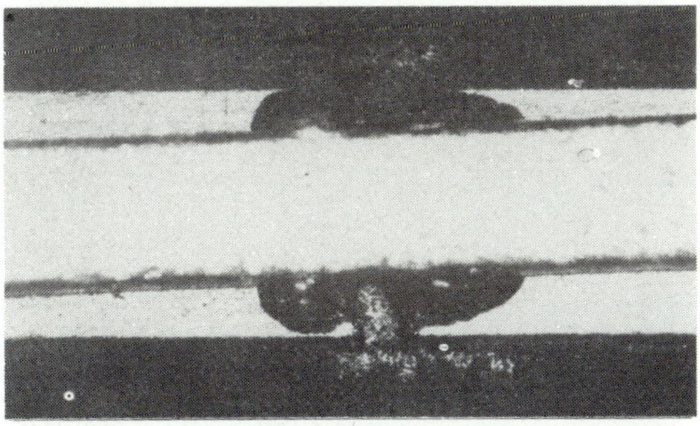
(f)

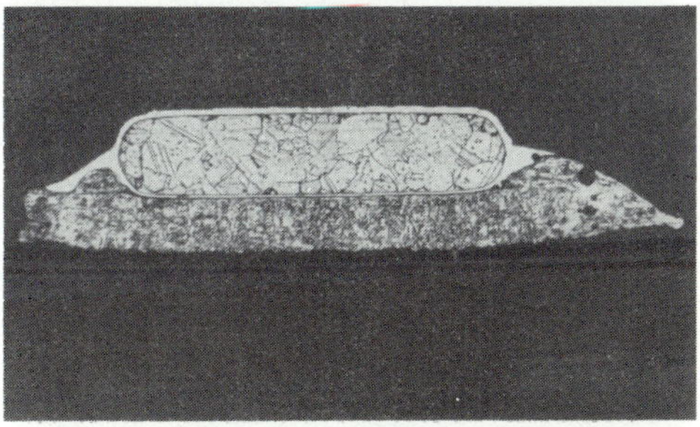
(g)

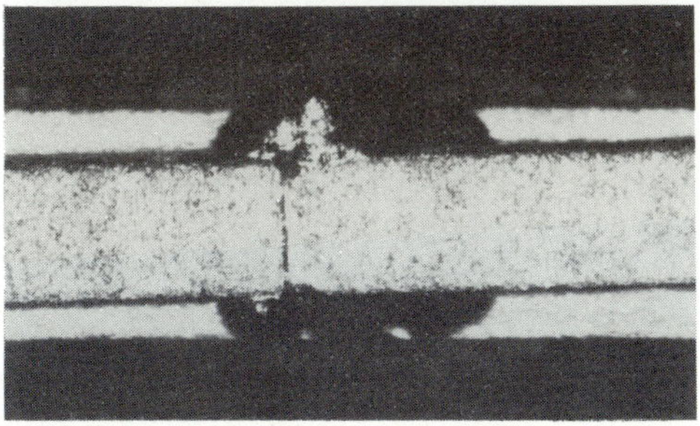
(h)

Fig. 20.8 (continued)
(e) Parallel gap weld with rather too high heat, showing melting of gold plating and alloying with copper circuit track, and some melting of latter. Transverse section. (80 ×).
(f) As (e), large fillets and track melting visible. (30 ×).
(g) Parallel gap welding with far too high heat, showing alloying of gold and copper and excessive melting of latter. Transverse section. (80 ×).
(h) As (g), showing large fillets. (30 ×).

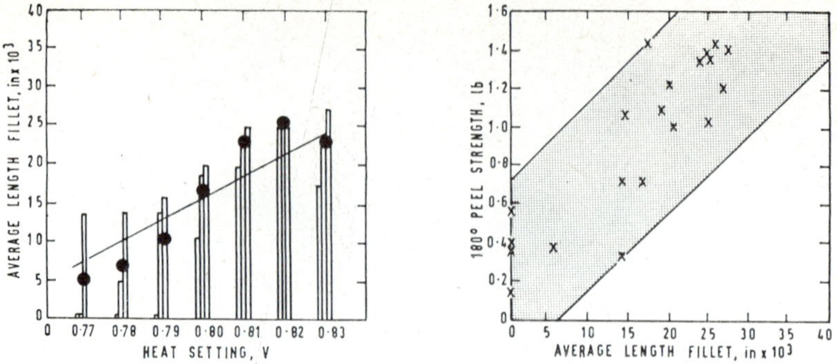

(a) (b)

Fig. 20.9 Parallel gap welds between gold plated KOVAR and copper circuit track. (a) relationship between heat and fillet length. (b) relationship between fillet length and peel strength.

REFERENCES

1 Philips L. S., *Microelectronics & Reliability*, **5**, 3, 197 (1966).
2 Baker D., Jones R., *Microelectronics & Reliability*, **5**, 3, 229 (1966).
3 Sawyer H. F., "Handbook of Electronic Packaging", C. A. Harper (Ed.), Chap. 4, McGraw-Hill, New York (1969).
4 Knowlson P. M., *Br. Weld. J.*, **14**, 7, 398 (1967).
5 Knowlson P. M., *Microelectronics & Reliability*, **5**, 3, 203 (1966).
6 Hansen M., "Constitution of Binary Alloys", McGraw-Hill, New York (1968).
7 Bernstein L., *Semicond. Prod.*, **4**, 7, 29 (1961); **4**, 8, 35 (1961).
8 Warner R. M., Fordemwalt J. N., "Integrated Circuits—Design Principles and Fabrication", McGraw-Hill, New York (1965).

Chapter 21

WEAR, FRICTION, AND LUBRICATION

M. ANTLER

The sliding properties of gold deposits are critical to the performance of many parts. Foremost in both technological importance and in the value of the gold consumed are current carrying devices, examples of which include electrical connectors, instrument slip rings and switches. Gold is plated on these components, particularly for service in low voltage circuits, because of its chemical stability and excellent resistance to the formation of insulating tarnish and corrosion films. Modern computer and communication systems involve large numbers of contacts. Therefore, the requirement of high reliability, consistent with reasonable cost, has stimulated extensive basic and applied studies of the wear and friction of gold for components containing contact surfaces[44].

Another important category of gold plated articles which require wear resistant surfaces are those in which the coating is decorative. Included in this category are jewellery, watchcases and tablewear. Unlike contacts, however, there is little published work which describes the durability requirements of the deposit for these uses or of current commercial plating practices.

Finally, in aerospace bearing applications, where conventional lubricants fail, thin gold coatings, deposited by either physical or chemical means, are used as solid lubricants[24]. Gold shears easily—an essential lubricant requirement.

WEAR

WEAR VERSUS TRANSFER

Wear is the loss of material as loose particles from solid surfaces due to mechanical action. Transfer is movement of material from one surface to another to which it becomes attached. Both processes may occur simultaneously with one dominating although transfer may be followed by a second stage which results in wear.

TESTING FOR WEAR

Testing for wear often involves the parts themselves in a real or simulated application. For screening, and almost always in basic studies, the test apparatus is designed to facilitate measurement of wear and employs contact members with simple shapes. Such apparatus also usually permits the load and sliding velocity to be varied. Figure 21.1 illustrates the contact elements in some simple test devices.

Abrasive wear is often determined with equipment in which sandpaper is rubbed over the gold plate at a known load. The plated article can also be rotated in a slurry of loose abrasive[12,32].

Wear and transfer are commonly determined by change in weight of the object, by alteration of dimensions such as length, and by changes in surface roughness. Profile meters are especially useful for gauging roughness. Of special importance for gold coatings are methods which assess the porosity in the deposit, such as the electrographic technique and by tests involving corrosive gases (see Chapter 27). The development, or a marked increase in porosity of deposits having only a few pores initially, signifies the end of reliable life of the component.

TYPES OF SLIDING DAMAGE EXPERIENCED BY GOLD DEPOSITS

Sliding damage to gold deposits occurs primarily by adhesive, abrasive, or brittle fracture mechanisms.

Adhesive damage occurs when surfaces experience metal transfer. Adhesive bonds form which are stronger than the cohesive strength of the metal and involve small areas of the mating surfaces. This leads to transfer and wear as sliding continues.

Abrasive wear is produced by ploughing of the surface by a member which is rough, hard, and to which the softer wearing material does not adhere. Alternatively, loose hard particles between sliding surfaces can abrade material from one or both members.

Fracture wear occurs with brittle materials on deformable substrates. The surface develops cracks during sliding and is often followed by catastrophic loss of the coating.

WEAR OF GOLD PLATED CONTACTS

Wear is a complex process and depends on many factors involving the materials, the environment and operating conditions, such as contact geometry and load. Thus, comparative wear rates for gold platings cannot be understood without considering their mechanisms of wear.

Stages in Wear

Gold contacts wear mainly by an adhesive process, particularly when unlubricated. The details of the adhesive wear mechanism, called "prow

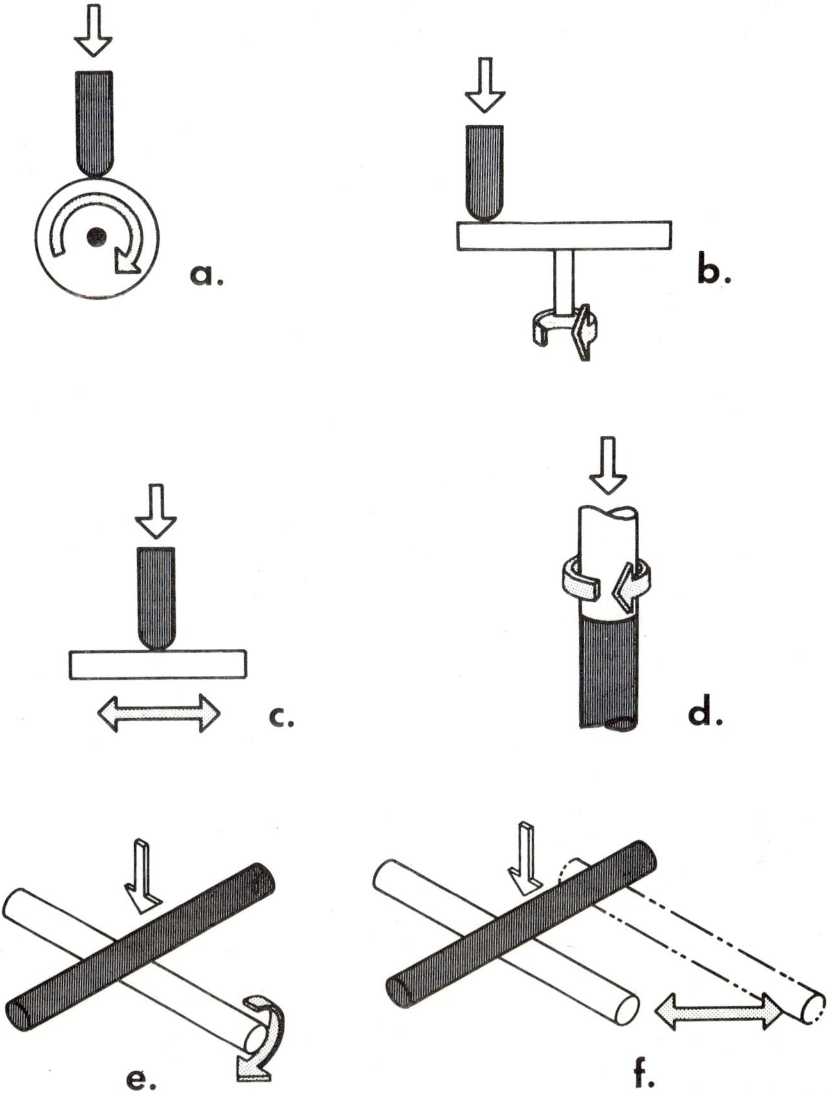

Fig. 21.1 Contact members in some simple test devices. (a) rider on ring (b) rider on flat (undirectional) (c) rider on flat (reciprocation) (d) flat on flat (e) crossed cylinders (f) crossed cylinders (lower member moves back and forth at 45° angle). Open arrow indicates load. Solid arrow indicates direction of motion. Solid member is stationary.

formation", have been determined for solid gold and other wrought metals[16], and for thick[31] and thin[41] gold electrodeposits.

In most cases, the contact members are not identical in shape and size. Furthermore, action is generally localised to a small area on one surface and spread out on the other contact. In such situations, the unlubricated sliding of gold, silver, palladium, platinum, copper and many other metals is characterised by unsymmetrical transfer, virtually all of the metal moving initially from one member to the other and not in both directions. In a second step, the transferred metal is removed by back-transfer to the original part or is lost as loose debris.

These processes are readily observed with test equipment in which the sliding members have the idealised geometry of a hemispherically ended rider contact that is pressed against a flat contact attached to a turntable. At the onset of rubbing a lump of metal, called a prow, which comes from the flat, forms between and separates the specimens, as shown in Figure 21.2 and schematically in Figure 21.3. Sliding now continues at the junction between them. The prow projects against the direction of sliding and gouges the opposing member. Routed solid becomes attached to the prow so that it grows in length.

It is not certain how the prow originates or why it is always located on the rider. A plausible explanation is that this is related to the difference in size of rubbing areas of the members. The population density of transfer particles (formed by breaking at other than the original interface of asperities of the specimens that have cold welded together) is greater on the smaller part. Since the particles are clean and the compressive forces are large, the particles readily weld to each other and to the rider to form a prow. The prow is harder than either original surface because of the extreme degree to which it is cold worked in its transfer and growth.

Prows form on the smaller part in any arrangement of sliding members. For example, with slip ring contacts, prows form on the brush; it has the smaller wearing area and corresponds to the rider in rider-on-flat systems. Gold plated slip rings have been found to wear by prow formation at loads of a few grams[23]. This mechanism occurs regardless of speed or type of motion and is unrelated to the passage of current through the contact.

Table 21.1 lists, for various electrodeposited golds as well as for the wrought metal, the hardness of wear debris and of the flat from which it originates. A fivefold increase in hardness of the debris is not unusual. One of the most striking features of prow formation is that it results in wear of the large member even though the small surface may be considerably softer. The data in Table 21.1 were obtained in runs in which the solid gold riders had a hardness of 60 kg/mm^2.

When the rider traverses the same track repetitively, prow formation eventually ceases and is replaced by rider wear, a mechanism in which the member with the smaller surface involved in sliding wears (Figure 21.3)[16]. This is due to the accumulation of sufficient back-transfer prows on the flat

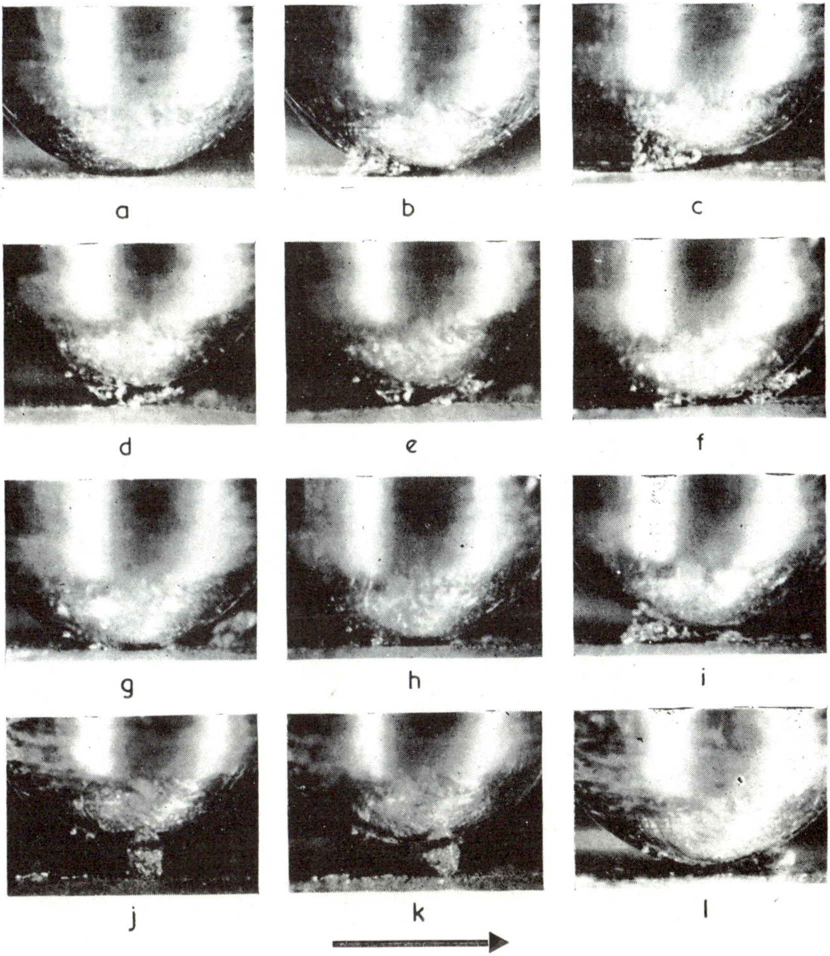

Fig. 21.2 Prow formation mechanism. Gold rider (0.125 in. diameter) on gold flat, 500 g. (a) start of test (b) well developed prow (c–h) loss of portion of prow by back transfer to flat (i–l) new prow formed and breaking to give roller-shaped wear debris. Note continuously changing separation of rider and flat during prow growth and loss stages. Prow metal originates in flat. Arrow indicates direction of movement of flat.

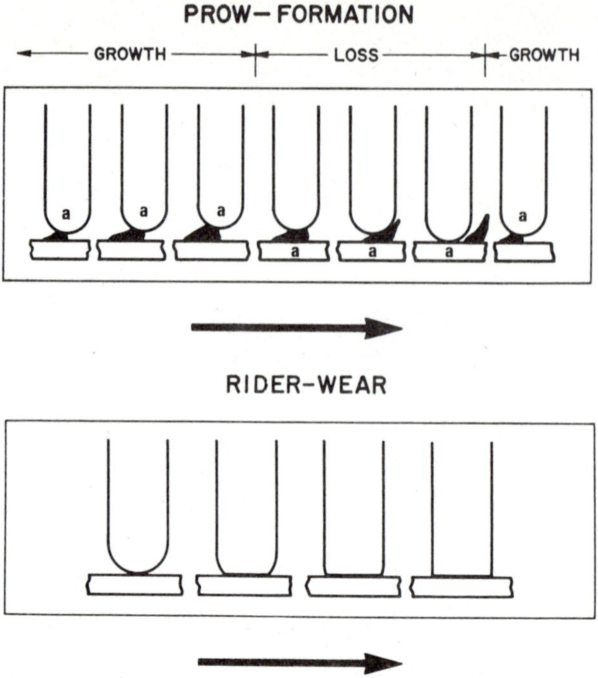

Fig. 21.3 *Top*: schematic representation of prow formation mechanism of wear. One of the processes by which rider loses prow and welds to flat is shown. Letter "a" designates surface to which prow adheres. *Bottom*: schematic representation of rider wear mechanism.

Table 21.1
Hardness of Wear Debris and of Unworn Flat, KHN_{25} (kg/mm^2)

Metal		*Flat*	*Debris*
Wrought gold,	<0.01% impurities	30–86	106–165
Plated gold[a],	<0.01% impurities	45–55	106–154
	1% Cadmium-gold	110–135	183–290
	1% Silver-gold	120–140	197–222

[a] Deposit thickness approximated 0.01 in., copper substrate. Hardness of plating determined in the outermost 0.003 in. thick region of microsectioned samples.

which increase its hardness in all places to the level attainable by extreme work hardening. When the surface of the flat reaches the hardness of the prow on the rider, routing of the flat ceases and a wear land appears on the rider.

Additional studies[31] indicate that there is a load-dependent roughness of the flat above which prow formation does not occur, pure gold being 25–50 microinches centre line average (CLA) at 100 grams[31]. Above this roughness, sliding is by rider wear.

Another investigation[38] has shown that the number of passes to the transition from prow formation to rider wear is related to both track length (in undirectional or reciprocating sliding) and to load. For pure gold this number equals

$$17{,}000 \; \frac{\text{track length (cm)}}{\text{load (g)}}$$

Past the transition, the flat gains mass and the rider loses metal by transfer to the flat or as loose debris. The combined wear of both members for a given total sliding distance lessens with decreasing track length, both before and after the transition.

Figure 21.4 illustrates some of these phenomena for pure gold. It also shows the common occurrence of a prow stuck to the rider past the transition. Eventually, however, the stuck prow wears away and the rider begins to show rapid weight loss as sliding continues.

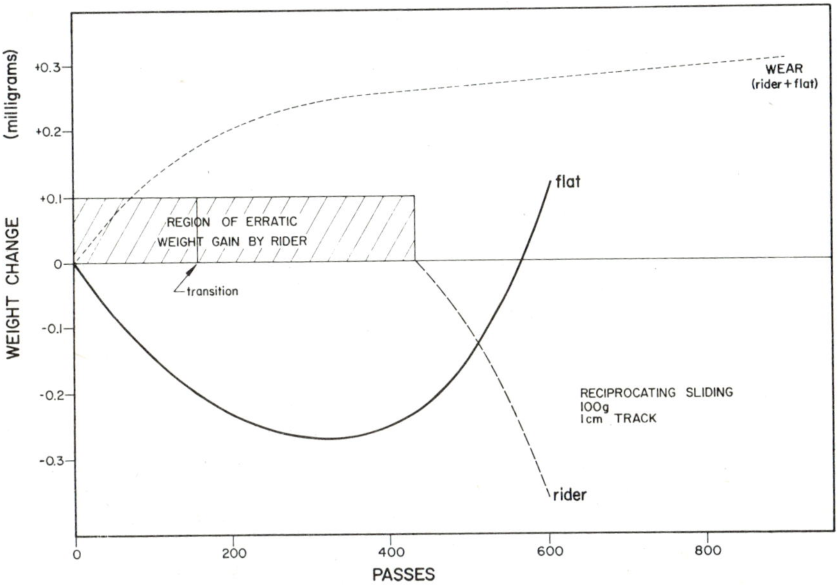

Fig. 21.4 Weight change vs numbers of passes for pure wrought gold members. The wear of rider and flat individually, and the combined wear of both members are plotted.

In general, the wear rate of a sliding system changes as sliding progresses. Characteristically, for unlubricated metals, a high initial (break-in) rate is replaced eventually by a lower equilibrium level. The individual wear rates of the members may be different. The prow formation—rider wear mechanisms are a striking example of this fact.

Separable connectors slide only short distances in their lifetimes, typically 10–500 insertions and withdrawals for a pass length of 2–20 mm., depending on design. Their contacts generally do not operate beyond the break-in process. Prow formation is undesirable and when it occurs, the connector will be unsatisfactory for most applications. Instrument slip rings slide long distances and equilibrium wear rates are more important, provided catastrophic damage by prow formation does not occur during the initial stages of sliding.

To illustrate these mechanisms, two investigations concerning the wear of gold plate will be described. A rider-on-flat apparatus was used. Table 21.2 lists the weight changes of flats measured in an experiment with several electrodeposited golds, using the pure wrought metal for comparison[31]. The mating riders were solid gold. Sliding was unidirectional. To eliminate the effects of the substrate, relatively thick deposits of gold, about 0.01 inch, were used. The surfaces were also normalised by abrasion to 10 microinches CLA, which was less than critical roughness.

Table 21.2

Wear of Solid and Electroplated Gold[a]

Alloy Metal Content of Flat (nominal)	*Hardness of Flat*[b] *KHN_{25} (kg/mm^2)*	*Weight Changes (mg)*		*Wear Mechanism*	
		Rider	*Flat*	*Prow Formation*	*Rider Wear*
Wrought gold, <0.01% impurities	30–85	−0.2 to +1.5	−5 to −10	▲[c] ⟶	△[d]
Plated gold, <0.01% impurities	45–55	−0.2	−11.5	▲ ⟶	△
1% Cadmium–gold	110–135	+1.1	−5.2	▲ ⟶	△
1% Silver–gold	120–140	+0.1	−32.4	▲	
Various hard gold plates (nickel-, cobalt-, indium-hardened)	190–234	0	−0.2	△ ⟶	▲

[a] About 10 μin. CLA finish; wrought gold riders: 100 g; 500 rev. on 1 in. circle.
[b] Hardness values for wrought gold from many samples having different degrees of cold work; for plated samples, hardnesses are of top 0.003 in. of metal from microsectioned specimens.
[c] ▲ Primary mechanism which predominated during run.
[d] △ Secondary mechanism.
Arrow signifies that sliding converted during the run from one process to the other.

Prow formation occurred in all cases, although it was less significant than rider wear with the harder electroplated golds. Wear of pure (soft) electrodeposited and wrought golds was practically identical. Of greater interest, the weight changes of the deposits were not simply related to their hardness. The deposit which contained 1% silver wore more severely than the sample with 1% cadmium. The nickel-, cobalt-, and cobalt plus indium-hardened golds showed the least weight change.

Figure 21.5 illustrates the durability of about 0.1% nickel-gold deposits in a range of thicknesses on oxygen-free high conductivity (OFHC) copper flats. The mating riders were similarly plated. Sliding was by reciprocation. Lifetimes of the deposits were determined by electrography, the part being considered a failure when porosity appeared in the wear track of the flat. The resistance to wear-through of the plate increases dramatically when its thickness exceeds a value which depends on load. This is because wear rates are significantly less after running-in, whether sliding is by prow formation or rider wear. Running-in requires a certain thickness of plate. Pure (soft) and 5% silver-gold plates are less durable than 0.1% nickel-gold. Figure 21.6 compares the durability of these deposits. The pure and silver-gold coatings wore by the prow formation mechanism during running-in, whereas the nickel-gold slid by rider wear.

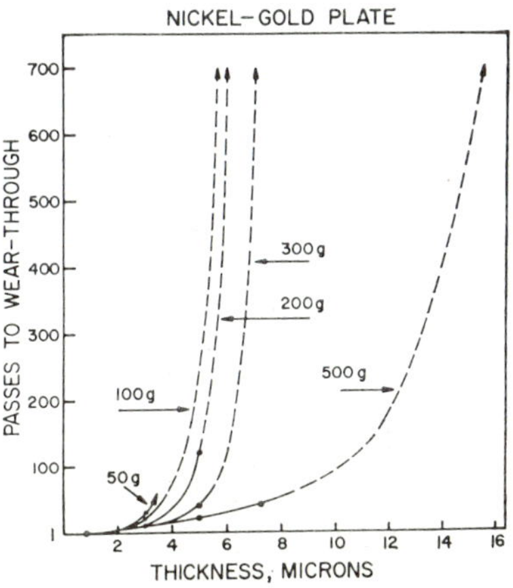

Fig. 21.5 Durability of 0.1% nickel–gold plate in range of thicknesses on OFHC copper. Mated to thick nickel–gold plated riders (0.125 in. diameter). Rider-flat apparatus, reciprocating sliding, 1 cm track length. Sliding velocity, 0.3 cm/sec. Load range, 50–500 g. Arrows at ends of dashed curves indicate runs at 700 passes without wear-through.

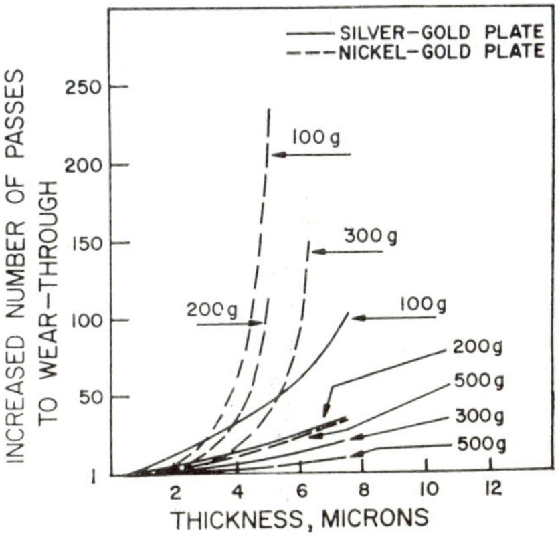

Fig. 21.6 Comparison of silver- and nickel–golds with pure gold plate. The increased number of passes to wear through of the alloy golds is plotted for various deposit thicknesses and loads.

Factors Controlling Wear

Many factors determine the durability of metals. Undoubtedly the most important for gold plate is its surface cleanliness. If adventitious contamination is absent and lubricants have not been used, gold plated surfaces readily cold weld when slid against each other[48], particularly if the deposits are soft and pure.

The closely related properties of composition, hardness, ductility and structure also control wear. Golds alloyed with large amounts of base metals which can produce superficial inorganic films, or gold surfaces contaminated by substrate metal which has diffused to the surface, resist cold welding and resulting transfer and wear.

The contact of mated surfaces occurs at their high spots. The real area of contact between surfaces is, therefore, determined primarily by their hardness and by load. Various metals are often codeposited with gold during plating to increase its hardness. Many authors recommend[1,4,5,6,9,10,11,13,15,17,19,21,36,41] this practice for wear resistance, particularly for electrical contact applications. Commonly used alloying elements include cobalt, nickel, silver, copper, cadmium, antimony and indium plus cobalt. However, it is also claimed[18] that very hard golds (over 200 kg/mm^2) wear poorly compared to those of intermediate hardness (say, 100–150 kg/mm^2). These values are nominal and depend on the application[22]. Ductility and hardness of gold deposits, in general, vary inversely[43]. It is likely that examples of high wear with hard golds were due to it having been deposited on deformable substrates which allowed brittle fracture to occur[29,41].

A combination of high toughness and hardness are considered necessary for wear resistance[36]. Toughness is a parameter which comprises both strength and ductility. Certain electroplated alloy golds having a hardness of 300–450 kg/mm^2, a tensile strength of about 90 kg/mm^2, and an elongation of 5% have been found[36] to be unusually wear resistant.

Thin contaminant films are able to lubricate hard surfaces better than soft ones[7]. Improved wear life observed with harder golds in practical applications may be attributable in part to effective lubrication by adventitious contamination.

An organic polymer is codeposited[37] with gold during electroplating from acid solutions of $Au(CN)_2^-$ operated at low cathode efficiency. It has been hypothesised[34] that friction and wear may be lowered by the presence of this material. However, silver-gold from alkaline solutions of $Au(CN)_2^-$ and excess CN^- also contains codeposited polymer[47], yet it is far less wear resistant than the widely used cobalt and nickel acid golds[44]. A recent study[48] could not find any significant lubricating effect attributable to intrinsic polymer.

The structure of gold deposits is dependent on the composition of the electroplating solution and on operating conditions. Deposit surfaces usually have a preferred crystallographic orientation and it has been suggested[31] that orientation can control sliding behaviour.

A rough, hard deposit can induce wear of a mating contact by abrasion. This is important for thick coatings having a roughness little related to that of the substrate. Prow formation, for example, does not occur above a load-dependent roughness, characteristic of the metal. Nodular deposits induce wear of the mating member[20,31].

The wear of a coating is adversely affected if it is not adherent to its substrate[2].

What is a Good Wearing Gold for Electric Contacts?

Good wearing golds for contacts, in thicknesses to several microns, are obtained from cyanide-based acid plating solutions producing deposits containing from 0.1 to several tenths of a percent of either nickel or cobalt and having a hardness of about 140–200 kg/mm^2 (Figure 21.5). Metallic impurities, especially iron, which in small amounts cause brittleness, should be low. The impurity level of each metal should preferably not exceed 0.1% in the deposit[43]. The substrate should be as hard as possible for the application and consistent with adequate ductility, particularly if the article is to be deformed.

Pure nickel is a preferred[25,33,39,41] underplate for gold. A three layer deposit of 5 microinches (maximum) of soft gold on 20 microinches of rhodium on 300 microinches of nickel has been found[29] to be very wear resistant.

Some high alloy golds are also wear resistant, but there is little information about them. One of these has a nominal composition of 70–80% gold, 15%–20% copper and 5% cadmium[40] and was described earlier[36]. The desirable properties of a gold having this composition have been confirmed[46].

ABRASIVE WEAR OF GOLD PLATE

A study of numerous types of gold coatings, by a method in which the plated specimens were rotated in sand, showed no clear relationship between wear rate and various properties of the deposits. The abrasion resistance of golds having a hardness of about 180 kg/mm^2 from cobalt-gold electrolytes was significantly greater than those having a hardness of approximately 80 kg/mm^2 from cobalt-free solutions. These 180 kg/mm^2 golds did not contain cobalt, their hardness being attributed[5,6] to a finer grained structure. Improved abrasion resistance has also been reported[15] when gold plating solutions incorporating additions of nickel have been used. However, the hardened deposits contained nickel. Decorative gold deposits containing from 10–40% copper are reported[9,19] to be harder and more abrasion-resistant than pure gold. Indium-cobalt-gold alloy plates are claimed[17] to be particularly abrasion resistant.

Stainless steel tableware is sometimes plated with gold. Current practices involve a deposit of 100–200 microinches of bright nickel, followed by an acid gold plate containing several tenths of a percent of nickel or cobalt[42]. The underplate is desirable for levelling, which improves the appearance of the product, rather than for wear. Nickel- and cobalt-golds are generally more wear resistant than pure gold deposits. A minimum of 7 microinches of the metal is required in order to meet United States government standards for the designation, "Gold Electroplated".* In commercial practice, an average of 15 microinches of gold is deposited to meet the minimum thickness. These products have reasonable durability (they last for 90–120 servings). Thicknesses of up to 100 microinches are used on better quality lines.

Systematic studies of the abrasion resistance of alloy and pure gold electrodeposits should be made to achieve a better understanding of the basic phenomena of this mode of wear.

FRICTION

Friction is the resistance to relative motion between two solid bodies in contact. In general, the force necessary to induce motion is larger than the force required to maintain it at constant velocity. Friction force is proportional to the normal load for most materials over wide ranges of conditions. The coefficient of friction is friction force divided by load; thus, there is a coefficient of sliding friction (μ_k) and a coefficient of static friction (μ_s). The former (μ_k) depends on velocity, especially at very low speeds, while the latter (μ_s) depends on the time of contact for short times, i.e., less than one second.

Stick-slip is intermittent motion and is experienced with devices which are free to vibrate. The stick originates in the higher value of μ_s and slip in the lower value of μ_k.[3]

Metallic friction originates primarily in the adhesion of surfaces to each other. However, when one member is much harder and deforms the surface to which it is mated, ploughing becomes a major cause of friction, because work is done in producing the groove.

Although the coefficients of friction are not constants, it is, nevertheless, useful to cite[31,41] some values in Table 21.3. The metal-hardened wear resistant golds have lower friction coefficients than the pure metal. They are lower still for contacts which have been exposed to the atmosphere or handled because their surfaces are no longer clean.

LUBRICATION

Transfer, wear, and friction originate largely in the interaction of surfaces with one another. Lubrication minimises this interaction by interposing a material between them having low shear properties. Gold plated electrical contacts are often purposely lubricated, because organic contamination from the atmosphere is variable and frequently ineffective as a lubricant.

---

* Specifications are dealt with in the Appendix by Mills.

Table 21.3

Typical Values of Coefficient of Sliding Friction for Gold Electrodeposits (in air, unlubricated)

Alloy Content	*Against Solid Gold*[a]	*Against Electrodeposited Gold*[b]
< 0.01% impurities	2.0–2.2	1.1–2.1
1% Silver	0.5–0.7	1.1–1.9
0.1% Cobalt	0.4–0.7	
0.15% Nickel	0.3–0.5	0.7–1.3
1% Cadmium	1.3–1.8	
2.5% Cobalt + 1% Indium	0.3–0.5	

[a] Riders: solid gold, 99.99% pure. Flats: electrodeposited golds, several thousandths of an inch thick, on copper substrates (ref. 31).

[b] Riders: electrodeposited golds, approximately 0.0008 in. thick, on copper. Flats: same gold deposits as riders, several hundred millionths of an inch thick, on copper substrates (ref. 41).

Lubricants may be classified according to their physical state, i.e., solids, liquids, and greases. Those that work well with solid gold are also effective on gold platings. Corrosion inhibitors can be incorporated in lubricants to protect the surfaces should the gold plate wear through in service. They are also used to minimise film formation in agressive atmospheres which originates in as-plated porosity in the deposit[33]. N-oleoyl sarcosine and alkoyl sarcosinate salts are corrosion inhibitors for gold plated nickel or copper alloys[35]. It has been suggested[30] that solid lubricants, such as niobium diselenide, could be codeposited with gold to make a totally inorganic self-lubricating contact material.

Contact lubricants in bulk are insulators. However, as thin films, a properly chosen lubricant does not cause a significant increase in contact resistance. The reason for this behaviour originates in its liquid or plastic character. A few asperities always make metallic contact through the film and only a few asperity contacts are required for good conduction. The contact resistance of a small number of widely scattered spots is little more than the resistance of a single large contact area bounded by their envelope[27]. On the other hand, wear is closely related to the actual area of metallic contact. A reduction in the area produces a corresponding reduction in wear rate.

The problem in lubrication is durability. Relatively few materials are able to continue to lubricate on repeated engagements of a connector, particularly over an extended period. Topography plays a key role in lubrication. Thin films of solid lubricants, for example, are effective only on rough contacts. Surface scratches act as reservoirs for the lubricant[28]. Although lubricated gold plated contacts become burnished on repeated cycling, the solid lubricant which is pushed aside is continuously replenished. When the reservoirs disappear, lubrication ceases. Provided the contact has a minimum roughness of about 10 microinches CLA, films less than 40 microinches thick of octadecylamine hydrochloride, a wax-like solid lubricant, will lubricate gold plated contacts for hundred of insertions and withdrawals.

When the substrate is rough, the surface of a thin coating is also rough. However, porosity in the deposit is directly related to substrate roughness[26] (see Chapter 23). It is desirable, therefore, to lubricate contacts with fluid materials which are less dependent on topography than solid lubricants.

A comprehensive study[14] of fluid lubricants for solid gold has established the principles by which such materials function. Experiments were made with a rider-on-flat apparatus in unidirectional sliding, and contact resistance and friction were continuously recorded. Various fluids were applied to the flat, either as 1% solutions in a volatile carrier which on evaporating gave thin films, or by flooding. The roughness of the flats was 10 microinches CLA.

Fluids differed markedly in their ability to lubricate gold. Figure 21.7 shows worn riders and flats and also track profiles obtained with several of the fluids, arranged in order of poor to good. Poor lubricants allowed prows to form on the riders with accompanying severe roughening of the flats. This is similar to damage obtained in the unlubricated case. Good lubricants gave burnishing of the flats, no visible transfer or wear, and polishing of the rider tip with cold flow of metal in the direction of sliding (opposite to the direction in which prows grow). Much of the experimental data is summarised in Figure 21.8, where a plot of coefficient of friction after run-in versus viscosity of the fluid shows that:

1. The more viscous fluids are generally the better lubricants.
2. Halogen-containing fluids as a group are better than those incorporating silicon. Hydrocarbons, fatty acids, fatty esters and other chemical types demonstrate intermediate behaviour.
3. Very viscous fluids cause electrical noise by inducing hydrodynamic lift of the members.
4. Fluids having low viscosity give noisy operation, this being associated with stick-slip and the formation of wear debris. For a given viscosity the halogen-containing fluids produce less electrical noise than other fluids.
5. Most lubricants have similar effectiveness provided their viscosity is adequate, i.e., 100–1000 centistokes at 25°C. In this study[14], particular conditions gave low wear, low friction and negligible electrical noise with several types of fluids in a certain viscosity range. At lower loads, with increasing velocity, or in sliding with smoother flats, the viscosity of fluids required for both low friction and noiseless performance would also diminish.
6. A few unusual lubricants, notably those containing organometallic additives such as tetraethyl lead, can cause electrical noise by "non-corrosive film formation"[8,14]. This is a mechanism by which a trace of the lubricant decomposes during sliding, either thermally or catalytically, to generate insulating inorganic films. However, 5% of di-n-butyltinsulphide

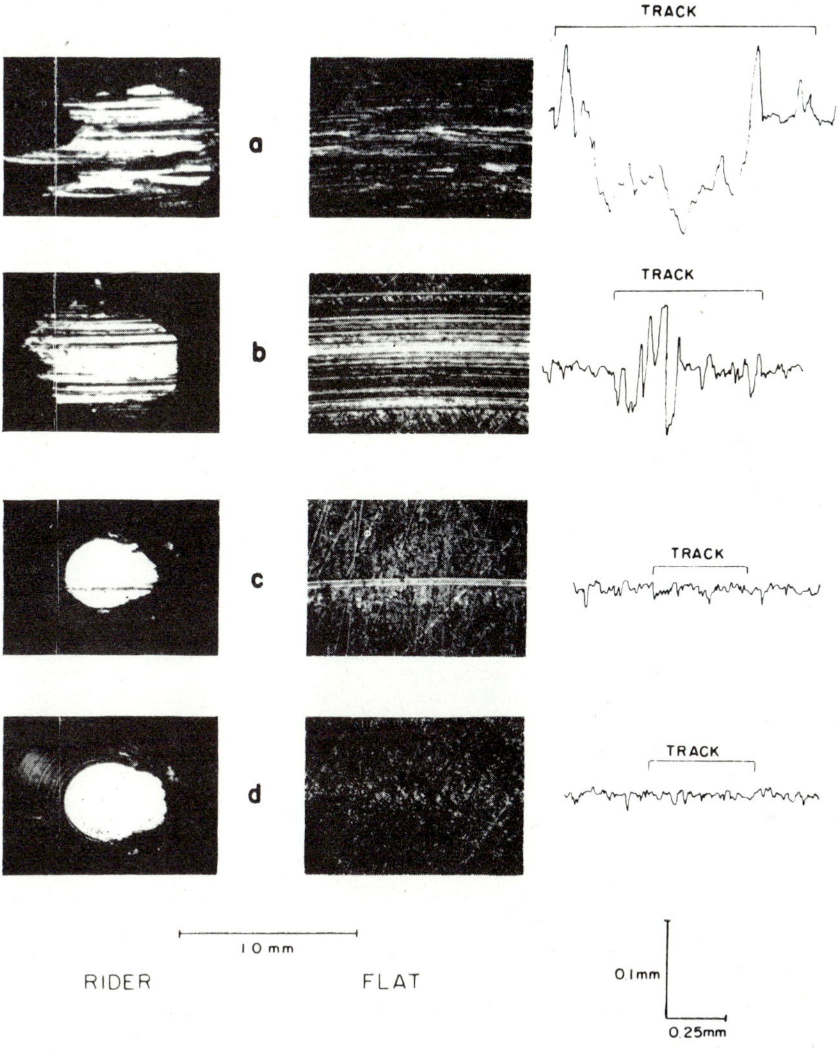

Fig. 21.7 Gold riders and flats from lubricated sliding. Photomicrographs of representative worn riders and flats from 500 revolution runs at 100 g and 1 cm/sec. Rider diameter: 0.125 in. TALYSURF* traverses made across tracks. Arranged in order of poor to good sliding. Note prows in (a) and (b), and polishing of rider and flat in (d). Lubricants: (a) alkyl silicate ester (b) dimethly-polysiloxane, 500 centistoke grade (c) mixture of fluorinated esters (d) polychlorotrifluoroethylene.

in mineral oil, which forms films of stannous oxide and stannous sulphide on gold, did not give electrical noise under these conditions. It is, therefore, a good lubricant for gold. However, organometallic-containing lubricants are difficult to use in contact applications because of their toxicity.

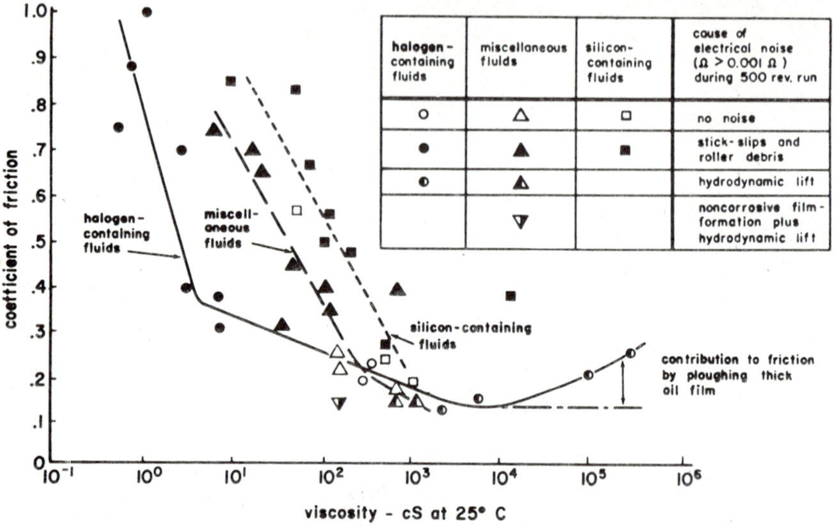

Fig. 21.8 Coefficient of friction vs viscosity for fluid lubricants. Rider-flat machine with wrought gold specimens. Friction determined near end of 500 revolution runs at 100 g and 1 cm/sec. Friction is related to both fluid type and its viscosity. Silicon-containing fluids are poorest. Electrical noise with poor lubricants is due to processes which generate loose debris and rough surfaces. Noise with good lubricants is due to hydrodynamic lift or sliding-generated insulating solids which adhere to surface.

Associated studies of the volatility of thin fluid films of a few of the better lubricants showed that a polyphenyl ether and a fluorinated ester, among others, persist without degradation for over 3000 hours at 65°C in a moving air stream. Fluids such as these are preferred to those having greater volatility. Liquid polyphenyl ethers are used commercially as lubricants for gold plated contacts in separable connectors.

Instrument slip rings having wrought or electroplated gold surfaces are generally lubricated. Polyphenyl ether, however, is too viscous for these lightly loaded devices, and thin films of di-2-ethylhexyl sebacate have been recommended[23].

The mechanisms of lubrication with liquid contact lubricants have been identified[14] as mobility and hydrodynamic lift. Mobility is the ability of the lubricant to self-heal after being displaced in sliding, a quality which depends on the viscosity of the fluid. The ideal lubricant for connectors would be one that could completely separate surfaces in sliding so that there is no metallic contact. At the cessation of sliding, hydrodynamic effects disappear and contact would be re-established.

Probably the best lubricants for gold plated contacts in connectors and other sliding devices are greases made with a liquid from one of the effective chemical types in a viscosity range suitable for the application. The solid portion of the grease should also be an easily sheared material, such as microcrystalline wax. The filler helps retain the lubricant on the surface.

Finally, good lubricants can to a large degree compensate for the inherently poor wearing properties of some electrodeposited golds. Coatings which have marginal sliding characteristics, but other good points such as processing advantages and lower cost, can often be upgraded with lubricants to make them successful in many applications.

REFERENCES

1 Rinker E. C., *Plating*, **40,** 861 (1953).
2 Raub E., *Metalloberfläeche*, **7,** 2, 17 (1953).
3 Bowden F. P., Tabor D., "The Friction and Lubrication of Solids", Vol. I, pp. 105, Clarendon Press, Oxford. (1954).
4 Fedot'ev N. P., Ostroumova N. M., Vyacheslavov P. M., *Zh. prikl. Khim.*, **27,** 1, 43 (1954).
5 Fedot'ev N. P., Vyacheslavov P. M., Ostroumova N. M., Leningradskogo Tekhnologicheskogo Inst. imeni Lensoveta, 33 (1955).
6 Fedot'ev N. P., Ostroumova N. M., Vyacheslavov P. M., *Zh. prikl. Khim.*, **29,** 4, 489 (1956).
7 Feng I-M., Chang C. M., *J. appl. Mech.*, **23,** 3, 458 (1956).
8 Antler M., *Ind. Engng Chem.*, **51,** 6, 753 (1959).
9 Fedot'ev N. P., Kruglova E. G., Vyacheslavov P. M., *Zh. prikl. Khim.*, **32,** 2014 (1959).
10 Parker E. A., *Plating*, **46,** 621 (1959).
11 Vasil'eva G. S., *Med. Prom.*, **13,** 5, 48 (1959).
12 Schumpelt K., *Am. Jewelry Mfr*, 10th Feb. (1960).
13 Bacquias G., *Galvano*, **31,** 310, 757 (1962).
14 Antler M., *Wear*, **6,** 1, 44 (1963).
15 Fedot'ev N. P., Vyacheslavov P. M., Kruglova E. K., Teoriva i Prakt. Blestyashchikh Gal'vanopokrytii, Akad. Nauk Lit. SSR, Inst. Khim. i Khim. Tekhnol., Osnovnye Materialy Vses. Soveshch., Vilnyus 1962, 349 (1963).
16 Antler M., *Wear*, **7,** 2, 181 (1964).
17 Danemark M. A., *Metal Finish. J.*, **10,** 12, 483 (1964).
18 Fischer J., Weimer D. E., "Precious Metal Plating", Draper, Teddington, England (1964).
19 Korovin N. V., *Electroplg Metal Finish.*, **17,** 5, 151; **17,** 7, 249; **17,** 8, 269 (1964).
20 Angus H. C., Trans. Inst. Metal Finish., **43,** 4, 135 (1965).
21 Day B. H. S., *Plating*, **52,** 228 (1965).
22 Lee W. T., *Electroplg Metal Finish.*, **18,** 2, 40 (1965).
23 Morris C. G., Hensley W. D., Reed P. L., Proc. Symp. on Precision Sliding Contact Devices Poly-Scientific, Blacksburg, Virginia (1965).
24 Missel L., *Metal Finish.*, **63,** 3, 50 (1965).
25 Tweed R. E., "Manufacturing Methods for Electroplating Silver, Gold, and Rhodium on Electrical Connector Contacts", Interim Engineering Progress Reports III, V, and VI, under Air Force Contract AF33 (657)-9752, Nu-Line Industries, Inc. (Sept. 1963–Dec. 1963; July 1964–Jan. 1965; and Jan. 1965–Feb. 1965) AD425933; AD457143; AD458539.
26 Garte S. M., *Plating*, **53,** 1335 (1966).
27 Greenwood J. A., *Brit. J. appl. Phys.*, **17,** 12, 1621 (1966).

28 Steinberg G., Proc. 3rd Int. Res. Symp. on Electrical Contact Phenomena, pp. 95, Univ. of Maine, Orono, Maine (1966).
29 Holden C. A., Proc. Engng. Semin. on Electrical Contact Phenomena, pp. 1, Illinois Inst. of Technology, Chicago (1967).
30 Van Auken R. L., Hensley W. D., Cole S. R., Proc. Engng Semin. on Electrical Contact Phenomena, pp. 117, Illinois Inst. of Technology, Chicago (1967).
31 Antler M., *Trans. Am. Soc. Lubric. Engrs*, **11,** 3, 248 (1968).
32 Duva R., Foulke D. G., *Plating*, **55,** 1056 (1968).
33 Krumbein S. J., Antler M., IEEE Trans. on Parts, Materials and Packaging, PMP-4, 1, 3 (1968).
34 Abbott W. H., Bartlett E. S., NASA CR-1447 (1969).
35 Antler M., Krumbein S. J., U.S. Pat., 3,484,209 (1969).
36 Flühmann W., *Plating*, **56,** 1351 (1969).
37 Munier G. B., *Plating*, **56,** 1151 (1969).
38 Antler M., *Trans. Am. Soc. Lubric. Engrs*, **13,** 2, 79 (1970).
39 Antler M., *Plating*, **57,** 615 (1970).
40 Flühmann W., private communication (1970).
41 Solomon A., Antler M., *Plating*, **57,** 812 (1970).
42 Zobbi R., International Silver Co., private communication (1970).
43 Baker R. G., Palumbo T. A., *Plating*, **58,** 791 (1971),
44 Antler M., Proc. Sixth International Research Symposium on Electrical Contact Phenomena, Illinois Inst. of Technology, Chicago (1972).
45 Horn G., Merl W., Proc. Sixth International Research Symposium on Electrical Contact Phenomena, pp. 65, Illinois Inst. of Technology, Chicago (1972).
46 Nobel F. I., Thomson D. W., Leibel J. M., Proc. Fourth Plating in the Electronics Industry Symp. pp., 106, Am. Electroplaters' Soc., Indianapolis (1973).
47 Antler M., *Plating*, **60,** 468 (1973).
48 Antler M., Proc. Engineering Seminar on Electrical Contact Phenomena, Illinois Inst. of Technology Chicago (1973).

Chapter 22

CONTACT PROPERTIES

M. ANTLER

DEFINITION OF TERMS

Contact resistance (R_t) is the electrical resistance between two touching bodies. It consists of constriction resistance (R_c) and film resistance (R_f). In some measurement techniques it is not possible to eliminate bulk resistance (R_b), the resistance of the metal pieces comprising the contact and the resistance of the wires and connections used to introduce the test current into the samples. In these cases, the measurement is actually of an overall resistance, which is often confused with contact resistance (see Chapter 26).

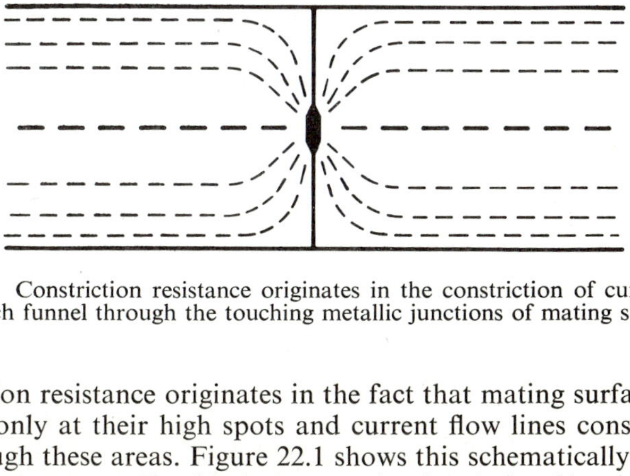

Fig. 22.1 Constriction resistance originates in the constriction of current flow lines which funnel through the touching metallic junctions of mating surfaces.

Constriction resistance originates in the fact that mating surfaces touch in most cases only at their high spots and current flow lines constrict as they funnel through these areas. Figure 22.1 shows this schematically for surfaces that touch at one spot. If oxide layers or other insulating films interfere with metal-to-metal contact, the contact resistance will be higher.

These relationships are simplest for a single circular contact spot with identical metals having a uniform film so that

$$R_t = R_c + R_f$$

$$R_t = \frac{\rho}{2a} + \frac{2\sigma}{a^2}$$

where a = radius of the contact spot

ρ = bulk resistivity of the contact material (ohm-cm.)

σ = film resistance (ohms per square).

Film resistance is important only when little or no metallic contact exists. Gold plated contacts are ordinarily free of films when new. Most processes that degrade gold deposits produce insulating layers which increase film resistance.

Contact rarely involves a single spot. Mated surfaces touch at a multitude of points having a total area that is considerably less than that of the macroscopic (geometric) contact, as illustrated in Figure 22.2. However, it has been shown theoretically that the constriction resistance of a cluster of many small spots is similar to that of a single spot having the same geometric area as that of the cluster. This has practical implications inasmuch that if an insulating film on the surface is broken in many places, measured resistance may be as low as that for a clean surface. The low contact resistance of lubricated gold contacts is due to metallic interaction through breaks in the lubricant film[13,33].

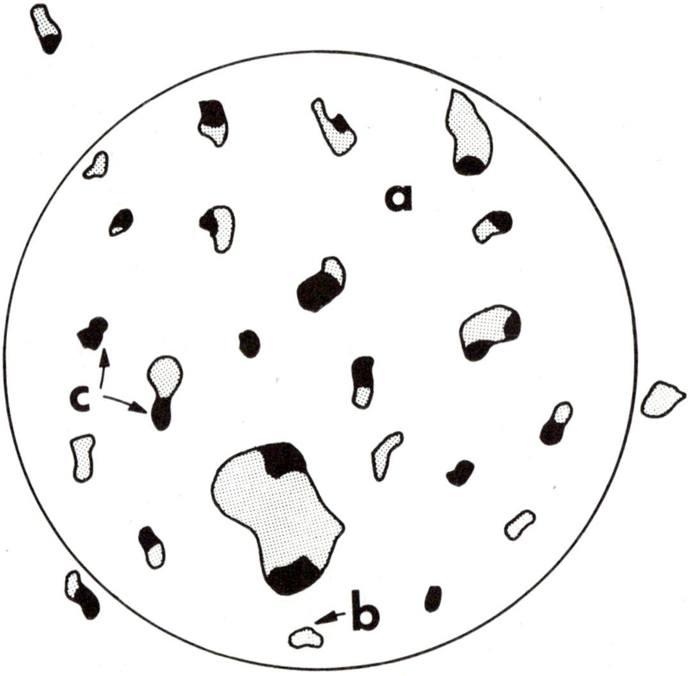

Fig. 22.2 Schematic illustration of a contact surface. The area of metallic contact, "c" (solid regions) in most cases is only a small fraction of the apparent area. Contact at "b" (shaded regions) is with insulating contaminant. Region "a" does not touch. Circle is envelope which encloses most of the metallic contact spots.

The higher the bulk resistivity or the hardness of a material, the higher will be R_c. Conversely, the higher the load, the lower the resistance. There is, however, one additional complication. Gold platings are often so thin that the resistivity and the hardness of the substrate affect R_c. When deposit thickness is less than about 200–300 microinches for connector contacts of conventional design with common substrate metals, this effect on R_c can be significant.

Table 22.1 gives some bulk resistivities of pure and of alloy gold platings. Additional data[31] for deposits from −40°C to 300°C are listed in Tables 22.2 and 22.3.

Table 22.1
Resistivity of Gold Electrodeposits

Gold Deposit	*Microhm-cm*	*Reference*
High purity	2.44 at 20°C	(10)
High purity	2.97 at 100°C	(10)
High Purity	2.195 at 0°C (after heat treatment)	(17)
10–15% copper–gold	150–250 before heating	(15)
10–15% copper–gold	10–20 after heating to 150°C or 200°C	
Cobalt–gold, bright acid	16.75 after heating to 120°C	(22)
Cobalt–gold, bright acid	9.70 after heating to 300°C	(22)
Low cobalt–gold, bright acid	10.01 after heating to 120°C	(22)
Low cobalt–gold, bright acid	6.08 after heating to 300°C	(22)
Low nickel–gold, bright acid	7.01 after heating to 120°C	(22)
Low nickel–gold, bright acid	5.56 after heating to 300°C	(22)
Bright cyanide gold	7.49 after heating to 120°C	(22)
Bright cyanide gold	6.17 after heating to 300°C	(22)
Acid gold, 24 carat	2.34 after heating to 120°C	(22)
Acid gold, 24 carat	2.20 after heating to 300°C	(22)

Table 22.2
Gold Plating Baths Investigated
(See Tables 22.3 and 22.4)

Bath	*Type*	*Deposit Purity (% Gold)*	*Deposit Hardness (kg/mm²)*
A	"Pure" alkaline cyanide matte	99.999+	47–86
B	"Pure" alkaline non-cyanide bright	99.955	131–185
C	"Pure" neutral semi-bright	99.999+	45–82
D	"Pure" neutral matte	99.999+	44–78
E	"Pure" acid soft matte	99.999+	52–78
F	"Pure" acid hard matte	99.999+	78–129
G	Silver "alloyed" alkaline cyanide bright	99.020	121–137
H	Cadmium "alloyed" alkaline non-cyanide bright	98.576	176–236
I	Cadmium & copper "alloyed" alkaline non-cyanide	76.350	204–287
J	Nickel "alloyed" acid bright	81.260	312–392
K	Nickel "alloyed" acid bright	99.898	135–167
L	Nickel "alloyed" acid bright	99.165	167–206
M, N	Cobalt "alloyed" acid bright	99.905	137–196
O	Cobalt & indium "alloyed" acid bright	98.275	245–292

Table 22.3
Resistivity for Deposits in Table 22.2 (Microhm-cm)

Bath	*Temperature*					
	−40°C	*0°C*	*50°C*	*100°C*	*200°C*	*300°C*
A	1.8	2.2	2.6	3.1	3.9	4.7
B	2.2	2.6	2.9	3.6	4.6	5.6
C	2.0	2.3	2.8	3.2	4.1	5.0
D	1.9	2.3	2.7	3.2	4.1	4.9
E	1.8	2.2	2.6	3.0	3.9	4.8
F	2.9	3.5	4.2	4.9	6.4	7.0
G	6.5	7.0	7.6	8.3	9.3	10.8
H	3.9	4.2	4.6	5.0	5.7	6.4
I	13.5	14.1	15.0	15.9	17.7	19.5
J	*	*	*	*	*	*
K	3.4	3.8	4.4	5.0	6.2	7.3
L	3.6	4.1	4.7	5.3	6.4	7.3
M	9.3	9.7	10.1	10.6	11.5	12.5
N	4.4	4.0	5.4	6.0	7.3	8.5
O	8.3	8.8	9.8	10.7	12.5	14.3

* not determined.

CONTACT RESISTANCE OF CLEAN SURFACES

The contact resistance of gold plated contacts is frequently cited in the literature and in specifications for connectors, but few systematic studies have been made of this property and of the factors on which it depends. The components of the resistance that is measured, R_c, R_f, and R_b, are not always recognised and there are no methods for determining the contact resistance of coatings which have industry-wide acceptance.

Since the substrate can influence R_c, particularly with thin deposits or at heavy loads, a study was made of the R_c versus load characteristic curves with 0.01 in. (effectively infinitely thick) deposits from a variety of commercial gold plating solutions[35]. The results are shown in Figure 22.3. The constriction resistances of as-plated and of wrought gold of comparable purity and hardness are not identical, the resistance of the former being somewhat higher. Gold electrodeposits alloyed with as little as 0.1 % of another element may have 5–10 times higher resistance. The structure-property relationships which determine the hardness and bulk resistivities of electrodeposits, hence R_c, have not been systematically explored.

The dependence of R_c on thickness of deposit[31] is clearly shown in Table 22.4. A variety of deposits from different solutions were plated on 0.030 in. copper wire and R_c was determined at 200 grams. The relatively pure gold platings, A and C-F, had constriction resistances similar to that of pure, clean copper. Hence, they changed little with deposit thickness. On the other hand, the hardened gold platings showed a marked increase in R_c with increase in

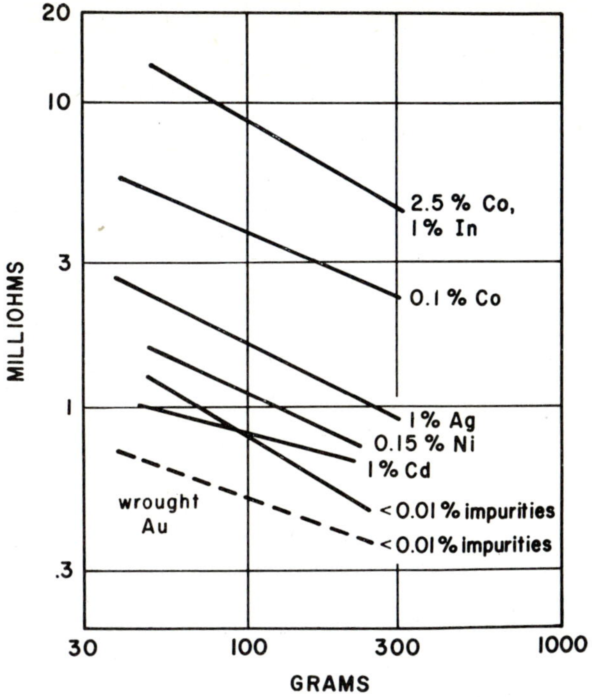

Fig. 22.3 Constriction resistance of a variety of gold plates. Specimens were OFHC copper in the form of a hemispherically-ended rod pressed on a flat, both members being plated with about 0.01 in. thick gold. The diameter of the rod was 0.125 in.

Table 22.4
Constriction Resistance at 200 Grams for Deposits in Table 22.2 (Milliohm)

Bath	*Deposit Thickness* (μ*-in.*)		
	50	*200*	*1000*
A	0.22	0.25	0.20
B	0.24	0.32	0.50
C	0.23	0.24	0.20
D	0.24	0.24	0.20
E	0.23	0.24	0.20
F	0.22	0.25	0.25
G	0.20	0.43	0.39
H	0.24	0.30	0.45
I	0.54	1.18	1.74
J	1.74	4.90	*
K	0.27	0.39	0.41
L	0.32	0.48	0.55
M	0.41	0.75	1.05
N	0.34	0.63	0.80
O	0.57	1.28	1.75

* not determined.

thickness because of a diminishing effect on R_c by the substrate. These trends are consistent with those in Figure 22.3.

The role of substrate in determining R_c has been discussed[26]. For thin deposits, less than 100 microinches, the R_c of gold at 100 grams is significantly higher when plated on solid nickel than when plated on solid copper. Copper is softer and more conductive than nickel, hence the measured R_c of gold plated copper is less. Typical data for approximately 50 microinches of gold are 0.4 and 1.4 milliohm when plated, respectively, on copper and on nickel[26]. Another way of showing this is in Figure 22.4, the R_c versus load characteristic curves for very thin (20 microinches) pure gold on copper with various thicknesses of nickel underplate (these data were obtained by mating the samples to pure wrought gold)[39]. Pure nickel underplate had a hardness (KHN_{25}) of 445 kg/mm^2 versus 60 kg/mm^2 for the gold. Its bulk resistivity was, perhaps, 7×10^{-6} ohm-cm.[1], compared to an estimated value of 2.5×10^{-6} ohm-cm. for the gold (Table 22.1).

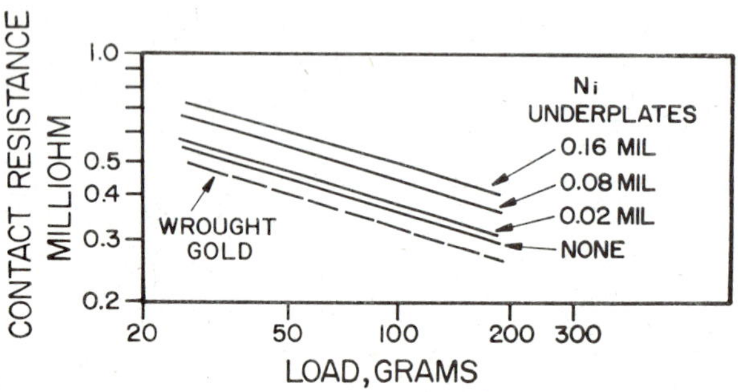

Fig. 22.4 Contact resistance vs load characteristic curves for unheated 0.02 mil pure gold plate with no underplate, and with various thicknesses of nickel underplate. Curve for wrought pure gold shown for comparison. Probed with hemispherically-ended solid gold rod, 0.125 in. diameter.

CONTACT RESISTANCE OF FILM-COVERED SURFACES

The widespread use of gold as a contact material[43] originates in its nobili ty it being the least reactive metal to oxygen and to common air pollutants. However, there are several mechanisms of degradation which involve the formation of solid insulating surface contaminants on gold plate. These are as follows:

1 Corrosion and tarnishing reactions when the gold coating is imperfect, either inherently through porosity or because a mechanical stress, such as sliding wear or mechanical deformation during manufacture, has exposed basis metal.
2 Diffusion to the surface of basis metal or underplating, or of codeposited base hardening elements in the plating.

3 Golds alloyed with high concentrations of base metals, e.g., more than 25 % by weight, are chemically less stable than golds of higher purity.
4 Codeposited organic material from $Au(CN)_2^-$ plating baths that are improperly maintained or from carbon electrodes.

In addition to these film-forming processes, "friction polymer" has been found[4] on gold contacts when rubbed in the presence of low concentrations of organic vapours. The quantity of these solids is, however, so small that it does not interfere with the functioning of most practical contact devices, such as relays, unlike metals of the palladium and platinum groups which form polymer readily. In fact, a composite of about one thousandth of an inch of a 22 carat hard gold alloy over palladium on one contact versus solid palladium on the mating contact is used[4] in relays that serve some telecommunications systems. This combination resists both erosive wear and sliding damage. If the contact involves arcing the gold is quickly lost, leaving palladium which has excellent erosive characteristics. If the contact is used where rubbing predominates, the gold reduces the formation of friction polymer. Occasionally, however, contact failures in lightly loaded instrument slip rings made of gold have been attributed[42] to friction polymer.

Finally, contaminants can be physically deposited on the contact during its manufacture, or from the environment during service.

CORROSION AND TARNISH FILMS

The origin of films formed by corrosion and tarnishing is in deposit porosity. This has been emphasised by many authors[9,14,19,20,23,24,27,32,34,36,37], and, indeed, common methods of determining porosity involve exposure of the contact to corrosive atmospheres (see Chapter 27).

Film formation is related to the numbers and sizes of the pores, the reactivity of the substrate metal, and to environmental factors like relative humidity, temperature, and types and quantity of air pollutant[14].

The contact properties, however, of film-covered gold are determined by:

1 Mechanical load normal to the surface—which can break insulating films.
2 Extent of wipe of the contact—which also can break insulating films.
3 Open circuit voltage—films can be punctured electrically to establish metallic contact[29].
4 Geometry of the contact members—needle point and knife-edged contacts facilitate film fracture.
5 Surface roughness.

The effect of surface roughness on the contact resistance of film-covered gold is especially important, because topography also affects porosity. The rougher the substrate the more porous the deposit (see Chapter 23) and the greater the amount of corrosion and tarnish films such surfaces will have after exposure to polluted atmospheres. On the other hand, it has been shown by studies of base metals having uniform films that roughness leads to lower contact resistance[18,21,28].

The cracking of corrosion, tarnish and oxide films is shown schematically in Figure 22.5. Most contaminating films are less ductile than the metal from which they originate. When two surfaces touch, strains are created which can rupture the films, a necessary condition for metal-to-metal contact[25].

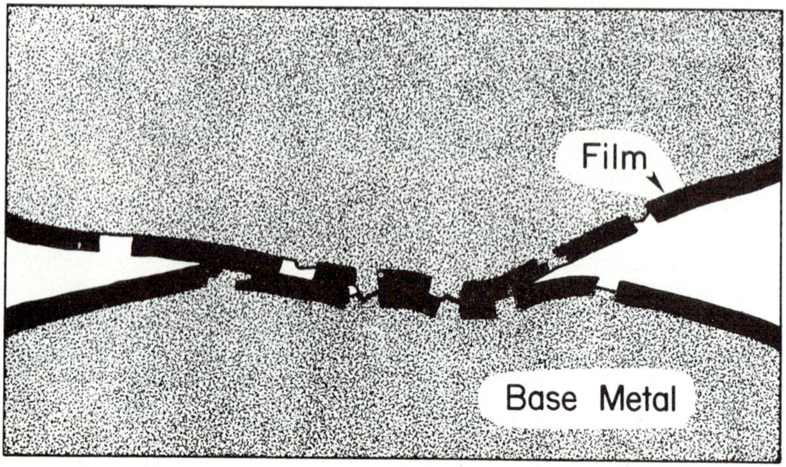

Fig. 22.5 Schematic representation of one mechanism of producing metal-to-metal contact between surfaces, both of which are covered with a uniform oxide or corrosion film. The touching spots of the surfaces are plastically deformed and strains induced in the films cause them to crack. The extent of film disruption increases as roughness or load is raised. Wherever cracks in the films on both surfaces match and metal is extruded through these breaks, metallic junctions are created.

It is not obvious whether thin (hence, corrosion susceptible) gold coatings on surfaces should be rough or smooth. In order to answer this question, gold plated nickel, silver and copper in a wide range of substrate roughnesses was studied[34]. Porosity and contact resistance determinations were made with specimens that had been exposed in polluted atmospheres. The measuring circuit for contact resistance impressed a low (or "dry") potential across the film, limited to 50 mV. In this way electrical breakdown of the films could be avoided. Figures 22.6 and 22.7 and Table 22.5 illustrate the critical dependence of contact resistance on roughness and porosity, and the effect of load in facilitating film breakdown.

It was found that the reliability of contacts fabricated from thin gold plate on corrodible substrates can be markedly improved if the substrate is prepared so as to be smooth. A decrease in the porosity of the gold plated surface

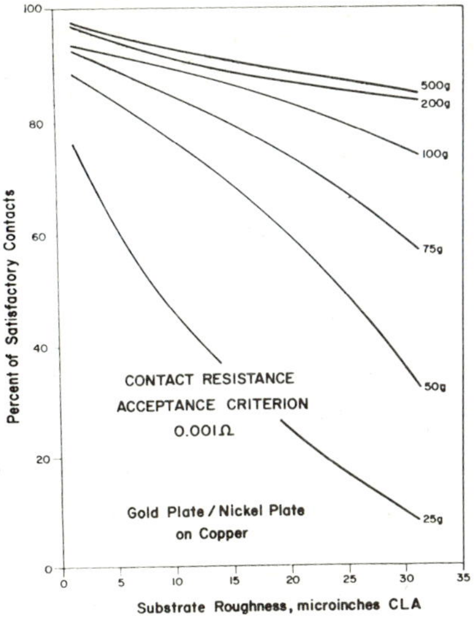

Fig. 22.6 Percentage points probed at various loads that gave contact resistance of 0.001 ohm or less vs substrate roughness. Samples exposed to 1% SO_2 for 120 hours at 95% RH and 45°C. Contacts probed with 0.125 in. diameter hemispherically-ended gold rod. Samples described in Table 22.5.

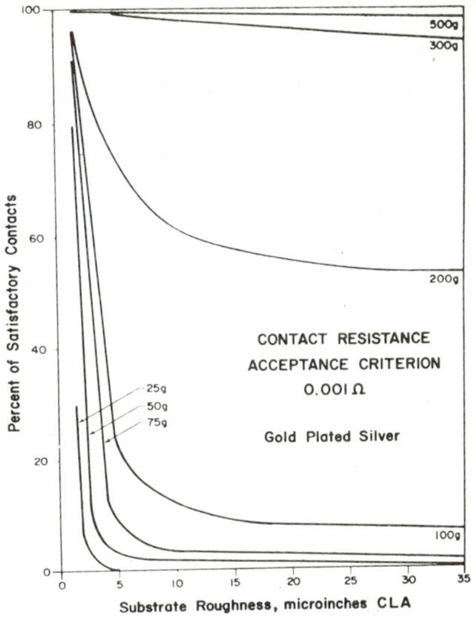

Fig. 22.7 Percentage points probed at various loads that gave contact resistance of 0.001 ohm or less vs substrate roughness. Samples exposed to 1% H_2S for 120 hours at 60–90% RH and 25°C. Contact probed with 0.125 in. diameter hemispherically-ended gold rod. Samples described in Table 22.5.

Table 22.5
Description of Samples
(See Figures 22.6 and 22.7)

System	*Basis Metal*	*Basis Metal Roughness*[a]	*Under-plate*	*Gold Plate*	*Roughness After Plating*[a]	*Average Pores/cm²*
Au/Ni	OFHC Cu	smooth 1.5 medium 5 rough 31	Watts nickel, 100 μ-in.	pure acid citrate[b] 42 μ-in.	smooth 2.5 medium 8 rough 36	60 100 160
Au/Ag	Wrought Ag 99.99% pure	smooth 1.5 medium 5 rough 35	none	pure acid citrate[b] 42 μ-in.	smooth 2.5 medium 6 rough 35	

[a] Roughness in microinches Centre Line Average, determined with a TALYSURF.* Roughness wavelength cutoff, 0.01 inch.
[b] See Reference 6.
* Rank–Taylor–Hobson, Ltd.

reduces the tendency of the contact to develop insulating films. This is more significant than any enhancement attributable to roughness in the ability to fracture films on making contact.

Several investigations of the effect of porosity of gold plate on its behaviour in electric contacts after exposure to air pollutants have been reported[9,14,16,20,24,37]. These studies entail exposure to the field or to synthetic environments produced in the laboratory. The latter generally involves individual pollutants that are common and particularly corrosive to gold on various substrate or underplate metals used on electrical contacts. Thus sulphur dioxide[11,22,30,44,45], hydrogen sulphide[34], sulphur vapour[21] and sulphur dioxide followed by hydrogen sulphide are often employed. Generally, relative humidity and temperature are controlled in these tests, in addition to the concentration of pollutant.

A problem with many laboratory tests is that they do not involve conditions which allow films to grow on contact surfaces having the same chemical compositions as films that form in the field. Field environments are complex and are difficult to simulate in the laboratory. The chemistry of corrosion is highly specific to the environment. Furthermore, most contacts must function at many different field conditions and it is generally not practicable or possible to simulate a wide spectrum of field environments. But even beyond film composition, it is the mechanical properties and morphologies of films that determine whether they will be broken on making mechanical contact to give low contact resistance. These characteristics may depend, for example, with chemically identical films having the same thickness on the rate at which they grow.

Laboratory tests are usually intended to accelerate the degradation of contacts, relative to the field. Ideally, the acceleration rate is known. Laboratory tests need extensive study to be meaningful. Nevertheless, even incompletely developed laboratory procedures may be useful to permit a comparison to be made of the relative behaviours of several contact materials, or to enable a determination to be made for quality control purposes as to whether a contact meets a defined quality standard.

The most common substrates for gold plated contacts are copper, copper alloys and nickel. The substrates are chosen for their current-carrying ability, ease of fabrication (copper, brass), spring properties (beryllium-copper and phosphor-bronze) and strength (nickel). Common underplatings are copper, nickel and silver, although usage of the latter is decreasing.

Corrosion occurs above a critical humidity in the presence of traces of gaseous or solid pollutants. A film, usually heterogeneous, grows rapidly on many metals under these conditions. For example, Figure 22.8 illustrates corrosion solids on porous gold plated nickel after exposure to sulphur dioxide at high relative humidity. The contact behaviour of corrosion solids from this system has been studied in detail[37], the critical relative humidity being about 70%. Porous gold, plated directly onto copper, also produces discrete corrosion spots in atmospheres containing sulphur dioxide.

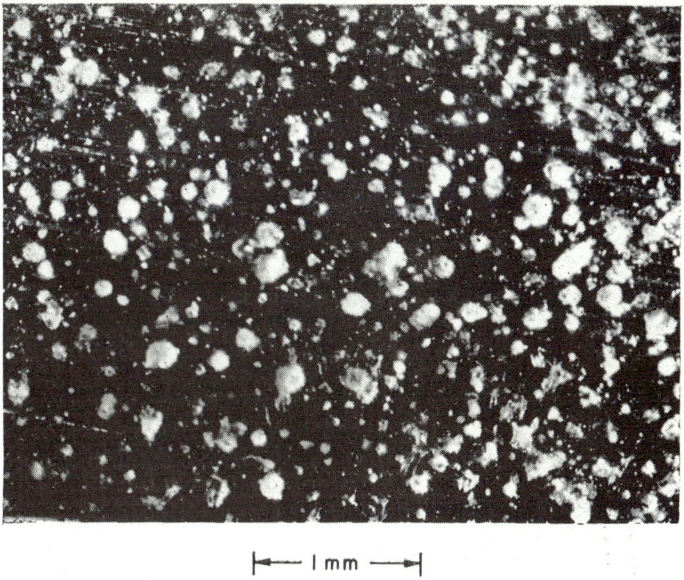

Fig. 22.8 Corrosion solids on porous gold plated nickel after exposure to moist sulphur dioxide.

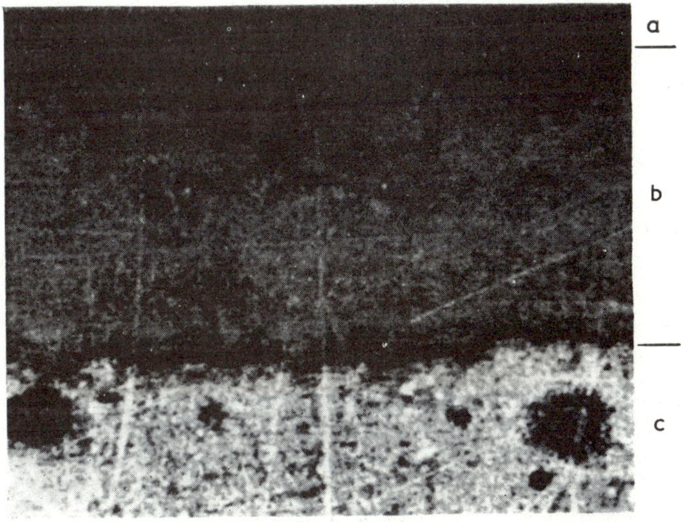

Fig. 22.9 Gold plated silver after exposure to hydrogen sulphide. Unplated area (a) is severely tarnished, and silver sulphide film (b) has advanced over the interface between the silver and gold. Note sulphide creep from pores in region (c).

Tarnish films on porous gold plated silver formed in hydrogen sulphide or sulphur-containing atmospheres are generally thin. However, silver sulphide is able to spread from pore sites and can quickly contaminate the surface[7], an example of which is shown in Figure 22.9. Spreading originates at the pores in gold on copper and its alloys[14], although at a slower rate than for gold plated silver. Silver sulphide spread on gold has particularly bad contact properties, being tougher to break than silver sulphide on solid silver[7,24]. Also, the film readily transfers to and accumulates on the wiping contact in common connector designs, e.g., edge-type connector contacts for printed circuit boards[14]. Typical data in Table 22.6 shows that on wiping, contact resistance may increase after an initial decrease.

Table 22.6
Contact Resistance (ohms)—Wipe Properties of Sulphides at 150 Grams Load

Wipe Distance	*Sulphide Migrated Onto Gold*		
	Silver	*Phosphor–Bronze*	*Beryllium–Copper*
A			
Before wipe	> 100 Ω	0.004	0.004
0.010 in.	0.032	0.003	0.004
0.025	0.058	0.003	0.003
0.050	1.95	0.003	0.010
0.100	0.0057	0.003	0.010
B			
Before wipe	> 100 Ω	0.005	0.005
0.010 in.	0.155	0.004	0.004
0.025	0.350	0.004	0.004
0.050	0.007	0.003	0.006
0.100	0.082	0.041	0.002
C			
Before wipe	1.32 Ω	0.002	0.006
0.010 in.	0.020	0.002	0.004
0.025	0.008	0.002	0.003
0.050	0.016	0.002	0.003
0.100	0.014	0.006	0.004

Recently, it has been found that silver sulphide is formed on porous gold plated silver on exposure to high concentrations of sulphur dioxide produced from a mixture of sodium thiosulphate, sulphuric acid and water in a closed vessel[30]. This reaction will not proceed, however, unless the relative humidity is nearly 100 % at the initiation of the exposure[40].

Another form of corrosion from porous gold plated metals is produced at high relative humidity by hygroscopic dust particles. A synthetic dust has been proposed for laboratory studies of contacts[14].

FILMS FORMED BY THERMAL DIFFUSION

The thermal degradation of gold plated contacts involves diffusion of base metals to the surface, where they oxidise to give insulating films. The base metals originate in codeposited hardeners in the gold or in the substrate. These mechanisms have long been recognised to be significant above room temperature[8,9], and barrier platings which retard the diffusion of copper and zinc from substrates containing these metals are used. Although rhodium is excellent, nickel is the most cost-effective for this purpose.

Diffusion coefficients for copper through gold plate have recently been experimentally determined[41] in the temperature range, 50–750°C. Below about 250°C diffusion was found to occur more rapidly than could be predicted by extrapolation of older high temperature results.

The rate of diffusion through gold plate depends on time, temperature, the compositions of basis metal and underplate, and on the thickness of the gold and the underlayer. It may also depend on the structure of the deposits. The compositions (and mechanical properties) of the insulating films that form from the diffused metal at the surface depend on the environment. In clean air copper oxides are found on the surface of gold plated copper, while other compounds can be expected to form in polluted atmospheres. Finally, the contact resistance of film-covered gold depends on surface roughness and on design factors, such as contact load and length of wipe on mating.

A study of the reliability of gold plated contacts when heated has been reported[39]. Gold platings as thin as those in common use, 20 microinches, and thick platings (200 microinches) of pure and of certain alloy golds, were considered. The thickness of nickel underplate was varied between 20–160 microinches. The substrates were smooth.

A preliminary study was made with pure gold plated directly on both solid copper and on 200 microinches of copper underplate. There was no significant difference in the rates of degradation of these surfaces as determined by contact resisting probing.

Studies were made at these temperatures: 65°C, a normal maximum for systems which serve at room temperature, such as business computers; 125°C, the temperature at which general purpose connectors must be able to operate and is often specified for military applications; 200°C, which is becoming increasingly important for aircraft equipment; and 250 or 350°C, to determine any unusual high temperature behaviour.

Reliability data were obtained by probing the surfaces, without wipe, using a clean, smooth, hemispherically-ended gold rod. Contact resistance with a circuit having a maximum open circuit potential of 50 mV was determined at 25–200 grams, the load range for which the contacts of most connectors are designed. Many probings of presumably identical surfaces were made to eliminate the effects of film variability and to provide statistically relevant data. A satisfactory contact is defined as one that has a resistance not exceeding 0.001 ohm, which is approximately double the initial value for unheated specimens.

Gold Plated Copper

Figure 22.10 gives the reliability curves for 20 microinches of pure gold plated on copper at intervals up to 1300 hours. The resistances of samples heated at 65°C have not changed. However, reliability was seriously degraded in only 300 hours at 125°C and in 10 hours or less at the higher temperatures.

Table 22.7 lists the maximum times and various temperatures that the thick (200 microinch) coatings can be heated at before they become unreliable. It is evident that thick deposits of pure gold are superior to thin deposits and that alloy golds degrade more quickly than pure platings, especially at high temperature.

Table 22.7
Effect of the Alloy Content of Gold Plate on Reliability*

Temperature (°C)	*Maximum Heating Time (Hours) for 100% Reliable Contact*		
	Pure Gold Plate	*0.1% Cobalt–Gold*	*19 carat Gold***
65	>1,000	>1,000	50
125	500	500	2
200	300	2	—
300	25	—	—

* Deposits were 200 microinches thick. Criterion of failure = 0.001 ohm.
** Alloyed with 19.3% cadmium and 4.2% copper.

Gold on Nickel Underplate

Figures 22.11–22.13 are reliability curves for 20 microinches of pure gold plate on 20, 80 and 160 microinches of nickel underplate. Data from the samples without an underplate are given for comparison.

Several facts emerge from this study. An effective way to retard the diffusion of substrate copper through gold is to use a nickel underplate, the barrier metal itself diffusing only very slowly through gold plate. As little as 20 microinches of nickel is remarkably effective at 300°C. The increase in effectiveness is not proportional, however, to the increase in nickel thickness. Nevertheless, 80 microinches of nickel is needed at 125°C for thin gold plate to make the contact reliable for 1000 hours.

This work also shows that degradation of plated gold is least for a given thickness when the deposit is pure. This is because diffusion of substrate metal only contributed to film formation. If alloying elements are used, reliability may be sacrificed, particularly at temperatures exceeding 200°C, compared to the behaviour of pure gold deposits. The rate of degradation probably increases with increase of alloy content of the gold. Not all alloy metals are

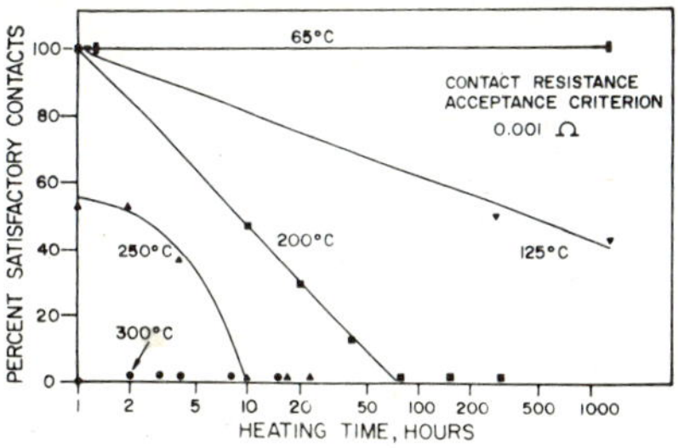

Fig. 22.10 Reliability vs heating time for 0.02 mil pure gold plate on copper; heated at 65, 125, 200, 250, and 300°C. Probed at 100 g with hemispherically-ended solid gold rod.

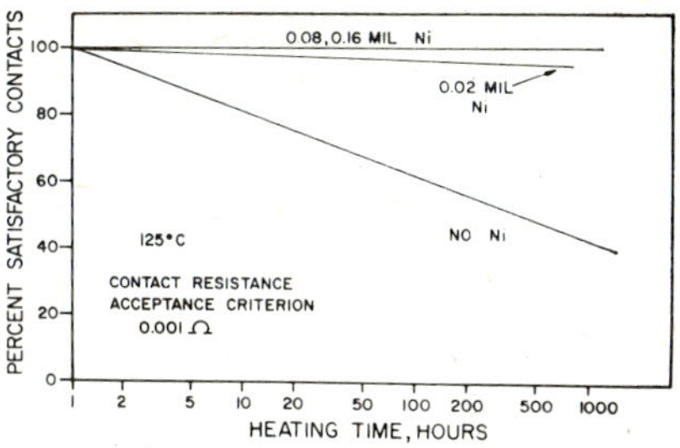

Fig. 22.11 Reliability vs heating time for 0.02 mil pure gold plate on copper with no underplate, and with various thicknesses of nickel underplate. Heated at 125°C.

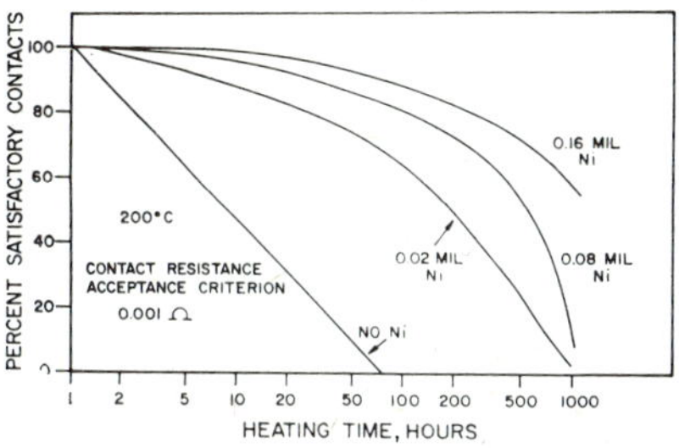

Fig. 22.12 Same as Figure 22.11, except heated at 200°C.

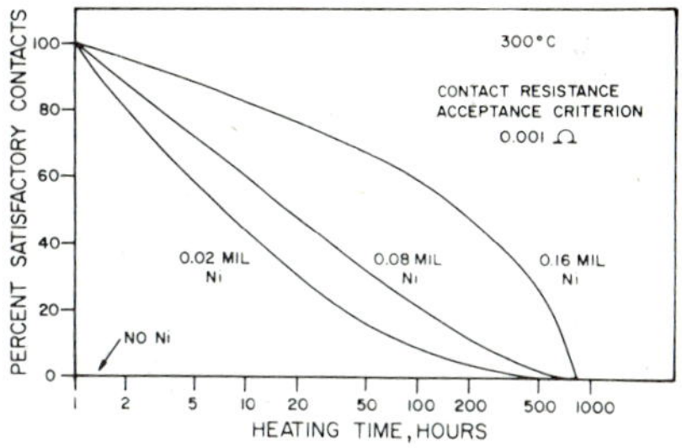

Fig. 22.13 Same as Figure 22.11, except heated at 300°C.

equivalent; for example, a noble metal hardener in gold would not produce oxide films. Gold platings that contain large amounts of base metal, such as the example of this study, may not be stable much above room temperature for extended periods.

It is likely that small amounts of base hardener metals, say, tenths of a percent, are acceptable for the majority of contacts, which are limited in service to 125°C. The value of small amounts of cobalt and nickel in the gold for wear resistance is discussed in Chapter 21.

FILMS FROM ALLOY GOLDS

Alloy gold platings may be susceptible to degradation in humid or polluted atmospheres because of their diminished nobility. This is of practical importance only with deposits on base metals where the deposit is so thick as to be practically free of pores. When porosity is present, films that originate from them are more significant. Little is known about the contact behaviour of deposits containing more than 1 % of base alloying elements.

The tarnish resistance of alloys having over 20% copper is low[2,5]. Bright acid golds containing less than 1% cobalt or nickel and detached from their substrates prior to exposure discoloured in an atmosphere of flowers of sulphur[31]. Those containing up to 2% nickel were unaffected by hydrogen sulphide, high relative humidity, or exposure to an industrial gas atmosphere[3]. Studies of the contact resistance of wrought gold alloys after exposure to polluted atmospheres have been reported[24], and may offer guidance in the selection of a deposit for contact applications. Interest in alloy golds is great, and several new systems containing 60–80% gold have been described[48,49].

CODEPOSITED POLYMER

Carbon-containing materials, presumed to be organic polymers, have been found to be an intrinsic constituent of many golds obtained from $Au(CN)_2^-$ solutions[38]. There is some conjecture regarding the location of these materials,

Fig. 22.14 Fracture surface of an acid cyanide gold containing 0.16% Co and 0.14% C. Heated at 150°C after fracture. Note dark protrusions of low atomic number materials, believed to be intrinsic polymers, emerging from the surface. Scanning electron microscope.

whether on the surface, in the bulk, or in both places of the deposit. Polymers were first observed[38] on the surfaces of gold plated telephone contact springs which had abnormally high contact resistance, but subsequent study also showed them to be in the body of the deposit. Probably polymers occur only in the interior of deposits from correctly operated plating baths.

Polymers are especially prone to deposit from acid solutions in the presence of cobalt, nickel, or indium cations and increase in amount as the cathode efficiency of the bath is reduced. Alkaline solutions containing free CN^- also produce deposits having codeposited organics.

The form of the polymers is of considerable interest. Fracture surfaces of polymer-containing electrodeposits from both alkaline and acid solutions show[50] low atomic number material to be present in discrete pockets. Gentle heating often caused the material to emerge from the fracture surface. The pockets vary in size, and generally range from 1000 Å downwards. Possibly the pockets are interconnected. Figure 22.14 is a photomicrograph of the fracture surface of a typical deposit. Earlier work[38] suggested that the polymers were layered in certain deposits, but this interpretation is probably incorrect. Grain boundaries are also believed[46] to be possible sites for the materials.

Most of the golds now used in the contact industry are obtained from potential polymer-forming solutions. Small amounts of polymers cause no contact problems. However, additional studies of the practical effects of polymer on contact properties are needed. It may be necessary to specify maximum carbon content in deposits for certain applications.[35]

REFERENCES

1 Zentner V., Brenner A., Jennings C. W., Am. Electropl. Soc. Res. Rep. No. 20, pp. 70 (1952).
2 Raub E., Sautter R., *Metalloberfläche*, **10,** 3, 65 (1956).
3 Atanasyants A. G., Kudryavtsev N. T., Karataev V. M., *Zh. prikl. Khim.*, **30,** 6, 876 (1957).
4 Hermance H. W., Egan T. F., *Bell Syst. tech. J.*, **38,** 5, 739 (1958).
5 Fedot'ev N. P., Kruglova E. G., Vyacheslarov P. M., *Zh. prikl. Khim.*, **32,** 2014 (1959).
6 Ehrhardt R. A., *Proc. Am. Electropl. Soc.*, **47,** 78 (1960).
7 Egan T. F., Mendizza A., *J. electrochem. Soc.*, **107,** 4, 353 (1960).
8 Frant M. S., *Plating*, **48,** 1305 (1961).
9 Angus H. C., *Trans. Inst. Metal Finish.*, **39,** 1, 20 (1962).
10 Dettner H. W., *Galvanotechnik*, **53,** 2, 87 (1962).
11 Fairweather A., Lazenby F., Parker A. E., Proc. I.E.E., 109 (Part B, Suppl. No. 22), 567 (1962); see especially pp. 585, 10.3 (vii).
12 Frant .M S., Symp. on Connectors, Los Angeles, California, 29-30th Nov. (1962).
13 Antler M., *Wear*, **6,** 1, 44 (1963).
14 Antler M., Gilbert J., *J. Air Pollut. Control Assoc.*, **13,** 9, 405 (1963).
15 Brenner A., "Electrodeposition of Alloys— Principles and Practice", Vol. 2, Chap. 37, pp. 494, Academic Press, New York (1963).
16 Baker R. G., Proc. 2nd Int. Res. Symp. on Electrical Contact Phenomena, pp. 545, Graz, Austria (1964).
17 Danemark M. A., *Galvanotechnica*, **15,** 6, 129 (1964).
18 Williamson J. B. P., Res. Rep. No. 4, Burndy Corp., Norwalk, Conn. (1964).

19 Alfiero J. M., paper presented at the 13th National Relay Conf., 26–29th April, 1965; summarised by Oliver, F. J., Electro-Technology, **76,** 69 (1965).
20 Chiarenzelli R. V., Proc. Engng Semin. on Electrical Contact Phenomena, pp. 63, Univ. of Maine, Orono, Maine (1965).
21 Chhabra D. S., Wenning G. T., Proc. Engng Semin. on Electrical Contact Phenomena, pp. 313, Univ. of Maine, Orono, Maine (1965).
22 Foulke D. G., *Metal Finish.*, **63**, 7, 42 (1965).
23 Antler M., *Plating*, **53,** 1431 (1966).
24 Chiarenzelli R. V., Proc. 3rd Int. Res. Symp. on Electrical Contact Phenomena, pp. 83, Univ. of Maine, Orono, Maine (1966).
25 Osias J. R., Tripp J. H., *Wear*, **9**, 5, 388 (1966).
26 Walton R. F., *Plating*, **53,** 209 (1966).
27 Antler M., *Plating*, **54,** 915 (1967).
28 Harada S., Mano K., Proc. Engng Semin. on Electrical Contacts, pp. 45, Illinois Inst. of Technology, Chicago (1967).
29 Holm R., "Electric Contacts", 4th Edn, pp. 135, Springer, Berlin & New York (1967).
30 Clarke M., Leeds J. M., *Trans. Inst. Metal Finish.*, **46,** 2, 81 (1968).
31 Duva R., Foulke D. G., *Plating*, **55,** 1056 (1968).
32 Krumbein S. J., Antler M., I.E.E.E. Trans., on Parts, Materials, and Packaging, PMP-4, 1, 3 (1968).
33 Aronstein J., Campbell W. E., Proc. Engng Semin. on Electrical Contacts, pp. 7, Illinois Inst. of Technology, Chicago (1969).
34 Antler M., *Plating*, **56,** 1139 (1969).
35 Antler M., *Products Finish.*, **34**, 10, 56 (1969).
36 Chow W., Stepke E., Proc. Engng Semin, on Electrical Contacts, pp. 99, Univ. of Maine, Orono, Maine (1969).
37 Krumbein S. J., I.E.E.E. Trans. on Parts, Materials, and Packaging, PMP-5, 2, 89 (1969).
38 Munier G. B., *Plating*, **56,** 1151 (1969).
39 Antler M., *Plating*, **57,** 615 (1970).
40 Antler M., unpublished work.
41 Pinnel M. R., Bennett J. E., Proc. Engng Semin. on Electrical Contacts, pp. 13, Illinois Inst. of Technology, Chicago (1971).
42 Jentgen R. L., I.E.E.E. Trans. on Parts, Hybrids, and Packaging, PHP-7, 2, 86 (1971).
43 Antler M., *Gold Bull.*, **4,** 3, 42 (1971).
44 Preston P. F., "An Industrial Atmosphere Corrosion Test for Electrical Contacts and Connections", British Post Office Branch, Materials Section, Birmingham, England (1971).
45 Leeds J. M., Such T. E., *Trans. Inst. Metal Finish.*, **49**, 4, 131 (1971).
46 Holt L., Stanyer J., *Trans. Inst. Metal Finish.*, **50**, 1, 24 (1972).
47 Silver H. G., *J. electrochem. Soc.*, **116**, 591 (1969).
48 Nobel F. I., Thomson D. W., Leibel J. M., Proc. 4th Plating in the Electronics Industry Symp., pp. 106, Am. Electropl. Soc., Indianapolis (1973).
49 Foulke D. G., Duva R., Proc. 4th Plating in the Electronics Industry Symp., pp. 131, Am. Electropl. Soc., Indianapolis (1973).
50 Antler M., *Plating*, **60,** 468 (1973).

Chapter 23

POROSITY

S. M. Garte

In the fabrication of articles, a thin gold layer on the surface provides a considerable degree of solid gold quality at greatly reduced cost. The advantages of gold plate, from both engineering and decorative aspects, may be lost if the layer is porous and the basis metal exposed. This is particularly true in those applications which do not permit the use of protective lacquers. Knowledge of the factors controlling porosity in gold plate enables the design engineer to obtain maximum cost effectiveness from the plated finishes which he specifies.

Porosity may be defined in a number of ways:

Bulk Porosity, P_b, is defined as the volume fraction of the deposit occupied by voids and discontinuties of all types.

$$P_b = \frac{V_v}{V_m + V_v} \tag{1}$$

where V_v and V_m are the volumes occupied by voids and by deposited metal respectively. Bulk porosity may be related to density by the following equation:

$$P_b = 1 - \frac{D_b}{D_m} \tag{2}$$

where D_b is the bulk or overall density of the deposit:

$$D_b = \frac{\text{Weight}}{V_m + V_v} \tag{3}$$

and D_m is the density of the deposited metal:

$$D_m = \frac{\text{Weight}}{V_m} \tag{4}$$

A method for the measurement of the bulk density of gold deposits has been described by Cooley[1] (see Chapter 24). Gold is deposited on to a polished panel, one half of which is masked off and thickness measured interferometrically at the resulting step. Thickness is also measured by a beta-ray backscatter instrument calibrated against accurate weight standards. The ratio of these measurements is D_b/D_m.

Equation (2) is strictly correct only if it is assumed that the voids in the deposit are empty. Fully enclosed voids may, however, be filled with residual liquids and precipitates from the various processing baths. Under certain conditions gold deposited from baths containing the gold as the cyanide complex may contain a significant quantity of codeposited polymeric material. This effect is described by Munier[2].

Transverse Porosity, P_t, is the fraction of the total area at which the basis metal substrate is exposed. Transverse pores form a continuous path from the substrate to the plate surface. P_t cannot be measured by the density method unless it is assumed that all the porosity is transverse, an assumption which is justified only for very thin coatings. Electrochemical methods for the measurement of exposed area are discussed in Chapter 27 on "Measurement of Porosity".

An alternative definition of transverse porosity which is very convenient for engineering purposes is:

$$P = \frac{\text{number of pores}}{\text{cm}^2} \tag{5}$$

This definition gives equal weight to all pores regardless of size. Arbitrary units for P are often used when an indirect method, such as the determination of corrosion product, is used in place of pore counting. Procedures for porosity measurement as defined by equation (5) are also described in Chapter 27.

THE CONSEQUENCE OF POROSITY IN GOLD PLATE

Gold forms no oxide or tarnish films. It is this property which makes it attractive for both cosmetic and engineering applications. Tarnish films and corrosion products which do form on porous gold plate, must originate from metallic impurities in the deposit or alloying metals added to it as hardening agents, or from the basis metal. At high temperatures, diffusion of metal to the surface from any of these sources may occur. Diffusion of copper to the surface of gold plate at elevated temperatures has been described by Frant[3]. Pinnel and Bennett[4] have recently reported a quantitative study of copper diffusion through electroplated gold. Based on their data, they predicted times required for the accumulation of 0.1 and 1.0 atomic % copper on the surface of pure gold plate at thicknesses of 1 micron (40 microinches) and 2.5 micron (100 microinches) at various temperatures from 25°C–250°C. These times are much shorter than previously predicted from bulk diffusion coefficients. For example, at 65°C, 8.3 years is required for the accumulation of 0.1 atomic % copper on the surface of a 40 microinch deposit. Extrapolation from diffusivities measured at high temperatures gives a time estimate of 6670 years for these conditions. Interdiffusion phenomena between various combinations of gold, silver, nickel, and copper electroplated bronze panels were studied by Modjeska and Kann[5]. Antler[6] has discussed the effect of high temperature on the engineering properties of alloy gold deposits. These effects may occur even in pore-free gold by a bulk diffusion process; it is reasonable to hypothesise that pores may accelerate diffusion processes by providing channels along which surface diffusion can occur. Pinell and Bennett[4] pointed out that the activation energy for defect diffusion mechanisms is much less than for bulk diffusion and that the former may predominate at lower temperatures, e.g., 25°C–250°C. In Frant's study the amount of surface copper oxide increased as plate thickness decreased. This can be attributed to the combined effect of two causes; the shorter diffusion path and the increase in porosity.

The special significance of porosity is that it permits the formation of tarnish films and corrosion products on the surface, even at room temperature. This can be understood only in terms of the interaction of three factors: (i) the pore density of the as-plated gold, (ii) the composition of the substrate, if gold is deposited directly, or any underplates which may be deposited between the substrate and gold, and (iii) the environment in which the part is used or stored. For example, porous gold over silver is highly sensitive to sulphide environments but will show little change in salt spray tests. Gold over nickel is unreactive in low humidity environments, but is highly sensitive to salt spray and to sulphur dioxide at relative humidities greater than 70%. Gold over copper occupies a somewhat intermediate position, sulphiding more slowly than silver and reacting more rapidly in salt spray.

It is usually the transverse porosity which is of major concern from both appearance and functional viewpoints. This follows from observations made by Frant[7] and later by the writer[8] that porosity decreases as the thickness of the deposit increases. As the plate grows it tends to bridge over transverse pores so that they no longer react with test reagents or with real environments. Nevertheless bridged, i.e., non-transverse pores, will continue to contribute to the bulk porosity as determined by density measurements. In some instances bridged pores may be of engineering significance. High temperatures such as occur in soldering may cause blistering and exudation of contaminants contained in the voids. Such effects may seriously interfere with the solderability of the plate.

In the electronics industry, which utilises the largest quantity of gold plate for engineering purposes, porosity is a major concern because of its effect on the electrical properties of the plated parts. Plate quality is often rated in terms of freedom from pores in functional areas, but the problem is more complex, however, and again the interaction between environmental factors and substrate composition must be taken into account. For example, silver sulphide spreads rapidly over a gold surface so that total blackening may occur within a short time[9]. Copper sulphide also spreads, but less rapidly.

Nickel forms a variety of corrosion products under different conditions[10]. At high humidities, nickel corrosion products will remain localised at the pore sites. Figure 23.1 illustrates the effects on gold plate over different under-plates when exposed to sublimed sulphur. Nickel does not react in this environment but gold-nickel coincident pores are revealed.

The effect of corrosion and tarnish on the contact resistance of electrical contacts is discussed more fully in Chapter 22 on "Contact Properties".

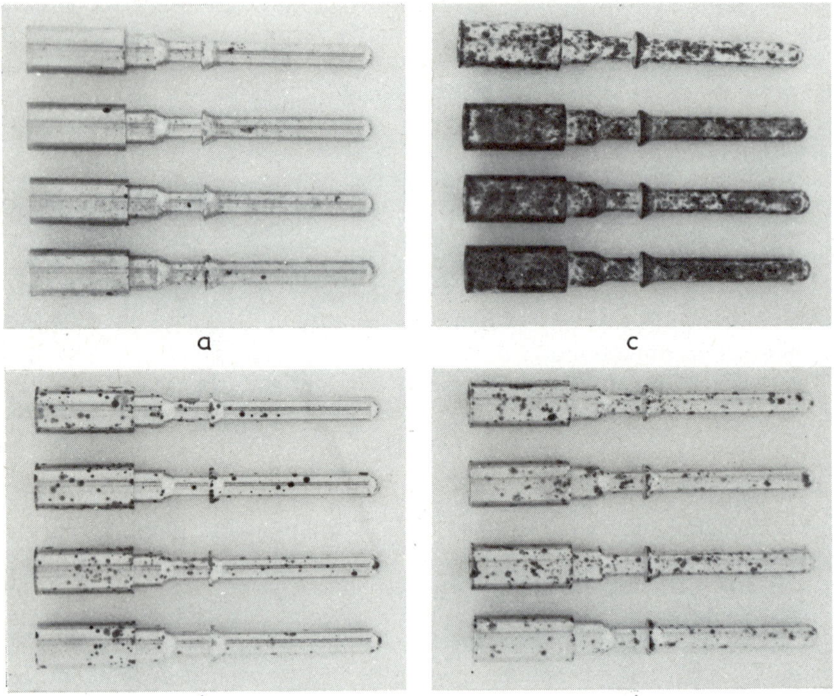

Fig. 23.1 Interaction of underplate types to a sulphiding environment. All were exposed to sublimed sulphur at 65°C, 100% RH for 24 hours. (a) 20 μ-in. gold over nickel and copper (b) 35 μ-in. gold over copper (c) 20 μ-in. gold over silver (d) 35 μ-in. gold over silver.

FACTORS AFFECTING THE AS-DEPOSITED POROSITY OF GOLD PLATE

This is dangerous subject indeed! If X is almost anything pertaining to the plating process, then the most probable answer to the question, "Does X affect the porosity of gold?" is yes. Nevertheless, quantitative studies have revealed many important relationships which are discussed in this section.

THE RELATIONSHIP BETWEEN GOLD THICKNESS AND POROSITY

The decrease in porosity with increasing thickness has been observed qualitatively for many years. The first quantitative study was made by Frant[7], who suggested that the data would fit an equation of the form,

$$P = AT^{-n} \tag{6}$$

where P is transverse porosity as defined earlier in this chapter, T is thickness, and A and n are constants, the values of which depend on the nature of the electrolyte and plating conditions. Why does porosity decrease with thickness? The simplest explanation is that as the plated layer grows it bridges over the smallest pores first, and then the larger ones.

This is indirectly supported by the observation, in porosity testing, that increasing the time of exposure to a gaseous test reagent often results in an increase in the amount of corrosion product generated, but very little increase in the number of pores that can be counted. This is particularly true for thick deposits of about 200 microinches. If the pores in a thick layer were simply deeper and not bridged, it would require more time for a reagent to penetrate into and corrosion product to diffuse out of the narrower channels than the wider ones. Increased exposure time would result in an increased pore count as the narrower pores were revealed.

More direct support for the idea that a growing electrodeposit bridges over pores is provided by a recent scanning electron microscope study of pore structure by Cooksey and Campbell[11]. With an X-ray spectrometer attachment they were able to detect gold growing into the small pores and in one particular case the beginning of a gold bridge which had not yet completely covered the pore.

Diehl[12] made a detailed study of the shapes of two transverse pores in a 125 microinch gold electrodeposit. One of them was cone shaped with the narrower end at the plate surface; the other, illustrated in Figure 23.2 from Diehl's paper, was narrowest part way down the channel rather than at the surface. It had, therefore, begun to flare outward again after an initial growth period during which it was becoming smaller since some pores do not get bridged over by a growing deposit; merely increasing plate thickness may not reduce the pore count to zero. Figure 23.2 b shows the narrowest part of the pore as a transmission shadowgraph, and the upper and lower surfaces of the same pore.

How do pores originate in the first place? The causes are numerous and may be conveniently grouped into two categories: (i) factors relating to the condition of the substrate upon which the gold is plated and (ii) factors relating to the plating baths and the conditions of deposition.

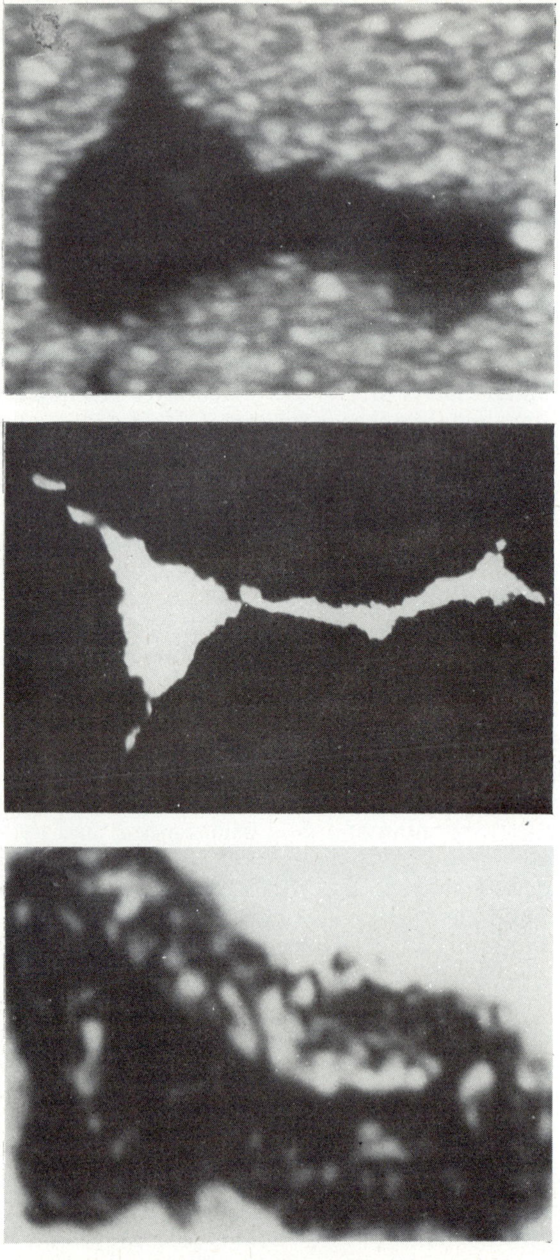

(a)

(b)

(c)

Fig. 23.2 Example of a pore in gold plate which at first decreased, then increased in size as plate grew. (a) Optical micrograph of pore at the plate-substrate interface (b) Shadowgraph by electron microscopy of narrowest portion of pore (c) Optical micrograph of pore at outer surface of the plate.

FACTORS RELATING TO THE SUBSTRATE

Substrate Roughness

A large increase in the porosity of gold plated copper with increased substrate roughness was noted by Ehrhardt[13] and by Noonan[14]. A quantitative study of this effect was also made by the writer[8]. Pure acid citrate gold was plated directly onto (i) OFHC copper discs and (ii) OFHC copper discs with a nickel underplate. The substrate roughness was varied by random direction abrasion with successive grades of metallographic paper and with 0.3 micron alumina under running water. The experimental data are illustrated in Figure 23.3, which clearly indicates a large increase in porosity with roughness.

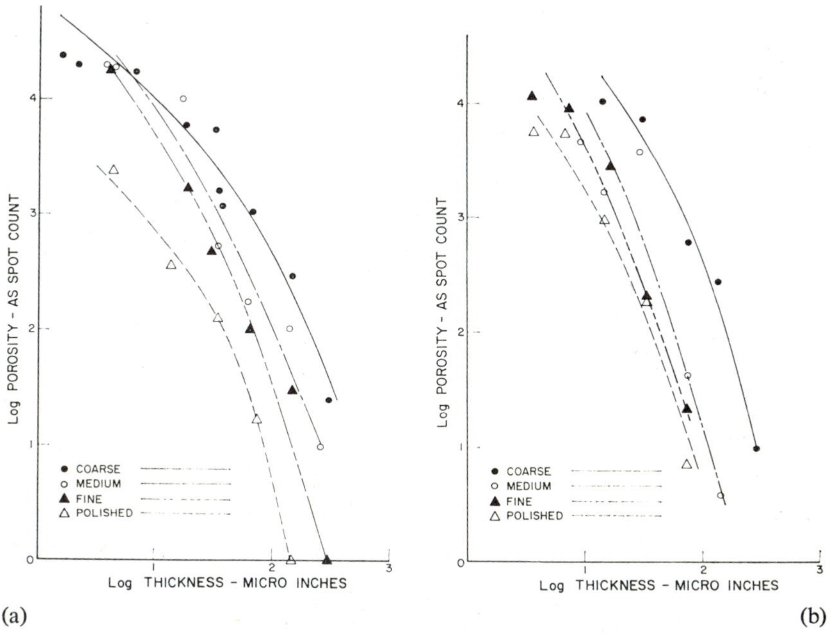

Fig. 23.3 Relationship between porosity-thickness-roughness for (a) Acid citrate gold on OFHC copper (b) Acid citrate gold on OHFC copper with a nickel underplate.

How are these results to be interpreted? Is roughness *per se* a cause of porosity? Do pores follow scratch lines in a clearly correlated fashion? This sometimes does happen, but in the writer's experience this is more often not the case. Roughness is characterised primarily by the geometric shape of the surface which can be quantitatively defined and measured. Surfaces which are geometrically similar may, nevertheless, differ in micro-scale features which may be chemical, physical, or crystallographic in nature, but which are always dependent on the method used for producing the surface finish. Since these differences can have a large influence on porosity, an attempt was made to keep the micro-scale effects constant by using comparable techniques for the preparation of the specimens with varying roughness.

For our purpose, the most convenient quantitative characterisation of roughness is the Roughness Factor which may be defined as

$$R = \frac{\text{true surface area}}{\text{projected surface area}} \tag{7}$$

Usually gold thickness, T, is measured by a weight per unit area method or by methods which are dependent on weight per unit area standards. If we consider a "true" thickness, t, to be the local thickness measured normal to the surface at any point, or the average value of many such measurements then

$$t = \frac{T}{R} \tag{8}$$

For thin plates, at least, equation (8) is strictly true, but it does not apply to plates several times thicker than the scale of the roughness. However, it is the thinner deposits which are of major concern. Measurements on photomicrographs of cross-sections gave a value of about 1.4 for R for the roughest surface of Figure 23.3. If it is postulated that the porosity-thickness relationship is the same for all roughnesses, taking the thickness as the true thickness, t, we can then calculate the value of R which would theoretically account for the increase in porosity with roughness, at constant apparent thickness, T. This value is about 3; details of the calculation may be found in the original paper. We think of the geometric aspect of roughness, not as a *per se* cause of porosity, but as an amplification factor. In addition to the effect on gold thickness, if the method of producing roughness also produces micro-scale surface changes, then there will be more such localised effects per unit of projected area than per unit of true area. The roughness effect on porosity is indeed a complex one, though nonetheless real. To understand it more fully we must turn to a detailed consideration of the micro-scale nature of the substrate surface.

Substrate Surface Defects

Do pores line up with scratches? In an attempt to explore this, the writer prepared abraded specimens and after gold plating, exposed them to nitric acid fumes. Examination of the surface at 70× magnification revealed no general tendency for the corrosion spots to follow the scratch pattern. However, when photographs were taken before plating, after plating, and following corrosion, a different kind of correlation was observed. Most corrosion spots on rough surfaces and all of the corrosion spots on smooth surfaces, can be exactly correlated to point defects on the original substrate surface. On a smooth surface, Nomarski interference contrast microscopy[16] renders such point defects clearly visible. Figures 23.4 and 23.5 are illustrations of these observations. Although the scribed lines in Figure 23.5 were intended as markers to identify areas for photographing, pores exhibited a tendency to line up with the scratches in a few cases. This is shown in Figure 23.6.

An experiment was designed to see whether this tendency can be related to the manner in which the scribing was done. Scratches were made at various loads, with and without lubrication, on OFHC copper discs prior to plating. Corrosion spots which fell on each type of line as well as those falling on the intermediate spaces were counted. Figure 23.7 summarises the observations. The effect of lubrication in reducing the spot count is particularly noteworthy. Lubrication does not reduce the width of the scratch, but within the groove the surface is smoother and less damaged.

The correlation between substrate surface defects and pores in the plate has been confirmed by Cooksey and Campbell[11]. In addition, they were able to identify silica inclusions within the defects. Silica would be expected as a result of the polishing procedures used to prepare specimens.

Zoning

A substrate surface effect on porosity which is of major importance with rolled strip materials, was reported by Clarke and Britton[17] for various base metal platings on strip steel. This study was later extended by Clarke and Leeds[18] to multilayer deposits on steel and to gold plate on copper foil rolled from OFHC copper. They found that the porosity varied in a periodic fashion when successive layers of the basis metal were removed prior to plating. Typically, the removal of the first 1.0×10^{-5} in. of steel gave a sharp reduction in porosity, but the removal of additional amounts gave alternating increases and decreases in porosity. A similar effect occurs with gold plated copper foil.

By plotting a graph of porosity against depth of metal removed, a zone diagram is obtained which is reproducible for different panels cut from the same batch of material, but differs from batch to batch. The effect of varying the method of metal removal is to alter the magnitude of the observed porosity changes, but the positions of the maxima and minima remain almost unchanged. These observations provide strong, albeit indirect, evidence for the existence of metallurgical zones in metals, which in rolled strip materials would be expected to exhibit a directionality related to the rolling process. These zones have not been detected or identified by metallographic means as they are extremely thin, yet they are able to exert a significant effect on the porosity of electrodeposits.

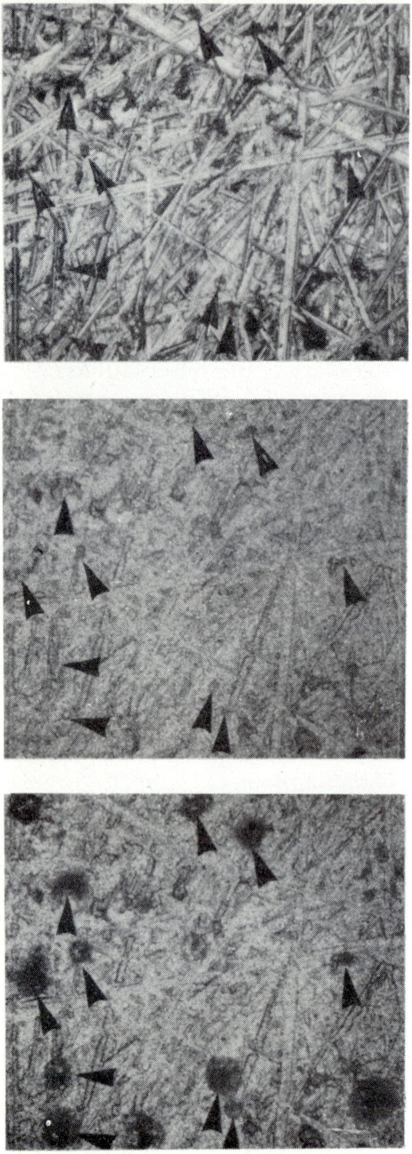

Fig. 23.4 Correlation between pores in gold plate and substrate surface defects on a rough surface. (a) Before plating (b) After plating (c) After exposure to nitric acid fumes.

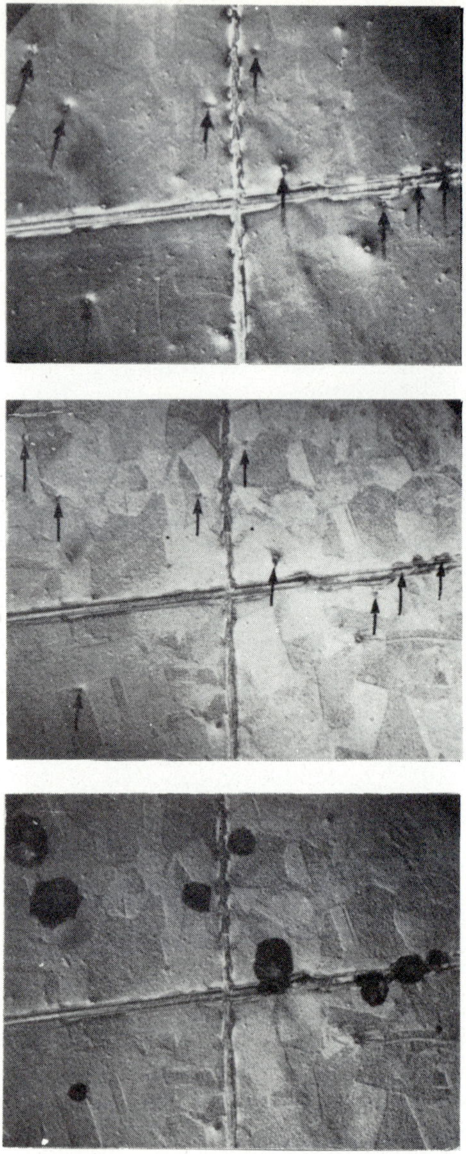

Fig. 23.5 Correlation between pores in gold plate and substrate surface defects on a smooth surface. (a) Before plating (b) After plating (c) After exposure to nitric acid fumes.

The outermost zone, i.e., no metal removed, has different properties from the deeper zones; the latter have the same effect on all deposits. A deep zone causing a porosity maximum on one type of electrodeposit will do so for all others. The untreated surface on the other hand may, for some electrodeposits, cause high porosity which falls sharply with removal of a small amount of metal. With others the same surface will give low porosities which show less change with surface removal. To explain these effects the authors postulated different mechanisms for porosity induction by the surface and by the high porosity inner zones. The surface of a rolled metal will have a highly disturbed structure and can give rise to "crystallographic porosity".

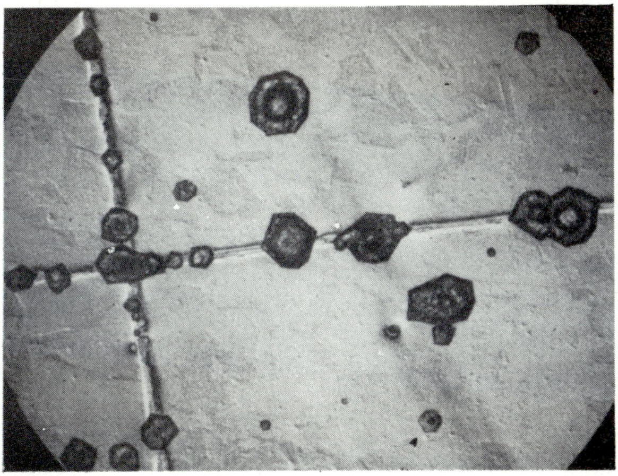

Fig. 23.6 Line up of pores along a scratch made in substrate before plating.

Some electrodeposits "class S", will tend to mimic the structure of the disturbed layer. This is referred to as pseudomorphism and is most pronounced in the initial stages of deposition. This class of deposit tends to be highly porous when deposited on untreated surfaces, an effect which is attributed to mismatching of grains.

"Class I" electrodeposits are relatively insensitive to the disturbed structure of the surface and thus have a lesser degree of crystallographic porosity. They show a porosity minimum on untreated surfaces. The inner zones are said to cause "inclusion porosity", a term which appears to suggest a mechanism similar to that illustrated in Figures 23.4 and 23.5.

These ideas are supported by an observation from the present writer's work which can be seen in Figure 23.7. On the polished copper specimen, etching with 50% nitric acid prior to scribing and plating greatly reduced the porosity on the intermediate spaces, but had no significant effect on the number of

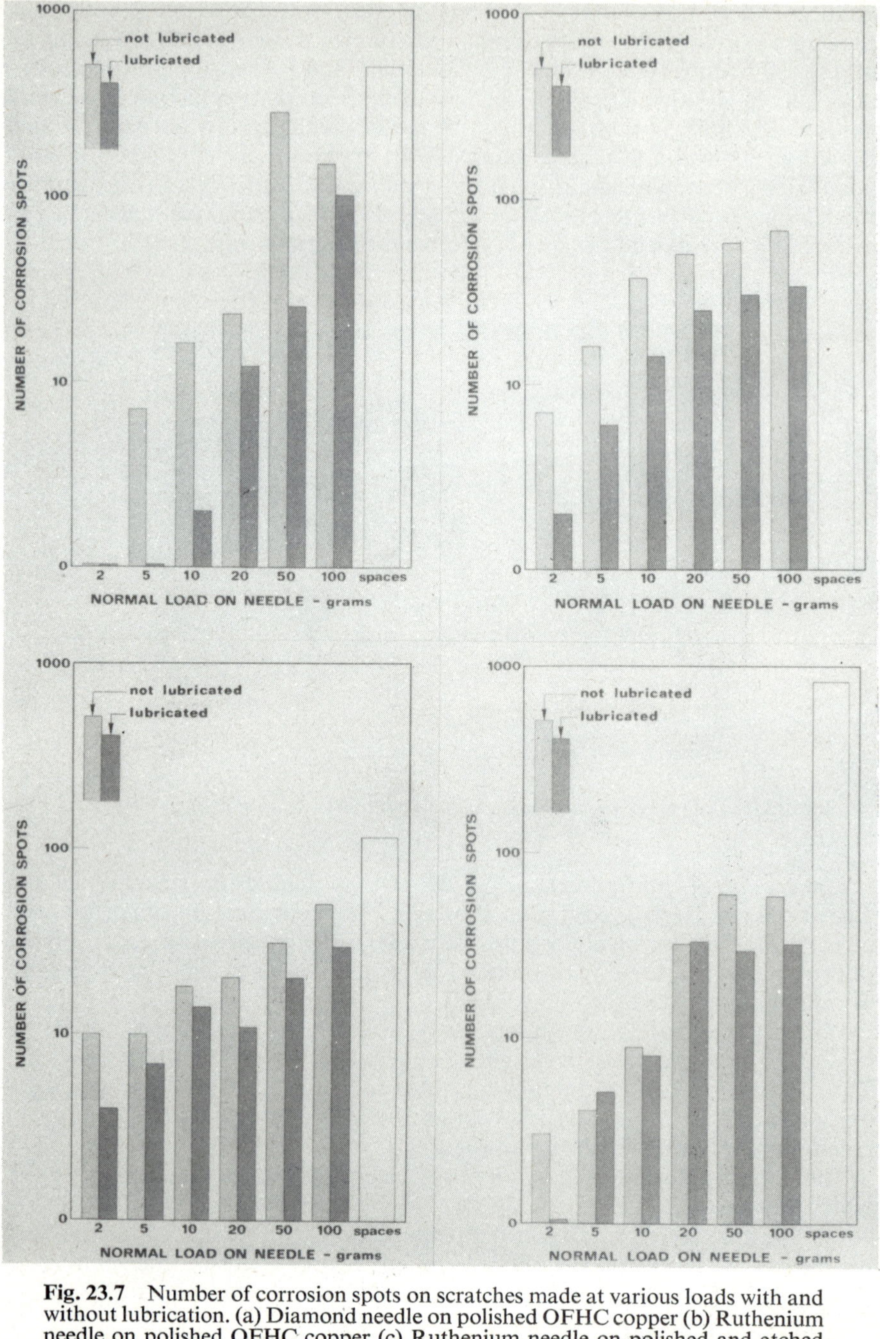

Fig. 23.7 Number of corrosion spots on scratches made at various loads with and without lubrication. (a) Diamond needle on polished OFHC copper (b) Ruthenium needle on polished OFHC copper (c) Ruthenium needle on polished and etched OFHC copper (d) Ruthenium needle on brass.

pores falling directly on the scratches where excessive mechanical working has taken place.

In summation, it is clear that the physical, chemical, and metallurgical structure of the substrate surface, and the interaction of this structure with the growing electrodeposit is a primary factor in controlling the porosity of the deposit. The inclusion porosity of Clarke and Leeds may for all practical purposes be equated with the writer's substrate surface defects. The evidence for this mechanism is the direct observation of exact correlation between such defects and individual pores[11,15]. The evidence for the crystallographic mechanism for porosity is more indirect. In the writer's opinion, it may be difficult to produce a surface which has a high degree of crystallographic distortion, yet is free of the inclusion porosity effect. The converse, a surface with high substrate surface defect density, but free of a distorted surface layer, is quite easy to prepare. An experiment of the latter type has been reported by Clarke and Chakrabarty[19] who used electroformed copper foil as a substrate. One side of this foil, that which was next to the electroforming drum, was smooth and shiny, while the other side was matte, quite rough and free of zoning effects.[20] The porosity of gold, plated on the rough side, decreased when the substrate surface was smoothed by chemical or electropolishing before plating. Mechanical polishing produced an increase in porosity. The initial surface could have produced little or no crystallographic porosity. Electropolishing of this substrate might reduce point defect or inclusion density, but would have no crystallographic effect. Mechanical polishing can increase the number of inclusions or point defects as well as introduce a crystallographically disturbed layer.

Underplates

This topic is included under substrate factors to emphasise the fact that a convenient way to modify the substrate as it is presented to the gold bath, is to use an intermediate deposit. The term underplate will be used when there is only one plate between the gold and the substrate. It will also be used to denote the layer just underneath the gold when there is more than one intermediate plating. In the latter case, the deposit which is first applied to the substrate will be referred to as the subplate.

Nobel, Ostrow, and Thompson[21] noted a decrease in the porosity of gold plate with a nickel underplate from both Watts and sulphamate baths. Ashurst and Neale[22] studied the effects of Watts and bright nickel, and of bright acid and matte copper. All the underplates reduced porosity but the nickel, both bright and matte, were most effective.

The writer[15] explored the effect of underplates on the porosity of pure acid citrate gold. Nickel was plated from a Watts bath and copper from an acid sulphate solution, both without any addition agents. Both underplates reduced the porosity of 30 microinches of gold, but in these experiments the copper was more effective. Two additional results were that very thin nickel plates resulted in increased gold plate porosity, and that when both platings were used in combination with the copper as a sub-plate, the porosity was decreased to a level lower than that obtained by either plate alone. This is illustrated in Figure 23.8.

The basic reason for the effect of underplates on porosity is that the underplate modifies the surface upon which the gold is plated. Favourable modifications are those which result in smoothing of the surface, and in covering substrate surface defects and metallurgically strained layers. A good rule is

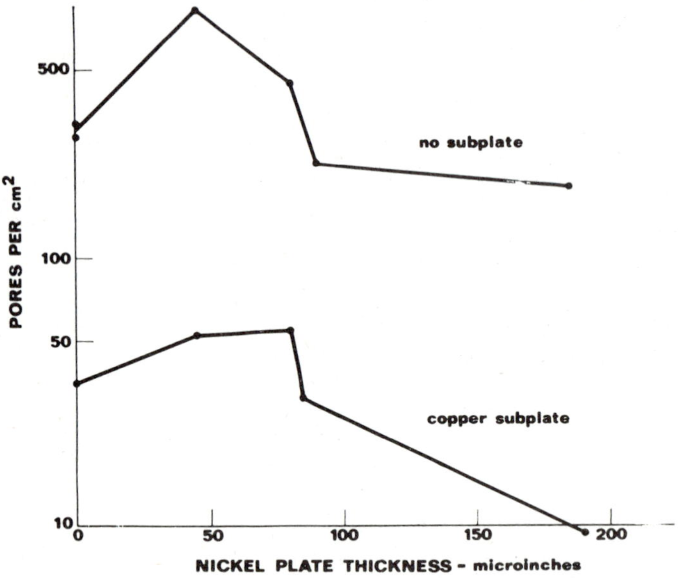

Fig. 23.8 Porosity of 30 μ-in. of acid citrate gold plated over various thicknesses of Watts nickel underplate with and without 75 μ-in. of acid sulphate copper subplate.

that electrolytes with good microthrowing power will best perform these functions. Recently Clarke and Chakrabarty[23] studied the effect of plating variables and different bath types for copper underplate upon gold plate porosity. Bright acid copper was the most effective in reducing porosity. Increased bath temperature and current density gave improved results. When thickness was above some value at which the undercoat became rough, porosity increased.

The choice of an underplate for gold is based on various and sometimes conflicting considerations. Adhesion may be improved in some cases. Conductivity is important in electrical contact applications. Some plates act as diffusion barriers to prevent substrate metals from diffusing to the surface of the plate. The mechanical properties of the underplate must also be considered on parts which are to undergo post-plating deformations or heavy wear. These matters are discussed in Chapter 21. The porosity of the gold will be affected by the underplate, often for the better, but sometimes for the worse.

One cannot regard porosity of gold plate as an isolated matter but must consider substrate, underplates and gold plate as a total system, the elements of which interact with each other to produce the final result.

FACTORS RELATING TO THE ELECTROLYTE AND CONDITIONS OF DEPOSITION

A perusal of the literature on the variation of porosity with plating conditions reveals three facts: (i) variation does exist and can sometimes be very great, (ii) observations by different workers have not produced agreement on the nature of the variations, and (iii) though speculative hypotheses have been offered, little progress has been made toward elucidating the mechanism of porosity variation with plating conditions. It will be instructive to compare the various findings for each of the parameters which have been examined.

Current Density

In his study of acid citrate gold, Ehrhardt[13] noted an increase in porosity of gold plated from a bath containing 30 g/l sodium gold cyanide and 100 g/l ammonium citrate, as the current density was increased. Points were taken at values of 1, 2, 4, and 8 A/ft^2. Leeds and Clarke[20] investigating a similar electrolyte showed porosity decreasing and then increasing with current density, with an optimum at about 5 A/ft^2. Ashurst and Neale[22], reported increasing pore counts with current density and correlated this to decreased current efficiency. Their bath was a cobalt-hardened acid citrate solution. Leeds and Clarke[20] found that addition of nickel to the acid citrate bath shifted the porosity-current density curve to lower porosity levels, but that the initial reduction to an optimum value still occurred.

Plating Bath Composition

Ehrhardt[13] found the acid citrate gold to be far superior to the cyanide bath with respect to the porosity of the deposit. Ashurst and Neale[22] compared a citrate-type cobalt brightened gold, with a mixed phosphate-citrate nickel brightened gold. Their data show a preference for the latter. The most comprehensive study of the effect of solution composition and all of the important plating variables, is that made by Leeds and Clarke[20,24]. They investigated the alkaline cyanide, the acid citrate and the phosphate bath. The porosity obtained depends not only on bath composition and current density, but on pH, temperature and anion concentration. These factors may interact so that the form or position of a current density-porosity curve can be changed or displaced by, e.g., a change in operating temperature, pH, agitation or anion concentration. These inter-relationships are so complex that to present them fairly would require the reproduction of nearly the whole of the original papers. Some of the more salient observations are presented in a condensed form in Table 23.1. The table should be interpreted as follows:

Table 23.1

Variation of dV/dI with Bath Type and Composition, and with Plating Conditions

Alkaline Cyanide Bath	*Citrate Bath*	*Phosphate Bath*
42–70°; 8 Au, 3 CD	32–60°; 8 Au, 3 CD, pH 4	50–80°; 8 Au, 3 CD, pH 6
—	45–pH 6; 16 Au, 3 CD, 50°	50–pH 6; 8 Au, 3 CD, 70°
22–0.5 CD; 8 Au, 20°	—	50–3 CD; 8 Au, pH 6, 70°
22–4 CD; 16 Au, 20°		47–2.5 CD; 16 Au, pH 6, 70°
30–3 CD; 24 Au, 20°		57–3 CD; 24 Au, pH 6, 70°
		Note 1
22–0.5 CD; 20°, 8 Au		
39–3 CD; 70°, 8 Au	—	—
—	51–5 CD; 16 Au, pH 6, 70°	47–2.5 CD; 16 Au, pH 6, 70°
14–32 CN^-, 8 Au, 3 CD, 20°	32–450 citrate; 8 Au, pH 4, 3 CD, 50°	50-at all PO_4; 8 Au, pH 6,3 CD, 70°
37–160 K_2CO_3; 8 Au, 3 CD 20°		
Note 2	Note 3	
22–0.5 CD; 8 Au, 20°	30–4 CD; 16 Au, pH 4, 50°	
80–0.5 CD; 0.16 Ag, 8 Au, 20°	46–6 CD; 1 Ni, 16 Au, 50° pH 4	—
34–3.5 SCO, 8 Au, 3 CD, 25°	37–1 Ni; 16 Au, pH 4, 50° 3 CD	—
60–0.5 CD; 2.5 SCO, 8 Au, 25°		—

Note 1—At current densities less than 2.5 or greater than 3 A/ft^2, the 16 Au bath gives lower porosity than the 8 Au bath.

Note 2—32 CN^- g/l was the lowest value tested; porosity increased at all higher values. The range of K_2CO_3 concentrations tested was 0–200 g/l.

Note 3—450 g/l citrate was the highest value tested, porosity increased at all lower values.

Each entry in the table is in the form, 42–70°C, which means that in the curve of dV/dI* vs temperature (or whatever parameter was varied for this entry) the maximum value of dV/dI, in this case 42, occurs at 70°C. Following each entry are given the values of other parameters held constant for the entry. A number followed by CD gives current density in A/ft^2. A number followed by a chemical symbol gives the concentration of this constituent in the bath in g/l; SCO is sulphated castor oil. Horizontal rows are spaced so as to

* *dV/dI* is a measure of the electrolytic resistance of the pore channels and an inverse measure of porosity. Hence, maximum *dV/dI* is minimum porosity (see Chapter 27).

facilitate comparisons between the three bath types. The last group of entries in the table deals with the effects of brightener additions.

Other interactions not listed in Table 23.1 are reported in these papers, as well as data on cathode efficiencies and alloy compositions. The interested reader should consult the original papers.

THE PRODUCTION OF LOW POROSITY GOLD DEPOSITS

Nowhere is the proverbial, luckless, "man in the middle" better typified, than by the Foreman Plater, whose job it is to bring all of this to practical fruition. On the one hand he is confronted by a mass of research data, much of it contradictory, some of it confused, and all of it prefaced with the warning, expressed or implied, "these results apply to the experiment as performed under the stated conditions and may not apply to other conditions". On the other hand he is faced with the specification writer, usually his customer, who knows what performance he wants and has his own, often immutable, ideas on how this performance is to be obtained. The practical plater will not find in this section any "how to do it" recipes and would be well advised to take with a pinch of salt any such recipes which he finds anywhere. The plater's first responsibility is to meet specifications and to do this he must have methods of testing. Chapter 27 deals with this subject in considerable detail.

GENERAL SHOP PRACTICE

A surface defect does not have to originate in the substrate; it can be acquired by the surface during plating if the solution is contaminated with insoluble particles of any description. Anode sludge, shop dirt, oils and precipitates due to improper rinse cycles are common offenders. Continuous filtration of all plating baths in the cycle including the underplates is highly desirable. Rinse cycles should be carefully designed to avoid cross-rinsing practices which could result in deposition of insoluble films on the surface. Every book on plating practice ever published has recommended adequate rinsing and this one will be no exception. Each bath in the cycle should have its own two-stage and preferably three-stage cascade rinse tank, with the water flowing counter-currently to the work. The argument is simple. All the effects detailed in the previous sections on surface roughness, plating conditions and solution composition, can be totally overwhelmed by dirty solutions and improperly cleaned and rinsed work entering the gold bath. Proper solution maintenance is absolutely essential for all aspects of plate quality.

One particular matter should be mentioned. Small amounts of iron in all acid gold solutions can codeposit with gold and cause stress cracking of the deposit. This cracking may often be invisible except in a cross-section under high magnification, but it is sufficient to cause very high porosity in the deposits. This effect was brought to the attention of the writer by R. G. Baker of Bell Telephone Laboratories. Tramp iron from overhead structures and stainless steel anodes in acid gold solutions are common causes. Many proprietary solutions have additives for the purpose of chelating metallic contaminants, but these can become saturated in time. Prevention is the best practice.

SUBSTRATE PREPARATION

Smooth substrates give lower porosity deposits; the writer has proclaimed this principle in the literature, but the exceptions to the rule must be clearly understood. An important example is grease buffing, which usually results in the highest porosity of all for the various methods of surface preparation. Any mechanical method of polishing is suspect and should be tested for resulting porosity before adoption. In general, chemical or electropolishing will be effective in the removal of highly strained surface layers and will not introduce inclusions into the metal. Poor machining practice resulting in coarsely finished surfaces can raise the porosity levels catastrophically. This can be seen in Figure 23.7. The Plater's oldest complaint, "I can't put a good finish on a bad surface", is well established.

PLATING CONDITIONS

The summary of the effects of plating conditions given can serve as a useful guide provided it is understood that proprietary baths may not necessarily behave in exactly the same way as the prototype solutions which were investigated. The ideal situation would be for supply houses to investigate their solutions and publish complete data for the effects of plating conditions on porosity. In the absence of such information, however, the plater can do this himself, at least well enough to avoid the most drastic pitfalls. It would be to guide the plater in making such an investigation that Table 23.1 would be most useful. In this connection a few comments will be made on the problem of current density in particular.

The effect of current density on porosity is well documented and has already been discussed earlier in this chapter. A problem arises in the interpretation of these results in the case of barrel plating. The investigations cited were carried out using rack plated specimens. Can the results be directly applied to barrel plated work? In the practical world there is always a variation in the average current density on individual parts whether they are plated on a rack or in a barrel. The average, however, arises in a different way. The current density on a part placed in a given position on a rack will be reasonably constant during the plating cycle although various positions on the rack may differ. An individual part in a barrel will receive a widely varying current during the plating cycle. A rack plated part set to be plated at a given current density will receive this current during all of the cycle. The barrel plated part may receive the same average current, but this will have fluctuated over a wide range during the cycle. What then happens to the current density-porosity relationship? In the writer's experience the relationship continues to hold, but the effects are much less sharp. The choice of an optimum current density is still possible and necessary. It is also important to use barrels designed to minimise current fluctuation and to observe proper loading practices so that parts will tumble properly.

Another problem concerns thickness control of the deposit. Deposit thickness is proportional to the product of the current and the plating time. Economic considerations dictate the use of the minimum plating period. With automatic equipment, the choice of current density is often based on equipment design considerations, which in turn may hinge on the space available for the plating machine. Platers who must plate different jobs at varying thicknesses may prefer to change the current rather than the time. These practices must inevitably result in porosity variations, the causes of which may be difficult to trace. When plate porosity is an important quality requirement it is essential that all elements of the plating cycle be chosen with consideration of their effects on porosity and that once chosen, they should be adhered to.

POST-PLATING OPERATIONS

The design of gold plated parts often requires fabrication operations which must be performed after plating. This is always undesirable and carries the risk of degrading the quality of the plate. When bending or forming after plating is unavoidable, it is essential to choose high ductility gold plates and to avoid contaminants, such as iron, which can increase the stress of the deposit. In particular, it is important to consider the ductility of underplates which are chosen. If nickel is used, the bath must be operated to produce the least internal stress in the deposit.

In many applications gold plated parts are assembled into a product. Such operations may involve handling, deformation, baking and mechanical or chemical cleaning. All handling and mechanical operations should be suspected of increasing the gold porosity until it is shown that they are safe, or that the effect is confined to a non-functional area. Chemical cleaning operations should be designed to avoid entrapment of insoluble residues. Vapour degreasing is generally safe provided that the equipment is properly maintained. Mild aqueous detergent solutions are safe for the removal of soils which are insoluble in the solvent vapours. Parts must be thoroughly rinsed and dried. Such contaminants are usually inorganic and may be corrosive. The best practice is to eliminate them at the source. If this is not possible the necessary chemical cleaning should be carried out as soon as is feasible in the process cycle. Never hold dirty gold plated parts in storage!

POROSITY SPECIFICATION REQUIREMENTS

In the United States, specifications for gold plate are promulgated by the government (M1L G45204-with revisions) and by ASTM, in particular

Committee B-8. These documents specify freedom from porosity, scratches, surface imperfections, etc., in a single catch-all sentence. Since one can, with sufficient diligence, find imperfections in any surface, or at least one pore in any "pore-free" plate, it is difficult to attach practical significance to such statements. Another ASTM Committee, B-4, has a sub-committee now studying various gold plate porosity test methods with the view to recommending practices for porosity testing of gold plated electrical contacts, but has not yet issued any document on the subject as yet. However, such a document has now been approvedand is scheduled to appear in ASTM Standards this year (1974).

The task of defining specification limits to porosity is therefore largely left to negotiation between buyer and seller. This is seldom a simple problem. It is complicated by (i) the variability of porosity levels in practice and their sensitivity to so many factors, and by (ii) the fact that what is at issue is not really the porosity itself, but its effect on performance. It will bear repeating that this results from the interaction between the level of as-deposited porosity, the composition of the substrate and the environment in which the plated component is used. The writer strongly believes that porosity specifications should be in something like the following format: "Porosity levels on functional surfaces, should be less than those which can (at a specified confidence level) degrade the performance of the part when it is subjected to (specified) environmental exposure". This would require establishing the relationship between porosity levels and performance, as well as agreement on relevant accelerated tests. The latter subject is covered in Chapter 27.

SUMMARY

Porosity in gold plate can, under some environmental stresses, negate both the cosmetic and engineering advantages of the gold. Its causes are numerous and complex. The condition of the substrate surface is a basic factor and individual pores can be directly related to the presence of some sort of visible point defects on the substrate surface. Surface roughness can be thought of as a geometric amplifier which acts to increase the surface defect concentration and to decrease the true gold thickness.

Although surface smoothness is desirable, the manner of achieving this must be carefully chosen. Chemical or electrochemical polishing methods which cannot introduce metallurgically strained surface layers are best. Underplates, especially of copper and nickel, can reduce gold plate porosity and their choice in a given application must take into account their effect on inherent porosity of the plate. Environmental factors which may affect different underplates in varying degrees and their effect on mechanical requirements of the plated part should also be considered. Finally, the conditions for deposition of gold must be carefully selected to give the best results. In making this choice it must be realised that these conditions will not be identical for all bath types. The production of low porosity deposits requires the judicious choice of underplate systems, gold electrolytes and plating conditions, as well as careful attention to the maintenance of good shop practice and adherence to the chosen conditions.

In the United States at least, porosity specification is still largely a matter for buyer-seller negotiation although progress is being made in ASTM toward the development of uniform testing standards and specifications for the porosity of gold plate.

REFERENCES

1 Cooley R. L., Lemons K. E., *Plating* **56,** 511 (1969).
2 Munier G. B., *Plating*, **56,** 1151 (1969).
3 Frant M. S., *Plating*, **48,** 1305 (1961).
4 Pinnell M. R., Bennett J. E., Proc. Holm Semin. on Electrical Contact Phenomena, Illinois Inst. of Technology, Chicago (1971).
5 Modjeska R. S., Kann S. N., *Proc. Am. Electropl. Soc.*, **50,** 117 (1963).
6 Antler M., *Plating*, **57,** 615 (1970).
7 Frant M. S., *J. electrochem. Soc.*, **108,** 774 (1961).
8 Garte S. M., *Plating*, **53,** 1335 (1966).
9 Egan T. F., Mendizza A., *J. electrochem. Soc.*, **107,** 353 (1960).
10 Krumbein S. J., Proc. Holm Semin. on Electrical Contact Phenomena, Illinois Inst. of Technology, Chicago (1968).
11 Cooksey G. L., Campbell H. S., *Trans. Inst. Metal Finish.*, **48,** 93 (1970).
12 Diehl R. P., *Plating*, 57, 51 (1970).
13 Ehrhardt R. A., *Proc. Am. Electropl. Soc.*, **47,** 78 (1960).
14 Noonan H. J., *Plating*, **53,** 461 (1966).
15 Garte S. M., *Plating*, **55,** 946 (1968).
16 Nomarski G., Weill A. R., *Rev. de Metallurgies*, **52,** 121 (1955).
17 Clarke M., Britton S. C., *Trans. Inst. Metal Finish.*, **37,** 110 (1960).
18 Clarke M., Leeds J. M., *Trans. Inst. Metal Finish.*, **43,** 50 (1965).
19 Clarke M., Chakrabarty A. M., *Trans. Inst. Metal Finish.*, **48,** 99 (1970).
20 Leeds J. M., Clarke M., *Trans. Inst. Metal Finish.*, **46,** 81 (1968).
21 Nobel F. I., Ostrow B. D., Thompson D. W., *Plating*, **52,** 1001 (1965).
22 Ashurst K. G., Neale R. W., *Trans. Inst. Metal Finish.*, **45,** 75 (1967).
23 Clarke M., Chakrabarty A. M., *Trans. Inst. Metal Finish.*, **50,** 11 (1972).
24 Leeds J. M., Clarke M., *Trans. Inst. Metal Finish.*, **47,** 163 (1969).

BIBLIOGRAPHY

1 Harding W. B., "Tarnish Resistance of Gold Plating Over Silver", *Plating*, **47,** 1141 (1960).
2 Hoar T. P., "Electrochemistry of Protective Metallic Coatings", *J. Electropl. Depos. tech. Soc.*, **14,** 33 (1938).
3 Hothersall A. W., Hammond R. A. F., "Causes of Porosity in Electrodeposited Coatings", *Trans. electrochem. Soc.*, **73,** 449 (1938).
4 Kutzelnigg A., "Porosity of Electrodeposits, Causes, Classification and Assessment", *Plating*, **38,** 382 (1961).
5 Leeds J. M., "A Survey of the Porosity in Gold and Other Precious Metal Deposits", *Trans. Inst. Metal Finish.*, **47,** 222 (1969).
6 Niehoff R. T., Faust C. L., Batelle Memorial Inst. & Univ. S. Calif., "Properties of Electrodeposited Gold and Gold Alloys for Electronic Uses", Bibliography with Abstracts Am. Electropl. Soc. Res. Project No. 25, A.E.S. Res. Rep. Serial No. 58 (1970).
7 Ogburn F., *et al.* Natn. Bur. Stand., "The Nature, Cause and Effect of Porosity in Electrodeposits", Am. Electropl. Soc. Res. Project No. 13, A.E.S. Res. Rep. Serial Nos. 36, 40 and 47 (1956).
8 Thon N., "Porosity of Electrodeposited Metals", *Proc. Am. Electropl. Soc.*, **36,** 241 (1949).

Chapter 24

DENSITY

J. M. Leeds

In common with other metals, the density of gold in the electrodeposited condition is less than the generally accepted value of 19.32 g/cm^2 pertaining to the pure metal in the cast or wrought condition. Relatively little work has been carried out on density measurement and reported results are conflicting, showing densities ranging from only slightly less than theoretical to as low as about 50% theoretical. Such divergence may well be attributable to the different techniques employed and to the general experimental difficulties in measurement in view of the very small amounts of material usually involved, more especially if determinations are to be made on deposits in the rather low thickness range of normal application.

The density of gold plating is important from two main viewpoints. Firstly, an excessively low value may serve as a pointer to the presence in the deposit of porosity or to undue contamination by base metals or by non-metallic or gaseous inclusions. Secondly, and of more immediate importance in day-to-day operations, density is closely concerned in the accurate measurement of plating thickness by radiation methods, e.g., beta-ray backscatter, which are already widely used and increasingly included in standard specifications where they may complement or replace traditional techniques based on weight gain or optical microscopy.

For example, measurement of thickness by the beta-ray method, which is discussed in more detail in Chapter 28, is based essentially on the comparison of backscatter intensity from a sample of unknown thickness with that from known standards. The intensity of backscatter depends on the depth of penetration of the gold deposit by the beta-ray particles, which in turn is dependent on the mean energy of the latter and the density of the coating. Clearly the accuracy of measurement will be affected if the density of the test coating differs significantly from that of the calibration standards.

This point has been well illustrated by the work of Cooley and Lemons[1] who stress the desirability of some knowledge of the effects on density of processing variables such as electrolyte pH and temperature, and deposit composition in terms of alloying constituents and codeposited foreign matter, all of which factors may influence the structure of the coating. These authors also comment on the lack of general published information on these apsects, particularly as applying to the relatively thin gold deposits of commercial importance, whose structures may be strongly influenced by the nature of the basis metal.

MEASUREMENT TECHNIQUES

The traditional type of procedure for determining the density of a solid by comparison of its specific gravity with that of water or an organic reference liquid such as di-iodemethane or bromoform, is fully described in the standard literature[2], together with the various corrections to take account of possible variation in air buoyancy effect, temperature, etc. In applying the method to gold plating, a thick coating is prepared, then stripped from the substrate, finely crushed and sieved, coned and quartered for duplicate determinations. It is important that any air entrapped with the powder should be excluded from the specific gravity bottle, to which end a trace of wetting agent may be added to the liquid to reduce surface tension, and ultrasonic vibration may be applied during filling. Weighings to constant weight are made at a standard temperature, usually 20°C.

This technique was used by Craig and his associates[3] in a study of 24 carat gold coatings, but Cooley and Lemons[1] considered it unsuitable for measurements on thin deposits of commercial importance. In their technique, density of thin coatings was measured without stripping, by comparison of the real thickness of coating, as measured by profilometry, with the thickness determined by beta-ray backscatter using evaporated gold films of theoretical density as calibration standards. Since the beta-ray thickness represented that which would pertain to a pure gold coating of theoretical density containing the same mass of gold per unit area as the coating under test, a value for the density of the latter could be simply derived from the ratio of the thicknesses as determined by the two techniques. A possible source of error in this method could arise through grooving of the gold coating by the diamond stylus of the profilometer. This was not detected in the present case, though it has been reported to occur with copper deposits of similar hardness in studies by Arrowsmith, Dennis and Fuggle[4], when some ploughing through of surface peaks by the stylus was observed by electron microscopic examination. It is possible that some unidentified factor may have influenced Cooley's results, which are remarkably lower than any other reported values.

EXPERIMENTAL OBSERVATIONS

Density measurements have been reported by Craig and coworkers[3], Foulke[6], Duva and Foulke[5], and Cooley and Lemons[1]. The results are presented in Tables 24.1–24.4.

All the observations by Craig (Table 24.1) are much higher than any others reported, most of the values obtained being within about 0.2% of theoretical. This may be because specific gravity comparison was made using di-iodomethane, with careful corrections for air buoyancy and temperature fluctua-

Table 24.1
Density of Gold Deposits Reported by Craig *et al.*[3]

Bath No.	*Type*	*Density g/cm³*
1N	Hot cyanide, phosphate, carbonate	19.22
1U		19.13
2N	Citrate, phosphate in equal amounts	19.26
2U		19.25
3N	As 2 plus 25 g/l polyamino acid complex	19.27
3U		19.27
4N	As 2 plus 10 g/l metallic brightener	19.26
4U		19.26
5N	As 4 plus 20 g/l potassium chromium sulphate hardener	19.23
5U		19.26
6N	As 5 plus 25 g/l boric acid inhibitor	19.25
7N	Citrate plus 6 g/l hydrazine sulphate redox agent	19.26
7U		19.25
8N	Phosphate plus 6 g/l hydrazine sulphate	19.27
8U		19.26
9N	Ammonium citrate plus 20 g/l potassium chromium sulphate hardener	19.25
9U		19.27
10U	Borocitrate plus tetraethylene pentamine	19.26
11U	Phosphate	19.27
12U	Phosphate	19.27
13N	Gold sulphite with excess sulphate and sulphite	19.23
13U		19.25
14N	As 13 plus 15 ppm arsenic	19.13
14U		19.24

* "N" and "U" signify new and used baths respectively.

Table 24.2
Density and Purity of Gold Deposits (Duva and Foulke[5])

	Type	*Deposit Purity (%)*	*Density (g/cm³)*
A	"Pure" alkaline cyanide (matte)	99.999	18.9
B	"Pure" alkaline non-cyanide (bright)	99.955	19.2
C	"Pure" neutral (semi-bright)	99.999	19.1
D	"Pure" neutral (matte)	99.999	19.1
E	"Pure" acid soft (matte)	99.999	19.2
F	"Pure" acid hard (matte)	99.999	19.1
G	"Alloy" (Ag) alkaline cyanide (bright)	99.020	16.7
H	"Alloy" (Cd) alkaline non-cyanide (bright)	98.576	18.9
I	"Alloy" (Cd & Cu) alkaline non-cyanide (bright)	76.350	15.7
J		81.260	16.0
K	"Alloy" (Ni) acid (bright)	99.898	17.9
L		99.165	17.4
M	"Alloy" (Co) acid (bright)	99.080	17.3
N		99.905	17.8
O	"Alloy" (Co & In) acid (bright)	98.275	16.4

tion. The low density of the deposit from bath 1 was thought to be associated with the presence of large pores within the deposit, and the low values for baths 12 and 14 were attributed to codeposition of sulphur and arsenic respectively. No other comments were offered on the differences in values for deposits from plating solutions of different compositions, nor on the slight differences observed for deposits from new and used baths of similar composition.

The results shown in Table 24.2 were also determined by specific gravity comparison and were corrected to constant temperature (20°C), but no correction was made for air buoyancy. The authors state that baths containing free cyanide will always yield deposits of low density, probably due to occluded organic matter and/or water, but no analysis for contaminants was presented.

The figures for purity of coatings in Table 24.2 refer to codeposited metallic impurities only, as determined by DC arc emission spectroscopy, and take no account of carbon codeposition. The density figures in this Table are fairly consistent, where bath types can be matched, with some results published some 5 years earlier by Foulke[6], as shown by comparison of the figures for Baths A, G, E and J, K, L, M, N (Table 24.2) with those for Baths 1, 2, 5 and 6 (Table 24.3).

Table 24.3

Density Measurements of Foulke[6]

No	*Type*	*Gold Density (g/cm^3)*
1	"Pure" alkaline cyanide	19.2
2	"Alloy" (Ag) alkaline cyanide (bright)	15.6
3	"Alloy" (Sb) alkaline cyanide	18.1
4	Neutral	18.0
5	"Pure" acid	19.2
6	"Alloy" acid (bright)	17–18

Table 24.4

Density Observations of Cooley and Lemons[1]

No	*Substrate*	*Gold Density (g/cm^3)*
3	Au/Cu, mechanically buffed	9.19
6	Au/Pd/Cu, chemically polished, Cu mechanically buffed	7.65
9	Au/Pd/Cu, chemically polished, Cu mechanically buffed	11.28
11	Au/Cu, mechanically buffed	16.50
12	Au/Pd/Cu, chemically polished, Cu mechanically buffed	11.86
13	Au/Pd, chemically polished, Cu mechanically buffed	12.50
14	Au/Pd/Cu, chemically polished, Cu mechanically buffed	10.33
15	Au/Pd/Cu, chemically polished, Cu mechanically buffed	8.62

EFFECT OF CODEPOSITED SECOND METAL AND CARBON ON DEPOSIT DENSITY

In order to improve hardness, wear resistance and brightness of gold deposits it is common practice to codeposit a second metal such as nickel or cobalt. Codeposition has a marked effect on density; for example, the addition of nickel to Bath E (Table 24.2) so that 0.1% is incorporated in the deposit (Bath K) lowers the density from 19.2–17.9 g/cm^3. Further codeposition, up to 18.7% nickel (Bath J) lowers the density to 16.0 g/cm^3. Codeposition of 0.98% silver also has a marked effect (Bath G), lowering the density to 16.7 g/cm^3.

The above observations are very important. For any gold plating bath there is a wide range of electrolyte and deposition parameters, such as pH, temperature, current density, agitation and so on, which may influence the amount of codeposited base metal and hence the density, so ultimately the accuracy of thickness measurement.

Lowering the pH of an acid gold bath by 0.5 pH unit, from 4.5–4.0 can change the amount of codeposited nickel by as much as 0.6% and influence density[7] by as much as 1.6 g/cm^3.

In discussing the low densities observed for electrodeposited gold all authors have postulated the codeposition of organic material as a contributory factor. Quantitative studies of this aspect are of comparatively recent origin, having been greatly stimulated by the observations of Munier[8] in the course of studies aimed at determining the cause of deterioration of gold coatings on sealed reed switch blades due to the formation of dark insulating films. It appears that quite large amounts of carbon can be codeposited, e.g., from 0.2–0.6%, the actual amount, as shown by Munier and more recently by Holt and Stanyer[9], depending on the plating conditions, such as temperature, pH and cathode efficiency, and also being strongly influenced by the amount of base metal, e.g., nickel or cobalt, deposited. The carbon is thought to originate from the cyanide ion, of the potassium cyanide complex used as the gold salt. Polymeric material is distributed throughout the deposit and various types of material have been isolated by Munier and more recently by Davis[10]. The effect of codeposited organic material has not been measured, though it must have contributed to the results reported in Tables 24.1–24.4.

EFFECT OF SUBSTRATE STRUCTURE

It is widely recognised that the structure of most electrodeposits, and hence properties such as hardness, porosity and so on, is influenced by epitaxial and pseudomorphic growth on the substrate in addition to the conditions of deposition. The latter, however, which control the metal ion concentration at the substrate surface and the degree of inhibition of growth, are primarily responsible for determining whether or not the substrate will exert a strong or subtle influence on deposit structure. In the case of a strong substrate influence, pseudomorphism ensures that the substrate grain boundary structure is continued by the deposit, which is most clearly seen when the grain size is large, as in the case of an annealed substrate. For a heavily worked substrate it is more difficult to follow the main grain structure of deposit/substrate due to the many small sub-grains of the highly disorientated substrate. Since most commercial deposits are thin, and almost invariably applied to mechanically worked basis metals, Cooley and Lemons[1] attempted a study of the effect of substrate pretreatment on coating density.

Their results, summarised in Table 24.4, lack consistency. For example, deposits on a heavily worked substrate produced by mechanical buffing gave two separate density figures of 9.19 and 16.50 g/cm^2. Similarly, a mechanical polished substrate subsequently treated by chemical polishing, followed by the application of acid copper and palladium undercoats, led to density figures of 8.62 and 10.33 g/cm^2.

SUMMARY

As pointed out in the opening paragraphs of this chapter, the density of a gold coating is of great practical significance in the measurement of thickness, and equally so as an indication of contamination of the coating by base metals and/or organic occlusions. The latter are particularly interesting, since it has been suggested that the presence of small amounts of polymer may exert a beneficial lubricating effect, whereas large amounts are likely to adversely affect certain properties of the deposit, such as mechanical strength, solderability and bondability, and contact performance.

In view of this, and having in mind the increasingly stringent functional requirements placed on gold coatings, the need for further studies concerning the effect of electrolyte and deposition parameters on the nature and amount of contaminants present in gold plate is evident, and these could reasonably include density measurements for purposes of correlation. In the meantime, the value of the relatively few measurements reported is limited by the wide variations recorded by different observers.

REFERENCES

1 Cooley R. L., Lemons K. E., *Plating*, **56,** 511 (1969).
2 Density of Solids and Liquids, National Bureau of Standards Circular 487. Density of Fine Wire and Ribbon for Electronic Devices. A.S.T.M. Standard F 180-50 Part 8, pp. 808–810, Nov. (1968). Brenner A., Zentner V. W., Jennings C. W., *Plating*, **39,** 865 (1952).
3 Craig S. C., Harr R. E., Henry J., Turner P., *J. electrochem. Soc.*, **117,** 1450 (1970).
4 Arrowsmith D. J., Dennis J. K., Fuggle J. J., *Electroplg Metal Finish.*, **22,** 19 (1969).
5 Duva R., Foulke D. G., *Plating*, **55,** 1056 (1968).
6 Foulke D. G., *Metal Finishing J.*, **9,** 487 (1963).
7 Leeds J. M., Clarke M., *Trans. Inst. Metal Finish.*, **47,** 163 (1969).
8 Munier G. B., *Plating*, **56,** 1151 (1969).
9 Holt L., Stanyer J., *Trans. Inst. Metal Finish.*, **50,** 24 (1972).
10 Davis M. V., *Am. Electropl. Soc.*, 4th Plating in the Electronics Industry Symp., Indianapolis, Feb. (1973).

PART 5

Testing

Chapter 25

SOLDERABILITY TESTING

C. J. THWAITES

BASIC REQUIREMENTS

The process of soldering involves the wetting of the surfaces of the components being joined by the molten solder, followed by flow of the solder into all the joint spaces. The speed at which the initial wetting process takes place is largely dependent on the nature of the surface, solder composition, flux and temperature. The second stage of joint formation in which crevices are penetrated by the molten solder involves joint geometry and the rate at which solder spreads over the surface, so that the forces due to surface tension also play a major role in the filling of the joint space. For a complete assessment of solderability therefore, the rate of wetting and the surface tension of the solder should be measured. The latter factor is most easily indicated by the contact angle of the molten solder.

The contact angle between a meniscus of liquid solder and the surface on which it rests is controlled by a balance of the interfacial tensions at the solder-substrate, substrate-flux and flux-solder interfaces. When metallurgical bonding of the solder and substrate occurs, with a consequent change in the interfacial forces, the contact angle is lessened towards zero to maintain the balance of forces. A contact angle of 180° represents no wetting of the surface, an angle of 90° represents wetting but no effective spreading of the solder, while 0° indicates perfect wetting conditions. In practice, an angle of up to 10° provides good solderability, but 20° and above indicates that there will be considerable difficulties in achieving a good soldered joint[1].

The contact angle as solder wets a surface is low (advancing contact angle is perhaps 5°) but when solder is drained from the surface, the receding contact angle is usually a little higher. Under certain circumstances, however, the angle may be as high as 80°, this condition giving rise to the phenomenon known as de-wetting. De-wetting produces an appearance similar to beads of water on unclean glass and can be caused by soldering to an inadequately prepared or contaminated surface[2].

TEST METHODS

It is rarely necessary to actually measure the contact angle on a surface, most solderability tests being designed not only to measure the rate of wetting, but also the extent of spread of the solder, which is only indirectly indicative of the contact angle.

SOLDER-SPREAD TESTS

Probably the most widely used solderability test is the solder-spread test in which a fixed volume of solder is allowed to spread over a horizontal surface under specific conditions and the area of spread of solder is determined after a given time. This area is directly related to the contact angle of the solder with the surface. The rate of increase in the area of spread may also be measured to determine the rate of wetting, but this involves a more complex procedure such as the use of cinephotography. A simple method for carrying out the solder-spread test is as follows:

Pellets of volume 20 mm^3, or some other suitably chosen volume or weight, of 60% tin–40% lead solder, are stamped from rolled sheet, made up from ingot metals of greater than 99.9% minimum purity. 25 × 25 mm test pieces are cut from the sheets to be tested and a solder pellet is placed at the centre, together with 0.025 ml of flux applied from a capillary tube dropper; this volume is not critical. The test pieces are placed one at a time on a cradle made of stainless steel or titanium, attached to a crank or hydraulic cylinder which permits them to be lowered at a predetermined rate on to a solder bath. The holder becomes submerged while the specimen remains floating on the surface of the solder (Figure 25.1) which acts solely as a source of heat. The test specimens may be heated for the required cycle, such as at 250°C for 10 seconds, after which they are removed from the solder without vibration by slowly raising the specimen support. After cooling, flux residues are removed if necessary and the area of spread of solder measured with a planimeter on a projection microscope or by comparison with circles of known area. Not less than five individual specimens of one material should be tested to obtain a meaningful average solder-spread value.

The area of spread test has been widely used for comparing the solderability of coatings as well as for comparing the efficiency of fluxes and solders. For assessing the solderability of gold electrodeposits it is suggested that this test provides the most valuable data. For example, Figure 19.4 (see Chapter 19) shows the spread of solder on a cobalt-alloyed gold compared with that on a pure cyanide gold coating. Although the area of spread is about the same, such differences as the uniformity of spreading on the two types of gold and the effects of gold dissolved in the pool of solder on it's surface appearance, are also demonstrated.

WETTING-TIME TESTS

The application of gold plating in the electronics industry often involves automatic soldering procedures. In many of these, such as the wave soldering process, all of the gold plated surface is brought in contact with the molten solder so that there is no requirement for the solder to spread over the surface. However, automatic processes involve very short soldering times so that a high rate of wetting is of prime importance. For this reason a wetting-time solderability test is of the greatest value for printed circuitry. One method of measuring wetting-time is the rotary-dip test[3] (Figure 25.2) which is included in BS 4025 and in IEC 68-2-20. A recommended test procedure is as follows:

Fig. 25.1 Titanium jig carrying test piece, flux and solder pellet about to be lowered onto the solder bath for a solder-spread test. [Tin Research Institute].

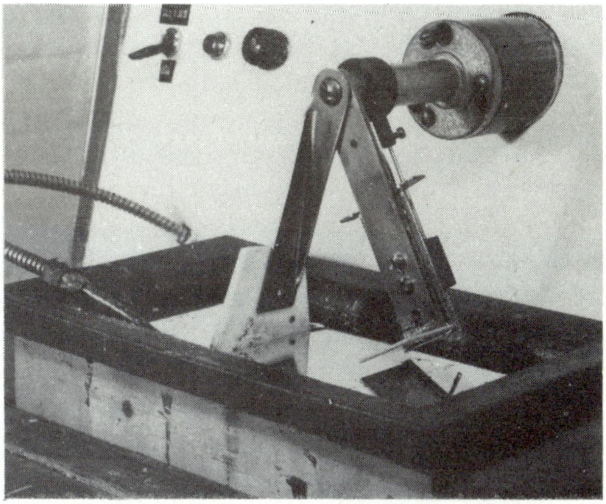

Fig. 25.2 Rotary dip solderability test for printed circuits. A succession of test pieces are moved by the rotating arm (right) across the surface of the molten solder and the time of contact which gives perfect wetting by the solder indicates solderability. The left hand arm carried a PTFE blade which precedes the test piece through the solder to remove oxide. [Tin Research Institute].

Test pieces measuring 25 × 25 mm are cut from the panels to be tested or from the regions of printed circuit boards having the greatest density of conductors. Flux solution is applied by immersion and draining, and the test piece is then placed in a clamp at the end of the radial immersion arm. This arm may be rotated by a horizontal shaft driven by a variable speed motor so that the period of contact of specimen with the solder may be varied over the range of about 1–10 seconds. A second radial arm carries a polytetrafluoroethylene strip to remove oxide from the surface of the solder just before the test is carried out. The soldering time is measured by an electromagnetic counter, energised when a probe behind the test piece is in electrical contact with the solder.

To determine the wetting-time, a succession of about twelve specimens are immersed for different times until the minimum time that produces an unbroken, smooth and reflective solder coating is found. Any tendency for de-wetting of the solder is checked by using contact times greater than necessary for perfect wetting, these being in the range of 5–10 seconds.

The solder bath should normally be 60 % tin–40 % lead and prepared from ingot metals of greater than 99.9 % purity.

In a simplified "go, no-go" form of the test, suitable for works control, two times are chosen such as 2 and 8 seconds, the test pieces immersed for these times being examined for perfection of wetting or de-wetting. If there are no visible defects, it may be safely assumed that good wetting occurs over the whole of this range of soldering times. It is clear, however, that times obtained in this test do not necessarily indicate an identical soldering time on the production line because of other factors involved, such as preheating before soldering. It is essential that the operators of this test be completely familiar with the defects that may be encountered such as non-wetting and de-wetting[4]. To facilitate inspection, it is recommended that standard photographs of these phenomena (Figure 25.3) should be available at the test station for comparison with the samples being tested. Colour reproductions of such photographs are obviously preferred for this purpose.

THE SURFACE TENSION BALANCE

A delicate method of measuring the surface forces acting on a specimen as it is immersed and becomes wetted by molten solder has recently been developed[5,6,7]. This involves the use of a transducer which measures the change in force and produces a continuous record on a chart recorder. A typical trace is shown in Figure 25.4. The rate of wetting and the wetting force are easily measured, but de-wetting is not clearly indicated. It has been suggested that de-wetting of the solder produces a discontinuous force during withdrawal of the specimen in comparison with a perfectly wetted specimen which gives a constant withdrawal force. This is a particularly useful technique in the laboratory, and as from mid-1973 has been available as a commercial test method.*

GLOBULE TEST

The methods just described are of most use for flat specimens although the surface tension balance may also be used for round wire terminations. However, the widely accepted and commonly used solderability test for round wires is the globule test proposed by ten Duis[8] and developed commercially in conjunction with the Electronic Engineering Association. The method is covered by IEC Publication 68-2-20 (Test T) recommendation TC50 and by BS 2011, Part 2T. The procedure for the globule test is as follows:

* Enthoven Solders, Ltd.

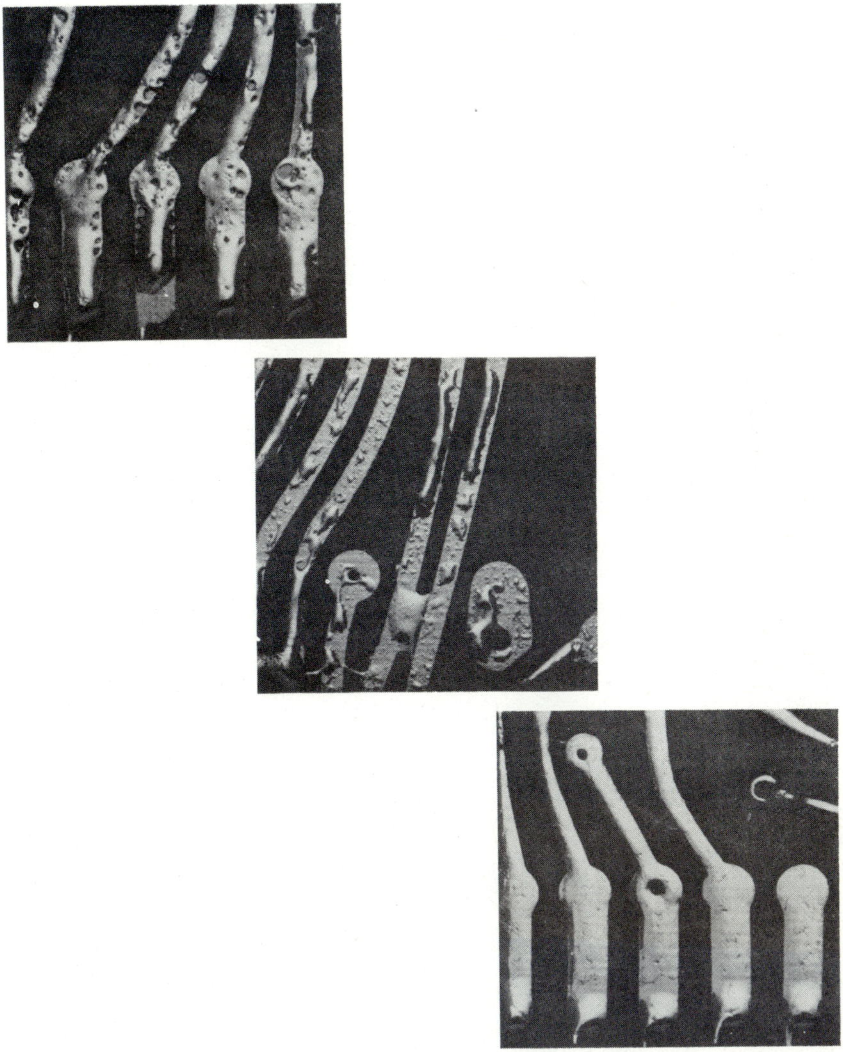

Fig. 25.3 Photographs suitable for use in a soldering quality control section illustrating (top to bottom) examples of non-wetting, de-wetting and complete wetting. [Tin Research Institute].

The wire to be tested is held in a horizontal position between clamps and is coated by brushing with flux. It is allowed to fall at a controlled rate into a small globule of molten solder resting on a heated iron pin set within a heated aluminium block. The globule volume is dependent on the diameter of the wire being tested. In the first instant the wire bisects the globule, but as the surface of the wire becomes wetted by the solder, the globule meniscus climbs and joins over the wire. The time is measured between the first moment of contact and the final submersion of the wire.

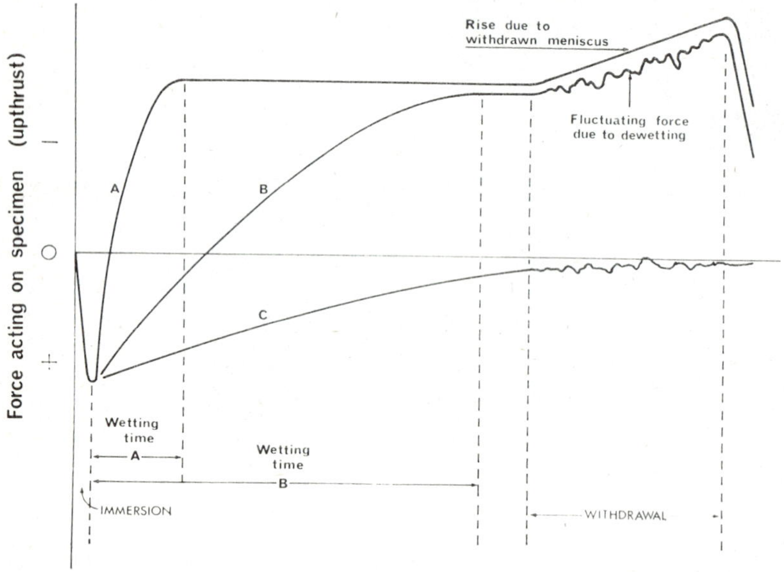

Fig. 25.4 A typical trace obtained from a surface tension balance showing the changes in magnitude and direction of the force acting on a test specimen as it is immersed and then withdrawn from a solder bath. "A" represents rapid wetting, "B" slow wetting and subsequent de-wetting on withdrawal, and "C" indicates no wetting occurring. [Tin Research Institute].

Although ambiguous results with this test procedure have been reported from time to time, there is little doubt that it represents a convenient method for the solderability testing of wires. It will distinguish readily between fluxes of varying activity and between coatings of different solderability. Becker[9] has demonstrated its use in a statistical quality control system for component terminations, although the test machine was slightly different in that it had fully automatic timing. The disadvantage is that any tendency for the solder to de-wet on the wire surface is not revealed in the normal test procedure. A second test carried out on exactly the same spot after removal of the first solder globule has been suggested as a means of detecting de-wetting.

DIP TEST

A simple procedure for determining solderability of lead wires is covered by the United States MIL-STD-202B, Method 208. Using a cam operated arm, the pre-fluxed component terminations are lowered vertically at a controlled rate into a solder pot and held there for 5 seconds before withdrawal. The solder coating acquired by the wire is examined and should have at least 95% of the surface perfectly wetted. The writer has found that small diameter wires tend to have excess solder coating due to rapid cooling which hides defects in wetting.

This system has also been adopted by the Institute of Printed Circuits for vertical edge dip tests on circuit boards, but experiments by the Electronic Engineering Association indicate that horizontal contact of the specimen with the solder (as in the rotary-dip test previously described) produces solderability results which correlate better with production line soldering.

FLOAT TEST

A test procedure devised by Fabish[10] for printed circuit boards and covered by MIL-P-55110A, should, perhaps, be mentioned. A horizontal board is lowered test-face downwards on to three circular solder meniscuses formed by depressing a plate having three holes, into a solder bath. The area of spread of the solder on to the test board along the conductors is measured.

STRENGTH OF SOLDERED JOINTS

It is generally considered that soldered joints should not be called upon to withstand stresses of any significant magnitude. This may be accomplished in many cases by the use of careful design and of such devices as mechanical clamps to support components. However, there are many other connections where it is not possible to design mechanical strength into the joint and the trend in modern electronics is for higher operating temperatures to well above 100°C. At temperatures as close to their melting point as this, solders have a low strength and are prone to creep under the influence of quite low stresses. Vibrations, which may be encountered in aerospace and other applications, may also result in complex stress systems which may lead to failure of soldered joints.

It is feasible to solder straight wires through holes in printed circuits normal to the surface and then to perform a tensile test. It is also possible to carry out a tearing or pull test on wires soldered along the length of conductors on printed boards and this method has been used in some investigations as indicated earlier in Chapter 19. There is some doubt, however, whether such tests have a meaningful value in assessing the quality of a soldered assembly.

Much of the previous work on the strength of soldered joints has been carried out with the specific object of studying the effect of certain variables on joint strength, such as joint gap, temperature and type of surface coating. Tensile strengths of joints have usually been determined on a specimen made by butt-soldering together the machined ends of two rods. During test this requires uniaxial stressing since, if the soldered interface is not normal to the stress axis, there is a tearing action which results in an abnormally low strength being obtained. In practice, the majority of joints are stressed in shear, hence tensile stressing is perhaps not the preferred test. Specimens for determining the shear strength of soldered joints may consist of a normal overlap joint between flat sheets which is simplest to produce or may be of the ring and plug variation with, in effect, a cylindrical soldered joint. The lap joint suffers from the difficulty that unless the joint members are of sufficient rigidity some bending will occur which leads to a combination of tensile and shear stressing. This problem may be overcome by using a double lap joint. On theoretical grounds the ring and plug system is favoured, but there is some evidence that it is easier to obtain a consistent joint quality with lap joints.

Pull tests in which a wire is torn away from the surface have been mentioned in relation to printed circuit boards. Tearing tests have also been used by Chadwick[11] for joints between sheet specimens. In this system two "L" shaped specimens are soldered to form a "T". The two vertically disposed unsoldered arms are pulled in tension so that the horizontal soldered members are slowly torn apart. The strength is measured as the load per unit width of joint, averaged over a specific length of joint pulled apart.

The effect of gold plating on the strength of soldered joints has been studied mostly with regard to the tensile and shear strength measured under rapid loading conditions and some of the results have been quoted in Chapter 19 (see, for example, Refs., 13, 20). The creep and fatigue strength of soldered joints are much less well documented and there seems to be no relevant data on these properties for joints between gold plated surfaces. Electromagnetic vibrators are currently used for vibration testing of printed circuit assemblies and the stresses imposed on a given soldered joint at a certain frequency can be calculated or measured. However, no systematic study of high frequency fatigue testing of soldered joints by this technique has been published.

TESTING THE QUALITY OF SOLDERED JOINTS

Since soldered joints in the electrical and electronics industry have the prime function of conducting electricity from one component to another, so a quantitative measurement of conductivity might be thought to be a suitable method of assessing the quality of joints. For example, a soldered connection made to a surface which exhibits de-wetting has low mechanical strength and as such, might be expected to give rise to electrical noise due to the imperfect bonding. However, resistivity measurements have been shown to be insensitive to the quality of a joint, a metal-to-metal mechanical contact without metallurgical bonding giving the same values as a strong, well soldered joint. Efforts have been made over a period of years to develop a non-destructive electrical method for assessing the quality of joints without success to date. One system which is of interest is to monitor the production of infra-red radiation from any dry joints where a temperature rise may occur,[12] but this does not seem to be applicable at the present moment. In a poorly made soldered joint where solder is in mechanical contact without bonding, although it exhibits a low resistivity when freshly made, the resistivity in service may increase to produce higher levels of electrical noise or the classical "dry joint". This is due to the ingress of oxygen and water-vapour along the solder-substrate interface.

The most reliable method of ensuring quality during production is to make sure that good solderability has been achieved. Neale[13] has described many of the factors to be considered in obtaining reliable production soldering. Careful thought in the choice of surface finish and design of assembly, together with adequate solderability testing, are the best means of providing trouble-free soldered assemblies. Visual inspection is normally essential to check for minor faults, but without doubt it is more economic to take all of the necessary steps to ensure perfect soldering in the first instance, than to try and repair faulty equipment at a later stage.

REFERENCES

1 Bailey G. L. J., *Sh. Metal Inds*, **32,** 47 (1955).
2 Thwaites C. J., Mackay C. A., *Metal Finish. J.*, **14,** 165, 291 (1968).
3 Thwaites C. J., *Electl Mfr*, **8,** 5, 18 (1964).
4 Tin Research Inst, Leaflet No. L52.
5 Duis ten J. A., Meulen E. van de, *Philips tech. Rev.*, **28,** 362 (1967).
6 Mackay D., Proc. Inter-Nepcon, Brighton (1970).
7 Budrys R. S., Brick R. M., *Trans. Met. Soc. Ass. Inst. Min. Engrs*, **2,** 103 (1971).
8 Duis ten J. A., *Philips tech. Rev.*, **20,** 6, 158 (1958–59).
9 Becker G., *Schweiss Schneid.*, **7,** 318 (1968).
10 Fabish J. P., *Weld. J. Res. Suppl.*, 400s, Sept. (1964).
11 Chadwick R., *J. Inst. Metals*, **62,** 277 (1938).
12 Parry N. C. J., *Trans. Inst. Metal Finish.*, **47,** 106 (1969).
13 Neale R. A., Proc. Inter-Nepcon, Brighton (1970).

Chapter 26

MEASUREMENT OF CONTACT RESISTANCE

M. Antler

INTRODUCTION

DEFINITION OF TERMS

Two solid surfaces in contact at moderate load touch at only a few isolated spots. Current flowing between the bodies is constricted through these small areas. This gives rise to contact resistance, which is the sum of (*a*) constriction resistance and (*b*) film resistance (when contaminated surfaces have to be taken into consideration). Resistance in the connection associated with the body of the contact, which often cannot be eliminated on making the measurement because of the geometry of the device, is (*c*) bulk resistance. The sum of these three is the overall resistance. These definitions are amplified in Chapter 22.

Contact noise is varying resistance and originates in a changing number and distribution of conducting contact spots. For lightly loaded static (i.e., connector) contacts, it may occur when there are severe external vibrations. Noise in sliding contacts is influenced also by wear particles, surface roughness, velocity, lubricants, and especially by the mechanical design of the system[8]. Noise is the most important characteristic of slip rings, particularly the instrument types which have gold plated or clad gold alloy contact surfaces. It is measured with oscilloscopes or with fast recorders and the maximum peak instantaneous resistance over a specified time interval, minus the static value, is generally described as the noise[7].

VOLTAGE BREAKDOWN OF FILMS

Contact resistance is higher than can be tolerated in most low voltage circuits when the normal load is small and the surfaces are extensively covered with films. However, when the circuit is capable of impressing a significant voltage across the contact, the film may break down electrically, depending on its thickness and composition[19]. This action, called the coherer effect or fritting, results in the formation of minute molten metallic bridges through the film which solidify to establish a viable conductive path. The puncturing level is 1 volt per 50–150 angstroms of film. The potential drop across the contact after puncturing, called melting point voltage, is 0.2–0.5 volt, depending on the metal.

The potential drop across contacts in a dry circuit is too low to affect the films.

RESISTANCE HEATING OF CONTACTS

There is a current-carrying limit to a contact, defined by the materials, contact area and its heat dissipating ability, and that of the structure of which it is a part. If excessive current is passed through the contact, it will reach the softening temperature. The area of metallic contact then increases and contact resistance is reduced. At still higher currents the mating junctions will melt. Both softening and melting occur at characteristic voltages. Typical values[20] are shown in Table 26.1.

Table 26.1
Some Softening and Melting Voltages

Metal	*Softening* *Temperature* (°*C*)	*Softening* *Voltage*	*Melting* *Temperature* (°*C*)	*Melting* *Voltage*
Gold	100	0.08	1,063	0.43
Silver	150–200	0.09	960	0.37
Copper	190	0.12	1,083	0.43

TECHNIQUES FOR MEASURING CONTACT RESISTANCE

FOUR-WIRE METHOD[31]

The most common method for measuring contact resistance is the four-wire method. A test current is introduced into the sample with two terminals, the voltage probes being positioned as close as possible to the point of contact. This is illustrated in Figure 26.1a for a pin socket contact. The resistance is the voltage drop divided by the test current. The current path between the voltage probes includes the bulk resistance of the contact members, in addition to contact resistance. Often, the bulk resistance is considerably larger than the contact resistance.

If the members can be mated in a crossed configuration, the four-wire method determines contact resistance alone, since bulk resistance will be zero. This is illustrated in Figure 26.1b. The only circuit common to the current path and the voltage probes is across the interface between the members.

In resistance measurements, it is important to specify voltage and current for the reasons just described. Most testing with gold plated contacts is carried out at dry circuit conditions and this is commonly defined at a maximum (open) circuit potential of 20 mV. Current is also limited by the test circuit, usually to 100 mA or less.

Testing may be done at "rated current", which is generally a high level for the device at which there can be significant heating. Ratings are related to wire size and are usually the same as specified by the wire manufacturers. The potential drop across the contacts at rated current is normally more than 20 mV.

Measurement of contact resistance can be made by either AC or DC methods. Direct current procedures permit positive or negative polarity control. Contact resistance may differ according to polarity with surfaces that

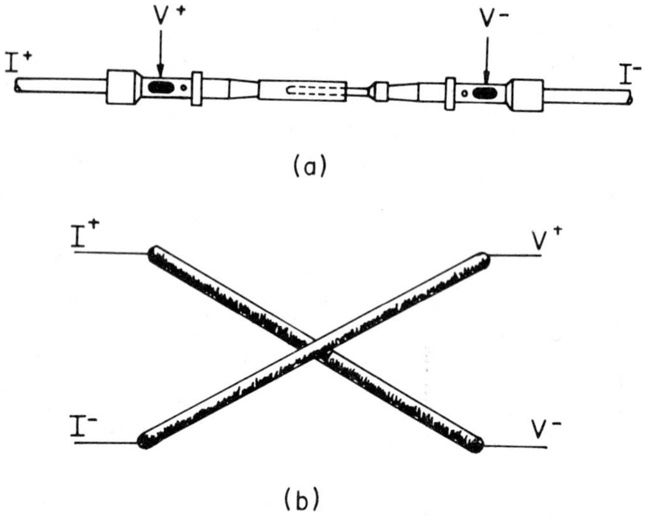

Fig. 26.1 (a) "Standard four-wire" technique for measuring the overall resistance of a contact. Circuitry between voltage probes includes bulk resistance of contacts and contact resistance. (b) "True four-wire" technique for measuring constriction or contact resistance directly. The only circuitry common to the current path and the voltage probes is across the interface between the rods.

are covered with films. The contact resistance of such members can change with time as current flows, again depending on polarity, due to physical changes in the film caused by the current[19]. Contact resistance determined by AC and DC methods is usually the same for clean surfaces when its level is low and currents are small. Direct current methods are preferred when obtaining the contact resistance of a material in physical property determinations.

Alternating current methods of measuring contact resistance are of special interest for contacts that serve in AC circuits. However, a special effort must must be made to eliminate electromagnetic coupling between stray fields in the vicinity of the equipment due to motors, power lines, etc. Capacitance in the samples may also affect the measurements. A number of self-contained AC four-wire resistance instruments are commercially available, including portable dry circuit meters.

The Appendix to this chapter describes a simple DC circuit suitable for determining dry circuit contact resistance of gold plated contacts. When specimens are in the form of crossed rods, it is convenient to contain one member in a V-shaped block and to press the other against it with suitable fixtures.

THERMAL PROCEDURES

Contact resistance can be isolated for measurement in a system involving a large bulk resistance in series with it, as with separable connector contacts at the end of a pair of long leads. The two procedures which have been devised are based on a difference from the rest of the system in thermal behaviour of the metallic contact spots when large currents are passed through them.

Pulse Methods[21,26]

Contact spots are small and have much lower heat capacity than the bulk metal. They will, therefore, heat faster than the bulk metal on passage of current. If successively higher currents of about a millisecond duration are pulsed through the contact, at some current level the spots will reach their softening point without affecting the bulk of the contact. Softening increases the contact area which in turn lowers contact resistance. By measuring the contact resistance before and after each pulse, the critical current, I_{crit}, which causes the softening is found. Contact resistance = K/I_{crit} where K is the softening voltage. To determine K, measurements of both contact resistance and I_{crit} are made with crossed rods for the material of interest, for instance, a particular type of gold plate. This K value is used for devices having contacts made from that metal.

Critical currents are high, of the order of 100 amps for gold at a contact resistance of 0.5 milliohm. The pulse method is, therefore, unsuitable where electron tunnelling through thin films contributes significantly to conduction, for contacts having very low contact resistances such as solder and crimp joints, and for devices where the magnetic field associated with the current pulse can affect contact load by interacting with an actuating coil, as in a sealed reed relay. On the other hand, this method is useful for detecting a tendency to early failure of connector contacts, such as those plated with porous gold over a silver underplate which is exposed to a sulphide environment[22]. This method is also recommended as a "go, no-go" test to insure that the contact resistance of a separable connector is below a specified value. The justification for making such measurements is that the contact resistance of unused contacts predicts their service life. When the contact resistance is above some level, the likelihood of early failure is indicated.

Non-linear Method[17,23]

The non-linear contact resistance technique is based on the voltage-current property of metallic conduction. This property is non-linear for metals, higher currents causing more heating and the resulting temperature rise increasing the resistivity. Under AC excitation, a measurable DC voltage is generated across the test contact which is a direct function of the constriction resistance portion of the overall measured resistance. Large currents are needed to obtain significant non-linearities for low contact resistances. For example, about 8 amps will produce a 10 microvolt DC signal with a constriction resistance of 1.0 milliohm, while 40 amps are required for the same DC output at a constriction resistance of 0.2 milliohm. Unlike the pulse method, the non-linear technique is based on equilibrium thermal behaviour. The current tends to heat the contact bulk which makes measurements difficult. Currents necessary for good DC measurements may be beyond the current-carrying ability of small devices containing gold plated contacts. The non-linear method has also been found to be of value in determining the resistance of thin (less than 50 angstroms) films which allow tunnel conduction when used in conjunction with the conventional method. The former method ignores tunnel films, while the latter does not[30].

Little practical work has been reported in the measurement of contact resistance by the non-linear technique.

CONTACT RESISTANCE PROBES

A probe is an apparatus for determining the contact resistance of a material. It generally includes the following:

1 Fixtures for holding specimens of varied shape and for clamping electrical leads to them.
2 A loading mechanism which applies a discrete load to the contacts or one which can be increased, decreased and measured.
3 A reference surface (the probe) that is pressed against the material and which is generally made of a noble metal. In the case of crossed rod contacts, the material of interest is often used for both members.
4 A current source with current and voltage measuring instrumentation for determining contact resistance.
5 A slide which permits the sample to be moved (wiped) small distances after loading.

Associated electrical circuitry can be simple or complex, depending on whether four-wire or thermal methods of determining contact resistance are used and on whether AC or DC contact resistance is desired. Circuitry may also be included to permit related measurements to be obtained, such as the voltage breakdown properties or the current versus voltage characteristic of film covered surfaces or semiconducting materials.

A few probes which have proven to be especially useful will be described to illustrate the range of instrumentation.

ZERO FORCE

Probes may be constructed so that the contacting member is a fine drop of mercury that is partially expelled from a capillary tip. Such probes are useful in exploring the surfaces of materials with which mercury does not amalgamate.

LIGHT LOAD

A light load probe may be defined as one having a member which contacts the test surface below 5 grams. Many types have been devised, with load limits as low as one milligram. The probe contact is usually a fine wire of platinum or gold. The value of such probes is that they permit the detection of insulating films on surfaces. Low loads are used to minimise the possibility that the probe will mechanically disturb any films.

Savage and Flom[1] describe a probe made of wire 0.002 in. thick and bent in the form of a loop. The load is determined from its compression and the loop is calibrated by pressing against the pan of an analytical balance. A wire loop probe is illustrated in Figure 26.2, where a micrometer is used to measure the compression. An X-Y table is convenient for indexing the specimen. This probe is generally used in experiments in which a multiplicity of sites of the contact surface are touched individually.

Flom[3] describes a modification involving a quartz fibre spring as the resilient member which permitted a wider range of loads to be explored than was possible with a metal loop. Chaikin[4] and coworkers devised an apparatus in which the fine wire could be slid slowly over the test surface and a chart record of instantaneous resistance versus time (or distance) obtained, together with cumulative resistance using a capacitance type integrator.

A probe can be constructed having the overall configuration of the instrument in Figure 26.2 with an electronic force transducer located between the micrometer and the probe for measurement and control of load. Some commercial transducers have considerable resilence and when they are used it is often convenient to employ a rigid member as the probe, such as a 0.020 in. diameter solid gold wire with a gold ball tip formed by fusion in a flame.

The ASTM "Standard Method of Test for Resistance Characteristics of Microcontacts"[28] describes a crossed rod contact tester for materials in the form of fine round wires. Its loading mechanism is made from a galvanometer coil.

Critical studies of the contact properties at low load of gold and other noble metals have been reported by Wilson[2] and Angus[5].

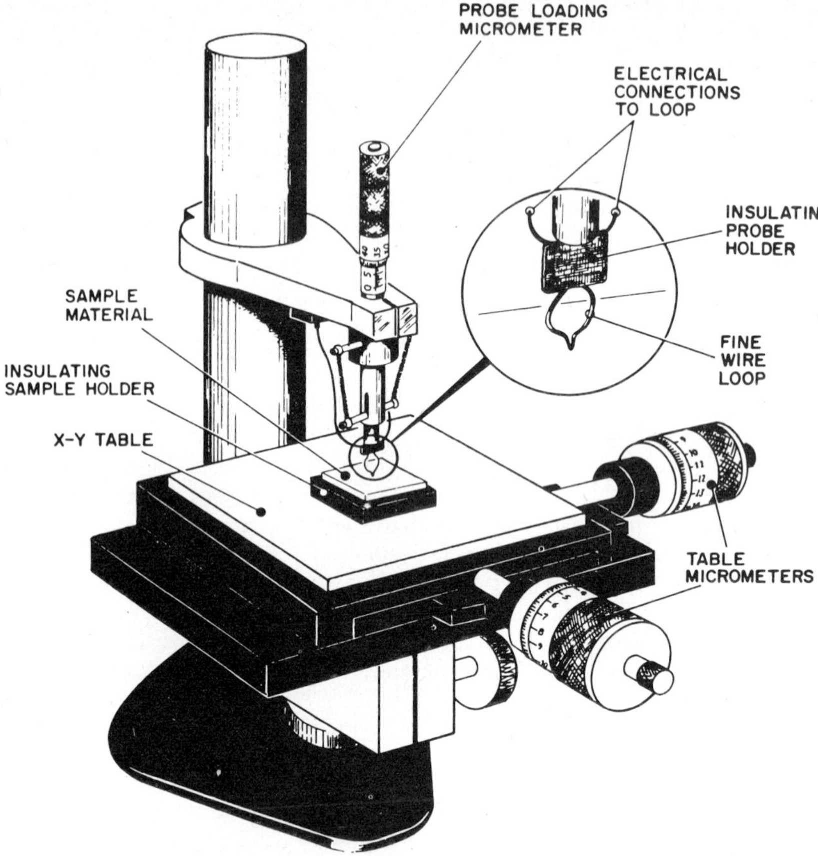

Fig. 26.2 Wire probe for determining contact resistance of film-covered metals at light load. Clean loop-shaped gold wire is pressed against surface. One loop connection is for current lead, the other for voltage. Second connections for current and voltage are made directly to the sample.

WIDE RANGE BEAM LOADING DEVICES

Dead weight beam loading is widely used for probes that operate at loads upwards of 1 gram. A simple device is described in the ASTM "Standard Method of Test for Surety of Make of Electrical Contact Materials"[29]. The probe has a counterweighted beam on a V-notched fulcrum and is used at 10, 40, and 150 grams.

Orso and the writer have designed an apparatus (Figure 26.3) for obtaining the contact resistance-load characteristics of surfaces while loading and unloading. The probe is usually a hemispherically-ended gold rod and the specimens are normally flat plates. Load is applied or removed by siphoning a fluid from one reservoir into another. Water is used for loads from 1–300 grams and mercury is used when operating within the 10-2000 grams range. Load is measured with a ring dynamometer fitted with wire strain gauges coupled to a carrier amplifier. Contact resistance is determined with a circuit similar to that described in the Appendix to this chapter. The voltage drop for resistance measurements and the signal from the carrier amplifier are fed into an X-Y recorder. The contact resistance dependence on wipe is determined by moving the sample with the X-Y table.

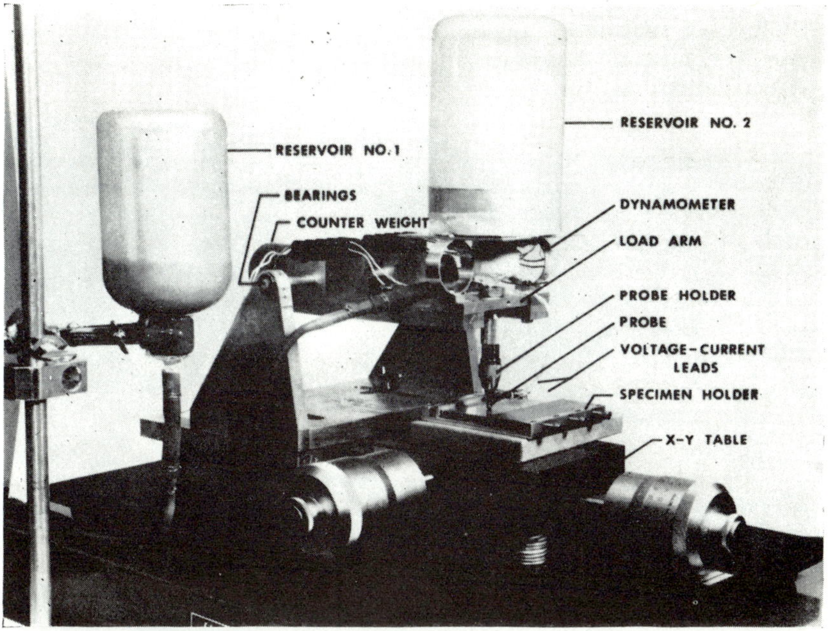

Fig. 26.3 Fluid siphoning probe. Water or mercury in the reservoirs is siphoned to load or unload probe against sample, mounted on X–Y table. Load is measured with dynamometer fitted with wire strain gauges.

AUTOMATED

The necessity for obtaining many measurements of contact resistance either of numerous materials or in statistical studies of the variability of the contact resistance of film-covered samples, has stimulated the development of automated probes. "Massacre"[6] determines contact resistance at a number of discrete loads, from 10–1000 grams, before and after wiping the contacts. The specimen is automatically repositioned for each measurement; numerous sites on a test surface can be explored.

A versatile probe, the "Autoprobe", has been described by Antler, Auletta and Conley[9]. It determines (*a*) the contact resistance-load characteristics of samples to 1000 grams (*b*) contact resistance-wipe properties at fixed load, and (*c*) the current-voltage properties of films to 200 volts. Specimen positioning is automatic and programmable. The probe locates the test surface on an X-Y plane, operates in the selected mode, and then automatically repositions and repeats the measurement according to the test schedule. The resistance span is divided into six ranges which change automatically when the resistance decreases to a preset percentage of the scale. Resistance, current, voltage, load, wipe distance and location of the specimen are read out on a multi-channel oscillographic recorder. Open circuit voltage is 10 mV or 50 mV, depending on resistance range. Improved versions of the "Autoprobe" (Figure 26.4) measure resistance on decreasing as well as increasing load at dry circuit conditions.

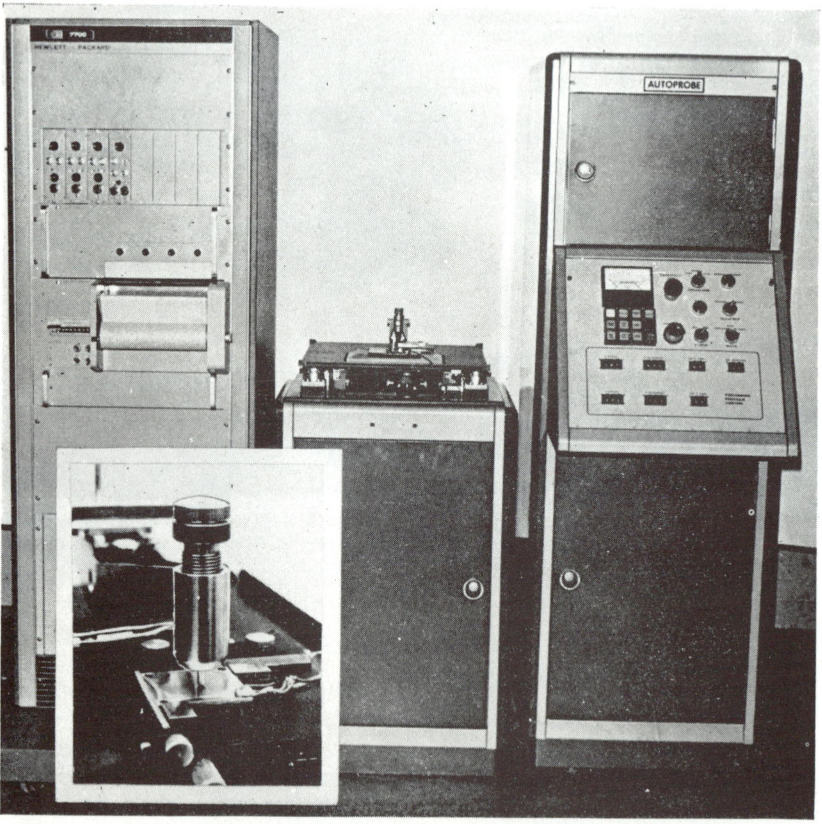

Fig. 26.4 "Autoprobe". Left to right: four-channel recorder, probe unit, control console. Inset shows hemispherically-ended probe in contact with sample.

The "Autoprobe" has been used as the key experimental tool in several studies of the contact properties of gold plate[10–16,18,24,25,27] and other materials.

APPENDIX

CIRCUIT FOR MEASURING CONTACT RESISTANCE

A circuit suitable for the measurement of contact resistance at dry circuit conditions imposes a maximum open circuit potential across the contacts of 0.02 V, with the current limited to 0.1 A. This circuit, shown in Figure 26.5, is

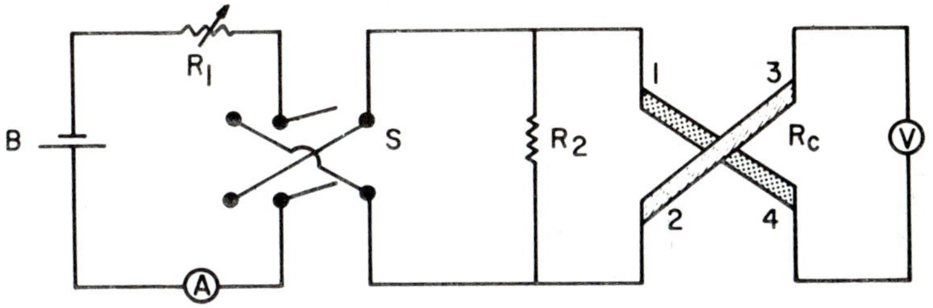

Fig. 26.5 Circuit for measurement of contact resistance.

recommended for resistances from 0.0001–0.1 ohm, the range ordinarily encountered in devices made with gold contacts. The parts of the circuit are:

B = 1.5 volt dry cell

A = DC ammeter, 0.15 A full scale, 2% accuracy.

V = electronic voltmeter, high impedance, multi-range, 0.0001–0.1 V full scale, 2% accuracy. Null centre meter preferred to eliminate necessity for changing leads when resistance is measured with both forward and reverse currents.

S = reversing switch with "off" position

R_1 = current-limiting resistor, variable, 0–25 ohms

R_2 = 0.2 ohm

R_c = contacts (crossed rod configuration shown).

It is sometimes necessary to include a capacitor across R_2 to eliminate voltage spikes when S is actuated. This is determined with an oscilloscope across contacts 3 and 4.

The experimental procedure is as follows:

(i) Short circuit test leads to contacts 1 and 2. Adjust R_1 to 0.1 A.
(ii) Attach leads to samples at 1, 2, 3, 4.
(iii) Measure resistance in both directions.
(iv) Calculate resistance in both directions from Ohm's Law.

REFERENCES

1 Savage R. H., Flom D. G., *Annals N.Y. Acad. Sciences*, **58,** 6, 946 (1954).
2 Wilson R. W., *Proc. Phys. Soc.* (*London*), **68B**, 625 (1955).
3 Flom D. G., *Rev. Scient. Instrum.*, **29,** 11, 979 (1958).
4 Chaikin S. W., Anderson J. R., Santos G. J., *Rev. Scient. Instrum.*, **32,** 12, 1294 (1961).
5 Angus H. C., *Brit. J. Appl. Phys.*, **13,** 2, 58 (1962).
6 Fairweather A., Jury R. L., Lazenby F., Parker A. E., Thrift D. H., Wright L. J., Proc. I.E.E., 109A, Suppl. No. 3, 210 (1962).
7 Glossbrenner E. W., Sun J. K., Proc. Engng. Semin. on Electrical Contacts, Univ. of Maine, Orono, Maine (1962).
8 Antler M., *Wear*, **6,** 1, 44 (1963).
9 Antler M., Auletta L. V., Conley J., *Rev. Scient. Instrum.*, **34,** 9, 1317 (1963).
10 Antler M., Gilbert J., *J. Air Pollut. Control Assoc.*, **13,** 9, 405 (1963).
11 Blake B. E., Proc. 2nd Int. Symp. on Electrical Contact Phenomena, pp. 531, Graz, Austria (1964).
12 Antler M., Krumbein S. J., Proc. Engng Semin. on Electrical Contact Phenomena, pp. 103, Univ. of Maine, Orono, Maine (1965).
13 Chiarenzelli R. V., Proc. Engng Semin. on Electrical Contact Phenomena, pp. 63, Univ. of Maine, Orono, Maine (1965).
14 Chhabra D. S., Wenning G. T., Proc. Engng Semin. on Electrical Contact Phenomena, pp. 313, Univ. of Maine, Orono, Maine (1965).
15 Antler M., *Plating*, **53,** 1431 (1966).
16 Chiaranzelli R. V., Proc. 3rd Int. Res. Symp. on Electrical Contact Phenomena, pp. 83, Univ. of Maine, Orono, Maine (1966).
17 Whitley J. H., Proc. 3rd Int. Res. Symp. on Electrical Contact Phenomena, pp. 63, Univ. of Maine, Orono, Maine (1966).
18 Walton R. F., *Plating*, **53,** 209 (1966).
19 Holm R., Electric Contacts, 4th Edn, pp. 135–152, Springer, New York (1967).
20 Holm R., Electric Contacts, 4th Edn, pp. 436–438, Springer, New York (1967).
21 Russakoff R., Snowball, R. F., *Rev. Scient. Instrum.*, **38,** 3, 395 (1967).
22 Snowball R. F., Lawrence J. W., Trans. tech. Conf. Am. Soc. Quality Control, pp. 213 (1967).
23 Whitley J. H., *Electron. Instrum. Digest*, **3,** 7, 7 (1967).
24 Krumbein S. J., Antler M., I.E.E.E., Trans. on Parts, Materials, and Packaging, PMP-4, 1, 3 (1968).
25 Antler M., *Plating*, **56,** 1139 (1969).
26 Landis J. M., *Electl Des. News*, **14,** 17, 65 (1969).
27 Antler M., *Plating*, **57**, 615 (1970).
28 "Standard Method of Test for Resistance Characteristics of Microcontacts", A.S.T.M. Designation: B 326–66, Annual Book of A.S.T.M. Standards, Part 8, pp. 273, Am. Soc. for Testing and Materials, Philadelphia (1970).
29 "Standard Method of Test for Surety of Make of Electrical Contact Materials", A.S.T.M. Designation: B 340–61 (Reapproved 1968), Annual Book of A.S.T.M. Standards, Part 8, pp. 283, Am. Soc. for Testing and Materials, Philadelphia (1970).
30 Bock E. M., Whitley J. H., Proc. Holm Semin. on Electrical Contact Phenomena, pp. 45, Illinois Inst. of Technology, Chicago (1970).
31 "Standard Methods for Measuring Contact Resistance of Electrical Connections (Static Contacts)", A.S.T.M. Designation: B 539–70, Annual Book of A.S.T.M. Standards, Part 8, Am. Soc. for Testing and Materials, Philadelphia (1970).

Chapter 27

MEASUREMENT OF POROSITY

S. M. GARTE

In dealing with a physical or chemical property of a material it is desirable to be able to quantify it; to place some numerical value on it. To be useful, such numbers should be reproducible by any one method of measurement and determination by different methods should agree. A more subtle requirement is that such values should be "true" or at least proportional to the true values. An analytical chemist can prepare samples containing known amounts of a given substance and check his analytical procedures against these standards. Such "umpire" methods do not exist for the measurement of porosity.

Many measurement techniques have been reported and many cross-checks of different methods have been made. The major difficulty has been that all previously known methods for rendering pores visible are destructive to some degree, so that measurements cannot be repeated on the same sample. Cooksey and Campbell[1] have applied the technique of scanning electron microscopy to the problem of detecting pores in gold plate in a truly non-destructive way, but they did not count pores by this method, or report on the feasibility of its use to provide specimens with "known" porosity. The writer[2] attempted to prepare standard porosity specimens by making synthetic pores using very small asphalt dots as a plating resist, thereby providing some reasonable basis for the technique of "Quantitative Electrography" described later on in this chapter, but there still remained a large size gap between the smallest synthetic pores and the real pores which occur in gold plate. Notwithstanding these difficulties, many methods of porosity measurement have been invented and all of them have served to provide useful information, both in scientific research on the problem of porosity and in the practical sense of providing some indication of the performance to be expected of various plate systems.

Since the proliferation of measurement methods makes their classification somewhat difficult, the system adopted here is therefore more convenient than logical. It would be advantageous for the reader to have previously read Chapter 23 on "Porosity in Gold Deposits", in which transverse porosity was defined in two ways: (i) as the ratio of exposed substrate area to the total specimen area, or (ii) as the number of pores per unit area of specimen. The second definition has the advantage that one can in many ways render pores visible and then count them. However, the first definition has had somewhat more scientific appeal. All methods purporting to measure the exposed area ratio, belong to the category of electrochemical methods and will be dealt with first. Bulk porosity is described in Chapter 23 and will not be discussed any further in the present chapter.

ELECTROCHEMICAL METHODS

THE SHOME AND EVANS CELL[3]

Although this method was originally developed for the measurement of porosity in nickel electrodeposits, it is described here because it has formed the basis for modifications for gold plate. In the original version the cell is a glass or acrylic cylinder cemented to the specimen surface with wax around the edges. The cylinder is filled with a solution of 3% sodium chloride and 0.01% sodium potassium tartrate. An auxiliary cathode consisting of a piece of copper gauze is inserted and connected by an external wire to the specimen. No external voltage source is used. The auxiliary cathode serves to increase the cathode area so that the cell current is no longer sensitive to variations in cathode area and is a function only of the anode area. The total bare area in the pores of the plate constitutes the anode. The solution must contain dissolved oxygen so that the cathode reaction is the reduction of oxygen and the formation of hydroxyl ions.

$$O_2 + 2H_2O + 4e = 4\ OH^-$$

In the presence of a small amount of hydrogen peroxide, the cathode reaction becomes reduction of the peroxide ion and the cell current is greatly increased. The cell can be operated in two ways: (i) after a measured time interval, the amount of iron dissolved in the solution is determined spectrophotometrically, or (ii) an ammeter may be inserted in series with the auxiliary cathode and the specimen, and the corrosion current measured directly. The two measurements were found to agree and were taken to be a measure of the bare area. This was not demonstrated by direct measurement on known areas. The method does not give either the number of pores or the actual areas associated with the pores.

Ehrhardt[4] modified the Shome and Evans Cell for use in determining the porosity of gold plated copper. He found that in a solution of 5% by volume of sulphuric acid and with an externally applied potential of 0.75 volt, the gold plate on the specimen, made the anode, passed no current except through any exposed copper basis metal. He calibrated the test by the use of heavily gold plated fine copper wires. When the ends were cut and polished this produced a synthetic pore. He calculated that a bare area due to porosity of 10^{-7} in.^{2} per in.^{2} of specimen surface could be detected. This is based on the assumption that the proportionality of corrosion current to bare area holds for sizes down to those of real pores, which are of course much smaller than the synthetic pore. The same assumption obviously underlies the original method of Shome and Evans. This assumption is discussed further under the heading of "Critique of Electrochemical Methods".

THE dV/dI METHOD

This method was developed by Clarke and Britton[5] for the measurement of porosity of tin-nickel and other platings on steel, and was later extended by Clarke and Leeds[6] to the study of porosity of gold on copper and its variation with bath type and plating conditions. The parameter dV/dI is the slope of the anodic polarisation versus current density curve for the specimen and is a measure of the resistance of the electrolyte in the pore channels provided (i) negligible current is passed by the gold coating, (ii) electrolysis does not affect the resistivity of the electrolyte in the pores and (iii) basis metal polarisation is constant so that the measured potential displacements are wholly resistance controlled. Since the pore channels are parallel resistors, dV/dI will increase with decreasing porosity. An electrolyte for gold over copper which meets these conditions was found to be a solution of 5% sodium sulphate containing 0.1% tartaric acid. The authors compared dV/dI measurements on various specimens with pore counts from sulphur dioxide exposure tests and found that the results obtained from the two methods gave good correlation.

CORROSION POTENTIAL MEASUREMENTS

A paper by Morrissey[7] describes a method for the measurement of porosity in gold plate by the determination of the corrosion potential of the specimen. He found the corrosion potential to be proportional to the logarithm of the exposed copper area fraction in a galvanically coupled copper wire to gold system. He also correlated the corrosion potential to porosity measurement on porous gold plate over copper by the ammonium persulphate etch method, which is described later on in this chapter.

CRITIQUE OF ELECTROCHEMICAL METHODS

All of the methods so far described require some sophistication in electrical measurements, but this ought not to be a real deterrent. All of them are virtually non-destructive, which is a considerable advantage. None of them give any information as to numbers and location of pores, which is from the engineering point of view at least, a disadvantage. It is fairly easy to correlate the electrochemical parameter being measured with pore count, and in fact Ehrhardt[4], and Clarke and his coworkers[5,6] have done so. The scientific attraction of these methods lies in the claim, made in some cases, that it is in fact the ratio of exposed base metal area to the total area which is being measured. Clarke and Britton[5] did not make this claim for their method and criticised its validity for the Shome and Evans[3] method. Ehrhardt[4] and later Morrissey[7], calibrated their methods on known areas of exposed copper wire and assumed that real pores would behave the same way.

Consideration of the assumptions made by the various authors indicates that the primary question is whether the electrode process during the measurement is (i) controlled by anode polarisation with pore channel resistance negligible or constant, or (ii) controlled by pore resistance with anode polarisation constant. The first case, assumed by Ehrhardt[4], leads to a relationship between the measurement and bare area. In actuality, Ehrhardt was measuring a single point on the potential-current curve. Since no gradient for the curve can be derived from his data, no judgement can be made as to the contribution of the pore resistance.

Case two, assumed by Clarke and Britton[5], of necessity involves the abandonment of the concept of bare area measurement. Calibration with known areas of exposed wires cannot be used since no pore resistance is involved in such a measurement. In principle, it should be possible to obtain areas. The resistance of a pore channel is a function of both its cross-sectional area and its depth, the latter to a first approximation being the coating thickness. A mathematical model of pore closure by increasing plate thickness is described by Leeds and Clarke[8], but is incompletely developed. For the present, the dV/dI method should be taken as a measure of porosity in units which can be related to the number of pores per unit area. The authors established this relationship by comparing dV/dI to the pore count on identically plated panels, which had been exposed to sulphur dioxide. In its present state of development dV/dI cannot be used directly to measure the porosity-thickness relationship and this is perhaps its chief drawback.

A corrosion potential method was described by Shome and Evans[3] who did not, however, apply it to gold plate. The dV/dI method does in fact involve the rest potential measurement, since this is the base line potential for the determination of displacement potential at each point on the V-I curve. Clarke and Britton[5] noted that the rest potential varied with porosity. The measurement of the rest potential does not involve the passage of any external current, so that it may be argued that pore resistance is not involved. However, local galvanic currents in the pores do exist and their magnitude must control the measured corrosion potential. These local currents must in turn be influenced by the electrolyte resistance in the pores. Although Morrissey[7] did calibrate his method with known area fractions, these did not involve any contribution of pore resistance. Thus, the validity of extrapolating down to real pore sizes for the estimation of area fractions remains in doubt. Nevertheless, Morrissey's method, in common with all the others, does in fact measure porosity and the most likely case is that what is being measured is proportional to the number of pores. This does not rule out its also being proportional to exposed area, the point simply being that the former could be easily demonstrated, whereas proof of the latter remains unconfirmed.

ELECTROGRAPHY

This is also an electrochemical method, but with the special feature that it provides a print of the actual pores so that their number and location can be determined directly. Since much higher currents are used, the method is more destructive of the specimen than with the previously described electrolytic methods.

PAPER ELECTROGRAPHY

A review of the procedures for paper electrography of gold over copper, silver, and nickel has been published by Noonan[9]. An electrograph may be made by compressing between two aluminium plates a sandwich consisting of the specimen under test and a piece of dye transfer paper which has been soaked in a suitable electrolyte. The shiny, hard surfaced side of the paper is placed next to the specimen. Specimens must be flat and of sufficient thickness so that the pressures used cannot distort the plating and produce channel pores due to cracking. Suitable electrolytes are (*a*) for gold over copper or silver: sodium carbonate, 43 g/l and sodium nitrate, 7 g/l and (*b*) for gold over nickel: sodium carbonate, 43 g/l and sodium chloride, 7 g/l. After soaking for about 5 minutes the excess reagent is squeezed out of the paper with a rubber roller and the test assembled in a press. Upon compression to about 1000 psi, a constant current, which may be about 5 mA/cm^2, is passed for a measured time interval of about 30 seconds. The specimen is made the anode. This results in deposition on the paper of a small quantity of salts of the basis metal at each of the pore sites. Good resolution is obtained because of the low rate of diffusion of the transferred metal salts in the hard paper surface. By extraction of the salts from the paper and analysis of the extract, the writer determined that under the above conditions 100 spots gave 1.3 micrograms of copper for gold over copper, or 1.8 micrograms of nickel for gold over nickel deposits[2].

When taken from the press the spots are invisible, but they may be developed by soaking the print in a suitable reagent. Photographic developer is used for gold over silver. A 1% solution of dimethylglyoxime in alcohol containing 5% ammonium hydroxide, gives green spots with a copper underplate and red spots with nickel. A 5% solution of rubeanic acid in alcohol containing 5% ammonium hydroxide may be used for copper.

In an attempt to relate quantitative electrography to exposed bare area of basis metal, the writer prepared polished copper specimens on which asphalt dots of various sizes were applied before plating[2]. In this way, synthetic pores were produced ranging from 0.1 mm–5 mm in diameter. The pore areas were measured directly by a photographic technique. For the larger spots, quantitative electrography was a measure of the exposed area. By contrast, for the smaller spots, the quantity of corrosion product generated per unit area was greatly increased, but the amount of corrosion product per 100 spots was the same for the smallest synthetic pores as it was for real pores, which are still considerably smaller. Clearly, in this size range, quantitative electrography is a measure of the number of pores rather than of the bare area. For this electrochemical method at least, extrapolation from large to small spots would be incorrect.

From a quality control point of view, the search for a method of measuring the area fraction of exposed basis metal may be rather irrelevant. It is the consequences of porosity rather than the porosity itself which concerns us, and these consequences are determined by the number of pores and by their location with respect to the functional areas of the part. A full discussion of these matters is to be found in Chapter 23.

Electrography with Cadmium Sulphide Paper

In the method just described the print is prepared in two stages. In the first, basis metal ions are electrolytically transferred to the test paper, and in the second a characteristic colour is developed by immersion in a suitable solution. This method is suitable for quantitative analysis since the transferred metal is in solution prior to development and can be easily washed out of the paper. An alternative method, requiring only a single step, is to use a test paper impregnated with a reagent which will produce an insoluble coloured product by direct reaction with the basis metal ions. The use of a test reagent of limited solubility to still further restrict "bleeding" of the print, was suggested by Hermance and Wadlow[10]. A technique using cadmium sulphide paper was employed by Reid[21] in delineating crack patterns in rhodium plate, and this work led to its use and further development by Fairweather[22] in the porosity testing of gold and other precious metal coatings. Present day practice stems mainly from Fairweather's recommendations.

Cadmium sulphide test paper is available in prepared form, but it may readily be prepared *in situ* by soaking high wet-strength filter paper or duplicating paper in a cadmium salt solution, blotting to remove excess moisture, then immersing in a sodium sulphide solution for about 5 minutes, when a uniform yellow colour is developed by precipitation of cadmium sulphide. The paper is finally washed for about 30 minutes in running water and dried in an air circulating oven for 1 hour. A 10% solution of cadmium chloride is usually recommended for the first stage, and 5% sodium sulphide for the second. 0.25 M cadmium acetate solution has also been used and since the presence of residual chloride in the paper may adversely affect the test, there may be some advantage in using the acetate instead of the chloride.

Typical test conditions employ a current of 50 mA/in.^{2} for 30 seconds with an applied pressure of 200 psi. Moistening of the test paper with sodium carbonate/nitrate electrolyte as described in the previous section is often recommended, but very sharp porosity prints can be obtained with the dry paper supported on a moistened pad of clean blotting paper. If the test paper is too wet there may be some bleeding of the print despite the insolubility of the test reagent. The degree of moistening must therefore be controlled on the basis of experience. The duration of the test may also require adjustment to take account of coating thickness. 30 seconds is adequate for coatings in the normal thickness range up to 5 microns, but with thicker deposits slightly longer times may be needed to permit proper penetration of the pores and to obtain sufficiently well defined prints.

Fairweather developed a variant of the technique utilising a Plaster of Paris mix impregnated with cadmium sulphide, which was cast around irregularly shaped components, e.g., spring contacts. The test current was passed while the plaster was still damp and the cast was finally separated from the test surface to observe the porosity print on the area in contact with it. The procedure is tedious and more convenient methods for the examination of actual components by electrography are now available as described in the following section.

ELECTROGRAPHY IN GELLED MEDIA

A method of electrography which permits the location of pores on real hardware rather than flat panels has been reported by Bedetti and Chiarenzelli[11] and involves the use of gelled media rather than paper. The pore spots are developed at the surface of the part which is immersed in a gelled solution of gelatine containing the electrolyte and indicator solutions. The composite reagent contains 94.6% by volume of gelatine solution, 3.4% electrolyte solution and 2% indicator solution. The gelatine solution is 9% by weight in water and may be stored under refrigeration. A suitable electrolyte for copper or nickel underplates is 4% sodium carbonate with 1% sodium nitrate. A saturated solution of rubeanic acid in alcohol may be used as the indicator. Other formulations are given by the authors.

A test is prepared by melting a quantity of the gelatine, sufficient to fill a glass beaker, and adding the required electrolyte and indicator solutions. The specimen, which may be any shape, is immersed in the gelatine supported by a gold wire. Another wire is placed in the medium to act as a cathode. After the medium has cooled and solidified, a current of about 3.8 mA/cm^2 is passed for about 15–20 seconds with the specimen as the anode. Porosity is developed as coloured spots on the specimen surface. The authors claim higher sensitivity for this method than for paper electrography. The method is suitable for electrography on specimens with complex shapes.

Gel Film Electrography

This procedure has been described by Noonan[9]. The article to be tested, which may be any shape, is attached to a gold or platinum wire and is immersed in the hot liquid gelatine solution containing both electrolyte and indicator, then withdrawn. Upon cooling, a thin film of gelatine adheres to the test specimen. The gel coated article is immersed in a suitable aqueous electrolyte and made the anode using current density and times similar to the original procedure. Some bubbling of the film may occur and it may be convenient to wash the film off the specimen with hot water. Pore centres containing sufficient corrosion product to be easily visible will remain on the specimen surface and can be readily counted.

POROSITY TESTING BY CHEMICAL REACTIONS

These methods do not involve the use of current or any electrical measurement. All of them, with the exception of the ammonium persulphate method, reveal the actual pore sites on the metal surface. All except three involve exposure to a gaseous reagent.

POLYSULPHIDE TEST[12]

The reagent is prepared by adding sulphur in excess of 250 g/l to a saturated solution of sodium sulphide. After 24 hours, filter, and dilute to an SG of 1.142. The solution may be stored in a plastic bottle and reused.

The specimen is immersed in the solution for 1 minute, rinsed, dried and examined for black spots under 4–10 × magnification. The test is sensitive to both copper and silver underplates. With a nickel underplate, the test will detect only those pores which are continuous from the gold surface, through the nickel, to the copper or copper alloy substrate.

AMMONIUM PERSULPHATE TEST[13]

This test is quantitative in the sense that it gives a numerical value based on the amount of corrosion product extracted from the porous plate. It is applicable only to gold over copper or copper underplate.

Test specimens of known area are treated with a mixture of equal volumes of ammonium hydroxide (SG 0.880) and 1 M ammonium persulphate, with agitation for 30 minutes. The volume used may be from 10–50 ml and is chosen for convenience with reference to specimen size. The quantity of copper in the extract as determined spectrophotometrically at 580 nm, is taken as a measure of the porosity.

Ultrasonic Ammonium Persulphate Test[14]

In this procedure the tubes containing the test specimens in the ammonium persulphate solution are held in a circular rack and rotated in an ultrasonically agitated water bath at about 10 rpm. The bath temperature is controlled to 30 ± 3°C. Under these conditions a soak time of 4 minutes is sufficient for gold over copper. The quantity of copper extracted may be determined colorimetrically, or by a spectrophotometric method.

SILVER ION TEST[15]

Porosity of gold plate on nickel, copper, copper alloys, iron, or ferrous metals is detected by immersing the article in 1 M silver nitrate solution for 24 hours at 70°C, and an additional 10 hours at room temperature. Crystals of silver will be deposited at the pores in the plate.

POROSITY TESTING BY EXPOSURE TO GASEOUS REAGENTS

As a class, gas testing methods are the most versatile, and for engineering purposes, the most relevant of all porosity testing methods. They all have in common the following advantages:

1 They are applicable to any size and shape of specimen and do not involve any mechanical distortion of the plating which could lead to porosity induced by the test itself.
2 They are the most convenient methods for the location of porosity with respect to the functional areas of the part.
3 Although some complex and sophisticated test apparatus have been described in the literature, it is usually possible to perform reliable testing with very simple equipment which can be operated by non-professional personnel.
4 Most real corrosive environments are gaseous. Laboratory gas tests with appropriate reagents are therefore most relevant to actual performance in the field. By comparison with field exposure testing, acceleration factors may be determined for the laboratory tests. They may then be utilised as accelerated ageing procedures to predict the change with ageing of properties, such as contact resistance, which are sensitive to porosity.
5 Although gas testing is generally destructive to some degree, it does provide a measure of pore density. Increasing exposure time beyond that necessary to reveal all the pores, results in an increase in the amount of corrosion product generated from the pores. The major disadvantage of over-testing is the obscuring of individual centres by running together of excess corrosion product.
6 Gas testing is the most efficient way to assess the efficacy of any corrosion inhibitors which may be used. In this connection, it is essential that the test reagent is closely related to the actual environmental stress which the parts will undergo.

Many variations of gas test procedures for porosity have been used. Only those which have come into some prominence will be described here.

Nitric Acid Vapour Test[16]

This test may be used for gold over copper, copper alloys, or nickel. Place about 100–300 ml of concentrated nitric acid in a shallow dish in the bottom of a large desiccator. Test specimens may be conveniently suspended with a nylon fishing line or other non-absorbent thread, and should be supported in such a manner that they are at least 3 inches above the liquid surface and 1 inch from the walls of the vessel. The rack must be of glass or other non-metallic material not attacked by the acid. Specimens should be degreased in solvent vapour before testing, except those which are coated with an organic inhibitor.

After placing the rack of test parts in the vessel, cover the vessel tightly and allow to stand at room temperature for 2 hours. Longer times may be used to increase the sensitivity of the test. The use of a thermostatically controlled chamber at 25°C is desirable to increase reproducibility of the test conditions, but is not absolutely essential. Remove parts at the end of the test period and, without rinsing, place them in an oven at 105°C for 1 hour. Cool and examine for porosity under 4–10 × magnification. Pores will be revealed as green or blue corrosion spots.

Sublimed Sulphur* Test

This is perhaps the simplest of all gas test procedures to operate. It is applicable to gold over copper and gold over silver. Gold over nickel gives no reaction at pores which terminate at the nickel, but porosity through the nickel to copper or copper alloy basis metal is revealed. The test reproduces accurately the effects of a sulphide environment on gold plated parts, but does not reveal all the pores in gold over nickel. These effects are illustrated in Figure 23.1 (see Chapter 23).

A desiccator similar to that used for the nitric acid test is used. Two dishes, one containing distilled water and the other sublimed sulphur, are placed in the bottom of the vessel. Since little or no odour is generated during the test, the reagents need not be changed between tests. Arrangement of specimens is similar to that in the nitric acid test. The vessel is covered and placed in an oven at 60°C for a period which may be chosen between 4–24 hours. Specimens may be removed, dried and examined for porosity at 4–10 × magnification. For testing porous gold over silver, shorter test times may be used to reduce the amount of silver sulphide creep which obscures the individual pore sites.

Sulphur Dioxide Test

Of all the procedures which have been described for porosity testing with sulphur dioxide, the simplest to operate is that of Clarke and Leeds[6]. Originally intended to test porosity of tin and tin-nickel coatings, it was later modified for use with gold plate. Reagent quantities suitable for a 10 litre vessel will be given although larger vessels may be used by appropriate

* Sublimed sulphur is often referred to as "flowers of sulphur".

alteration of the amounts. Preparation and arrangement of test specimens is exactly as described for nitric acid testing. The test operates on the principle that in a closed vessel containing a solution of sulphuric acid and sodium thiosulphate, both the humidity and sulphur dioxide content will reach an equilibrium content which will remain constant in spite of depletion by reaction with the metal and some slow leakage. The test is applicable to gold over copper and over nickel. For gold over silver, the procedure described may be used, except that before beginning the test a dish of warm water is placed in the covered vessel for a period of time sufficient to provide a condensed film of water on the vessel walls, thus giving 100% relative humidity in the vessel during the initial test period. The dish of water is removed just prior to commencing the test.

The test reagent is prepared for each test by mixing 4 parts by volume of a solution made by dissolving 250 g sodium thiosulphate crystals in 1000 ml of water, and 1 part of a solution comprising of equal volumes of sulphuric acid (SG 1.84) and water. The combined volume of the test reagent is 1/40 that of the vessel and the ratio of the chamber volume in cubic centimetres to the solution surface in square centimetres should not exceed 50 : 1. For a 10 litre vessel, 200 ml of the sodium thiosulphate solution is placed directly on the bottom and 50 ml of the sulphuric acid solution added. The rack with specimens is placed in the chamber which is then covered with a well fitting greased cover. The vessel should be kept in a draught-free cupboard for the test period which is usually 24 hours, but may be shortened when very thin deposits are tested. The test chamber should be opened in a hood or well ventilated area. Specimens are removed and after drying may be examined for pores under 4–10 × magnification. Under the test conditions described, the sulphur dioxide concentration will be about 10% and the relative humidity approximately 86%. A later paper by Clarke and Sansum[23] describes a two hour test procedure.

Two Day Gas Test[17]

The two day gas test consists of 24 hour exposure to sulphur dioxide in a closed vessel of 10 litre capacity to which 0.5 ml of water and 100 ml of sulphur dioxide gas are added, the latter by means of a gas metering device. After the first 24 hour period the vessel is ventilated and 100 ml of hydrogen sulphide injected, the vessel resealed and held for a second 24 hour period. This procedure is longer, more difficult to operate, and is inherently less reproducible since a fixed amount of gas is added and the concentration must fall as the gas is consumed in the reaction with the metal or by leakage. Results obtained are comparable to the Clarke and Leeds procedure.

Industrial Atmosphere Test

British Standard 2011: Part 2Kb: 1970, describes an ageing procedure for electrical contacts and connectors which calls for 20 days of exposure to an atmosphere consisting of: 25±5 ppm sulphur dioxide, 3000±500 ppm carbon dioxide and a relative humidity of 75±5% at 25±2°C. The prescribed conditions are obtained by burning town gas and carbon disulphide vapour, and mixing the combustion products with humidified air. An improved cabinet for carrying out the test has been described by Leeds and Such[18].

The test is intended as an accelerated exposure test for connectors. The results are generally evaluated in terms of electrical performance (contact resistance). For porosity measurements *per se* the test appears to offer little or no advantage over the much simpler, faster and less expensive methods outlined in the preceding paragraphs of this section.

FIELD EXPOSURE TESTS

No general description of test procedures can be attempted here. Such tests have been used for many years in the evaluation of paints and other various systems, e.g., nickel and chromium electrodeposits. There has been much less activity in the case of gold. Walton[19] reported on the change of contact resistance of various gold plates on different substrates when these were subjected to 3.8 months exposure in a field site close to a chemical plant, and compared these results to the effect of exposure to a water-sulphur slurry at 55°C for 204 hours. Exposure times after measurement were extended for a total of one year at the field site and 400 hours in the sulphur-water test. His data were not published in a form which permits direct comparison of pore counts by the two test methods, except for gold over silver, but the effect on contact resistance does show some correlation. In this environment the peak levels of gas contaminants were highest for H_2S, NO_2, NH_3, O_3, HF and SO_2 in that order, but the median levels were in the order, O_3, HF, NH_3, with NO_2, H_2S and SO_2 each at about 8 parts per billion in air. The sulphur compounds were of major importance and in the case of gold over silver, reasonable agreement in pore counts for comparable samples was obtained between the field and laboratory exposures.

Baker[20] reported on studies of gold plated metals which were exposed at field sites in New York City (industrial-marine), Kure Beach (marine) and Steubenville, Ohio (industrial). He studied the effects on contact resistance, due to corrosion at the pores in the plate, but gave no estimates of the amount of porosity found.

CRITIQUE OF GAS TESTS FOR POROSITY

It is obvious that a large number of test methods have been invented. Not all the variants have been described here, but clearly, the repertoire is sufficient for most foreseeable testing requirements. The writer is strongly of the opinion that future research and engineering effort should be directed towards correlating the results of the test methods now available, with the object of choosing the most relevant and reproducible procedures, rather than to the invention of more tests and variants on the old ones. Such an effort is now under way for gold plated electrical contacts in ASTM Committee B-4 and it is to be hoped that industry-wide accepted standard practices will be developed.

Pending this highly desirable outcome, some remarks on the problem of choosing a proper test procedure will be offered. The first point to be made is that the laboratory test gas reagents react differently on the various base metal substrates and underplates over which gold is plated. Table 27.1 is a rough guide to these effects, for the most commonly used basis metals, silver, copper, and nickel.

No single gas reagent can be considered as universal for all cases. The two day gas test in which sulphur dioxide exposure is followed by hydrogen sulphide would perhaps serve this purpose, but it is in fact two separate tests which could more easily and reliably be carried out using the Clarke and Leeds sulphur dioxide procedure for gold over nickel and gold over copper, and either the sublimed sulphur test or the Clarke and Leeds procedure with 100% initial relative humidity for gold over silver or copper. These tests, together with the nitric acid fume test, form a battery of gas tests from which a relevant and effective test procedure can be chosen for gold, plated over nickel, silver, copper or copper alloy. The Clarke and Britton[4] sulphur dioxide test procedure originally developed for tin-nickel on steel will work as well for gold on steel when no underplate is used. If underplates of copper or nickel in excess of 100 microinches are present, the appropriate test from Table 27.1 may be used.

Table 27.1
Effect of Various Test Gases on Base Metal Substrates
With Porous Gold Plate

Test Gas	*Gold over Silver*	*Gold over Copper*	*Gold over Nickel*
S, 60°C 100% RH	Reveals all pores, Ag_2S creep severe for long times, or with very porous plate.	Reveals all pores, resolution good.	Nickel not affected, Reveals Au–Ni coincident pores to copper basis.
H_2S 100% RH	Reveals all pores, Ag_2S creep severe, resolution poor.	Reveals all pores, resolution good.	Reveals Au–Ni coincident pores.
SO_2 85% RH	Variable, often no effect depending on initial ambient conditions.	Reveals all pores, resolution good.	Reveals all pores, resolution good, but high initial ambient RH may cause blurring.
SO_2 100% initial RH in test chamber.	Reveals all pores, resolution good.	Reveals all pores, resolution good.	Reveals all pores, resolution poor due to excessive corrosion product.
HNO_3 100% RH	—	Reveals all pores, resolution good.	Reveals all pores, resolution good.

When the test objectives are to reveal all pores in a plating for research or engineering studies, then the relationships given in Table 27.1 will be adequate for a proper choice. However, porosity testing is also used to predict performance under particular environmental conditions. This is a more difficult problem. If one is concerned with industrial, or industrial marine conditions, then the Sulphur Dioxide, the Industrial Atmosphere and/or the Sublimed Sulphur tests would appear to be the most relevant. It has generally been accepted that salt spray is relevant to marine environments; however, in a real situation, marine environments, particularly on board ship, are seldom represented by a purely salt atmosphere. Sulphide contamination is usually also present. A combination of sulphur dioxide and sublimed sulphur tests, would in fact give a more realistic appraisal of performance.

All gas tests can be varied in their severity by altering exposure times and in some cases by higher or lower temperatures. The rates of pore corrosion will be affected by temperature in a complex manner. The reactions are rarely representable by a single chemical equation, but are more often the result of a number of intermediate processes. Each of these may be affected differently by temperature changes. Variation in exposure time rather than temperature would seem to be more likely to lead to better correlation of tests. Variation of test severity is a proper practice when dealing with testing problems which involve different thicknesses of gold plate. A test procedure should, ideally, be carried to the point at which all pores are revealed. To go beyond this generally results in the copious evolution of corrosion product at the pores, which will totally obscure the porosity pattern. This can only be justified when it is known that the actual environment of use is so aggressive that such copious corrosion would in fact occur.

SUMMARY

Electrochemical test methods have been widely used for research purposes. Most of them are rapid and non-destructive, and give no visual indications of porosity. Electrography is rapid and convenient, and does provide a visual indication. Paper electrography is useful for panels of thick materials and porosity can be assessed quantitatively. Electrography in gelled media is useful for real hardware. Gas testing procedures are the easiest to operate and are most relevant to real environments. They are useful for conditioning tests, particularly for gold plated electrical contacts. General agreement on test methods has not yet been achieved, but this is likely to happen within the next few years as a result of the efforts of industry associations, especially of ASTM.

REFERENCES

1 Cooksey G. L., Campbell H. S., *Trans. Inst. Metal Finish.*, **48**, 93 (1970).
2 Garte S. M., *Plating*, **53**, 1331 (1966).
3 Shome S. C., Evans U. R., *J. Electropl. Depos. tech. Soc.*, **27**, 45 (1951).

4 Ehrhardt R. A., *Proc. Am. Electropl. Soc.*, **47,** 78 (1960).
5 Clarke M., Britton S. C., *Trans. Inst. Metal Finish.*, **36,** 58 (1959).
6 Clarke M., Leeds J. M., *Trans. Inst. Metal Finish.*, **43,** 50 (1965).
7 Morrissey R. J., *J. electrochem. Soc.*, **117,** 742 (1970).
8 Leeds J. M., Clarke M., *Trans. Inst. Metal Finish.*, **47,** 163 (1969).
9 Noonan H. J., *Plating*, **53,** 461 (1966).
10 Hermance H. W., Wadlow H. V., "Electrography and Electro Spot Testing", Monograph No. 1809, Bell Telephone Laboratories, Murray Hill, N. J. (1951); also in W. G. Berl "Physical Methods in Chemical Analysis" Vol. II. Academic Press, New York (1951).
11 Bedetti F. V., Chiarenzelli R. V., *Plating*, **53,** 305 (1966).
12 Nobel F. I., Ostrow B. D., Thompson D. W., *Proc. Am. Electropl. Soc.*, **52,** 49 (1965).
13 Frant M. S., *J. electrochem. Soc.*, **108,** 774 (1961).
14 Tweed R. E., tech. Rep. AFML-TR-321, Project 7-960, U.S.A.F. (1965).
15 Ciambrone D. F., *Metal Finish.*, **67,** 3, 60 (1969).
16 Baker R. G., Holden C. A., Mendizza A., *Proc. Am. Electropl. Soc.*, **50,** 61 (1963).
17 Khan A. A., *Plating*, **56,** 1374 (1969).
18 Leeds J. M., Such T. E., *Trans. Inst. Metal Finish.*, **49,** 131 (1971).
19 Walton R. F., *Plating*, **53,** 209 (1966).
20 Baker R. G., Proc. Int. Symp. on Electric Contact Phenomena, Graz. Austria (1964).
21 Reid F. H., *Trans. Inst. Metal Finish.*, **33,** 105 (1956).
22 Fairweather A., Lazenby F., Parker A. E., Proc. I.E.E. 109 (Part B, Suppl. No. 22), London (1962).
23 Clarke M., Sansum A. J., *Trans. Inst. Metal Finish.*, **50,** 211 (1972).

BIBLIOGRAPHY

1 Burns R. M., Bradley W. W., "Protective Coatings for Metals", Reinhold, New York (1955).
2 Galitzine N., Ashley S. E. Q., "Examination of Plated and Protective Coatings by Electrographic Analysis", A.S.T.M., Special tech. Publ. 98, 61 (1949).
3 Leeds J. M., Townley J. R., "Techniques for Measuring the Porosity in Precious Metal Electrodeposits" *Metal Finish J.*, **18,** 210, 190 (1972): **19,** 217, 36 (1973).

Chapter 28

THICKNESS MEASUREMENT

F. H. Reid

In functional applications the essential requirement of a gold coating is reliability. This depends on a number of inter-related properties such as hardness, ductility, wear resistance, porosity, etc., and assignment of quantitative values to these in relation to specific applications can only be made in the light of experience based on the systematic correlation of laboratory measurement and assessment tests with field performance records.

Despite an increasing appreciation of the need, there is little information available concerning the detailed effects of such factors in relation to performance in particular circumstances, and their relative importance may vary considerably in specific service conditions. A common denominator, however, is that all may be affected by plating thickness; some, like porosity, in a fairly clearly defined way while others, like hardness, ductility and wear resistance, less simply, but nevertheless significantly. This emphasises the overall importance of gold thickness measurement from the technical aspect.

Since gold is a precious metal, economic considerations are also strongly involved. To reduce cost, plating thickness must be restricted to a minimum consistent with reliability and when this has been specified, it is clearly important to the user that it should be strictly controlled. This is equally important to the plater, not only to the end of avoiding costly rejects, but also because the use of other than a reasonable marginal excess of gold over that required to achieve the minimum thickness on significant areas represents wastage. As such, this is to the detriment of profitability, and the customer is not likely to be sufficiently broadminded to accept an inflated charge which takes this contingency into account.

The serious specialist plater must therefore be in a position at least to carry out thickness checks on a sample basis. Ideally, he should be able to extend this facility to larger numbers of measurements, using non-destructive techniques, to permit some statistical quality control of his production. The user will usually be more concerned with spot checks of thickness on random samples and for this purpose may have at his disposal a wide variety of sophisticated procedures. In this situation it is highly desirable that the same method of measurement should be employed by both parties or, if this is not practicable, that tests should be carried out to establish the closest possible correlation between the methods adopted. This is necessary because some procedures (direct, or primary methods) give the true linear thickness of a coating directly, irrespective of other factors, whilst others (indirect, or secondary methods) measure thickness in terms of weight of coating per unit area, either directly, or through the response of the deposit to some form of external radiation. In the latter cases the linear thickness must be derived indirectly, either by assuming the density of the coating, or by suitable calibration procedures. The work of Cooley and Lemons[1] has shown significant variation in density of different types of gold deposit which can obviously affect the derived thickness figure. Similarly, in radiation methods changes in composition of substrate or coating may cause a change in the apparent thickness of the coating if careful attention is not paid to these factors. These aspects are discussed more fully in the following section, with reference to individual methods.

METHODS

Numerous methods are available for measuring the thickness of electrodeposits in general, but whilst in principle most of these, with the possible exception of tests of the jet type, could be used with reference to gold plating, practical restrictions are imposed by requirements of sensitivity and accuracy in relation to the relatively low thickness range commonly involved (0.5–5 microns). Thus, non-destructive procedures based on differences in thermal EMF and thermal conductivity between coating and basis metal, and on the generation of eddy currents in the substrate/coating system, are not sufficiently sensitive in this range. This also applies to the various magnetic pull-off tests. However, considerable developments have been made in electromagnetic methods as a result of which commercial instruments are now available which are capable of measuring gold coatings on ferromagnetic substrates down to about 0.5 μm with an accuracy of approximately ± 0.15 μm relative to the standards used. As this technique can be applied to quite small areas this method could readily satisfy certain gold measurement applications. Within the scope of the present chapter, therefore, main attention is devoted to methods which have become particularly well established in gold plating technology. Brief reference is also made to certain other techniques which, whilst not specifically relating to gold, could be of potential usefulness in this field.

There are a number of possible ways in which methods can be broadly classified, viz., direct (primary) and indirect (secondary): destructive and non-destructive. However, it is convenient for the present purpose to categorise them according to the basic nature of the measuring technique employed, e.g., optical, chemical/electrochemical, irradiation, electrical, mechanical, etc.

OPTICAL METHODS

Microscopic

PRINCIPLE

Thickness is measured directly on a carefully prepared normal or oblique section of the coating.

METHOD

It must be assumed that the reader is familiar with the general techniques of mounting and polishing used in the preparation of microsections in conventional metallographic practice. Special precautions, however, are necessary for optimum accuracy in measurements on plated coatings, as discussed by Wilson[2], and by Cullen and Petruna.[3]

For instance, it is advisable to apply a thick deposit of copper or nickel (20–50 microns) to the specimen before sectioning to protect the coating during initial cutting or sawing, and to prevent excessive bevelling of the edges during polishing. For mounting, a clear medium should be used so that the position of the specimen may be visually checked. It is also advantageous to employ a cold setting resin to avoid possible diffusion effects between coating and substrate which may occur if heating is used.

Whatever the sequence of abrasive papers used in preliminary grinding, final polishing is preferably carried out with diamond paste of particle size not exceeding 1 micron, supported on a selvyt covered polishing wheel. The polishing process should not be prolonged beyond the time required to produce a reasonably scratch-free finish, in order to minimise slight bevelling of the section, which may occur even in the presence of a supporting deposit. This is particularly important when there is a significant difference in hardness between the coating and substrate or supporting deposit.

Some drag-over may occur between the various layers during polishing which can lead to difficulty in delineating boundaries and to error in measurement (Figure 28.1). This can be removed and the contrast between layers improved by lightly etching the section for a short time in a suitable solution. For this purpose a 5% solution of nitric acid in ethanol is of fairly general applicability. Other etchants have been described by Baldwin.[4] Experience is necessary in judging the optimum etching time, which should be just sufficient to produce a clear delineation of boundaries. Over-etching may lead to uncertainty in the precise location of these and, in the case of oblique sections, to serious errors in measuring thickness as indicated in Figure 28.1.

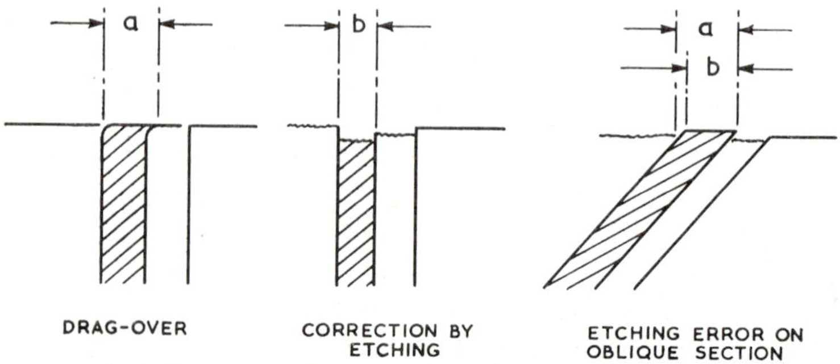

Fig. 28.1 Sources of error in microsection measurement[3].

Actual measurement may be carried out by projecting an image of the section on to the screen of a metallurgical microscope at a known magnification (up to 1000 ×) and measuring coating thickness by a linear scale, or by using an accurately calibrated eyepiece graticule or drum micrometer. To eliminate possible error due to movement of the eyepiece during measurement, Cullen and Petruna[3] have described an image splitting eyepiece by means of which two superimposed images of the coating can be moved so that the inner edge in one coincides with the outer edge in the other, the movement being recorded on a micrometer drum. This eliminates the need for crosswires or graticules in the field of view and results in greater accuracy in measurement on very thin sections.

ACCURACY

Sources of error in microscopic measurement of thickness may be associated with:

1 Incorrect mounting, leading to deviations of the section from the normal, or from the tilt angle in the case of oblique sectioning. The error will be greater in the latter case.
2 Poor preparation technique, leading to excessive tearing of the coating during grinding, or excessive drag-over in polishing. This will necessitate excessive etching, with attendant possibilities of error as indicated in the preceding section. As before, errors from this cause will be more serious for oblique sections, but if grinding scratches are not completely removed there will also be general difficulty in location of boundaries.
3 Over-etching, even in the case of properly polished sections.
4 Errors in nominal magnification at which measurements are made.
5 In the case of very thin coatings, thickness of crosswires or graticule scales may be appreciable in relation to coating thickness, leading to uncertainty in lining up.
6 Instrumental errors such as movement of eyepiece, backlash in micrometer movement, etc.

Even when instrumental errors are discounted, it is clear that a high degree of operator skill and experience is necessary in order to obtain optimum accuracy, and even so, definite limitations are imposed on this as deposit thickness decreases. On coatings, nominal thickness of which was determined by weight gain, Cullen and Petruna[3] report an accuracy of 2% over an optimum range of 2.5–5 microns, while Plog[5] quotes only ±4% at 10 microns and ±20% at 2 microns. Wilson[2] records discrepancies of up to 50% in measurements on the same section of a nominal 1 micron coating by three different observers. This thickness corresponds to only a few micrometer drum divisions and it is likely that error under (5) above was mainly operative. Cooley and Anderlee[6] assessed the error involved in thickness measurement by microsectioning due to variations in mounting and preparative technique by different operators, on submitting standard panels carrying 0.35, 1.1, 2.1 and 4.6 microns of plating to six different laboratories. They found mean errors of +62%, +24%, +8% and +6% respectively, in the reported thicknesses, which confirms the general view that 2 microns is the minimum thickness that can be measured with acceptable accuracy (±10%) by the microsectioning method.

ADVANTAGES AND LIMITATIONS

Advantages

1 Measurement is direct and independent of coating density.
2 Both local thickness and distribution along the line of cut can be determined.
3 Applicable to any coating/substrate combination. The thickness of undercoat or of individual layers of composite deposits may be determined.
4 Not limited by sample geometry. Suitable for thickness measurements on internal surfaces.
5 Section can give qualitative indication of adhesion and also serve for hardness testing, structural examinations, and general assessment of quality of coating and basis metal in terms of freedom from gross pores, inclusions, nodules, etc.
6 Best accuracy is obtained at thicknesses greater than 2.5 microns (ASTM recommendation).

Limitations

1 Destructive
2 Time consuming. Preparation and measurement can occupy 60–90 minutes per microsection but multiple mounting can reduce measurement time per single component.
3 Operator skill and experience is essential.
4 Poor accuracy obtained on thin coatings.

Chord Method

PRINCIPLE

Thickness is derived from microscopic measurement on an oblique section exposed by grinding a shallow groove on the coated surface.

METHOD

For optimum results a fairly large flat area of coating is necessary. The specimen is supported in a suitable holder to restrain lateral movement and a shallow cut is made using a fine grinding wheel (600 grit) of 1–2 in. diameter or by feeding fine abrasive on to a high-duty cast iron wheel, to just expose the surface of the basis metal, when oblique sections of the coating are revealed at each end of the cut (Figure 28.2). The thickness of the coating is obtained by measuring the lengths Cx and Cy and applying the formula:

$$\text{Thickness} = \frac{2xy}{D}$$

where D is the wheel diameter.

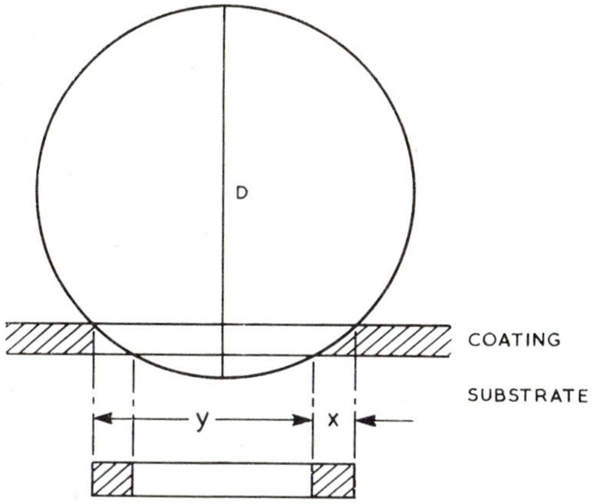

Fig. 28.2 Principle of "chord" method for thickness measurement. Coating thickness = 2x.y/D.

As in microsectioning it is advantageous to apply a protective copper or nickel deposit, the latter being preferred in view of colour contrast which aids in boundary definition. Since no polishing is involved, drag-over between the various zones is usually insignificant. If it is necessary to improve contrast between them, some form of heat tinting or staining of the basis metal may be used.

ACCURACY

There is little information available. Although the technique appears relatively crude in comparison with microsectioning, accuracy may nevertheless be comparable under the best conditions. This may be illustrated by the results recorded in Table 28.1 of comparative measurements on gold coatings on silver plated brass made by microsectioning and by the chord method, using a 600-grit wheel.[7] Appropriate skill and experience was available for both techniques.

Using 0.25 micron diamond powder as the grinding medium, this type of technique has been adapted for measuring the thickness of diffusion layers on silicon[8] where it is claimed that layers as thin as 0.25 micron can be readily measured. Whilst conditions for plated coatings are not immediately comparable, this appears to indicate that the technique could be of wider use in the measurement of very thin gold deposits.

Table 28.1
Comparison of Thickness Measurements by Chord and Microsection Methods

Sample	*Thickness* (*μm*)	
	Chord Method	*Microsection*
1	2.2	2.3–2.5
2	2.4	2.8–3.0
3	1.8	1.8–2.0
Silver undercoat (Sample 2)	5.2	4.2–4.8

ADVANTAGES AND LIMITATIONS

Advantages

1 Low cost method which is adaptable as a simple "workshop" test
2 Rapid
3 Measurement of thickness is independent of coating density.
4 Reasonable accuracy may be achieved
5 Undercoating deposits may also be measured.

Disadvantages

1 Destructive
2 Application limited by sample geometry
3 Accuracy limitations for thin deposits, as for microsectioning method.

Interference Microscopy

PRINCIPLE

Plating thickness is measured directly as the difference in level between unplated and plated areas of a surface, in terms of the displacement of interference fringes across the boundary.

METHOD

Application of simple double-beam interferometric techniques to the measurement of plating thickness have been described by a number of authors, notably by Thomas and Rouse[9] and by Saur,[10] the chief interest being in decorative chromium coatings which are readily stripped selectively from a nickel base to leave plated and unplated areas for measurement. Such a procedure could equally be applied to gold plating where the substrate is not subject to attack by the stripping medium, but in general the interferometric technique is more useful as a special method for the unequivocal determination of gold thickness on reference specimens for use in the calibration of non-destructive procedures.

For this purpose, specimens are prepared by masking a carefully prepared flat surface with a strip of masking tape or a thin lacquer coating. This is removed after plating the exposed surface with the nominal thickness of gold to leave a well defined boundary between plated and unplated portions. This is viewed through a microscope fitted with an interferometry attachment and adjusted to produce a system of interference fringes aligned normal to the boundary. The difference in surface level on crossing the boundary is manifested as a displacement of the fringe system by one fringe-to-fringe distance for each increment of $\lambda/2$ of the difference in height between plated and unplated areas (where λ = the wavelength of light employed for the observation). The thickness of the coating is then given by the simple relationship:

$$\text{Thickness} = \text{Fringe displacement (in units of fringe-to-fringe distance)} \times (\lambda/2)$$

As an alternative to a special interferometry attachment, Saur[11] has described a simple and inexpensive arrangement for the production of fringes by placing a piece of flat glass, e.g., a microscope cover glass, in contact with the bright plated surface. This method is suitable for use with an ordinary microscope equipped with vertical illuminator and a monochromatic light source.

The appearance of the field of view under various conditions is illustrated diagrammatically in Figures 28.3 a–28.3 d. Optimum sharpness of fringes for measurement purposes depends on the use of illumination of good monochromaticity such as that supplied by a sodium lamp, or low pressure mercury lamp with an appropriate filter. Under these conditions the fringes are of uniform appearance as shown in 28.3 a. White light gives a fringe system of finite width, the central, or first-order, fringe of which is dark, with red and green fringes on either side (28.3 b). If a tapered boundary can be formed it is simple to trace individual fringes and measure the deviation using monochromatic light only (28.3 c), but in the case of a sharp boundary it is necessary to make a preliminary observation in white light and to note the deviation between the first order fringes on an eyepiece reticle. The white light is then replaced by a monochromatic source and the number of fringes counted in the same reticle interval (28.3 d and 28.3 e).

The interference technique has been utilised by Fluhmann and Saxer[12] in thickness measurement of very thin gold coatings for the calibration of beta-ray backscatter equipment. They describe a simple device for transfer of specimens from the interference microscope to the beta-ray measuring table

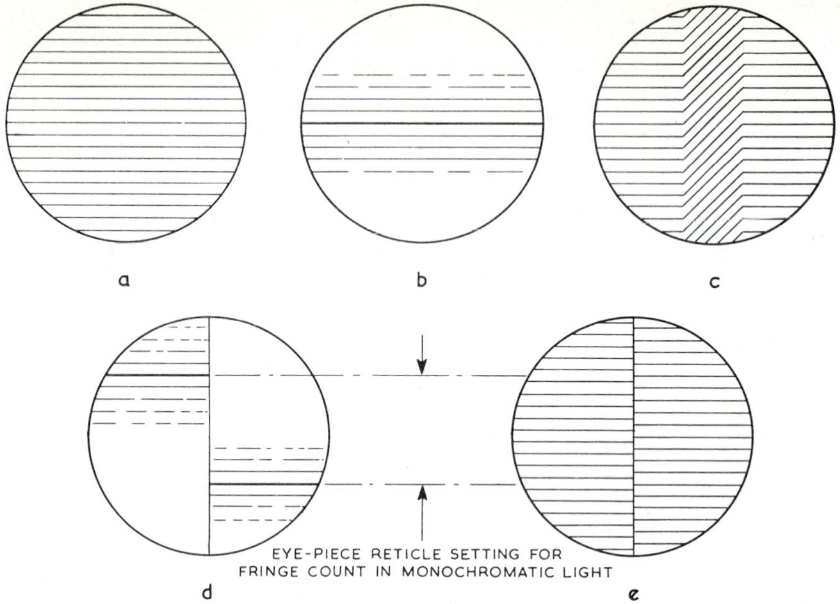

Fig. 28.3 Appearance of interference fringes under various conditions (schematic)
(a) Plane surface—monochromatic light (b) Plane surface—white light.
(c) Half-plated surface with tapered boundary—monochromatic light.
(d) Half-plated surface with sharp boundary—white light (e) As (d) monochromatic light.

to ensure coincidence of the areas examined by each method (Figure 28.4). For the highest precision in measuring very thin coatings it is also necessary to apply a thin coating of gold or other metal over the whole surface before measurement. This takes into account possible errors arising from differences in phase change of incident light on reflection from the basis metal and the coating respectively, due to differing optical properties of the metals.

ACCURACY

The accuracy obtainable by double-beam interference methods depends on the precision with which fringe deviation can be measured. Using mercury light of wavelength 5460 Å, fringe separation corresponds to 0.273 micron, but since fringe width may itself be comparable to this, Saur[10] considers that fringe deviation cannot be interpolated to better than one fifth of a fringe interval, i.e., about 0.50 micron. This estimate agrees with that proposed by Cooley and Lemons and applies to coatings of up to 10 microns, above which modifications of technique are necessary with a decrease in accuracy.

Where necessary, much greater accuracy can be attained by the use of multiple-beam interferometry[13] in which fringes are formed by multiple

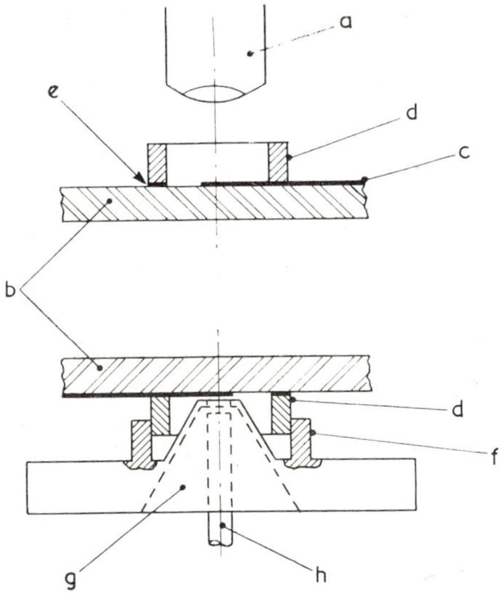

Fig. 28.4 Mechanical positioning and transfer system for defining specimen observation area in calibration of beta-ray backscatter standards by interferometry (a) Objective mount of microscope (b) Calibration standard (metal base) (c) Plating (d) Guide ring (e) Point of adhesion (f) Guide ring of beta-radiation backscatter apparatus (g) Diaphragm of beta-radiation backscatter apparatus (h) Radiation source (radionuclide).

reflection between the specimen and a reference mirror. In this case fringe width is much narrower in relation to fringe separation, permitting greater precision in the measurement of fractional fringe deviations. Using a simple multiple-beam interferometer as described by Tolansky,[14] Saur has achieved a sensitivity of 17 Å (0.0017 μm) in measurements of this type.

ADVANTAGES AND LIMITATIONS

Advantages

1. Since this is a direct method, no calibration is required
2. Time requirement not excessive
3. High accuracy on thin coatings makes this suitable as a referee method.

Limitations

1. Destructive, if stripping is involved
2. Limited application to plated samples due to difficulty in stripping gold without attack on basis metal or undercoat
3. Limitation of thickness for optimum accuracy
4. Applicable only to surfaces of good surface quality and brightness.

Depth of Focus Method

Where adjacent plated and unplated areas are available, thickness can be measured using a microscope with a short focal length objective by focusing successively on the surface of the coating and the bare substrate. The difference in level is noted from the movement of the calibrated fine focusing drum. Saur[10] has described a modification of this technique utilising interference fringes, which permits a wider field of view, The method is not highly accurate, accuracy being about half the minimum calibration, usually either 0.5 or 1 micron.[10] In the measurement of 12 micron gold coatings on beam leads, Hodgson and Szkudlapski[15] report a significant difference in readings by different operators.

Light Profile Method

This technique, originated by Schmaltz,[16] is primarily a method for general surface examination, but can be applied to measurement of step height between plated and unplated areas. The image of an illuminated optical slit or a hair line, is projected onto the surface at an angle of 45° and the specular reflection viewed through a microscope at the opposite 45° angle, when the line or slit image appears as a contour of the surface. In a modification due to Tolansky,[17,18] a grid of equally spaced fine lines is placed at the field iris of the microscope objective to give a profile of the surface with the grid image showing height contours at regular intervals. Saur[10] gives an accuracy of $\pm 6\%$ for this type of method in measurement of groove depths of 12.5–125 microns. A specific application relating to gold plating is in the measurement of 3 and 12 micron coatings on beam leads.[15]

A similar technique, based on electron microscopy, is described by Halliday.[19]

CHEMICAL/ELECTROCHEMICAL METHODS

Strip and Weigh

PRINCIPLE

The coating is separated from a known area by dissolution of the basis metal and weighed. Thickness is derived on the basis of coating density, usually assumed theoretical. If the coating separates as a handleable foil, thickness may be measured directly with a precision micrometer.

METHOD

A small area is cut from the plated sample or, in the case of very small samples, several complete items may be taken and the basis metal dissolved by treatment with dilute nitric acid (SG 1.2). If the residual coating is sufficiently coherent it may be washed by decantation with distilled water, with a final rinse in alcohol or acetone before drying and weighing. Otherwise, it is recovered by filtration through a small filter paper, which may be of fairly coarse texture, but should be of the ashless variety. After thorough washing, the paper and contents are dried and the coating finally transferred carefully to the pan of an assay balance and weighed. Should particles adhere tenaciously to the paper, the latter is ashed in a small crucible and the residue added to the balance pan after cooling.

$$\text{Thickness (microns)} = \frac{w \times 10^4}{a \times d}$$

where w = weight of coating (g)

a = area of coating (cm^2)

d = density of coating (g/cm^3).

ACCURACY

The accuracy of this procedure is in principle very high provided that a sufficient sample area can be taken. As an indication, a 1 micron coating of gold on an area of 1 cm^2 weighs 0.00193 g, and since a good assay balance can weigh to 0.00002 g, an accuracy in the order of 1% is to be expected. This level can be maintained in the case of thinner coatings by taking correspondingly larger areas, but in all cases achievement of optimum accuracy is dependent on sufficiently precise measurement of the area and a high degree of manipulative care and skill on the part of the operator, both in the isolation and weighing of the gold.

ADVANTAGES AND LIMITATIONS

Advantages

1 High accuracy over wide thickness range. Suitable as a calibration method.
2 Independent of sample size and geometry.

Limitations

1 It is difficult or impossible to measure discrete areas on sharply curved, complex surfaces.
2 Destructive and indirect.
3 Unfavourable time factor for routine control.
4 High operator skill required.
5 Not applicable to gold coatings on tin-lead alloys, stainless steel, or other substrates not completely soluble in nitric acid.

6 Not applicable to highly alloyed gold coatings which may be selectively attacked by nitric acid.
7 Gives average thickness only.

Chemical Assay

In cases where the strip and weigh procedure is inapplicable due to resistance of the substrate metal to attack by nitric acid, it is usually possible to dissolve the gold coating selectively. This may be accomplished by anodic attack in sodium cyanide–sodium hydroxide solution or by immersion in sodium cyanide solution with the gradual addition of hydrogen peroxide.[20] The amount of gold in the solution may then be determined by conventional analytical methods (see Chapter 33) and the deposit thickness derived from this data.

Gain in Weight

Though not strictly a chemical method, the well known procedure for determining gold plating thickness by recording the gain in weight of components or test plates before and after plating is mentioned here in view of its analogy in reverse to the strip and weigh procedure.

Anodic Dissolution

PRINCIPLE

The coating over a small defined area is dissolved anodically at constant current and 100% efficiency in a small test cell, the end-point being indicated by a change in cell voltage when the basis metal is exposed. Coating thickness is proportional to the time required for dissolution, and the procedure is calibrated accordingly.

METHOD

The anodic de-plate method has been described by Anderson and Manuel[21] and by Waite[22] with reference to decorative coatings of chromium on nickel. Francis,[23] and Narayanan and Venkatachalam[24] have described electrolytic thickness testers based on the same principle. The former author makes passing reference to its successful use on gold plating, but offers no details on electrolyte formulations or precise parameters of the de-plating process, although a 10% solution of sodium cyanide is referred to as of general applicability to cyanide-soluble coatings.

Baldwin[25] has dealt with the application of the technique in considerable detail to the measurement of gold plating thickness. He recommends an electrolyte containing 200 g/l of magnesium chloride (hydrated) and 100 g/l of sodium chromate, for use on preferred substrates of silver, copper and nickel.

Commercial instruments are available for the measurement of gold thicknesses up to about 5 microns on copper, nickel, electroless nickel, brass and silver. They comprise essentially a small test cell which may itself form the anode (Monel or stainless steel), an adjustable source of constant current, and suitable electronic timing or voltage recording devices for following the progress of dissolution. For measurements on plated components the cell is pressed in contact with the surface, a small area of which is exposed through an accurately apertured gasket in the base. Alternatively, very small components or fine wires may be completely immersed in the cell.

The electrolyte is then introduced and the de-plating current, usually in the order of a few milliamperes, is switched on. Mechanical stirring or continuous circulation of the electrolyte is employed to assist in maintaining 100% anode efficiency. If electronic timing is employed the timer, which may be calibrated directly in thickness units, is switched off by a relay, activated by the voltage change at the end-point.[26] The process can alternatively be followed by chart recording of the cell voltage, as described by Baldwin. It is claimed that this method of recording permits differentiation between different types of gold deposit in duplex deposits.[27] Mathur[28] has described an instrument in which the end-point is detected visually by the deflection of a voltmeter, this being rendered very sensitive to small voltage changes at the end-point by using a potentiometer arrangement to balance out the cell voltage during the stripping process.

In at least one commercial instrument* which has recently become available there is direct digital readout in thickness units. This particular instrument has two pre-selectable deplating time cycles and is fitted with a presettable compensator (±15%) for use in testing alloy coatings.

ACCURACY

Little precise information is available on the accuracy of the anodic dissolution procedure in the thickness measurement of gold coatings. It has been claimed[25] as an advantage of the technique that, since de-plating time is directly proportional to coating thickness, calibration is simplified because it can be based on a de-plating rate established on relatively thick reference coatings which can be accurately checked by microsectioning. This is only valid if the de-plating rate remains constant, even if less than 100%. Since this may vary with the type of gold, it is clearly essential, as in all calibration procedures, that the reference coatings should be of the same type as those to be tested. Other possible sources of error may arise from slight variations in exposed area from test to test and uncertainty in end-point detection as may

---

* Fischer Instrumentation Ltd.

arise, for example, due to premature exposure of basis metal through a porous coating before the latter is completely removed.

Cooley[29] refers to "good" results in measuring gold coating thickness, using an area of 0.040 in.^{2} and a de-plating rate of 0.13 micron/second. Thickness could be measured to an accuracy within 0.25 micron on small intricate shapes. Later,[6] on the basis of a more systematic study of the procedure in comparison with other "production" methods for thickness measurement comprising gain in weight, beta-ray backscatter and micro-sectioning, as applied to pure gold and some alloy coatings covering the thickness range 0.5–7.5 microns, it was concluded that the accuracy of the de-plate procedure was only comparable to that obtainable by micro-sectioning. There was, however, a suggestion by the manufacturer of the equipment used that this could probably be improved by adjusting conditions to give longer dissolution periods.

ADVANTAGES AND LIMITATIONS

Advantages

1 Procedure is rapid
2 Applicable to measurements on very small components and fine wires.

Limitations

1 Destructive
2 Need to take into account possible effects on accuracy of coating quality and composition.

Depolarisation Kinetics

Agarwal[30] has related the thickness of thin gold coatings on platinum to the time required for depolarisation after cathodically polarising specimens at 3 volts in 1 *N* sulphuric acid. The effect appears to be related to a progressive decrease in porosity with increasing thickness, and while this technique might be useful in porosity studies it seems to have little relevance to thickness measurement for control purposes.

ACTIVATION TECHNIQUES

X-ray Methods

PRINCIPLE

Depending on conditions, a metal can respond in two ways to irradiation by a beam of X-rays:

1 The metal lattice acts as a diffraction grating, giving rise to a characteristic X-ray diffraction pattern, as utilised in examination of metallic structures.

2 The metal atoms can be excited to produce emission of a characteristic secondary X-ray spectrum (fluorescence) which is related uniquely to atomic structure, and independent of crystalline structure or metallurgical condition.

Up to certain limits, the intensities of diffraction or fluorescent lines increase with the mass of metal producing the effect, hence, in the case of a metal coating, thickness can be related to the measured intensity of a selected line in the emitted radiation from the coating (emission method) or, alternatively, to the attenuation of intensity suffered by a line from the substrate emission due to absorption on traversing the coating (absorption method).

Lines in the fluorescence spectra are produced mainly as a result of electron transitions between the K and L shells of the atoms, hence they are designated as $K\alpha$, $K\beta$, $L\alpha$, $L\beta$ etc.

Measurement of diffraction lines from the coating itself is little used, though historically this was one of the earliest applications of X-rays to thickness measurement.[31] Attenuation of substrate diffraction lines has not been used on gold coatings, a general disadvantage of this technique being the sensitivity of diffraction line intensity to the metallurgical condition of the substrate, e.g., crystal structure, grain size, stress, etc. However, Keating and Kammerer[32] eliminate these effects by measuring the intensities of two orders of reflection, or by measuring intensities of reflection for two different incident radiations. Such techniques are most useful in measurements on coatings which contain major amounts of substrate elements.

By contrast, X-ray fluorescence is now often employed in thickness measurement and control of gold plating. Of particular relevance to the measurement of gold thickness on electronic components are papers by Mohrnheim[33] and Heller.[34]

METHOD

A typical arrangement for X-ray fluorescence measurements is indicated schematically in Figure 28.5. The fluorescent radiation emitted from the specimen passes through a collimator to an analysing crystal which is set by the spectrogoniometer to the specific angle for the characteristic line to be measured. Intensity, detected by a Geiger counter with appropriate electronic counting equipment, is expressed in counts per second or counts over a fixed period. More detailed descriptions of instrumentation and techniques are to be found in the literature, covering both emission and absorption techniques,[32–42] and a useful summary review of X-ray methods is given by Bertin and Longobucco.[43] Cook, Mellish and Payne[44,45] have used gamma-radiation from radioactive isotopes in place of a normal X-ray tube for excitation of fluorescence spectra, the fluorescent rays being analysed by the resolving power of a proportional or scintillation counter, which is sufficient to distinguish rays from the basis metal and the coating provided that the metals differ by at least 2 units in atomic number. Achey and Serfass[46] have avoided the need for precise collimation or crystal reflection by using a differential filter system to isolate the required narrow wavelength band of radiation.

Choice between the emission and absorption techniques depends on the nature of the sample to be examined. The absorption method is more sensitive than emission[33, 42], and hence to be preferred for measurements of coatings on single substrates, but since in practice one or more undercoating deposits may be present, the emission technique is more generally useful.

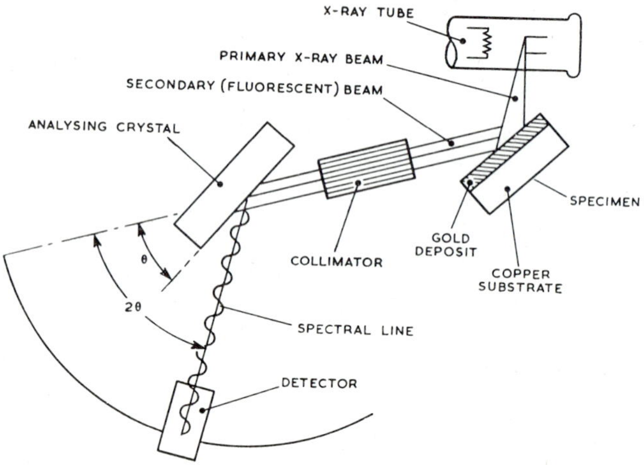

Fig. 28.5 Schematic diagram of X-ray fluorescence geometry for determining thickness of gold deposit on copper substrate.

For the preparation of calibration curves relating intensity of selected lines to coating thickness, standards are usually prepared by plating large test panels and punching out small test coupons on which thickness can be established by referee methods, such as strip and weigh, dissolution and chemical analysis, or optical interferometry. Measurements are most conveniently made on flat areas, 0.25 in. or more in diameter, but where this is not practicable, smaller areas, down to about 0.06 in. in diameter can be defined by masking.[33, 34, 42] Special devices and techniques have also been developed to produce measurable intensities from areas as small as 0.008 in. diameter.[47, 48] Techniques have also been described for dealing with samples of irregular shape where it is impossible to define a small flat area by masking.[49]

In the construction of calibration curves the use of intensity ratios rather than simple intensities avoids the need for separate working curves for each combination of sample area, sample support and masking material. In each case the selected ratio after correction for background radiation is plotted against thickness on a semi-logarithmic scale to give a straight line relationship. Ratios commonly used are:

For absorption I_t/I_0

For emission (i) $(I_{00}-I_t)/I$
(ii) $(I_t-I_0)/(I_{00}-I_0)$

where suffix "t" refers to the specimen under examination, "0" to unplated basis metal and "00" to a sample of the coating which is greater in thickness than the saturation thickness with respect to penetration by the incident radiation.

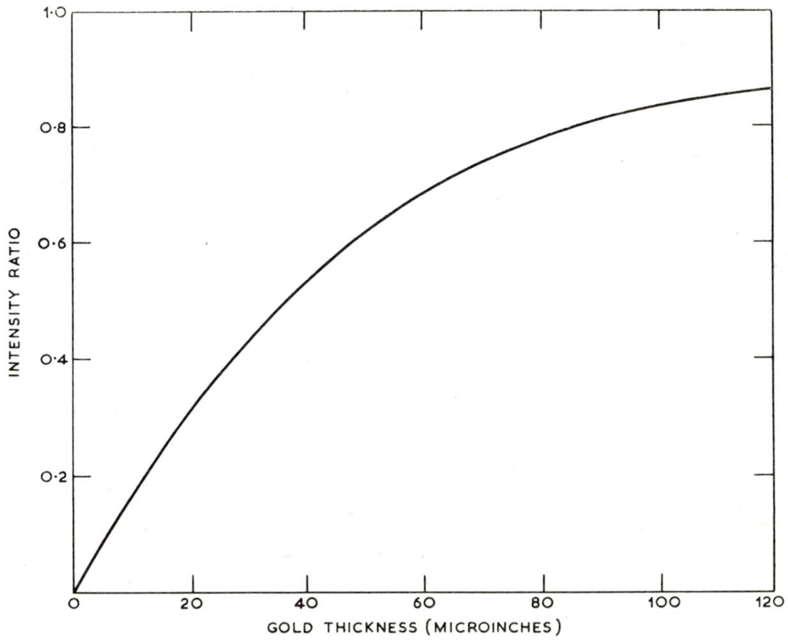

Fig. 28.6 Analytical curve for the determination of gold plating thickness by X-ray fluorescence. [Heller].

Since the ratio of the intensities of two different spectral lines is less influenced by instrumentation variables than the intensity of a single line, the ratio of $L\alpha$ to $L\beta$ lines of the gold spectrum is sometimes used. Mohrnheim[33] used a simple plot of net $L\beta$ line intensity against thickness (semi-logarithmic) and reported no appreciable difference in accuracy between the results obtained by this simple method, and by the various intensity ratio procedures.

A typical "straight" plot of intensity ratio (ii) against gold thickness, using the emission technique, is shown in Figure 28.6.[34] A simple plot of net intensity would have the same general form. The curve shows an almost linear increase in intensity or intensity ratio, with coating thickness for thin coatings. As coating thickness increases, this proportionality is lost due to increasing absorption of the primary X-rays and self-absorption of the fluorescent radiation by the coating, a limiting thickness being reached, above which change in intensity with thickness is not sufficient for measurement purposes. Similar considerations apply to the absorption technique. The limiting thickness will depend on the energy of the primary X-ray beam which is limited by the maximum tube voltage. Under normal conditions it is in the order of 5 microns.[33]

ACCURACY

The precision of X-ray measurements increases with the number of counts. Using a reasonable value for gold emission based on a 2-sigma value of 1% (40,000 counts), Zimmerman[42] quotes an accuracy of ± 0.025 microns on 0.75 microns, ± 0.05 on 1.5 and ± 0.075 on 2.25, but points out that this error could double with positioning errors and curvature of samples. Mohrnheim[33] obtained an arithmetic mean of 79 microinches from 10 measurements on a standard sample of thickness in the range 75–80 microinches, the mean sum of the differences from the squared average being 0.52, giving a coefficient of variation of 4.5. Heller[34] has also presented precision data on X-ray methods in comparison with beta-ray backscatter procedures. Cullen and Petruna[3] assign an accuracy of 0.5% over an optimum thickness range of 0–3 microns. Foulke,[50] however, considers that while X-ray methods are inherently highly accurate, the practical level achieved in day-to-day control may be in the order of only 5%.

ADVANTAGES AND LIMITATIONS

Advantages

1 Non-destructive, rapid.
2 Good precision and accuracy on very thin coatings.
3 Emission method free from substrate interference.

Limitations

1 High equipment cost and space requirements.
2 Susceptible to error if coating composition or surface condition changes during run.
3 Thickness limitation.

Beta-Ray Backscatter

PRINCIPLE

Beta-rays are high energy electrons emitted by various radioisotope sources.

"Backscatter" refers to the re-emission from a metal of a proportion of electrons in an incident beam as a result of successive collisions with the electron shells of the metal atoms. For a fixed source, the number of electrons backscattered increases with the mass of metal accessible to the incident radiation up to a limit determined by the penetration depth (saturation thickness) and with increasing atomic number of the metal concerned. In the case of a plated coating thinner than the saturation thickness on a bulk substrate of lower atomic number, the intensity of backscatter will be greater than that pertaining to the bare substrate and will increase with plating thickness up to the saturation thickness of the coating.

METHOD

The basic principles and techniques relating to the method are described very comprehensively in a paper by Joffe and Modjeska.[51] A number of

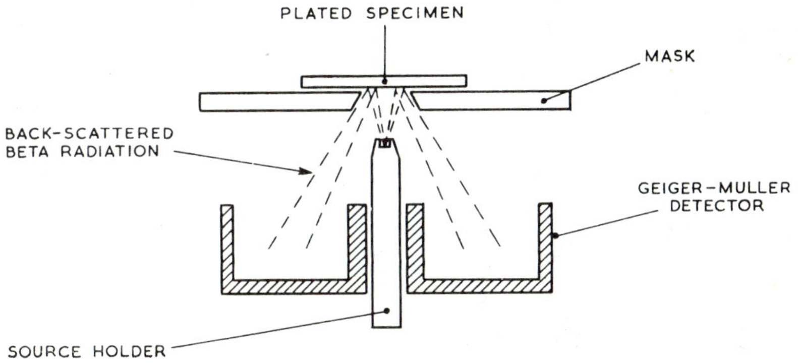

Fig. 28.7 Schematic arrangement for thickness measurement by beta-ray backscatter.

papers of applicational interest have since appeared, the most relevant to gold plating being contributions by Cooley,[52] Cooley and Anderle,[6] Cullen and Petruna,[3] Melrose and Cooper,[53] and Heller.[34]

A typical arrangement for backscatter measurements is illustrated schematically in Figure 28.7. The source is a long lived radioisotope producing beta-radiation of appropriate energy in relation to the range of coating thickness to be measured. Whilst it is essential that the maximum depth of penetration of the incident radiation should exceed the upper level of thickness to be measured, excessive penetration into the substrate reduces sensitivity. Several sources are therefore necessary to cover various thickness ranges, the most common being promethium-147 (up to 2.5 microns), thallium-204 (up to 13 microns), and strontium-90 (up to 35 microns).[35] The area of sample for irradiation can be controlled by interchangeable apertures which fit into the measuring table and are selected according to the geometry of the sample to provide the largest flat area for measurement, though curved surfaces can also be accommodated. Flat specimens or other suitable components may be simply placed over the measuring aperture, but smaller items such as contact springs may be held in position by a locating probe which lines up with the aperture. For measurements on large flat areas like printed circuit boards, remote self-contained measuring heads may be employed, the area for measurement being selected by a low power cross wire lens of the same diameter as the head, supported in a heavy guard ring to facilitate exact positioning of the head when this is substituted for measurement.

The latest developments[63] in the measuring head system has been to combine the table system with optical location devices such that small complex shaped parts can be measured in one mode, whilst in the second mode an optical system locates the area to be measured, a lever is pulled and the measuring head is moved on to the located area. An extension of this system is used for continuous measurement and control of plated strip and for automatic sequential measurement of the latest spot and selective plating processes.

Curved cylindrical surfaces down to 0.010 in. and flat areas down to 0.018 in. are now measureable with commercially available instrumentation. The problems associated with mechanically positioning, normally associated with handling small parts, have largely been solved using duplex platen aperture systems.

For calibration, backscatter counts are made on a relatively thick specimen of the substrate and on a series of reference coatings covering the thickness range of interest. Various methods are used in constructing calibration curves. A direct plot of increase in backscatter against coating thickness yields a curve of typical form indicated in Figure 28.8. The initial portion A-B is almost linear and may be used in this form for measuring very thin coatings. Over the range B-C the relationship is logarithmic and this region is more commonly used for measuring purposes, based on the linear calibration obtained by plotting on semi-logarithmic paper (Figure 28.8). In this case only two standards, one towards each end of the gold thickness range, are necessary to fix the slope of the line, but it is important to note than the line cannot be extrapolated beyond the linear range (B-C) for measurement purposes. In this context, Cullen and Petruna[3] have presented data for various sources as shown in Table 28.2.

It is sometimes preferred to use an intensity ratio as in X-ray methods to avoid the need for multiple calibration curves.[34]

The semi-logarithmic plot shown in Figure 28.8 is particularly useful in connection with modifications of the calibration procedure which are necessary in dealing with special cases, e.g., where the thickness of substrate is less than the saturation thickness (as in measurements on printed circuit foil) or where an intermediate deposit may be present. Errors involved in graphical methods of the above type are eliminated or greatly reduced by the use of computerised back-scatter instruments which give direct thickness readout. Manufacturers'

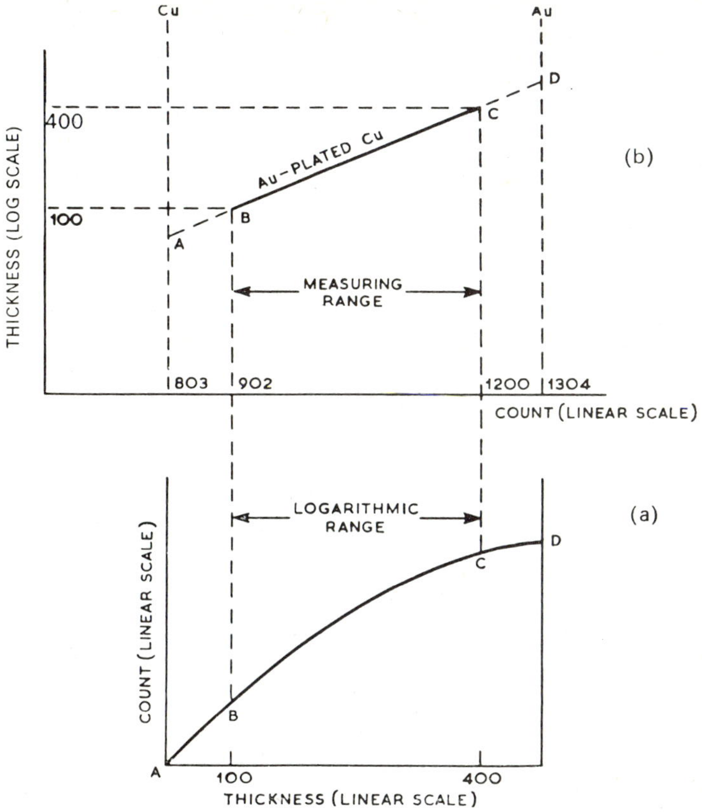

Fig. 28.8 Relation between linear (a) and semi-logarithmic (b) calibration curves for beta-ray backscatter method. *Standard B* 100 μ-in. gold on copper; *Standard C* 400 μ-in. gold on copper.

Table 28.2
Measurement Ranges of Beta-Ray Sources

Source	*Energy (MeV)*	*Logarithmic Range (range of highest sensitivity) (μm)*	*Maximum Range of Measurement (μm)*
Carbon-14	0.16	0.3–1.2	0–1.6
Promethium-147	0.22	0.4–1.6	0–2.5
Thallium-204	0.77	3.0–10	0–12
Radium D+E	1.17	5.0–13	0–18
Strontium-90	2.18	5.5–30	0–35

literature should be consulted for details and for specific recommendations concerning selection of sources, apertures and count times in relation to thickness ranges and component types.

ACCURACY

Although inherently accurate provided that sufficient counts are taken, the beta-ray method is less so than the X-ray methods because incident electrons have a range of energies and backscatter is a random effect to which all components of the coating/substrate system contribute. In the measurement of gold thickness in the range 30–80 micro-inches over nickel, Joffe and Modjeska[51] report an accuracy of better than 2%, while more generally, Cullen and Petruna[3] assign an accuracy of 1% over the whole range of 0.25–25 microns. While such levels may be achievable under the best conditions, it is unlikely that they will be attained in production measurements without very rigorous control over a number of factors.

Since substrate effects must always be included, errors may arise in practice due to variation in substrate composition and, more significantly, to variation in thickness of an undercoating deposit when present. Errors may also be associated with change in composition of the coating, though these are likely to be significant only in the case of highly alloyed deposits. A more serious source of error in relating backscatter measurements to gold coating thickness, which applies equally to the X-ray methods, arises from variation of coating density, since the beta-rays "see" only the mass of gold over the irradiated area and take no account of bulk porosity, inclusions or other incorporated non-metallic material. This factor has been underlined in work by Cooley and associates,[1,52] which stresses the inadequacy of reliance on prepared manufacturers' standards and the importance of establishing calibration curves on standards prepared under relevant "in-house" conditions if maximum accuracy is to be achieved.

Cooley also points out some more subtle effects of instrument variables and measuring environment on the accuracy of beta-ray methods, referring to significant effects of changes in line voltage, temperature, humidity and barometric pressure. The latter was operative by affecting the fragile end window of the Geiger-Muller counter, and it was reported that even the opening and closing of a door of the inspection room could cause "considerable" variation in readings. Accuracy in production measurement of coating thicknesses was improved from within 15–20% to about 3% by:

1 Use of a constant voltage transformer to maintain line voltage at optimum value.
2 Use of "in-house" standards and more frequent calibration.
3 Environmental control.

Precision data on the beta-ray method has also been presented by Heller.[34]

ADVANTAGES AND LIMITATIONS

Advantages

1 Rapid, non-destructive procedure.

2 Relatively low cost and compactness of equipment as compared to X-ray methods.
3 Good accuracy in measuring thinnest coatings.
4 Wide range of thicknesses measurable.
5 Simple operation.
6 Applicable to continuous measurement.
7 Readily used as "go, no-go" test.
8 Any combination of metal to metal, metal to non-metal, or non-metal to non-metal may be measured, subject to there being sufficient density difference between coating and substrate.
9 Due to the insensitivity of the Beta backscatter technique in differentiating between nickel and copper, gold thickness measurement of gold on an unknown thickness of nickel deposited over copper, are little affected by variations in nickel thickness.

Limitations

1 Low sensitivity cf. X-ray methods, since selective response cannot be isolated.
2 Susceptibility to error with variation in substrate or coating composition and coating density.
3 Need for frequent calibration for optimum accuracy in uncontrolled environments. In this context it should, however, be noted that continuing improvements in instrument design and components including, for example, the provision of built-in voltage stabilisers and the use of temperature-controlled Geiger–Muller tubes or compensating electronic supply circuitry to reduce temperature effects, is now claimed[63] to permit thickness measurement to within one fifth of the coefficient of variation of the items tested without special environmental control.
4 Due to closeness of atomic numbers, method cannot generally be used for measuring thickness of nickel on copper or brass, a frequent requirement in gold plating practice.

Electron Probe Method

PRINCIPLE

An electron beam of variable energy is directed into the plated sample. Coating thickness is related to the acceleration voltage of the electron beam at which electrons penetrate the coating, as indicated by the appearance of characteristic lines of the substrate material in the X-ray spectrum from the sample.

METHOD

The procedure was developed by Schumacher and colleagues at the Ontario Research Foundation. Technical papers from this source cover detailed theory of the method[54] and a simplified description of the technique as applied to plated coatings.[55]

For calibration purposes, two methods may be employed. In the first, the ratio of intensity of characteristic lines from the substrate and coating is plotted against electron energy, the value of the latter at which penetration of the coating occurs being indicated by a fairly sharp increase in the slope of the curve. Alternatively, the intensity ratio may be plotted directly against coating thickness using a beam of constant energy. The method then resembles the X-ray fluorescence procedure, with excitation by an electron beam in place of a primary X-ray source. Advantages claimed for the electron probe method are greater rapidity and the capability for measurements on a micro-spot.

The method may also be used for determining individual thicknesses of layers of a multiple coating,[55] which suggests a strong advantage. While the electron probe procedure is clearly of potential application to thickness measurement of gold plating, no reference has yet been made to its use in this specific context.

Radiographic Method

Coating thickness can be measured by the incorporation of radioactive tracers either in the substrate or the coating, thickness being related respectively to decrease or increase of activity from the substrate/coating as measured by the blackening of a photographic plate. The technique is useful for research purposes, but so far as the writer is aware, has not been used in practical gold thickness measurement.

ELECTRICAL

Eddy Current Method

PRINCIPLE

An eddy current is induced in the surface of the plated coating by a small probe coil energised by a radio frequency alternating current. Flow of current in the specimen is confined to a thin surface layer due to the well known skin effect, and the energising frequency is selected so that the current flow is partly in the basis metal and partly in the coating. Provided that these differ in conductivity and/or magnetic properties, a change in coating thickness will affect the induced current which, by interaction of its magnetic field with the probe coil, will alter the impedance and hence the current in the latter. Coating thickness can therefore be related to the change in probe current as measured, for example, by imbalance of a bridge circuit.

METHOD

Principles and techniques involved in the original development of this procedure by Brenner and associates are fully described in a paper by Brenner and Garcia-Riviera,[56] and further studies in operating characteristics, circuitry and calibration procedure by Brodell and Brenner.[57] A number of commercial instruments based on this principle are available, but though in fairly wide general use for other purposes, the method does not appear to be often employed for the measurement of gold thickness. This is probably because of lack of adequate sensitivity at the very low thicknesses generally involved, together with the advent of alternative non-destructive procedures now available.

DC Resistance Measurement

In circumstances where current flow between two electrodes placed in contact with a coating or substrate/coating composite is essentially laminar, and provided that the electrical conductivities of coating and substrate differ sufficiently, coating thickness may in principle be correlated with change in resistivity of the system to an applied direct current. A technique of this kind has been used by Davidson and Rahal[58] for measuring silver plating thickness on the internal surfaces of stainless steel waveguides using an "in-line" 4 probe system, the two outer probes, about 2 in. apart, carrying current and two spring loaded steel needles, about 1 in. apart, measuring potential difference. Their instrument could be used on either a constant current or constant potential basis, the latter being preferred since the current required to maintain constant potential between the probes is directly proportional to the conductance of the system, and hence varies directly with coating thickness.

This type of procedure is clearly applicable to measuring the thickness of gold under similar conditions, and could be especially useful in the case of coatings on non-conducting substrates. There are, however, obvious limitations in respect of component size and sensitivity in relation to more general application.

MECHANICAL METHODS

Profilometry

As an alternative to optical interferometry, the step height at a boundary between a plated and unplated surface can be measured by the use of a profilometer. The stylus is used as one arm of an electromechanical pick-off and electrical signals generated by vertical displacement relative to a reference slider or skid are amplified to actuate a recording device, usually a pen recorder. Commercial instruments differ in the design of the diamond stylus tip (which is not critical in simple thickness measurement) and in the means of mechanical-electrical transduction employed. A good general description of principles and use of various instruments of this type is given by Judge.[59]

Vertical magnification of stylus displacement up to 50–100,000× can be achieved and, on the basis of a comparative study, Timms and Scoles[60] have concluded that the accuracy of this method is comparable with that achieved by double-beam interferometry. A disadvantage in relation to the latter method, is the possibility of damage to soft surfaces by the stylus[61] to give erroneous readings. This is minimised by keeping the stylus reading as low as possible, e.g., 0.1 g in the TALYSURF* instrument.

TESTING ENVIRONMENT

As a general principle, cleanliness of the environment and freedom from vibration of benches are two essential requirements in obtaining optimum accuracy in physical measurement, and thickness measurement of gold plating is no exception. Wherever possible a separate room should be assigned to this purpose, remote or sheltered from the plating shop to provide a clean area for metallographic preparation, production of special calibration samples and actual physical measurements. Benches should be attached to outside walls wherever practicable, otherwise measuring equipment should be provided with effective anti-vibration mounting.

The ideal arrangement is described by Cooley,[52] who considered it worthwhile to construct a special clean room with control of barometric pressure, temperature and humidity, and a constant laminar flow of filtered air over the work area. The room is also provided with an air lock to prevent entry of contaminants.

SELECTION OF METHOD

In considering the selection of a method for thickness measurement a number of inter-related factors must be taken into account, viz.,

Technical

(*a*) Purpose of measurement, i.e., production control, random inspection, research.

(*b*) Range of thickness to be measured.

(*c*) Accuracy required.

(*d*) Validity of the technique for the substrate/coating system involved and geometry of the specimen.

(*e*) Type of method, i.e., destructive or non-destructive: direct or indirect.

---

* Registered Trade Mark of Rank-Taylor-Hobson.

Economic

(*a*) Prime cost of equipment.
(*b*) Level of operator skill or experience.
(*c*) Time required.

In Table 28.3 an attempt has been made to provide guidance on some of these aspects in summary form for the main methods described. Assessment of time and accuracy requires considerable qualification. Time per test has been quoted for "one-off" measurement, assuming the availability where necessary of calibration data. Where very small items are concerned and average thickness is acceptable, the strip and weigh procedure could be applied to a number of samples in a single batch. Similarly, the time per test in microsectioning can be substantially reduced by packing as many samples as possible in the same mount, thus spreading the "overhead" of grinding and polishing time.

Referring to accuracy, a distinction can be made between methods which are inherently accurate, such as the interferometric procedure with its built-in calibration in terms of light wavelength, methods which depend for optimum accuracy on the skill and experience of the observer, as for microsectioning, strip and weigh and chemical assay, and those which, whilst inherently accurate, are subject to error from factors other than the actual measuring technique, as for X-ray and beta-ray procedures, with their susceptibility to substrate and coating compositional variations, etc.

For research purposes, technical requirements will usually be the sole consideration. All of the methods listed, with the possible exception of the chord technique, could be applicable, including various extensions and modifications of certain of the procedures which have not been detailed in the present review because of their highly specialised nature (see, for example, the review by Wright,[62] which deals specifically with methods for thin film measurement).

Methods used by large industrial consumers of gold plating will often be dictated by the immediate availability of specialised equipment and skills. As for research, economic factors are relatively unimportant in this case, though they must be considered when sophisticated techniques such as X-ray fluorescence are used in production control of in-house plating.

The biggest problem with regard to selection of a method faces the specialist outplater, usually with limited resources of finance, time and specialised experience to devote to this aspect, but who, nevertheless, must in his own and his customer's interest, have some facility for checking production. For this purpose relatively inexpensive techniques based on gain in weight of components or test panels or on stripping and weighing of coatings can be extremely valuable, particularly when the results are systematically correlated with those obtained in checks by the customer, often by more sophisticated means. When large numbers of small items are being plated the destructive

Table 28.3
Comparison of Methods for Thickness Measurement of Gold Deposits

Method	*Type**	*Thickness Range* (μm)	*Accuracy*	*Approx. time per test (min)*	*Operator Skill & Experience*	*Commercial Availability*	*Ref.*	*Remarks*
Microsectioning	P, D	2 →	2% (5 μm) 10% (2 μm) 50% (1 μm)	60	High	Yes	2–6	Referee method for thicknesses > 2 μm. Wide applicability. Suitable for plating on internal surfaces
Chord Method	P, D	2 →	Comparable to microsectioning	5	Moderate	Yes	8	"Workshop" method—could possibly be extended to thinner coatings
Interference Microscopy	P, N(D)	0–20	0.5%	5	High	Yes	9–14	Reliable calibration method for thinnest coatings
Depth-of-focus method	P, N	1–1700	0.5–1 μm	5	High	Yes	9, 15	Main use for surface topography studies
Light profile	P, N	2–125	6%	5	High	Yes	10, 15–19	(Main use for surface topography studies)
Chemical assay	S, D	0 →	0.5–1%	20	High	—	20	Variant of "strip and weigh" for coatings on Sn, Sn/Pb, stainless steel, etc.
Strip and weigh	S, D	0 →	0.5–1%	20	High	—		Calibration method—wide applicability. Gives average thickness only
Gain in weight	S	0 →	0.5–1%	3–5	Moderate	—		Approx. Control method for average thickness only

Anodic dissolution	S, D	0.5–7.5	Comparable to microsectioning	3–5	Moderate	Yes	21–29	Suitable for thickness measurements on fine wires —gives local thickness
X-ray fluorescence	S, N	0–5	0.5%	3–5	High	Yes	32–50	Suitable for production control
Beta-ray backscatter	S, N	0.25–25	1–2%	2–3	Moderate	Yes	3, 6, 34, 35 51–53	Suitable for production control
Electron probe	S, N	0 →	—	—	High	Yes	54, 55	
Profilometry	P, N(D)	0–10	Comparable to interference microscopy	3–5	Moderate	Yes	59–61	Suitable as calibration method

* P—Primary (direct) method S—Secondary (indirect) method D—Destructive N—Non-destructive

nature of the strip and weigh method may present no serious disadvantage, a greater limitation being its inability to measure local thickness.

Where this is necessary, the thickness concerned will often be in the range of acceptable accuracy by microsectioning. Despite the increasing recognition of accuracy limitations at low thicknesses, the microscopic method is still well established as a "referee" procedure, and having in mind the ancillary advantages already enumerated, the purchase of a good metallographic microscope must represent a sound investment for the plater who wishes to hold his own in discussion or dispute with a customer on thickness and general quality of his product. The addition of a simple double-beam interferometry attachment would provide coverage of the whole range of thickness likely to be of interest. Although the interferometric method is currently regarded primarily as a calibration procedure, it could assume greater importance in routine control with the increasing tendency to partial plating and reduction in gold thickness, both of which factors favour the application of interferometric procedures. Optical and interference microscopy share the common advantage of being direct methods and therefore independent of calibration standards.

Local thickness measurement is possible by anodic dissolution methods. Although destructive, they offer the advantage over strip and weigh of being adaptable to fairly rapid measurement on individual small components, including fine wires, on a one-off basis. This procedure could lend itself to production control, but the cost in equipment and instrumentation approximates to that involved in beta-ray backscatter which is non-destructive and more generally useful in this context. The relatively sparse literature on the anodic dissolution method in specific relation to gold plating leaves some doubt concerning the accuracy obtainable. Where the technique is already established for measurement on base metal coatings it could no doubt be readily utilised for gold in many cases.

The non-destructive X-ray and beta-ray procedures are in principle ideally suited to large numbers of routine measurements required to support meaningful statistical production control. Although the X-ray emission method is less susceptible to substrate effects, its over-riding drawbacks from the plater's viewpoint are the very high prime cost, space requirements and general level of experience required in operation. For these reasons its use and that of related techniques such as the electron probe method, seems likely to remain restricted to the large specialist organisations, for research or "in-house" production control.

The beta-ray backscatter technique on the other hand, offers compact instrumentation at a much lower prime cost, with virtually negligible radiation hazard, and minimal requirements in terms of operator skill. As such, this provides the most generally convenient non-destructive means for thickness measurement in relation to production quality control of gold plating, having a good level of accuracy as well as the capability of dealing with a variety of sample geometries. The main danger here may well arise from indiscriminate use of the technique without a proper appreciation of its principles and limitations, which have been discussed earlier. In particular, the importance of correct calibration procedures, based wherever possible on "in-house" standards, is again emphasised.

REFERENCES

1 Cooley R. L., Lemons, K. E., *Plating*, **56,** 510 (1969).
2 Wilson G. A., *Metal Finish.*, **58,** 6, 50 (1964).
3 Cullen T., Petruna P., *Plating*, **54,** 1039 (1967).
4 Baldwin P. C., *Plating*, **50,** 1006 (1963).
5 Plog H., *Galvanotechnik*, **56,** 240 (1965).
6 Cooley R. L., Anderlee J. G., *Plating*, **54,** 1029 (1967).
7 Private communication, International Nickel Ltd.; Ericssohn Telephones Ltd.
8 Anon., *Metals Mater.*, **4,** 2, 41 (1964).
9 Thomas J. D., Rouse S. R., *Proc. Am. Electropl. Soc.*, **42,** 49 (1955).
10 Saur R. L., *Plating*, **52,** 663 (1965).
11 Saur R. L., *Plating*, **45,** 1232 (1958).
12 Flühmann W., Saxer W., *Metal Finish. J.*, **17,** 195, 84 (1971).
13 Tolansky S., *J. Electropl. Depos. tech. Soc.*, **27,** 171 (1951).
14 Tolansky S., "Introduction to Interferometry", Longmans, New York (1955).
15 Hodgson R. W., Szkudlapski A. H., *Plating*, **57,** 693 (1970).
16 Schmaltz G., "Technische Oberflachenkunde", Springer, Berlin (1936).
17 Tolansky S., *Nature*, **169,** 445 (1952).
18 Tolansky S., "Properties of Metallic Surfaces", Inst. of Metals Monograph, No. 13.
19 Halliday J. J., *Proc. Inst. mech. Engrs*, **169,** 777 (1955).
20 "Electroplaters Process Control Handbook", D. G. Foulke and F. R. Crane (Eds), Reinhold, New York (1963).
21 Anderson S., Manuel R. W., *Trans. electrochem. Soc.*, **78,** 373 (1940).
22 Waite C. F., *Proc. Am. Electropl. Soc.*, **40,** 117 (1953).
23 Francis H. T., *J. electrochem. Soc.*, **93,** 79 (1948).
24 Narayanan U. H., Venkatachalam K. R., *Plating*, **48,** 1211 (1961).
25 Baldwin P. C., *Plating*, **57,** 927 (1970).
26 Manufacturers literature, Kocour Co., Chicago, Ill., U.S.A.
27 Manufacturers literature, Baldwin Engineering Co., E. Segundo, Calif., U.S.A.
28 Mathur P. B., *Plating*, **47,** 1274 (1960).
29 Cooley R. L., *Quality Assurance*, **5,** 26 (1966).
30 Agarwal H. P., *J. Sci. Industr. Res.*, **21D,** 108 (1962).
31 Clark G. L., Pish G., Weeg C. E., *J. Appl. Phys.*, **5,** 193 (1944).
32 Keating D. T., Kammerer O. F., *Rev. Scient. Instrum.*, **29,** 34 (1958).
33 Mohrnheim A., *Plating*, **50,** 725 (1963).
34 Heller H. A., *Plating*, **56,** 277 (1969).
35 Birks L. S., Friedman H., *Phys. Rev.*, **69,** 49 (1946).
36 Friedman H., Birks L. S., *Rev. Scient. Instrum.*, **17,** 99 (1946).
37 Eisenstein A., *J. Appl. Phys.*, **17,** 874 (1946).
38 Beeghley H. F., *J. electrochem. Soc.*, **97,** 152 (1950).
39 Liebhafsky H. A., Zemany P. D., *Analyt. Chem.*, **28,** 455 (1960).
40 Zemany P. D., Liebhafsky, H. A., *J. electrochem. Soc.*, **103,** 157 (1956).
41 Sellers W. W., Carroll K. G., *Proc. Am. Electropl. Soc.*, **43,** 97 (1956).
42 Zimmerman R. H., *Metal Finish.*, **59,** 5, 66 (1961).
43 Bertin E. P., Longobucco R. J., *Metal Finish.*, **60,** 8, 42 (1962).
44 Cook G. B., Mellish C. E., Payne J. A., *Analyt. Chem.*, **35,** 590 (1960).
45 Cook G. B., Mellish C. E., 2nd United Nations Conf. on Peaceful Uses of Atomic Energy (Geneva) 19, 127 (1958).
46 Achey F. A., Serfass F. J., *Proc. Am. Electropl. Soc.*, **43,** 41 (1956).
47 Heinrich K. F. J., "Advances in X-Ray Analysis", W. M. Mueller (Ed.), pp. 516, Plenum Press, New York (1962).
48 Loomis T. C., Vincent S. M., "Handbook of X-rays", E. F. Kaelbe (Ed.), Ch. 37, pp. 10, McGraw-Hill, New York (1967).

49 Zimmerman R. H., *Iron Age*, **186,** 84, Oct. 13 (1960).
50 Foulke D. G., *Metal Finish.*, **63,** 7, 42 (1965).
51 Joffe B. B., Modjeska R. S., *Metal Finish.*, **61,** 12, 44 (1963).
52 Cooley R. L., *Plating*, **57,** 111 (1970).
53 Melrose S. P. G., Cooper B. S., *Trans. Inst. Metal Finish.*, **45,** 5, 199 (1967).
54 Schumacher B. W., Mitra S. S., *Electron Reliab. Micromin.*, **1,** 321 (1962).
55 Krieglee R., Schumacher B. W., *Plating*, **54,** 393 (1960).
56 Brenner A., Garcia-Riviera J., *Proc. Am. Electropl. Soc.*, **40,** 106 (1953).
57 Brodell F., Brenner A., *Plating*, **44,** 591 (1957).
58 Davidson M., Rahal N. S., *Tele-Tech and Electron. Ind.*, **13,** 9, 76 (1954).
59 Judge A. W., "Engineering Precision Measurements", 3rd Edn, Chap. 13, Chapman & Hall, London (1957).
60 Timms G., Scoles C. A., *Metal Treatment*, **18,** 450 (1951).
61 Dennis J. K., Fuggle T. T., *Trans. Inst. Metal Finish.*, **4,** 4, 177 (1969).
62 Wright P., *Electron. Reliab. Micromin.*, **2,** 227 (1962).
63 Latter T. D. T., private communication, October (1973).

Chapter 29

HARDNESS TESTING

F. H. REID

The hardness of a metal is most commonly measured by the resistance offered to penetration when a hardened steel or diamond tipped indentor of precise geometry is pressed on to the surface under a known load, hardness being expressed as the ratio of the load to the area of the impression produced. No single property is measured in this way, since the response of the metal to deformation is affected by a number of factors, such as tensile strength and the relative extent to which plastic and elastic deformation occur under the particular loading employed. For a detailed discussion of these effects the general literature on hardness tests should be consulted (see, for example, Lysaught[1] and Mott[2]).

An important requirement in measurements on plated coatings is the restriction of the loading in relation to coating thickness, thereby avoiding possible effects of the basis metal. This dictates the use of microhardness testing techniques, which differ from macrohardness testing mainly in respect of the low loadings employed, which are seldom greater than 100 g and more commonly in the order of 10–25 g. Correspondingly greater care is also necessary in specimen preparation, conditions of indentation and measurement of the impression to ensure reasonable precision of measurement.

In the case of bulk metal the hardness may sometimes be fairly clearly related to other mechanical properties of engineering interest such as tensile strength and wear resistance. However, in the case of electrodeposits, relationships are much less defined, since there is generally a marked difference in structure from that of wrought materials. Additional factors of brittleness and internal stress are also frequently introduced, which are less commonly encountered in the bulk condition.

Hardness may, however, be correlated to some extent with the wear resistance of a coating on the basis of experience and with due regard to specific service conditions, particularly loading. It can also be of value in providing a fairly sensitive indication of coating purity with respect to contamination by organic material. Hardness measurement can therefore serve as a useful control on both coatings and electrolytes. Careful inspection of a hardness indentation may also yield useful qualitative indications concerning other factors such as adhesion and ductility of deposits.

Although scratch hardness tests have been developed, notably by

Bierbaum[4], these have found relatively little use in comparison with indentation methods. For a given loading, scratch width is considerably less than that of a diamond impression and the accuracy of measurement is further affected by a tendency to furrowing, which may lead to poor definition of scratch edges. The present chapter will therefore be confined to tests of the indentation type.

MEASUREMENT

Various designs of microhardness tester are available, either as self-contained instruments or as attachments to direct or inverted metallurgical microscopes. A representative, but not comprehensive, list is given in Table 29.1. A detailed description of individual instruments is not within the scope of the present text since such information is readily available from the manufacturers concerned. It seems more useful to comment generally on the testing procedure, with particular reference to the common features which affect measurements on electrodeposited coatings and to which due attention must be given if meaningful results are to be obtained. These factors have been discussed in some detail in relation to gold plating by Wilson[5], and by Cullen and Petruna[6]. They comprise essentially:

1 Surface condition
2 Types of indentor
3 Selection of load in relation to coating thickness
4 Conditions of indentation
5 Measurement of impression, or of penetration depth.

SURFACE CONDITION

The surface to be tested should be as smooth as possible to ensure maximum clarity of the impression for measurement. In the case of cross-sections, these will normally be mounted and prepared in the usual manner (see Chapter 28), using a copper or nickel overplate to provide lateral support of the section. Grinding and polishing may produce some work-hardening of the surface and since etching of the gold is not generally practicable to remove this disturbed layer, the load used in testing should be the maximum consistent with the requirements discussed later under "Selection of Load".

For measurements on the surface, surface condition is equally critical. Some polishing may be necessary and permissible in the case of deposits exceeding 10 microns, but thinner coatings will usually be bright and the surface finish will depend on that of the initial basis metal. In the 5–10 micron range polishing is undesirable. At the lower end of the range metal removal may reduce the thickness below that necessary for a satisfactory measurement. In the case of surface measurements, any work-hardening during polishing may contribute unduly to the measured hardness in view of the very small permitted depth of the indentation. If polishing is required to remove surface

Table 29.1
Representative Types of Microhardness Testing Instruments

Instrument	*Type of Indentor*	*Load Range (g)*	*Method of Loading*	*Remarks*
Akashi	Vickers, Knoop	25, 50, 100, 200 500,1000	Dead weight—manual application	Self-contained instrument
Bergsman*	Vickers, Knoop	1–200	Dead weight—manual application	For use with inverted metallographic microscope. Can be used for scratch-testing at loads of 5 g.
Dawe Sonodur	Vickers	150 (fixed)	Dead weight, with dash-pot control	Measurement based on depth of indentation—direct reading hardness scale. Load limits application to coating.
Eberbach*	Vickers	7.5, 22.5, 65, 110, 190, 500	Loadings by calibrated springs	For use with inverted or direct metallurgical microscope.
GKN	Vickers		Load applied through lever arm—manual application	For use with non-inverted microscope.
Kentron*	Vickers, Knoop	1–100, 100–1000	Dead weight, with dash-pot control	Self-contained instrument.
Leitz 'Miniload'*	Vickers, Knoop	25, 50, 100, 200, 300	Dead weight, with dash-pot control	Self-contained instrument.
Newage*	Vickers	1000 (fixed)	Liquid loading by mercury column	Hardness measured by depth of penetration, with direct dial reading. High load limits application to coatings.
Tukon*	Vickers, Knoop	1–1000, 25–3600, 25–1000 } various models	Weight applied to lever arm between indentor and knife-edge. Automatic application and removal of load in some models, giving fixed dwell time.	Self-contained instrument.
Vickers	Vickers	5, 10, 20, 50, 100, 200	Constant rate loading by pneumatic pressure	Attachment for M 12a and M 55 metallurgical microscopes.
Zeiss Hanemann (D. 32)	Vickers		Load applied by deflection of disc springs supporting indentor objective-manual application	For use with inverted microscopes type "Neophot" and "Epitype 2". Visual load-indicating scale in field of view.
Zeiss MHT Device H	Vickers, Knoop	1.25, 2.5, 5, 10, 20, 40, 80, 160	Dead weight—manual application	For use with non-inverted microscopes.

* For general description and discussion, see H. J. Read[3].

bloom, it should be carried out lightly by hand using a damp selvyt cloth, dressed with polishing alumina or diamond powder, as used in the final polishing of microsections.

TYPES OF INDENTOR

The two types of indentor principally used are the Vickers[7] and Knoop[8] diamond pyramids, the characteristics of which are compared in Table 29.2.

Table 29.2
Characteristics of Vickers and Knoop Diamond Indentors

	Vickers	*Knoop*
Shape of indentor	Square Pyramid 136° 	Rhombic Pyramid 172°.30′ 130° 
Shape and relative sizes of impressions produced under similar loading	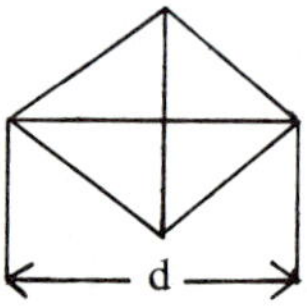	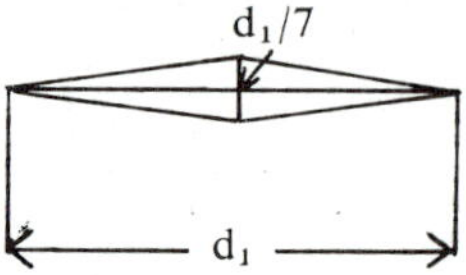
Approximate depth of penetration	$d/7$	$d_1/30$
Microhardness formula (Kg/mm^2) "d" in μm "P", loading in g	$MHV = \frac{1854.4P}{d^2}$	$MHK = \frac{14230P}{d_1^{\,2}}$

The Vickers diamond was developed primarily for the testing of steel and was designed to provide a relatively simple means of deriving the approximate tensile strength from hardness measurements by taking into account the mode of metal flow under indentation conditions. Though both types of indentor are widely employed, the Knoop diamond offers a number of advantages in the case of measurements on electrodeposited coatings, as follows:

1 Elastic recovery is restricted almost completely to the short diagonal, hence measurement of the long diagonal is little affected.
2 Depth of penetration for a given loading is less (Table 29.3 Figure 29.1).

3 Length of indentation is about three times that of a Vickers diagonal for the same loading, giving more accurate measurement at low loads (Table 29.3).
4 Elongated shape of impression is more convenient for measurements on relatively thin cross-sections.

Table 29.3
Diagonal Lengths and Penetration Depths for Vickers and Knoop Diamond Indentors

Hardness	*Load*	*Vickers Indentor*		*Knoop Indentor*	
	(g)	*Length of Diagonal (μm)*	*Depth of Indent (μm)*	*Length of Long Diagonal (μm)*	*Depth of Indent (μm)*
50	1	6.1	0.9	16.9	0.6
	5	13.6	1.9	37.7	1.3
	10	19.3	2.8	53.4	1.8
	20	27.2	3.9	84.6	2.8
	50	43.1	6.2	119.3	4.0
	100	60.9	8.7	168.7	5.6
100	1	4.3	0.6	11.9	0.4
	5	9.6	1.4	26.7	0.9
	10	13.6	1.9	37.7	1.3
	20	19.3	2.8	53.4	1.8
	50	30.5	4.4	84.4	2.8
	100	43.1	6.2	119.3	4.0
150	1	3.5	0.5	9.8	0.3
	5	7.9	1.1	21.8	0.7
	10	11.1	1.7	30.8	1.0
	20	15.7	2.2	43.6	1.5
	50	24.3	3.5	68.9	2.3
	100	35.2	5.0	97.5	3.3
200	1	3.1	0.44	8.4	0.3
	5	6.8	1.0	18.9	0.6
	10	9.6	1.4	26.7	0.9
	20	13.6	1.9	37.7	1.3
	50	21.5	3.1	59.7	2.0
	100	30.5	4.4	84.4	2.8
300	1	2.5	0.36	6.9	0.2
	5	5.6	0.8	15.4	0.5
	10	7.9	1.1	21.8	0.7
	20	11.1	1.7	30.8	1.0
	50	17.6	2.5	48.7	1.6
	100	24.9	3.6	68.9	2.3

A third type of indentor is the Berkovich three sided diamond pyramid, the geometry of which is such that the same area of impression is obtained as for a Vickers diamond at the same load. This has the advantages that more precise geometry of the indentor is obtainable in manufacture, particularly relating to a well defined apex. Measurement of the sides of the indentation, which lie in a single plane, may be easier than that of the Vickers diagonals. These are, however, very marginal in practice and this indentor appears to be little used in comparison with the main types.

SELECTION OF LOAD

Measurement on Surface

It is generally accepted that the coating thickness should not be less than ten times the depth of the impression in order to avoid substrate effects, though Wilson[5] suggests that where the hardnesses of coating and basis metal are of a similar order, a 30% penetration may be permissible. In Appendix E to BS 4292: 1968, covering electroplated coatings of gold and gold alloy, it is stated that figures of reasonable accuracy can be obtained with a coating thickness of five times the depth of indentation. For approximate guidance in this context, Table 29.3 lists depths of penetration of Vickers and Knoop indentors at various loads over a range of hardnesses, as derived from the formulae given in Table 29.2 and the same data are presented graphically in Figure 29.1. The advantage of the Knoop in terms of shallower penetration and greater diagonal length at comparable loadings is apparent. It is also clear that, if the 1 : 10 ratio of indentation depth to coating thickness is to be maintained, measurements on coating thicknesses of 3–5 microns in the common hardness range of 100–150 can be made, in principle, only at the lowest loading.

In practice, figures obtained under these conditions are of little significance. Measuring accuracy is greatly reduced due to the small dimensions of the impressions, and there is a more fundamental difficulty in that at the lower loadings the measured hardness is not invariant with load, probably due to the proportionately greater effect of elastic recovery after indentation. This effect applies equally to electrodeposited and wrought gold, as illustrated by the results quoted by Wilson[5] and recorded in Table 29.4.

In neither case does hardness approach a reasonably constant value at loads below about 25 g. This emphasises the importance of stating the loading at which tests are made whenever microhardness figures are quoted, viz.; $MHV_{25} = 106$. Taking 25 g as a minimum loading, the limiting thicknesses needed for realistic hardness determinations at various levels, as derived from Table 29.3 would be as shown in Table 29.5.

These figures apply to a 1 : 10 penetration/coating thickness ratio and may be correspondingly reduced if a lower ratio is accepted.

Measurement on Sections

In this case, basis metal effects are less critical and it is generally regarded as sufficient if the clearance between the coating edges and the corners of the

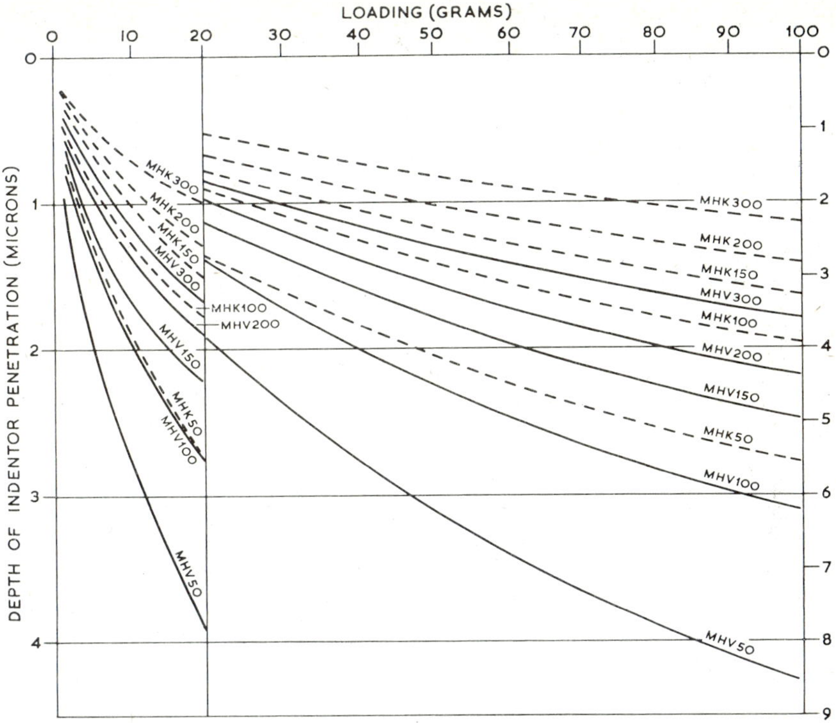

Fig. 29.1 Indentation depth vs loading for Vickers and Knoop indentors.

Table 29.4

Variation of Microhardness with Load for Wrought and Electrodeposited Gold

Electrodeposited	Load (g)	2	4	5	10	25	50	100
	MHV	61.1	70	76	82	106	102	105
Wrought	Load (g)	1	5	10	25	50	100	
	MHV	29	40	54	65	68	72	

Table 29.5

Minimum Thicknesses Required for Microhardness Measurement on Surfaces of Gold Coatings Using a 25 g Load

Hardness	*Minimum Thickness* (μm)	
	Vickers	*Knoop*
50	43	30
100	30	20
150	24	16
200	21	14
300	18	11

indentation is not less than half the length of the impression diagonal, which, for the Knoop indentor, may be the short diagonal. The variation of measured hardness with loading is still operative and as before, on the basis of a 25 g minimum load, coating thicknesses required at various hardnesses are as indicated in Table 29.6.

Table 29.6
Minimum Thicknesses Required for Microhardness Measured on Cross-Sections of Gold Coatings, Using a 25 g Load

Hardness	*Minimum Thickness* (μm)	
	Vickers	*Knoop*
50	69	26
100	42	18
150	33	14
200	28	12
300	25	10

It is clear that there are considerable limitations in the measurement of hardness of thin coatings, both in section and on the surface. If hardness is to be specified, this is best done on the basis of measurements on specially prepared coatings of adequate thickness.

CONDITIONS OF INDENTATION

In making the actual indentation the conditions for optimum accuracy are as follows:

1 Vibration-free environment.
2 Specimen surface exactly normal to axis of indentor travel.
3 No lateral or rotational movement of indentor.
4 Controlled rate of approach of indentor to avoid inertial effects.

Bulk specimens may be supported directly on the microscope stage. Mounted cross-sections are usually affixed to microscope slides. Small samples for surface measurement may require special jigging arrangements to ensure correct positioning and orientation of the surface.

Various means are employed by manufacturers to restrict lateral and rotational movement of the indentor, which usually presents no serious problem. Different methods are also adopted for making and measuring the impression. In some cases the indentor is fixed and the specimen is transferred to the measuring objective by a sliding table. In others, the indentor and one or more objectives are brought into position as required by rotation of a turret holder. In the Bergsman tester the indentor is supported in the objective holder and replaced by an objective for measurement. In the Vickers instrument the indentor is mounted in the centre of a special "search" objective, which enables the indentation area to be selected and the impression made without otherwise disturbing the instrument. A separate objective is, however, used for measurement.

In all cases a "sighting" impression is usually made initially, which is then centred in the field of view of the measuring objective to enable subsequent indentations to be pin-pointed on selected areas.

The most critical factor in making the indentation is the rate of loading, which depends on the method used to control the approach of the indentor. Various means employed for this purpose are indicated in Table 29.1. Manual application utilises the fine focussing screw of the microscope and great care is necessary to avoid impact effects when the indentor initially contacts the test surface, as usually indicated by a pilot lamp. If the rate of approach is too rapid, impact of the indentor will produce too large an impression and lead to a low hardness figure. For the same reason, significant discrepancies may be registered between measurements made by different observers on the same coating.

To avoid this difficulty, several instruments incorporate automatic control of the load application. This is commonly by dash-pot control under gravity, but a variety of other methods are used including liquid, pneumatic and electromagnetic loading. The time during which the load is applied, normally in the order of 15 seconds, and the withdrawal of the indentor may also be automatically controlled.

MEASUREMENT OF INDENTATION

In most cases the diagonals of the indentation are measured by means of a filar micrometer eyepiece at a suitable magnification. At least three impressions should be made a few microns apart and the average diagonal length used in determining the hardness from calibration tables provided with the instrument. Using the Dawe Sonodur and Newage testers, however, hardness is measured by the depth of penetration, with the advantage that elastic recovery effects are eliminated. In the former case, the indentor probe is excited at its natural resonant frequency by the magnetostrictive effect produced by a surrounding coil; the frequency changes in proportion to the depth of penetration of the indentor, and suitable electronic circuitry is employed to provide a direct reading on a meter which can be calibrated in terms of a standard hardness scale. In the Newage tester, direct reading of hardness in terms of penetration depth is obtained through a device which couples the indentor to a transparent capillary tube in which movement of a fluid meniscus indicates depth of penetration. Such instruments provide very rapid readings, but in both cases the use of a fairly high fixed load (Dawe, 150 g; Newage, 1000 g) severely curtails their usefulness in hardness measurement of gold plating.

SOURCES OF ERROR IN MICROHARDNESS MEASUREMENT

There is little information available concerning the accuracy of hardness measurement, since hardness is not a well defined physical property. The value obtained on any particular material will depend to some extent on the loading employed and the instrument used for measurement. Test blocks, usually steel, of standard hardness can be provided with most instruments for reference purposes, but since the mode of deformation may differ significantly for an electrodeposit, the main use of the former is to check the correct functioning of the instrument, rather than to provide standards in terms of which hardness may be expressed in any absolute way.

Gross errors are most likely to arise from the use of too great a loading in measurements on a plated surface, most often due to departure of the coating thickness from the nominal. It is wise, therefore, to confirm the thickness on any sample of unknown origin before carrying out a hardness determination.

Assuming that the loading is correctly selected, other factors affecting the precision of measurement may be mechanical (associated with the indentation conditions) or optical (associated with measurement of the impression). Mechanical errors have already been briefly discussed. Undue impact between indentor and coating when manual operation is employed is probably the most serious cause of low and variable hardness reading. It may be desirable, where this is indicated, to gear down the movement of the microscope fine focus screw to still further reduce the rate of indentor approach.

External vibration of the instrument can of course lead to major errors at the low loadings employed and a vibration-free environment is therefore essential. Apart from its possible effect on the dimensions or clarity of the impression, vibration will add greatly to the difficulty of placing an impression in a precise location, which is particularly troublesome in measurements on cross-sections.

The importance of surface finish has already been pointed out. Correct orientation of the surface with respect to the indentor is likewise essential. Incorrect orientation or undue surface irregularity will lead to the production of kite-shaped or generally distorted impressions which confuse measurement.

As mentioned earlier, errors due to lateral or rotational movement of the indentor during penetration are not likely to be encountered with commercial instruments. It should be emphasised however, that if a testing attachment is purchased for fitting to an existing microscope, care should be taken to ensure that the latter is in sufficiently good condition for the purpose in terms of tube stability, smooth movement of stage without undue backlash, etc., otherwise problems of impact and in placing and locating impressions are likely to be aggravated.

With due attention to the above factors, precision of hardness measurement is dependent to a large extent on the accuracy with which impression diagonals can be measured. Limitations of optical microscopy, which have been discussed in relation to thickness measurement (see Chapter 28) apply equally in the present case. From Table 29.3, considering a loading of 20 g, the difference in diagonal length on coatings of 100 and 150 hardness is approximately 3.6 microns for the Vickers indentor and 9.8 for the Knoop. Thus an error in length measurement of 0.25 microns, which is probably the lowest that could be achieved, would correspond to a difference in the measured hardness of approximately 4 units and 1.5 units respectively over this range, indicating a measurement precision that would certainly suffice for most practical purposes. No information is available on the precision obtainable with direct reading instruments where measurement is linked to depth of indentation but, as already pointed out, the usefulness of these in measurements on plated coatings is severely restricted by a high fixed loading.

REFERENCES

1 Lysaught V. C., "Indentation Hardness Tests", Reinhold, New York (1949); *Metal Prog.*, **78**, 2, 93 (1960); **78**, 3, 121 (1960).
2 Mott B. W., "Micro Indentation Hardness Testing", Butterworth, London (1956). Also available in German translation by K. F. Frank, "Die Mikrohärte Prufung" Berlin Union, Stuttgart (1957).
3 Read H. J., *Plating*, **49**, 602 (1962).
4 Bierbaum C. H., *Trans. Am. Soc. Steel Treating*, **18**, 1009 (1930).
5 Wilson G., *Metal Finish.*, **58**, 6, 50 (1960).
6 Cullen T., Petruna P., *Plating*, **54**, 1039 (1967).
7 Smith R. L., Sandland G. E., *J. Iron Steel Inst.* (*Lond.*), **111**, 285 (1925).
8 Knoop F. C., Peters C. G., Emerson W. B., *J. Res. Natn. Bur. Stand.*, **23**, 39 (1939).

Chapter 30

MEASUREMENT OF INTERNAL STRESS

F. H. Reid

As for most other metals, gold coatings show various degrees of internal stress in the electrodeposited condition. Stress is a complex phenomenon and although a number of theories have been advanced concerning its origin, the basic reasons are not fully understood.

In theoretical discussion distinction is made between microstresses and macrostress[1]. Microstresses, which may be orientated, are active over very small regions of a deposit and may be fairly closely related to changes in hardness and other physical properties. They can only be measured, however, by computer-assisted X-ray methods, based on the broadening of diffraction lines from the coating. Details of these very specialised procedures are outside the scope of the present text, but for those wishing to pursue this aspect further, a good review is given by Binder and Fischer[2].

Wherever the terms "stress" or "internal stress" occur in the literature without qualification, it may be assumed that they refer to macrostress, which may be regarded as the overall resultant of the various microstresses. Though less closely related than the latter to the physical properties of the coating, macrostress is of great importance since it governs the tendency of the coating to contract, if the stress is tensile, as is usual, or to expand if the stress is compressive. As such, it is closely associated with the various practical problems that may arise as a result of stress in deposits.

Such problems are well known. High stress may cause spontaneous cracking of deposits either during deposition or subsequently, leading to loss of protective value. When precleaning is incorrect or ineffective, stress will assist exfoliation of the coating but, it should be noted, is often wrongly identified as the prime cause of this effect. In electroforming, stress may lead to distortion of the electroform or to premature separation of the latter from the mould or mandrel. Even when cracking is not observed in the as-plated condition, stress may act as a contributory factor in the development of cracks in service, particularly when components may undergo flexure or be subjected to wear by sliding or impact. Fatigue strength of coatings may also be adversely affected by high stress.

The internal stress of deposits is usually increased markedly with contamination of an electrolyte by foreign metals or organic material, hence its measurement should be regarded as an important factor in the control both of coating characteristics and electrolyte condition.

MEASUREMENT OF MACROSTRESS

By comparison with the determination of thickness and hardness, measurement of internal stress is time consuming and cumbersome, a factor which probably accounts for the sparseness of quantitative data in the literature of gold plating. Instead, reliance is often placed on qualitative assessment.

QUALITATIVE

A satisfactorily low stress level is often inferred from the absence of cracks in the coating. From the serviceability point of view this is a reasonable method of assessment, particularly if coupled with bend or fatigue tests, though it is clear that other factors than stress alone are involved. At least a positive result indicates that whatever the exact magnitude of the stress, it is not likely, in conjunction with other mechanical properties of the coating, to adversely affect performance or protective value.

Another method commonly used for qualitative demonstration, is to dissolve the substrate from a plated strip and note any tendency of the unsupported deposit to curl or roll up. If the stress is sufficiently high to have caused cracking, the coating will disintegrate on removal of the substrate.

QUANTITATIVE

Qualitative techniques are useful for quick and simple assessment, but they clearly suffer limitations as serious control methods, for which quantitative measurement is necessary. The techniques most widely used for this purpose are based on measurement of the bending effect produced by the stress in the coating when it is deposited on to one side only of a flexible strip or disc specimen. The strip may be held rigid during plating or allowed to bend: in the latter case the amount of deflection may be measured at the end of the test or continuously during plating. Continuous measurement is also possible in the case of a rigid strip by measurement of the restoring force needed to restrain bending.

Strip Deflection (strip unrestrained)

Historically, this type of method was one of the first employed for stress measurement[3]. A known length of strip is plated on one side only and the change in deflection at the centre measured before and after plating. Stress is given by the formula:

$$S = \frac{4Et^2 D}{3dl^2}$$

where t = thickness of strip

l = length plated

d = coating thickness

E = Young's modulus of strip material

D = change in deflection at centre.

Alternatively, stress may be expressed in terms of curvature, thus:

$$S = \frac{Et^2}{6d}\left(\frac{1}{r_a} - \frac{1}{r_b}\right)$$

where the suffixes "b" and "a" refer to radius of curvature before and after plating. These are easily found by matching the strip curvature against a pattern of circular arcs of known radius.

Curvature may also be measured by a dial gauge[4] or, where very small, by a microscopic "depth of focus" method[5]. Interferometric techniques have also been employed[6,7], though these are mainly of interest in measuring the deformation of actual components due to stress in the applied coating. Strain gauge techniques have also been used for this purpose. It is often convenient to measure curvature by the deflection of the free end of a clamped strip. If the deflection is small relative to the plated length, the formula is reduced to:

$$S = \frac{Et^2 D}{3dl^2}$$

where D is now the deflection of the free end and the other symbols have the same significance as before.

It is not difficult to devise fairly simple arrangements for making measurements on this principle, the variants employed differing mainly in the method used to measure and record deflection. The writer[9] has described an arrangement in which the tip of the plated strip serves as an indicator by moving across a millimetre linear scale reproduced on a photographic plate, the enlarged image of strip and scale being projected on to a screen. This set-up permits the use of quite a crude lens system for projection, but yields good results for comparative purposes. Hammond[10] has described other simple methods used by earlier investigators which are suitable for continuous or "before and after" measurement of deflection.

Continuous observation of deflection may be useful in research work, but is not necessary for control purposes. In this case an extremely simple method of stress measurement which could be easily carried out in any plating shop is the double strip technique of MacNaughtan and Hothersall[11]. In this procedure, two suitably masked strips are clamped back to back and separation of their free ends is measured before and after plating using a travelling microscope or even a graduated linear scale.

Strip Deflection (strip restrained during plating)

On a rigid substrate, stress effects may be accommodated by modification of the mode of growth or structure of a coating, which may not occur in the case of a flexible strip where accommodation can take place via substrate deformation. An extreme example of this effect has been observed by the writer in the deposition of rhodium from a sulphate electrolyte containing aluminium salts[12]. Although the flexible strip method indicated a very high tensile stress, a deposit plated on to a rigid substrate showed no cracking and no tendency to curl when the substrate was dissolved, which indicated that residual stress was very low.

Though flexible strip methods based on free deflection are useful for control and investigational purposes, it is apparent that since most gold coatings are plated on to rigid substrates, more meaningful values of residual stress in relation to practical application should be obtained by techniques in which movement of the strip is restrained during plating.

In the type of method due to Soderberg and Graham[5], a flat strip is clamped against a glass or plastic supporting plate by a picture frame arrangement during plating. Residual stress is measured by the deflection of the strip at the centre when released after plating, using the formula:

$$S = \frac{4E(T+t)^2 D}{3TtL^2}$$

where S = residual stress in coating

E = modulus of strip

d = thickness of coating

t = thickness of strip

L = length of strip plated

D = deflection of strip.

Hoar and Arrowsmith[13] devised a neat method for continuous observation of stress even when the strip is not allowed to deflect. In this technique, deflection is restrained by the magnetic force applied through a solenoid acting on a small pin of mild steel carried by the free end, and the instantaneous stress is measured in terms of solenoid current required to maintain the strip in its original position. This method is capable of considerable refinement and can lend itself readily to automatic recording (see, for example, Stalzer[14]). Giles and Schrier[15] have described a sensitive stressometer for measurement of stress resulting from cathodic entry of hydrogen into steel, which incorporates the principles of the Hoar-Arrowsmith method and the spiral contractometer described in the following paragraphs.

COMMERCIAL INSTRUMENTS FOR STRESS MEASUREMENT

Spiral Contractometer

The principle of free deflection of a strip is utilised in the spiral contractometer originally developed by Brenner and Senderoff[16] and now in fairly general use for stress measurement. This is a portable and compact instrument in which the strip takes the form of a helix, usually of stainless steel, coiled around a cylindrical perspex guard tube. The upper end of the strip is clamped, and the lower end is attached to a torque rod passing up through the centre of the guard tube and engaging with a gear train to produce movement of a recording needle over a circular scale. Stress is given by the formula:

$$s = \frac{2KD(t+d)}{t^2pd}$$

where K = deflection constant of the helix, measured by subjecting the latter to a known torque

D = angular deflection of needle in degrees

p = pitch of helix

and "t" and "d" have their usual significance. The helix can be re-used after stripping the deposit.

The spiral contractometer has several attractive features:

1 Continuous reading
2 Suitable for research and control
3 Can be immersed directly in any bath of interest
4 Long length of strip provides high deflection for low stress.

The instrument has been used without special comment by Duva and Foulke[17] in measurements of stress in gold deposits, but Kushner[18] suggests that where very low stresses are involved, difficulty might arise due to friction in the gear train and as such, jewelled bearings might be necessary.

Stressometer

In this instrument, developed by Kushner[19], a flat metal disc 0.01–0.04 in. thick and 4 in. in diameter is used as the flexible test plate and is clamped to form the roof of a shallow cylindrical reservoir with the outer face exposed to the plating solution. The reservoir is filled with a manometric liquid and bowing of the disc due to internal stress of the electrodeposited metal is registered by movement of the meniscus in a vertical capillary measuring tube connecting with the reservoir. The stress formula is:

$$S = \frac{3r^2L}{4Kh}$$

where $K = kA^4/Eh^3$

L = change in height of capillary meniscus

r = capillary radius

A = disc radius

h = thickness of disc

t = coating thickness

E = Young's modulus of disc material

k = disc constant.

A useful practical advantage of this method is that no masking of the test disc is required: this can prove troublesome in bending strip techniques. Uniformity of coating thickness is also more readily achieved since there are no edge effects. The test disc may be fabricated from any metal of suitable mechanical properties, including particular substrates of practical interest, which can be subjected to the same pretreatment as used in the plating of actual components. Trapping of air bubbles in the reservoir in assembly can be troublesome and temperature effects may also be significant when tests are made on solutions at temperatures above ambient, though suitable corrections may be applied where necessary.

IS Meter

A novel method for stress measurement, now available commercially, has been described by Dvorak and Vrobel[20]. In this procedure a strip or wire specimen is supported in a special jig, the centre portion of which carries a device for pre-stressing, while the upper part supports a sensitive dial gauge. Stress is measured by the change in length of the strip during plating, as indicated by the gauge, and is calculated using the following formula:

$$S = \frac{Ed}{2lt} \times v$$

where S = internal stress

E = elastic modulus of test strip

d = thickness of strip

t = thickness of deposit

l = length of strip plated

v = change in length of strip.

Tensile stress produces a contraction, and compressive stress an elongation of the strip.

The authors report good sensitivity and reproducibility for the method in measurements on precious metal coatings using copper as the strip material, but no detailed results are presented.

Advantages of the method are that the strip can be immersed directly in a production bath, and since both sides are plated, no stopping-off problems arise. It is also claimed that the stress values obtained correspond more closely with those which would apply in the case of a rigid substrate, since no bending of the strip occurs.

FACTORS AFFECTING STRESS MEASUREMENT BY "DEFLECTION" METHODS

Since stress is not a well defined physical property there is no absolute standard by which to judge the accuracy of a determination. The value obtained in any particular case will depend to some extent on the method used for measurement because the mode of substrate deformation may differ, or the conditions implicit in the formula by which stress is calculated may be realised to a greater or lesser extent in practice.

This effect has been demonstrated in measurements on nickel deposits from a purified Watts electrolyte using the rigid strip method, the Kushner Stressometer and the spiral contractometer[21]. The level of precision of measurements by individual methods was, however, reasonably good, and it is this, rather than absolute accuracy, which is important for comparative and control purposes.

Optimum precision by any method requires that attention be paid to factors relating both to the substrate and to the conditions of deposition to ensure optimum compliance with assumptions made in deriving the stress formula, which are common to all methods based on substrate deformation.

SUBSTRATE FACTORS

A variety of substrate materials may be used, including stainless steel, tempered steel, brass, beryllium-copper, phosphor-bronze, or pure metals such as copper, nickel and silver, which may possess sufficient resilience in the as-rolled condition. The essential condition is that the strain developed in the substrate due to distortion during measurement should not exceed the elastic limit of the material. This may readily be checked by applying a mechanical deflection under a known load and noting whether the original curvature is regained when the load is removed. If this observation is carried out with sufficient precision in respect of load and deflection measurement it will serve as a measure of the elastic modulus, which should be checked from batch to batch of the metal from which strip or discs are cut or stamped. If pretreatment of the substrate involves the deposition of one or more intermediate layers, the composite material should be checked in the same way to provide a modulus value for use in the appropriate stress formula.

The thickness of strip is usually in the order of 0.01–0.02 in. whereas the width may vary from 0.2 in. upwards to about 1.5 in. This is important only insofar as possible edge effects in free deflection techniques may be enhanced with decreasing width of strip. Omitting the special case of the spiral contractometer, a strip length of about 5–6 in. is convenient for measurement purposes. These dimensions may be varied in particular cases in relation to the magnitude of the stress to be measured and the means available for accurate determination of deflection or curvature, particularly when the stress is relatively low.

In cases where deposition is carried out at elevated temperature, with subsequent measurements of deflection at ambient, a correction should strictly be applied to take account of change in curvature between the two temperatures due to differential thermal expansion of the substrate and coating (bimetal effect)[3].

DEPOSITION FACTORS

The most important of these is uniformity of coating thickness. Edge build-up can be particularly troublesome in plating long and relatively narrow strips and can lead to premature initiation of transverse cracking of the deposits. When the strip is clamped during plating, edge masking can be provided by the fixture itself. This applies also to the disc of the Kushner Stressometer. In the case of the spiral contractometer the close proximity of adjacent turns of the spiral assists in minimising edge build-up. Uniformity of thickness is also favoured by the disposition of the coil since a cylindrical anode may be used to achieve an even distribution of current density. In other cases some simple screening arrangement should be used to protect the strip edges from excessive current density concentration.

Thickness variation from top to bottom of a strip may also occur. This can be reduced by supporting the strip in a horizontal plane during plating, but this is not always convenient, particularly when continuous measurement is required on a freely bending strip. In this case the effect must be minimised by attention to anode size and disposition, and by agitation of the electrolyte. The latter is likely to cause fluctuation of the strip deflection and is therefore not advisable if strictly continuous observation is to be made. A practical compromise is to stop the agitation for short periods while intermediate measurements of strip deflection are made.

The average thickness of the coating is often conveniently determined from the gain in weight of the strip during plating and provided that precautions have been taken to ensure reasonable uniformity, this will usually suffice for comparative purposes. Alternatively, thickness measurements for averaging may be made at selected positions on the strip using any convenient method, with due attention being accorded to the factors affecting accuracy, as discussed in Chapter 28. In precise work it may be desirable to take into account the effect of coating density on determination of true linear thickness, but this would appear to be unnecessary for general purposes, since the error involved is not likely to be significant in relation to the general level of precision obtainable in stress measurement.

Since stress may vary with coating thickness, the thickness selected for measurement purposes should logically be that of maximum interest from the applicational viewpoint in specific cases. If the results are to be of general use in relation to those of other workers, however, it is important that both the measuring method and the coating thickness involved should be specified whenever stress figures are quoted.

REFERENCES

1 Binder H., *Metalloberfläche*, **17,** 263 (1963).
2 Binder H., Fischer H., *Metalloberfläche*, **17,** 295 (1963).
3 Stoney G. G., *Proc. R. Soc.*, **82,** 173 (1909).
4 Phillips W. M., Clifton F. L., *Proc. Am. Electropl. Soc.*, **34,** 97 (1947).
5 Soderberg K. G., Graham A. K., *Proc. Am. Electropl. Soc.*, **34,** 74 (1947).
6 Austen H. E., Fisher R. D., *J. electrochem. Soc.*, **116,** 2, 741 (1969).
7 Curkin L. E., Moeller R. W., *Proc. Am. Electropl. Soc.*, **41,** 196 (1954).
8 Journaud M., *Ind. Finish.*, **45,** 35 (1969).
9 Reid F. H., *Trans. Inst. Metal Finish.*, **33,** 105 (1956).
10 Hammond R. A. F., *Trans. Inst. Metal Finish.*, **30,** 140 (1954).
11 MacNaughtan D. J., Hothersall A. W., *Trans. Faraday Soc.*, **24,** 387 (1928).
12 Reid F. H., *Trans. Inst. Metal Finish.*, **36,** 74 (1958/9).
13 Hoar R. P., Arrowsmith D. J., *Trans. Inst. Metal Finish.*, **34,** 354 (1957).
14 Stalzer M., *Engrs Digest*, **25,** 11, 110 (1964); *Metalloberfläche*, **18,** 263 (1964).
15 Giles M. E., Shrier L. L., *Electrochimicia Acta* **11,** 2, 193 (1966).
16 Brenner A., Senderoff S., *Proc. Am. Electropl. Soc.*, **35,** 53 (1948).
17 Duva R., Foulke D. G., *Plating*, **55,** 1056 (1968).
18 Kushner J. B., "Electroplaters Process Control Handbook", D. G. Foulke and F. R. Crane (Eds), Chap. 14., Reinhold, New York (1963).
19 Kushner J. B., *Proc. Am. Electropl. Soc.*, **41,** 188 (1954).
20 Dvorak A., Vrobel L., *Trans. Inst. Metal Finish.*, **49,** 4, 153 (1971).
21 Borchert L. C., *Proc. Am. Electropl. Soc.*, **50,** 44 (1963).

Chapter 31

MEASUREMENT OF DUCTILITY

F. H. Reid

Ductility is defined as the ability of a metal to withstand plastic deformation without fracture or cracking. In the case of bulk metal it is measured by standard tensile tests in which strip or rod specimens are extended under an increasing load until fracture occurs. Ductility is expressed as the percentage elongation on a fixed initial gauge length of the specimen and/or the percentage reduction in cross-sectional area of the specimen at fracture.

In electroplated coatings some degree of ductility is an important requirement from several viewpoints. Lack of ductility renders a coating prone to cracking, either in the as-deposited condition under the influence of internal stress, or possibly later in the course of assembly operations, or under service conditions where mechanical stresses are encountered. The latter may arise, for example, through flexing of contact springs, sliding wear, impact, or differential expansion effects due to temperature variation. If cracking occurs the protective value of the coating is lost, or at least seriously impaired.

For optimum resistance to wear, brittle coatings are undesirable in view of the danger of fragmentation under load. Once initiated this is usually a cumulative process which leads to rapid penetration of the deposit. It has been suggested that the ideal requirement in this context is a combination of high tensile strength with maximum ductility, to confer "toughness" on the coating[1].

Ductility of gold deposits is affected very sensitively by the presence of foreign impurities in the coating which may be metallic, non-metallic, organic or gaseous in nature. Hence ductility measurement is valuable in the control both of electrolyte condition and plate characteristics.

METHODS OF MEASUREMENT

TENSILE TESTS ON STRIPPED COATINGS

The most unequivocal method for measurement of the ductility of an electrodeposit is by a tensile test on a specially prepared coating of at least 0.01–0.02 in. in thickness. This technique, however, suffers limitations of accuracy when ductility is less than about 2.5%. Since it also entails the time consuming preparation of special test coatings, it is of little value for production control and remains essentially a research method. It has been used recently by Wiesner and Distler[2] in a study of the physical properties of electrodeposited copper-gold alloys. In their work, tensile specimens were prepared by plating 0.01–0.02 in. of the coating under test on to highly polished copper panels. After heat treatment at 300°C for 3 hours to remove brittleness in the as-plated condition, tensile test specimens were cut and the copper backing removed by treatment in hot chromic acid (250 g/l) containing 20–50 ml/l of sulphuric acid.

These authors also carried out tests on ring samples, using a ring tensile test apparatus developed by Holman, Stiles and Fung[3] for measurements on tungsten. In this case samples were prepared by plating a copper cylinder (1 in. OD) over a length of 8 in. with deposit thicknesses ranging from 0.035–0.060 in. After heat treatment the plating was machined to uniform thickness and annular test rings were isolated by machining through to the copper, which was finally removed by dissolving as before. The ring tester gave good consistency of results and was preferred for the measurement of mechanical properties on coatings thicker than 0.02 in.

TENSILE TESTS ON SUPPORTED COATINGS

The preparation of thick coatings for tensile testing of unsupported foils is difficult and often quite impracticable due to the development or cracks in the as-plated condition. Even if cracks are not present initially, some form of stress relieving heat treatment may be necessary, as recommended by Wiesner and Distler in the preceding text, to avoid edge cracking of the coating when tensile specimens are cut from plated panels.

Apart, therefore, from the time consuming procedure involved, results obtained by this method, though useful for comparative purposes in investigational work, are of limited value with relatively thin coatings.

More meaningful from this viewpoint, and more convenient in practice, are tests applied to a composite of the coating on a suitable substrate. Although the interpretation of the results in terms of a true ductility measurement is confused by possible coating/substrate interactions, this type of test has the advantage of being applicable to coatings within a realistic thickness range. The elongation that can be sustained before cracking of the deposit occurs is a useful parameter of performance, particularly when the selection of substrate and pretreatment prior to plating of test specimens is made to correspond to practical cases.

The test of this type described by Such[4] for nickel plate on steel should be equally applicable to gold plating on copper or other substrates of sufficient ductility. In his procedure a strip specimen with central necked portion (Figure 31.1) is extended in a Hounsfield Tensometer until the first appearance of edge cracks in the coating in the necked area is observed under a magnification of 10X. At this point the load is slowly reduced until the elastic deformation has been recovered, the remaining permanent elongation being taken as a measure of the ductility of the coating.

The minimum coating thickness used in this case was about 0.001 in. For thinner coatings visual observation of the onset of cracking may be difficult, particularly if the surface is slightly rough. In this connection, a technique used by Angus[5] in comparing the ductility of relatively thin (5 micron) palladium coatings could be relevant to gold. Coatings on 3 mm diameter copper rod specimens were extended, also in a Hounsfield Tensometer, but the end-point was detected by an electrographic method. Porosity prints on cadmium sulphide paper were taken at regular increments of loading and the percentage blackened area (i.e., exposed basis metal) plotted against extension to give a typical curve of the form shown in Figure 31.2, from which the limiting elongation was taken as the extrapolated value indicated.

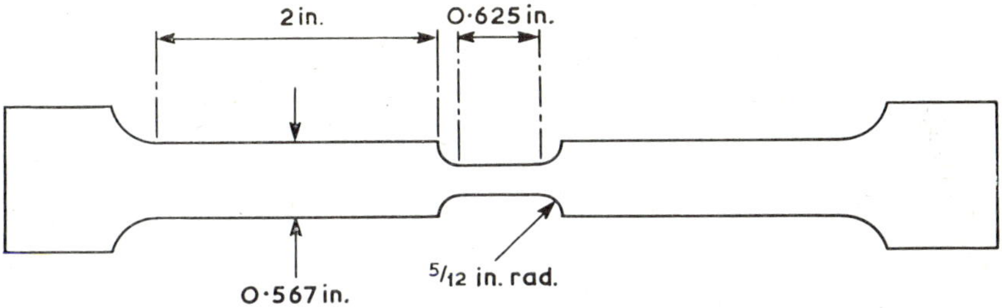

Fig. 31.1 Strip specimen for Such[4] ductility test.

It is interesting to note that in some cases failure of the coating by this criterion was established before the appearance of visual cracking under a binocular microscope, and in others, surface cracking was observed without exposure of the substrate. These observations emphasise the difficulty in interpreting "ductility" tests of this type, though in relation to noble metal coatings the practical significance of elongation figures related essentially to loss of protective value of the deposit can hardly be questioned.

BEND TESTS

Tensile tests require special equipment and are relatively time consuming, even on supported coatings, hence they are less useful in production control of coatings or electrolytes than fairly simple bend tests. These may be applied either to stripped foils or to coating/substrate composites and can be adapted to yield at least an approximate quantitative assessment of ductility for comparative purposes.

Tests on Stripped Foils

MICROMETER TEST

In a method described by Durbin[6], a coating stripped from a highly polished

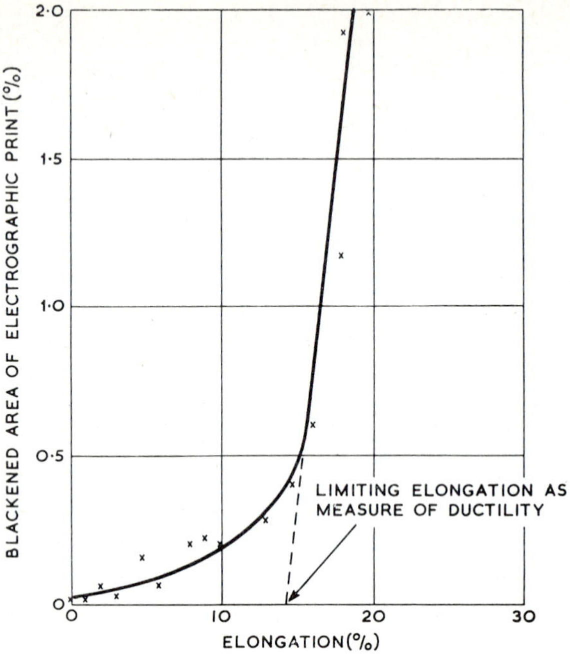

Fig. 31.2 Loss of basis metal protection with elongation of palladium plated copper rod[5].

stainless steel panel is cut into 0.25 × 2 in. strips, which are bent into a "U" shape and placed between the jaws of a micrometer, so that the bend remains between the jaws when they are slowly closed. The approximate elongation of the coating at which cracking or fracture occurs is given by the formula:

$$\text{Elongation }(\%) = \frac{100t}{d-t}$$

where d = separation of jaws at failure

t = thickness of coating.

REVERSE BENDING TEST

In this test, the stripped foil is clamped in a vice or other suitable holder and and bent through 90°, the free end being pressed flat against the jaws of the vice. A reverse bend through 180° is then made and this is repeated in opposite directions until failure of the coating occurs by cracking or complete fracture. The number of bends sustained is taken as a measure of ductility. A refinement of this test, in the form of the Jenkins Bend Tester, has been described by Foster[7].

Tests on Supported Foils

MANDREL TEST[8]

This is usually applied as a "go, no-go" test. A plated strip is bent around a mandrel until the two ends are parallel, the diameter of the mandrel being selected to produce a specific percentage elongation in relation to the total strip thickness, as derived by the formula:

$$D = \frac{100t(1-E/100)}{E}$$

where

E = % elongation

D = mandrel diameter.

The test is passed if no visible cracking of the coating occurs at the outer surface of the bend.

SPIRAL BEND TEST

A test of the simple mandrel type can be made semi-quantitative by utilising a series of mandrels of various diameters to cover a range of elongations, but it is more convenient to use a single mandrel of varying curvature, as in the spiral bend test described by Edwards[9]. In this version the mandrel profile takes the form of a logarithmic spiral, the polar equation of which is:

$$r = 0.4473\ e^{0.5\theta}$$

Between the values of θ of 0–2π, the radius of curvature (ρ) varies progressively from 0.5–11.6 cm, corresponding to elongations of 9.1 and 0.46% respectively for a 1 mm strip. In practice the mandrel is graduated directly in terms of the radius of curvature. The test strip (200 × 10 × 1 mm) is clamped firmly at one end, with the plated face outwards, and then slowly bent around the profile. The radius of curvature is noted at the point where cracking of the coating is observed and the corresponding elongation is calculated from the formula:

$$E(\%) = \frac{100t}{2\rho + t}$$

where t = the total thickness of the strip.

Due to the rapid change in radius of curvature with length, sensitivity is poor at the upper end of the elongation range, but the effective range of 0.4–6% is convenient in relation to the values commonly encountered in electrodeposits. In terms of precision, measurements on three similar strips gave elongation figures differing from the mean by a maximum of 37%. In over 100 tests of this kind, standard deviation from the mean was about 10%.

Elongation figures obtained by this procedure showed poor correlation with those obtained by the Such[4] tensile test on similar coatings. Two possible reasons for this are that edge cracking, which marks the end-point in the Such[4] test, is disregarded in the bend test and, secondly, that while the accuracy of tensile tests increases with increasing ductility, the reverse is the case for the bend test. These observations on nickel plated steel samples are equally applicable in the gold plating context.

TAB TEST

In this test a tab is formed in a plated panel by making two parallel saw cuts about 1.5 in. deep and 0.75 in. apart, and the tab is then bent away from the plated surface while the radius of the bend is observed for initiation of cracking. When this occurs, the angle of the bend from the original surface is taken as a measure of ductility.

CUPPING AND BULGE TESTS

Tests of this type are based on procedures used in general metallurgical practice for assessing the deformability of sheet. They involve the application of mechanical or hydraulic pressure over a circular area to produce a "cup" or bulge. Ductility is indicated by the depth of the cup or height of bulge at incipient fracture. In the case of plated coatings the end point is signified by the onset of cracking of the deposit on the outer surface of the bulge.

Erichsen Cupping Test

In this test, probably the best known of this type, a circular or square sheet specimen not less than 80 mm in diameter or width, is clamped between a die ring of 27 mm internal diameter. Deformation is produced by pressing a lubricated indentor of 10 mm radius or a hemispherical steel ball into the surface of the sheet. Use of this test has been proposed by several authors[10,11] and the test has been modified by Romanoff[12] for use in production control testing of nickel and copper plated steel sheet and strip. Such[4], however, found the sensitivity insufficient for quantitative measurement and also questioned the reproducibility.

Bulge Tests

Read and Whalen[13] have described a hydraulic bulge test, following Jovignot[14], for ductility measurement on both supported and unsupported coatings. Their apparatus is shown schematically in Figure 31.3.

A disc specimen is clamped between upper and lower plates A and B, the spherical seating between B and the head of the hydraulic jack J providing for uniform loading around the periphery of the disc. The disc is bulged by hydraulic pressure until cracking of the coating is observed. Ductility is expressed in terms of significant strain, S, which is related to bulge height by printing a standard 20 line per inch photogrid on the surface of a specimen and bulging to various heights. For typical electrodeposits, S has values in the range 0.010–0.050, which corresponds to elongations of less than 2.5%.

Bulge height may be measured by a high quality dial indicator, but for coatings thinner than 0.001 in. damage may occur through contact with the indicator probe. In these cases measurement is best made using a microscope with a very small vertical depth of focus (0.001 in.). By this technique, measurements have been successfully made on deposits as thin as 0.0002 in.

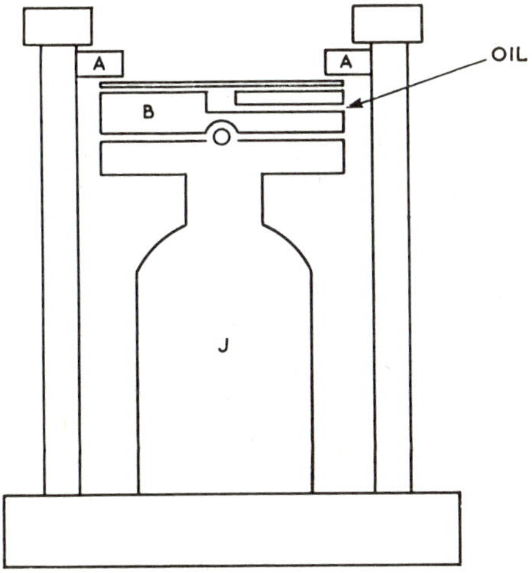

Fig. 31.3 Apparatus (schematic) for hydraulic bulge test[13].

Owen[15] has recently described a hydraulic ductility test applicable to 3 in. square samples of thin metal foil in which the basic unit is a Mullen Tensile Tester, more commonly employed in the testing of paper. Linearly variable differential transformers are used to monitor the applied hydraulic pressure and the height of hemispherical deformation, the two signals being fed to an X-Y oscilloscope or recorder to produce a tensile curve which can be fairly closely related to data from conventional tensile tests on foils using an Instron Tester (Table 31.1).

The Mullen Tester is a piece of precision built equipment and different testers are stated to produce closely comparable results, which is important for comparative purposes between different investigators.

Table 31.1
Comparative Elongation Data on Various Foils

Material	*Thickness* (μm)	*Elongation* (%)		
		Mullen Tester	*Instron Tester*	
			Longitudinal Direction	*Transverse Direction*
Clevite	35	2.2	3.1	2.3
Annealed ETP Copper	50	11.3	10.9	10.2
Brass Shim Stock	25	4.4	1.1	2.6
Pyrophosphate Copper	25	1.4	3.1	

SUMMARY

Notwithstanding the importance of ductility from a functional aspect, the technical and trade literature on gold plating is notably lacking in quantitative figures, though it abounds in qualitative statements.

Often the description of a coating as "ductile" appears to be based solely on the absence of spontaneous cracking in the as-deposited condition, which certainly represents a minimum criterion, but is of little value unless complemented by deformation tests of some kind. Where general metallurgical test facilities are available, direct tensile testing, cupping and bulge tests may be employed. Of these, cupping tests, in the form described by Romanoff[12], would appear most convenient for production control purposes, though tensile tests on supported coatings as used by Such[4] and Angus[5] would also be suitable.

Tensile and bulge tests on unsupported films are to be regarded essentially as research tools in the development of improved processes and coatings. In this context, there are severe limitations on tensile testing when coating ductility is low (say, less than 2.5%), both in terms of difficulty in preparing specimens of adequate thickness and in the accuracy of measurement obtainable. In this area, the bulge test, which measures ductility in terms of significant strain, would seem to have better possibilities.

For most practical purposes bend tests on supported coatings are convenient and adequate, with the merit of relative simplicity. Although the figures obtained are open to some doubt as true measures of ductility, they are useful for comparative purposes and of considerable practical significance. A test of the spiral bend type is the most readily adaptable to quantitative measurement and the precision quoted by Edwards[9] might be improved by the adoption of an electrographic method for crack detection, as utilised by Angus[5]. In this case, however, the procedure could be simpler, since it should not be difficult to devise a simple arrangement for electrography on the test strip in the fully bent position. This would eliminate the uncertainty in detecting the onset of cracking by visual means.

There is a need for systematic correlation of ductility figures obtained by the various tests cited, an aspect which is under study by an ASTM Committee (Section G, Sub-Committee III, Committee B-8). A useful preliminary survey of methods for ductility testing carried out under its auspices has been presented by Van Tilburg[16].

REFERENCES

1 Flühmann W., *Plating*, **56,** 1351 (1969).
2 Wiesner H. J., Distler W. B., *Plating*, **56,** 799 (1969).
3 Holman W. R., Stiles E. B., Fung E., *J. Scient. Instrum.*, **44,** 545 (1967).
4 Such T. E., *Trans. Inst. Metal Finish.*, **31,** 190 (1954); *Metallurgia* 56, 335, 121 (1954).
5 Angus H. C., *Trans. Inst. Metal Finish.*, **44,** 1, 41 (1966).
6 Durbin C., *Proc. Am. Electropl. Soc.*, **38,** 119 (1951).
7 Foster P. F., "The Mechanical Testing of Materials and Alloys", pp. 218, Pitman, London (1936).
8 B.S. 1224: 1959.
9 Edwards J., *Trans. Inst. Metal Finish.*, **35,** 101 (1958).
10 Fischer H., Barmann H., *Zeit. Metallkunde*, **32,** 376 (1940).
11 Phillips W. M., Clifton F. L., *Proc. Am. Electropl. Soc.*, **35,** 87 (1948).
12 Romanoff F. P., *J. electrochem. Soc.*, **65,** 385 (1934); *Metal Finish.*, **58,** 2, 55 (1960).
13 Read H. J., Whalen T. J., *Proc. Am. Electropl. Soc.*, **46,** 318 (1959).
14 Jovignot C., *Rev. Met.*, **27,** 443 (1930).
15 Owen C. J., *Plating*, **57,** 1012 (1970).
16 Van Tilburg G. C., *Proc. Am. Electropl. Soc.*, **50,** 51 (1963).

Chapter 32

MEASUREMENT OF ADHESION

F. H. Reid

From the functional viewpoint the most fundamental requirement of gold plating, as for any other electrodeposition process, is adequate adhesion of the coating to the substrate to which it is applied. The degree of adhesion obtained in any particular case is critically dependent on the correct choice of pretreatment, and as such, a detailed discussion of procedures suitable for most basis metals is to be found in Chapter 12.

For gold coatings on the common basis metals, such as copper and copper alloys, nickel and silver, true metal-metal adhesion can readily be achieved in the as-plated condition, in which case attempts to separate the coating from the substrate usually result in fracture within the weaker of the two rather than at the interface. This sets an upper limit on the strength of adhesion that can be recorded even in quantitative measurement. In addition to this limitation, most quantitative methods require the preparation of special test pieces, often involving very precise machining, which makes them time consuming and hence not readily adaptable to routine control.

Assessment of adhesion is most practical cases is therefore made by qualitative tests. These are based essentially on the development of tensile or shear stresses at the interface by severe local deformation, heating or mechanical working, one or all of which may be involved to varying degrees in individual tests. Such tests are quickly and simply applied and are sufficiently searching to provide the necessary assurance regarding adhesion in relation to satisfactory functioning of the coating in service.

QUALITATIVE TESTS

The literature on adhesion testing offers several excellent reviews of the subject, notably those by Ferguson[1], Polleys[2], and by Davis and Whittaker[3]. In the last mentioned, which deals generallly with the testing of metal coatings on metals irrespective of the method of application, qualitative methods are classified under the main headings of mechanical and non-mechanical, as follows:

Mechanical

1 Bend, twist and wrapping tests

2 Compression
3 Heating and quenching
4 Burnishing, buffing and abrasion
5 Scribing, chiselling and grinding
6 Impact and hammering
7 Cupping and indentation

Non-mechanical

1 Fluorescent
2 X-ray
3 Ultrasonic
4 Cathodic treatment

Not included in this listing is adhesive tape testing which involves a direct tensile pull on the coating.

Of the methods mentioned those most often used in the testing of gold coatings are based on gross deformation as in (1) and (2), heating/quenching, burnishing, scribing and adhesive tape testing. Cutting, grinding and indentation are involved incidentally in other tests on coatings such as thickness measurement by microsectioning and hardness measurement. Poor adhesion may often be detected without formal testing by careful examination of a microsection or by noting lifting or flaking of a coating at the edges or corners of a hardness impression.

MECHANICAL TESTS

Bend Tests

Under this heading may be included any test which involves gross deformation of the substrate. Such tests may be applied either to components themselves when these are expendable in sufficient numbers or to test strips of the same basis metal processed together with each batch. In the former case the nature of the deformation applied will depend to some extent on the shape of the component and may take the form of bending, crushing, twisting, etc.

Forming operations appropriate to specific components often provide convenient and practically significant adhesion tests as, for example, in wire-wrapping or crimping of connectors, and the forming of contact springs from preplated strip.

When a test strip is used, the thickness should be sufficient to produce a significant strain at the interface on bending, but not such that fracture of the test piece occurs too readily. A thickness of 0.5–1 mm is usually suitable. The test may be made in various ways. A fairly sharp bend of 90 or 180° will usually cause exfoliation of a poorly adherent coating at the inside or outside of the bend, either spontaneously or on careful probing. Cracking of the coating is disregarded. The results are generally unequivocal, but there may be some doubt of the interpretation in the case of coatings of unusually high ductility or of deposits on thin section strip when the elongation, and hence the shear strain, produced by bending is relatively small. In such cases, repeated reverse bending through 180° to fracture, as specified in US Military Specification MIL-G-45204, is particularly useful since work hardening will reduce the ductility of the coating and assist separation. An adhesive tape test at the site of the bend or fracture may also be helpful in confirming flaking of the coating.

Bend tests may be applied to plated wire by simple sharp bending or by wrapping the wire around a mandrel of the smallest diameter to permit continuous winding without fracture. The above comments on thin section strip and ductile coatings apply equally to fine wire. Examination at reasonably high magnifications under a binocular microscope may be desirable, assisted by careful probing to detect flaking or lifting of the coating. An adhesive tape test may also be applied after winding the wire on a flat former to offer a suitable surface for testing.

In all adhesion tests, if an intermediate deposit has been applied to the basis metal prior to gold plating, it is essential from the remedial viewpoint to ascertain in cases of failure whether separation has occurred at the gold-undercoat or undercoat-basis metal interface. This can usually be determined by careful inspection of the underside of the detached coating.

Heating Tests

These are widely employed in the testing of plated components in the electronics industry and in watchcase plating. The separation force at the substrate-coating interface is provided partly by the evolution of occluded gas or vaporised liquid from the basis metal, and partly by the shear stress developed as a result of differential expansion of coating and substrate. Failure is indicated by blistering or flaking of the coating.

Temperatures employed range from 120–400°C, longer heating times being used at the lower end of the scale. For example, MIL-G-45204, covering coatings up to 0.015 in. thick, specifies 120–150° for 1 hour, while BS 3315; 1960, on gold plated watchcases calls for 360–400°C for a minimum time of 5 minutes. Since heating may actually improve the adhesion of a coating by interdiffusion with a metallic substrate, short times and higher temperatures would seem to be generally preferable. It is also important that the components should be introduced into the oven or furnace at temperature to ensure the maximum effect of thermal shock. This is less essential in the case of deposits on non-metallic substrates which may themselves be damaged by thermal shock and where the possibility of coating-substrate interdiffusion does not arise. Heating is sometimes followed by quenching, though heating alone is usually sufficient to produce blistering if adhesion is not satisfactory.

In the case of alloy coatings, surface discolouration may occur on heating due to oxidation of base metal constituents. To avoid this it is necessary to heat in an inert atmosphere of argon, nitrogen or cracked ammonia. Alternatively, a hot oil bath may be used. In a procedure described by Marcovitch[4], components are immersed for 2 minutes at 250–260°C, quenched in kerosene and finally degreased in trichloroethylene and examined for blistering or peeling. The technique has been applied on a production basis to gold plating on aluminium and it is reported that parts which passed the test showed no tendency to failure on subsequent deformation. This is an important observation since heating provides one of the few adhesion tests which may be regarded as non-destructive.

Burnishing Tests

Although burnishing is a subjective operation and conditions are not readily definable with precision, this type of test figures commonly in gold plating specifications. A selected area of the coating is rubbed firmly for 15–30 seconds with a burnishing tool of agate or steel which may be hemispherical or chisel shaped, but with rounded edges in the latter case to avoid cutting into the coating. Poor adhesion is indicated by blistering of the deposit. This test is particularly suitable for the examination of thin coatings on reasonably flat substrates. Ferguson[1] suggests that local deformation is a significant mechanism, combined with a fatiguing effect. Local frictional heating also plays an important part.

Scribing Test

In this type of test a pattern of parallel or intersecting lines is made on the plated surface by cutting through to the basis metal with a suitable sharp tool. The coating is inspected for lifting or peeling along the scribe marks by visual observation assisted, if necessary, by careful probing or adhesive tape testing.

Adhesive Tape Test

This is generally useful as a complement to visual examination for lifting or peeling of coatings in other tests. It is also widely applied to coatings on flat surfaces where bend tests are not applicable or where the substrate might be affected by heating, as in the case of printed circuit boards.

In a typical specification[5] the adhesive side of a non-transferable adhesive tape (cellulose regenerated type) with an adhesion value of approximately 26–28 oz/in. width, is applied to the coating under test using a fixed weight roller, care being taken to exclude all air bubbles. Following a 10 second interval the tape is removed by a steady pulling force perpendicular to the surface, after which there should be no evidence of plating adhering to the tape. The test will readily detect gross defects of adhesion and is particularly suitable for the examination of thin ductile deposits which are not readily susceptible to simple bending tests.

NON-MECHANICAL QUALITATIVE TESTS

Hodgson and Szkudlapski[6] have reported the use of an ultrasonic method for checking the adhesion of thin (2 microns) gold plating on sputtered platinum layers on silicon wafers in the course of beam lead processing. The wafer is immersed in a small beaker of deionised water and placed in an ultrasonic bath. Poor adhesion results in a rapid detachment of the gold, as indicated by the appearance of gold flakes in the water after about 1 minute. An adherent coating shows no flaking even after 1 hour's exposure.

Ultrasonic, fluorescent and X-ray methods can also be used to detect a physical gap between coating and substrate. Such techniques have been employed in the examination of silver plate on bearing surfaces, but are not widely used in the gold plating field.

Tests based on cathodic treatment can only be applied to coatings which are permeable to hydrogen and hence are not suitable for gold plating.

QUANTITATIVE METHODS

As indicated in the introductory remarks, quantitative methods are not of great importance in the day-to-day technology of gold plating, but, nevertheless, will be briefly treated for the sake of completeness. They may be broadly classified into four types:

1 Direct pull-off or peel tests
2 Tensile or push-out tests applied to specially prepared test specimens
3 Scratch tests
4 Ultra-centrifuge tests.

Peel Tests

In the first category, the coating is pulled off a known area of the basis metal by means of some kind of grip which may be attached to it by various methods. Soldered attachments have been employed, but soldering is not permissible in the case of gold plating due to the interaction between solder and gold at soldering temperatures (see Chapter 19). Brenner and Morgan[7] applied nodules of nickel or cobalt to the coating by electroforming, and a similar technique has been employed in measuring the adhesion of silver plating to thin copper sheet[8]. There are, however, considerable difficulties in establishing reproducible conditions for the forming of nodules of suitable mechanical strength and in obtaining adequate adhesion to the coating over a well defined area, for which reasons the method appears to be little used at present. The most promising approach probably lies in the use of synthetic adhesives, as studied by Ferguson and Tsao[9], who obtained adhesive strengths of about

6000 psi and considered that this could be increased to 25,000 psi if formation of air bubbles could be avoided during curing. Cold curing adhesives are to be preferred to prevent possible interdiffusion effects on heating, which could affect the bond strength being measured.

Another method of this general type is the peel test originally devised by Jacquet[10] in which a second electrodeposited coating is applied to the coating under test, one zone of the latter having been treated to ensure poor adhesion. The second coating may then be lifted in this area to provide a tongue by means of which a separating force can be applied to a defined area of the coating. Somewhat similar to this is the "can-opener" test developed by Mesle[11].

Tensile Tests

Probably the most widely known test of this type is that due to Ollard[12], in which a heavy deposit of the coating under test is applied to one end of a cylindrical test rod and subsequently machined to form an annular lip through which a separating force can be applied by pushing or pulling through a die. The test has been modified and refined by a number of investigators including Hothersall and Leadbeater[13], Roehl[14] and Knapp[15].

When parallel sided test rods and die-holders are used, separation between the two is very critical. If too tight a fit is experienced, binding will occur, and if too loose, the results will be affected by stress concentrations at the outer edge of deposit and basis metal. Williams and Hammond[16], following Bullough and Gardam[17], avoided this difficulty by using a test rod in the form of a bolt with a tapered spigot which fitted accurately into a cap bored to fit the taper. The spigot end and cap were machined flush and the deposit was applied to the flush surface. The bolt and cap were then pulled apart using a special shackle assembly to ensure axial loading. Gugunishvili[18] used a similar arrangement with several tapered pins fitting into a common base plate.

These methods were developed primarily for evaluating the adhesion of nickel plate to steel and for other applications to base metal systems of more traditional engineering interest, but they could no doubt be adapted to measurements on thick gold plate. One technique which has recently been used in this specific context is a simple shear test initially described by Zmihorski[19]. In this procedure a number of annular rings of the coating under test are plated on to a suitably masked cylindrical rod, from which test specimens are separated on a cutting lathe. The samples are then forced through a die of larger diameter than the rod, but of smaller diameter than the coating, adhesion being measured by the shear force necessary to detach the coating from the rod. This method has been utilised by Dini and Helms[20] in a study of the effect of various pretreatments on the adhesion of gold plating to stainless steel. The total thickness of coating applied to the test cylinder was 0.060 in. (1.5 mm), but the thickness of gold was restricted to 0.003-

0.004 in., just sufficient to provide a diameter greater than that of the die, the coating being built up to the stated total thickness by a thick copper plate. A backing-up deposit could be similarly used in the other tests described.

Scratch Tests

Scratch tests are reported to be particularly useful for studies on thin films on hard, smooth substrates such as glass[21]. A smoothly rounded chrome steel point is drawn across the surface under a gradually increasing load until the coating commences to strip. At this point the shearing force can be calculated from the critical load, the tip radius and the indentation hardness of the substrate.

Ultra-Centrifuge Tests

The ultra-centrifuge test was first applied by Hallworth[22], but has since been developed largely by Beams[23] and co-workers. In this technique the test coating is applied to a small rotor which is spun at high speeds to develop sufficient centrifugal force to cause detachment of the coating. In the latest development of the test the rotor is partly constructed of ferromagnetic material and may be suspended *in vacuo* in a magnetic field and spun by a rotating magnetic field. This method has important advantages in that no attachment to the coating is required and, provided that longitudinal slits are made in the coating to eliminate hoop stresses, the separating force is purely tensile and applied normal to the interface. However, the equipment needed is elaborate and although the test is considered by some authors[3] to be the most reliable and satisfactory of the many that have been proposed for quantitative measurement of adhesion, it is likely to remain of academic interest only for some time to come.

REFERENCES

1 Ferguson A. L., *Monthly Rev. Electropl. Soc.*, **32,** 9, 894; 10, 1006; 12, 1237 (1946); **33,** 1, 45; 6, 620; 7, 760; 12, 1285 (1947).
2 Polleys R. W., *Proc. Am. Electropl. Soc.*, **50,** 54 (1963).
3 Davis D., Whittaker J. A., *Metallurg. Rev.*, **12,** 15 (1967).
4 Marcovith I. W., *Proc. Am. Electropl. Soc.*, **41,** 81 (1954).
5 B.S. 4025: 1966—General Requirements and Methods of Test for Printed Circuits.
6 Hodgson R. W., Szkudlapski A. H., *Plating*, **57,** 693 (1970).
7 Brenner A., Morgan V. D., *Proc. Am. Electropl. Soc.*, **37,** 51 (1950).
8 Schlaupitz H. C., Robertson W. D., *Plating*, **39,** 750 (1952).
9 Ferguson A. L., Tsao M. V., *Plating*, **35,** 724 (1948).
10 Jacquet P. A., *Trans. electrochem. Soc.*, **66,** 393 (1934).
11 Mesle F. C., *Proc. Am. Electropl. Soc.*, **27,** 152 (1939).
12 Ollard E. A., *Trans. Faraday Soc.*, **21,** 81 (1925).
13 Hothersall A. W., Leadbeater C. J., *J. Electropl. Depos. tech. Soc.*, **14,** 207 (1938).
14 Roehl E. J., *Iron Age*, **146,** 17, 26th Sept. (1940): 30, 3rd Oct. (1940).
15 Knapp B. B., *Metal Finish.*, **47, 12,** 42 (1949).

16 Williams C., Hammond R. A. F., *Trans. Inst. Metal Finish.*, **31,** 124 (1954).
17 Bullough W., Gardam G. E., *J. Electropl. Depos. tech. Soc.*, **22,** 169 (1947).
18 Gugunishvili G. G., *Metal Finish.*, **59,** 10, 67 (1961).
19 Zmihorski E., *J. Electropl. Depos. tech. Soc.*, **23,** 203 (1947–8).
20 Dini J. W., Helms J. R., *Plating*, **57,** 906 (1970).
21 Benjamin P., Weaver C., *Proc. R. Soc.*, A, **254,** 163 (1960).
22 Hallworth F. D., *Automotive & Aviation Ind.*, **95,** 2, 30 (1946).
23 Beams J. W., *Proc. Am. Electropl. Soc.*, **43,** 211 (1956).

Chapter 33

ANALYSIS

H. A. HELLER

Accurate analytical information is particularly important in the various phases of gold plating because of the relatively high cost of this metal. It is also very desirable to know whether or not additives are being codeposited in the intended amounts and that, equally important, impurities have not found their way into the electrolyte where they may adversely affect the properties of the deposit and thus defeat the purpose of using a noble metal. These considerations indicate the need for controlling the composition of the plating solution with respect to the concentration of major constituents, additives and inadvertently introduced impurities.

It is the purpose of this chapter to outline the analytical procedures which have been used to obtain the desired information in connection with the electrodeposition of gold. Recognising that there is a wide variation in the equipment available in different plating laboratories, every effort has been made to present as many different analytical approaches to specific problems as possible. In order to avoid a lengthy compilation of step-by-step procedures, however, it was decided to describe the various techniques in rather general terms, giving a few typical examples along with references to the original sources in all cases.

ANALYSIS OF GOLD ELECTROLYTES

GRAVIMETRIC METHODS

Determination of Gold Content

"BOIL-DOWN" PROCEDURES

Most gravimetric procedures for the determination of gold in plating solutions are based on a reduction to the spongy metal followed by careful washing, drying or ignition, and weighing. Because gold compounds are rather readily decomposed, it is possible to recover the gold content as the metal merely by acidifying with concentrated sulphuric acid and heating to copious fumes[1]. This operation must be carried out in an efficient fume chamber because of the very poisonous hydrocyanic acid vapours and corrosive sulphuric acid fumes which are evolved. Depending on the composition of the electrolyte, it may be difficult to eliminate all of the organic matter by this method. For this reason, most procedures call for the addition of an oxidising agent. Grassby, Gill and Bradford[2] add nitric acid along with the sulphuric acid. After the initial fuming, the solution is allowed to cool and a further addition of nitric acid is made after which the solution is again brought to fumes to complete the elimination of organic matter. To ensure the complete precipitation of gold, they then add hydrogen peroxide to the cooled and diluted solution after which it is again brought to fumes.

Another method of eliminating organic matter involves the dropwise addition of concentrated nitric acid after the sulphuric acid has reached fuming temperature and the organic matter is partially charred[3]. This would be less time consuming than the previous procedure although it is stressed that digestion and fuming must continue for a minimum of 15–20 minutes.

Miller[4] advocates the use of a perchloric acid-nitric acid mixture (1 : 1) to completely oxidise the organic constituents of any type of plating solution in a short time. After cooling and dilution, hydrogen peroxide is added to ensure complete reduction of the gold. If tin or antimony are present, a small amount of hydrofluoric acid is added to prevent contamination of the precipitated gold. Because of the potentially hazardous nature of hot perchloric acid and its fumes, this procedure should not be undertaken without appropriate safety precautions[5].

ALUMINUM REDUCTION

A rapid "control" method for the determination of gold in cyanide plating solutions is given by Rochat[6]. This procedure avoids the time consuming "boil-down" stage although a high degree of accuracy is not claimed. Aluminum foil is used to precipitate the gold after the sample has been made strongly alkaline with potassium hydroxide. The excess aluminum is soluble in the caustic solution. The precipitated gold is treated with hot dilute nitric acid to remove base metals such as copper and nickel.

HYPOPHOSPHOROUS ACID REDUCTION

Another gravimetric method in which gold is precipitated by an active reducing agent is advocated by Lemons[7] who uses hypophosphorous acid as the reductant in strong hydrochloric acid solution. Mercuric chloride is used to catalyse the reaction and to collect the gold as an amalgam. The mercury is removed from the gold by the drop-wise addition of concentrated nitric acid. After washing, the residue is ignited repeatedly until a constant weight is obtained. The author states that if interfering metals are known to be absent, the nitric acid treatment may be omitted and the amalgam fired directly in an efficient fume chamber.

ELECTROGRAVIMETRIC DETERMINATION

An electrogravimetric procedure is one in which a constituent to be determined is deposited on a weighed electrode by the process of electrolysis. The desired constituent is ordinarily deposited as the metal on a platinum cathode although lead may be determined through the deposition of lead peroxide at the anode.

Electrogravimetric methods are versatile in that a number of separations are possible by using the controlled cathode potential technique. Detailed discussions of these methods have been published by Sand[8] and later by Lingane[9].

The major advantage of the electrogravimetric technique over the "boil-down" procedures is that the time consuming fuming, filtration and ignition steps are avoided. In the version described by Langford and Parker[10] a small sample, the size depending upon the gold content, is diluted to approximately 120 ml and an excess of sodium cyanide is added. The solution is electrolysed at a current of 2–3 amps using a platinum anode and a weighed platinum gauze cathode. Any copper which might be present plates out only after all of the gold has been deposited. The copper is easily removed by rinsing the cathode in concentrated nitric acid before washing, drying and final weighing. The deposited metal may be removed from the cathode by immersing it in a sodium cyanide solution to which a few drops of a 30% solution of hydrogen peroxide have been added.

THIOGLYCOLLIC ACID REDUCTION

One of a number of organic reducing agents which can be used for the determination of gold in the lower concentration limits of approximately 6–60 mg is 2-mercaptoacetic (thioglycollic) acid. In the procedure reported by Mukherji[11], the precipitation is carried out in 6 *N* hydrochloric acid. Among the advantages claimed are a favourable weight factor, easy filtration of the precipitate and ready availability of the reagent.

Determination of Metallic Additives and Impurities

After gold has been removed by one of the "boil-down" procedures, minocationic constituents can be determined in the filtrate by any gravimetric procedure which is suitable for a given element at relatively low concentrations. Some representative gravimetric methods for the more common additives and impurities are summarised in the following text. It should, however, be understood that certain elements may be determined with greater accuracy and convenience by a more appropriate technique such as colorimetric, volumetric and so forth.

SILVER

Langford and Parker[10] give a systematic scheme of analysis for the common additive elements including silver. After gold has been removed, silver is precipitated as the chloride by the addition of sodium chloride to the filtrate. The precipitate may be weighed or it may be dissolved in an excess of sodium cyanide and then electrodeposited as the metal on to a weighed platinum cathode.

COPPER

The filtrate from the silver determination may be treated with hydrogen sulphide to recover any copper as copper sulphide. This precipitate may be ignited and weighed as copper oxide (CuO). If silver is absent, the hydrogen sulphide precipitation may be avoided by adding ammonium nitrate and sulphuric acid to the filtrate from the gold separation and subsequently electrodepositing the copper on to a weighed platinum cathode.

IRON

The filtrate from the sulphide precipitation of copper is boiled to expel hydrogen sulphide. Nitric acid is then added to oxidise the iron which is finally precipitated as the hydroxide by the addition of ammonium hydroxide. The precipitate is filtered, ignited and weighed as ferric oxide (Fe_2O_3).

NICKEL

If nickel is to be determined, dimethylglyoxime is added to the filtrate from the iron determination. The red precipitate is collected by suction filtration on a weighed, sintered glass crucible.

ZINC

The alkaline filtrate from the nickel filtration is treated with hydrogen sulphide to precipitate zinc as the sulphide. This may be collected by filtration, ignited and weighed as zinc oxide (ZnO).

COBALT

This ion may be precipitated from a dilute hydrochloric acid solution by the addition of α-nitroso-β-naphthol, although any copper, silver or iron must first be removed. It is reported that nickel, zinc and phosphate do not interfere in this precipitation[12].

INDIUM

After the removal of gold, indium may be precipitated by the drop-wise addition of ammonium hydroxide until the solution is nearly neutral[1]. After the addition of 5 g of ammonium sulphate, more ammonium hydroxide is added until all of the indium is precipitated. An excess should be avoided in order to prevent the indium hydroxide from redissolving. The precipitate is filtered, washed, ignited and weighed as indium oxide (In_2O_3). Iron and copper, if present, should be removed by a cupferron separation[13] prior to the indium precipitation.

Determination of Non-metallic Constituents

PHOSPHATE

Langford and Parker[10] use the classical pyrophosphate method for the determination of the phosphate ion, gold having first been removed by one of the "boil-down" procedures previously mentioned. To the filtrate a small amount of citric acid and 25 ml of "Magnesia Mixture"* are added. After making the solution strongly ammoniacal, it is allowed to stand overnight. The phosphate precipitates as magnesium ammonium phosphate which is subsequently collected by filtration, ignited and weighed as magnesium pyrophosphate ($Mg_2P_2O_7$). The purpose of the citric acid is to complex any iron and thus prevent its precipitation.

SULPHATE

This procedure was devised[14] to permit the determination of the sulphate ion, as barium sulphate, in gold plating formulations containing sodium thiosulphate. This method usually calls for precipitation from a hot, dilute hydrochloric acid solution which would, however, result in the decomposition of the sodium thiosulphate and possibly the precipitation of other constituents of the plating solution. It was found that a sufficiently accurate precipitation of the barium sulphate could be made in a cold, dilute solution which had been acidified with citric acid.

A 20 ml sample of the plating solution is pipetted into an 800 ml beaker and diluted to 500 ml with demineralised water. After the addition of 20 ml of a 10% solution of citric acid, 20 ml of a 5% barium chloride solution are slowly added with constant stirring. The solution is allowed to stand overnight. The precipitate is then collected by suction filtration after carefully decanting

---

* Magnesia Mixture—Dissolve 25 g of magnesium chloride and 50 g of ammonium chloride in 250 ml of water. Add a slight excess of ammonium hydroxide and allow to stand overnight. Filter off any precipitate which forms, make the solution just acid with dilute hydrochloric acid and then add 2 ml of the concentrated acid in excess. Dilute to a final volume of 500 ml and store in a polyethylene bottle.

most of the clear supernatant liquid. The precipitate is washed first with cold water and then with dilute hydrochloric acid. After a final washing with water, the precipitate is ignited and weighed as barium sulphate ($BaSO_4$).

VOLUMETRIC METHODS

Determination of Gold Content

IODOMETRIC METHOD

The most common volumetric method for the determination of gold is based on the liberation of iodine from potassium iodide by auric chloride (Au^{3+}). The reaction is generally considered to be as follows:

$$AuCl_3 + 3KI \rightarrow 3KCl + AuI + I_2$$

It is first necessary to destroy cyanides by the addition of hydrochloric acid in an efficient fume chamber. The solution is then carefully evaporated to a pasty mass, but not to dryness, to avoid any decomposition of the gold chloride. After cooling and dilution, potassium iodide is added and the liberated iodine is titrated with 0.01 *N* sodium thiosulphate using a few drops of starch solution as an indicator. Langford and Parker[10], and Hall[15] give the details of the basic procedure. The presence of Cu^{2+} or Fe^{3+} will lead to high results because they both liberate iodine from potassium iodide in hydrochloric acid solution.

ASCORBIC ACID METHOD

An indirect titrimetric method involves the addition of a measured excess of ascorbic acid[16] which reduces the Au^{3+} as follows:

$$2Au^{+++} + 3C_6H_8O_6 \rightarrow 2Au + 3C_6H_6O_6 + 6H^+$$

The excess ascorbic acid is back-titrated with a standard iodine solution. It is claimed that Cu^{2+} does not interfere.

POTASSIUM CYANIDE METHOD

Another titrimetric method which may be used is the potassium cyanide method[17] in which the precipitated gold is dissolved in an excess of standard potassium cyanide solution. The excess cyanide is then back-titrated with 0.01 *N* silver nitrate using potassium iodide as the indicator.

Determination of Metallic Additives and Impurities

Volumetric methods are usually employed for the determination of major constituents. For this reason they are not often used for the determination of brighteners and base metal impurities in high carat gold plating solutions.

Armet[1], however, gives a scheme of analysis for the rapid volumetric determination of silver, copper, nickel and free cyanide, as well as for the gold content, in alloy gold solutions.

Determination of Non-metallic Constituents

FREE CYANIDE

This is usually determined by titrating a diluted sample with 0.1 *N* silver nitrate, using potassium iodide as an indicator. The appearance of turbidity, due to silver iodide, indicates the end point.[2,10,15]

CARBONATE

The procedure most often used is based on the precipitation of barium carbonate by the addition of barium chloride. The washed precipitate and filter paper are macerated and the carbonate titrated with 1 *N* hydrochloric or sulphuric acid using methyl orange to indicate the end point.[10,15,18] A somewhat more accurate titration will result if an excess of standard acid is first added to react with all of the carbonate. The excess acid is then back-titrated with 1 *N* sodium carbonate.

It is to be noted that the sulphite ion, if present, will also react with barium chloride to give a precipitate which will consume acid. A correction is possible, however[10], after making a separate determination of the sulphite by a method such as that given below.

SULPHITE

The sulphite ion may be determined by precipitating it with barium chloride and then titrating the washed precipitate with a 0.1 *N* iodine solution[10]. Carbonates and sulphates will also be precipitated but these do not interfere with the iodometric titration.

THIOSULPHATE

This procedure was devised to permit the determination of thiosulphate in gold plating formulations containing sodium thiosulphate as an additive[14]. It is a straightforward iodometric determination involving the addition of an excess of standard iodine solution followed by a back-titration with sodium thiosulphate using starch as an indicator. One or more corrections may be necessary, however, to compensate for substances other than thiosulphate which consume iodine. The first of these is the cyanide ion. Thiosulphate is not ordinarily used in an alkaline cyanide type of plating solution although the gold is usually present as potassium gold cyanide even in the acid types of electrolytes. The correction, which should be determined experimentally for each type of plating solution, will depend upon the potassium gold cyanide content. Sulphite, if present, will also consume iodine. This correction may be based on a separate determination of the sulphite content by a procedure such as that just outlined in the preceding paragraph.

CITRATE

Since the introduction of the acid gold electrolytes, the determination of the citrate ion is sometimes required. Natarajan and Lalitha[19] have developed a method in which the citrates are complexed by the addition of an excess of Cu^{2+} as cupric sulphate. The excess Cu^{2+} is back-titrated with approximately 0.01 M sodium citrate using murexide as an indicator.

Silverman, et al.[20], report an oxidation method in which the citrate reacts with an excess of potassium iodate. After the addition of 30 ml of concentrated sulphuric acid the solution is heated to fumes of sulphur trioxide. The time of heating is important, this preferably being 13 minutes, but not exceeding 15 minutes. Carefully restricted heating time is to prevent decomposition of the iodate by the hot sulphuric acid. The cooled solution is diluted and heated until all free iodine is evolved. After cooling, 10 g of potassium iodide is added. This reacts with the excess iodate to liberate iodine which is then titrated with standard thiosulphate. The authors cite earlier work by Unger and Haynes[21] upon which the present method is based. Other studies which probably could be adapted to the determination of citrates in gold plating solutions are (*a*) the use of ceric sulphate as a volumetric oxidant by Willard and Young[22] (*b*) a study of cerate, periodate and perchloratocerate as oxidants for the determination of various organic compounds by Smith and Duke[23] who conclude that cerate is superior to periodate for this purpose (*c*) the use of potassium permanganate in acid medium as an oxidant as studied by Berka and Hilgard[24].

PHOSPHATE

Foulke and Crane[25] advocate, as a rapid control procedure, the precipitation of phosphate as ammonium phosphomolybdate. The washed precipitate is dissolved in an excess of standard sodium hydroxide solution followed by a back-titration with hydrochloric acid using phenolphthalein as an indicator. The yellow precipitate is of uncertain and variable composition, however, so that very accurate results are not to be expected by this method.

COLORIMETRIC METHODS

A colorimetric method is based on the relationship between the concentration of a substance in a solution and the depth or intensity of a characteristic colour associated with that substance. The colour may be that of the substance itself or, more often, one developed by the addition of a suitable reagent. This relationship provides a means for determining concentration if a series of standards are available for comparison. The simplest method of evaluating the intensity of the colour consists of visually matching it with that of a standard although more accurate results are obtainable by using a photoelectric device to measure quantitatively the intensity of a beam of light after it has passed through a definite depth of solution. The basic principles of these techniques are presented in detail by Snell and Snell[26] and by Sandell[27].

The term "colorimetry" usually implies measurement of the average intensity of a band of wavelengths in the visible region with or without the aid of optical filters for improved selectivity. It is thus distinguishable from spectrophotometry in which a dispersing element such as a prism or grating enables the measurement of transmitted light at essentially a single wavelength, not necessarily in the visible region. Colorimetric procedures are readily adaptable to spectrophotometry although the reverse may not be true.

These methods are most useful at relatively low concentrations. In the case of a major constituent such as gold in gold plating solutions, it is necessary to make a rather large dilution which is not desirable where high accuracy is required.

Determination of Gold Content

WITHOUT COLOUR REAGENT

Baldwin[28] gives a control procedure for gold in plating solutions based on the use of a photoelectric colorimeter with a series of colour filters. The gold is removed from a 5 ml sample by fuming with sulphuric acid and hydrogen peroxide. The precipitated metal is dissolved in aqua regia and ultimately diluted to 100 ml from which a 10 ml aliquot is taken for colorimetric determination using a blue filter. No colour reagent is used.

O-TOLUIDINE

For the determination of gold in alloy plating electrolytes Nell[18] uses o-toluidine as a colour reagent after a series of dilutions amounting to a net dilution of 1 : 500. He uses a photoelectric colorimeter with a purple filter (440 nm)* or a blue filter (425 nm).

STANNOUS CHLORIDE

A very old colour test for gold, said to be capable of good quantitative accuracy under controlled conditions, is the stannous chloride method[12,29] which produces a colour known as "Purple of Cassius". The test solution must be free of all reducible metals such as mercury, selenium, platinum, palladium, ruthenium and metals which form insoluble chlorides or coloured

---

* nm, nanometer (10^{-9} meter), formerly millimicron.

reduction products with stannous chloride. The test solution should contain no more than 0.5 mg of gold in 0.1 *N* hydrochloric acid. The colour reagent is a 10% solution of stannous chloride in 2 *N* hydrochloric acid. The colour develops immediately.

STANNOUS BROMIDE

Pantani and Piccardi[30] use a 0.5 *M* stannous bromide solution in approximately 1 *N* hydrobromic acid to produce a stable violet colour in the presence of 0.05% gelatin. The transmittance is measured at 540 nm. Beer's law holds in the concentration range from 5–50 ppm of gold.

BROMOAURATE

For gold concentrations in the range from 0.025–2 mg, the bromoaurate method of McBryde and Yoe[31] may be used. To the sample, which must be freed of oxides of nitrogen, 3 ml of concentrated hydrochloric acid and 5 ml of colourless hydrobromic acid are added. After dilution to 50 ml, the transmittance at 380 nm is measured. Interference from traces of iron may be avoided by adding a few drops of phosphoric acid.

MISCELLANEOUS METHODS

For gold concentrations in the range from 7–40 ppm the tetraphenylarsonium chloride method of Murphy and Affsprung[32] may be used. Other colorimetric methods which could have an application in the determination of low concentrations of gold would include the rhodamine B method of McNulty and Woolard[33], the improved o-tolidine* method of Clabaugh[34] and the rhodamine method of Cotton and Woolf[35].

Determination of Metallic Additives and Impurities

COPPER

For the colorimetric determination of copper, Grassby, Gill and Bradford[2] used as a colour reagent bis-cyclohexanone oxaldihydrazone as a 0.5% w/v solution in alcohol. After the removal of gold and silver, an aliquot was taken to contain 0.05 to 0.2 mg of copper. Ammonium citrate was added and the solution neutralised with sodium hydroxide. Ten minutes after adding the reagent the transmittance was measured at 600 nm. The anions and cations normally present in gold plating solutions do not interfere.

---

* It should be noted that o-tolidine exhibits carcinogenic properties and should be handled with care.

In a rapid control method, Nell[18] develops the blue copper-ammonia complex. The optical density is read at approximately 650 nm in an electrophotometer using a red filter.

NICKEL

The preferred reagent for the colorimetric determination of nickel appears to be dimethylglyoxime[2,18,28] which is added to an alkaline sample solution after it has been treated with a strong oxidising agent, usually bromine water. Sandell[27] recommends a minimum of ammonium hydroxide, a liberal excess of dimethylglyoxime and no more oxidising agent than necessary to avoid destruction of the dimethylglyoxime. Sodium hydroxide is used to raise the pH to 11, above which a stable form of the nickel complex is obtained. The colour may be read on a photoelectric colorimeter with a green or blue filter.

COBALT

For the colorimetric determination of cobalt in gold plating solutions, Baldwin[28] uses potassium ferricyanide which is added to an ammoniacal preparation of the sample. A reddish-brown colour develops which is read on a colorimeter using a green filter. Nickel does not interfere. Grassby, Gill and Bradford[2] use nitroso-R-salt as a colour reagent in a hot acetate solution of the sample. After cooling and dilution, the transmittance is read at 520 nm. The amount of cobalt which can be determined by this method is from 0.01–0.2 mg.

ZINC

Dithizone is reported to be a suitable reagent for the determination of zinc after the removal of gold[2]. A sample is taken to contain from 0.01–0.025 mg of zinc. This is acidified with hydrochloric acid and any copper is removed by extracting with 5 ml portions of a 0.01 % w/v solution of dithizone in carbon tetrachloride. After neutralising with ammonium hydroxide, any nickel is complexed with potassium cyanide. The solution is buffered to pH 6–7 with sodium acetate/acetic acid. Sodium thiosulphate is then added to prevent interference from silver and certain other metals. The zinc is extracted with 5 ml portions of dithizone in carbon tetrachloride, the excess dithizone being removed by washing with sodium sulphide solution. After volumetric dilution with carbon tetrachloride, the transmittance is measured at 530 nm. It is reported that this method can be used in the presence of any amounts of copper, nickel, cadmium, cobalt or indium.

SILVER

It appears that the best colorimetric procedure for silver is the "mixed colour" dithizone method[27,28] in which a carbon tetrachloride solution of copper dithizonate is used as a reagent. Increasing amounts of silver change the deep red colour of the reagent through various shades of orange to yellow because the silver dithizonate is more stable than the copper complex. Care must be taken that the capacity of the reagent is not exceeded by large amounts of silver. Copper must not be present in the sample solution. The reagent must be freshly prepared since it is not stable for more than a day. The colour is measured in a photoelectric colorimeter using a No. 59 (yellow) filter.

INDIUM

The fact that indium hydroxyquinolate has a strong yellow colour in chloroform solution[36] provides the basis for a colorimetric determination. Grassby, Gill and Bradford[2] give a method for determining indium in gold plating electrolytes using 8-hydroxyquinoline. The transmittance of the chloroform phase is measured at 400 nm.

METALLIC CONTAMINANTS IN EFFLUENTS

Stevens, Fischer and MacArthur[37] give colorimetric methods for the determination of most of the metallic contaminants which might be encountered in plant effluent waters from electroplating operations.

Determination of Non-metallic Constituents

PHOSPHATE

Baldwin[28] gives a colorimetric method for phosphates in gold plating solutions based on the formation of a deep blue colour when ammonium molybdate followed by hydroquinone are added to a diluted aliquot of the sample. The colour is measured with a photoelectric colorimeter using a No. 66 (red) filter. The concentration range covered is from 0–112 g/l. Nell[18] gives a similar procedure except that he adds a 1-amino-2-naphthol-4-sulphonic acid reagent to the sample after adding the molybdate reagent and sulphuric acid.

CYANIDE

Colorimetric methods are not ordinarily used for the determination of cyanide at the high concentrations encountered in plating solutions although they are useful for checking the cyanide content of plant effluents.

Stevens, Fischer and MacArthur[37] give a method based on the reaction of an *n*-butyl or *n*-amyl alcohol extract of the sample with a pyridine-benzidine reagent to give an orange colour when cyanides are present. It can be compared visually with those of a series of standards or it can be measured colorimetrically at 480 nm. The range covered is from 0.04–0.35 ppm of CN^- in a 10 ml sample.

SPECTROPHOTOMETRIC METHODS

Spectrophotometry may be regarded as a refined form of colorimetry in which monochromatic, rather than polychromatic, light is passed through the sample. Measurement of the transmitted light is always by a photoelectric detector rather than by visual means. Among other advantages, this permits the use of wavelengths beyond the visible range, extending usually to 200 nm in the ultra-violet. The essential parts of a spectrophotometer include a source of radiation such as an incandescent lamp for the visible region and a hydrogen discharge tube for the ultra-violet, a monochromator which isusually a quartz prism or a ruled grating and interchangeable photoelectric detectors to provide complete coverage of the near ultra-violet, visible and near infra-red regions. Sample cells with windows essentially transparent to the wavelengths of interest are positioned immediately in front of the detector enclosure.

The relationships between the several variables involved in spectrophotometry are expressed by the Beer-Lambert-Bouguer law:

$$\log (I_0/I) = abc$$

where I_0 = the intensity of the light after passing through a cell filled with solvent only, i.e., a "blank"

I = the intensity of the light after passing through a cell filled with a sample preparation

a = a constant usually called the "Extinction Coefficient"

b = the path length of the light through the sample

c = the concentration of the absorbing material in the sample.

The two terms commonly used in analytical applications of spectrophotometry are "transmittance" and "absorbance" which are defined as follows:

$$\text{Transmittance} = T = I/I_0$$

$$\text{Absorbance} = A = \log (1/T) = \log (I_0/I) = abc$$

In setting up a spectrophotometric procedure it is customary to draw a curve relating absorbance, at a specified wavelength, to concentration on the basis of measurements made on a series of standards which have been treated in exactly the same manner as the unknown samples to be prepared. Such a curve is referred to as a "working curve" or an "analytical curve". More detailed presentations of the principles of spectrophotometry are given by Harley and Wiberly[38], by Mellon[39] and by Schilt and Jaselskis[40].

Determination of Gold Content

WITHOUT COLOUR REAGENT

Spectrophotometric methods, like colorimetric methods, are most useful in the determination of constituents at relatively low concentrations. This technique, therefore, is not often used for the determination of gold in plating solutions although it is very useful for analysing rinse solutions. Conrad and Kenna[41], however, report the use of a spectrophotometric method for gold in plating solutions. Nickel and cobalt do not interfere at their ordinary concentrations, but when the nickel content is greater than 0.8 g/l it is separated in an ion exchange column. The cobalt complex, however, cannot be separated in this manner so that if much cobalt is present the gold must be determined gravimetrically. A 3 ml sample is diluted to 2000 ml and the transmittance is immediately read at 239 nm without the addition of a colour reagent. A gold determination can be made by this method in 5 minutes if the nickel content is less than 0.8 g/l. A mean error of approximately 0.014 g/l at the 9 g/l level is reported. The concentration range covered is from 0–35 g/l.

TETRAPHENYLARSONIUM CHLORIDE

Murphy and Affsprung[32] report a spectrophotometric method for gold in the concentration range from 7–40 ppm. Tetraphenylarsonium chloride is used to precipitate the gold chloride complex which is then extracted with chloroform. Traces of nitrates do not interfere. The transmittance of the chloroform phase is read at 323 nm.

Determination of Metallic Additives and Impurities

NICKEL

Doherty[42] gives a spectrophotometric method for the determination of nickel in gold plating solutions. The gold is first removed by fuming with sulphuric acid, adding sodium sulphite and fuming again. To the diluted sample the following are added: 2 ml of 1 : 1 sulphuric acid, 5 ml of 20% citric acid solution, 5 ml of bromine water, 3 ml in excess of the amount of ammonium hydroxide needed to decolourise the solution and 4 ml of a 1% solution of dimethylglyoxime in alcohol. The transmittance is read at 530 nm and the nickel content is determined from a previously prepared calibration curve.

OTHER METALLIC ELEMENTS

Any of the colorimetric methods described in the preceding sections are readily adapted to spectrophotometric use.

POLAROGRAPHIC METHODS

Polarography is a branch of electrochemistry which involves the measurement of the current-voltage relationships of a microelectrode in a supporting electrolyte containing relatively low concentrations of one or more ionic species whose identity and/or concentrations are to be determined. Polarography is thus a type of voltammetry in which the indicator electrode takes the form of mercury continuously dropping from a very fine glass capillary tube. The other electrode consists of a pool of mercury in the bottom of the cell. When the potential applied to the dropping mercury electrode is increased at a uniform rate, discrete increases in the current through the cell will occur as the applied voltage approaches and passes through the potential at which the various ions present are either oxidised or reduced. Modern instruments employ a strip-chart recorder on which is recorded the resulting "polarogram". The chart movement (horizontal axis) is synchronised with the applied potential while the pen deflection (vertical axis) indicates the cell current. The mid-point of a polarographic wave indicates the "half-wave" potential of the responsible ion while the amplitude of the wave is proportional to the concentration of the ion. The concentrations of the ions to be determined are usually from 10^{-5}–10^{-2} molar in the supporting electrolyte. Two very good sources of information on the principles and applications of polarography are Kolthoff and Lingane[43] and Meites[44].

Determination of Gold Content

Polarography is not a preferred method for determining the gold content of plating solutions. This is probably because of the large dilution required to reduce the gold concentration to approximately 0.005 molar. When polarography is used to determine minor constituents in plating solutions it is customary to first separate the gold by heating the sample to fumes with concentrated nitric and sulphuric acids. Knotowicz and Tatoian[45], however, give procedures for the determination of gold in a 1 *N* potassium cyanide supporting electrolyte. The gold wave coincides with that of nickel, however, so that a correction must be made if nickel is present.

Determination of Metallic Additives and Impurities

ALLOYING ELEMENTS

Grassby, Gill and Bradford[2] give a series of polarographic procedures for alloying elements in gold plating solutions. These procedures were developed to avoid mutual interferences which would occur if a single supporting electrolyte were to be used for all elements. For example, copper, in the presence of nickel, zinc, indium and cadmium, is determined in an ammoniacal ammonium chloride electrolyte while copper in the presence of cobalt requires a potassium chloride electrolyte. Conditions are also given for the determination of nickel in the presence of copper, cadmium, indium and zinc. These authors also give procedures for cobalt in the presence of copper, zinc, cadmium and indium with nickel absent; for zinc in the presence of copper, zinc, cadmium and indium with nickel absent; for zinc in the presence of copper, indium, cadmium and nickel; for zinc in the presence of cobalt, nickel, indium and cadmium; for indium in the presence of copper, nickel, cobalt and zinc; for cadmium in the presence of copper, nickel, zinc, indium and cobalt.

Knotowicz and Tatoian[45] investigated the determination of silver, copper, nickel and zinc in four different supporting electrolytes, viz., in both 1 *N* ammonium hydroxide and ammonium sulphate, 1 *N* potassium cyanide, 6.8 *N* calcium chloride/5 *N* sodium chloride, and 5 *N* sodium chloride. Craft and Schumpelt[46] give polarographic methods for the determination of nickel, cobalt and/or indium in acid gold alloy plating solutions after the removal of gold by fuming with nitric and sulphuric acids.

EMISSION SPECTROGRAPHIC METHODS

Optical emission spectroscopy is a particularly versatile analytical tool which is readily adaptable to the detection and quantitative determination of metallic additives and impurities in gold plating solutions.

The general term "spectroscopy" implies the measurement of radiant energy in any region of the electromagnetic spectrum. One of the regions which is of particular value in analytical chemistry is that which extends from the near ultra-violet through the visible and into the near infra-red, i.e., from approximately 2,000–10,000 angstroms. This is the region which is generally designated the "optical" region because it includes the visible wavelengths from about 4,000–7,000 angstroms.

Any form of "emission" spectroscopy involves three basic processes, viz., excitation, dispersion and detection. In the excitation process, energy is supplied to the atoms comprising the sample material, causing them to emit very definite characteristic wavelengths. The array of wavelengths characteristic of a chemical element therefore serve to indicate the presence of that element while the intensities of the characteristic emissions provide a means for determining the concentration of the element in the sample.

In the optical region, the exciting energy is usually in the form of an electric arc, a high voltage spark or a high temperature flame.

The emitted wavelengths are separated or dispersed by a quartz prism or a ruled grating. The individual wavelengths are often referred to as "lines" because of their appearance in an eyepiece or on a photographic plate.

The separated wavelengths are usually recorded, or "detected", on a photographic plate although some instruments employ a number of "photomultiplier" tubes, each positioned to intercept a selected wavelength characteristic of a particular element.

Optical emission spectroscopy is characterised by its ability to detect very low concentrations of most of the chemical elements. It is therefore most useful for the detection and determination of minor and trace constituents in gold electrolytes although it is not recommended for the quantitative determination of gold when it is the major constituent.

The basic concepts and operations of spectroscopy have been presented in a number of texts[47,48,49].

Determination of Metallic Additives and Impurities

SOLUTION EXCITATION TECHNIQUE

Optical emission spectroscopy originally involved the excitation of sample materials in the solid state exclusively, e.g., powders, metal surfaces and so forth. In 1935 Duffendack, Wiley and Owens[50] reported the direct excitation of a conductive solution by maintaining a high voltage spark discharge between a pointed graphite electrode and the surface of the liquid sample preparation. Gradually, other methods for the direct excitation of liquids were developed. Five of the most prominent of these techniques were evaluated by Baer and Hodge[51].

The determination of minor and trace constituents in gold plating solutions by liquid excitation is rapid and convenient in that little sample preparation is required and quantitative standards are readily prepared from reagent chemicals. The most widely used liquid excitation technique appears to be that which is based on the use of a "rotating disc" apparatus. A spectroscopically pure graphite disc, usually 0.5 inch in diameter and 0.125 inch thick, is mounted on a slowly rotating horizontal shaft. The lower portion of the disc is immersed in the liquid sample which is usually contained in a small porcelain combustion boat. A pointed graphite electrode is positioned a few millimeters above the top of the disc to form the analytical gap.

McLain and Wade[52] report the use of the rotating disc technique in the determination of cobalt, nickel, aluminium and boron in commercial acid gold baths. Palladium, as the cyanide, is added as an internal standard. Analytical curves are based on intensity ratios obtained by running standard solutions covering the range from 2.5–20 ppm for aluminium and boron, and from 20–140 ppm for cobalt and nickel. Excess potassium citrate is added to overcome the effects of varying concentrations of potassium and citrate ions in plating solutions of unknown composition.

A similar procedure for the determination of nickel, cobalt, indium and silver in gold electrolytes of any type has been described by the writer[53]. In this method the added internal standard is copper. Three parts, by volume, of a potassium citrate solution (250 g/l) are added as a matrix diluent to one part of sample. The concentration range for silver is from 0.01–0.2 g/l and for nickel, cobalt and indium it is from approximately 0.04–1.0 g/l. In order to obtain measurable densities for the analytical lines of all four elements in a single exposure, a neutral-density step filter is used in front of the slit of the spectrograph. The nickel and cobalt lines are "densitometered" at the unattenuated steps while the densities of the intense indium, silver and copper lines are measured at line segments corresponding to the 18% transmittance step of the neutral density filter.

Much greater sensitivities may be obtained if required by first removing the gold and organic anions by fuming with sulphuric acid. This technique is very useful when conducting general surveys for the purpose of identifying unknown contaminating elements.

ATOMIC ABSORPTION METHODS

The atomic absorption technique is a form of spectroscopy in which the concentrations of selected elements in a sample are determined by measuring the amount of characteristic radiation they absorb from a light beam of known initial intensity. This is in contrast to emission spectroscopy in which the characteristic radiations emitted by the sample elements are used as a measure of their concentrations.

The sample must be in the form of a solution which is sprayed into a fuel-gas mixture in a special burner which produces a long, ribbon-like flame. A beam of light which includes wavelengths characteristic of the element(s) to be determined is directed through the length of the flame which now contains atoms from the sample which are in a condition to absorb energy from the light beam at their characteristic wavelengths. The light source is a special "hollow-cathode" gas-discharge lamp which emits wavelengths characteristic of a particular element or group of elements. The light leaving the flame, now diminished by any "atomic absorption" which has taken place, enters a monochromator which is used to pass a particular wavelength, rejecting all others, The selected wavelength then enters a photodetector which is used to compare the intensity of the attenuated beam with the intensity observed when the absorbing atomic species is not present in the sample solution.

As an analytical technique, atomic absorption is highly selective and sensitive because a light source is used which emits only wavelengths characteristic of a single element or, in the case of special sources, from two to as many as six selected elements. Like optical emission spectroscopy, atomic absorption is most useful in the determination of minor and trace constituents.

Although the phenomenon of absorption of characteristic radiation by atoms has been known for some time, it was apparently first investigated and proposed as an analytical tool by Walsh[54] in 1955. Since that time other authors have discussed light sources[55], absorption wavelengths and detection limits[56], and the general application of atomic absorption to chemical analysis[57,58]. A comprehensive textbook on the subject of atomic absorption is by Elwell and Gidley[59].

Determination of Gold Content

Although atomic absorption would not be the method of choice for determining the relatively high concentrations of gold which are present in plating solutions, it has been used to a considerable extent for determining the gold content of rinse and waste solutions.

Mikhailova *et al*[60], determine gold directly in electrolytes in the concentration range from 0.01–2 g/l using the Au2428A line after diluting the original samples 1 : 25 or 1 : 100 to minimise interference. They use an air-acetylene fuel mixture and report coefficients of variation from 1–4%.

For the analysis of solutions containing gold at much lower concentrations, the final atomic absorption determination is often preceded by a pre-concentration step involving extraction of the gold into a suitable organic solvent. Strelow *et al*[61], for example, determine gold in the range from 0.01–0.5 ppm in cyanide waste solutions by extracting into methyl isobutyl ketone.

Groenewald[62] determines gold (I) in concentrations between 2.5×10^{-4} M–5×10^{-8} M by extracting into di-isobutyl ketone containing a quaternary ammonium salt. This large concentration range is covered by varying the ratio of the aqueous to organic phases from 1 : 1 to 100 : 1.

Yudelevitch *et al*[63], achieved 100% extraction of gold from a hydrochloric acid solution into dibutyl sulphide in benzene. Of the 23 elements studied as possible interferences, only silver was extracted to an appreciable extent. A sensitivity of 0.0015 micrograms/ml of gold was reported.

In all of the three preceding methods, the organic phase is aspirated directly into the burner of the atomic absorption apparatus.

Determination of Metallic Additives and Impurities

Atomic absorption spectrophotometry is readily adapted to the determination of minor and trace constituents in gold electrolytes.

Whittington and Willis[64] give a procedure for the determination of trace elements in electroplating solutions. Their preliminary investigations included work with copper, cadmium, silver, zinc, chromium and nickel plating baths in which copper, nickel, and zinc were the impurity elements of major concern. They give the details of a procedure for the determination of copper and zinc in a Watts-type nickel plating solution. The same principles would be applicable to gold electrolytes. In this case it is advantageous to carry out an atomic absorption analysis on the filtrate remaining after the precipitation of gold as a preliminary to its gravimetric determination.

X-RAY SPECTROMETRIC METHODS

X-ray spectrometry, sometimes referred to as "X-ray fluorescence" or "X-ray emission spectroscopy" is an analytical tool of particular value in the electroplating industry. Unlike optical emission spectroscopy, it is well adapted to the determination of the major elements in a plating solution as well as many additives and other minor constituents. It can also be used for the measurement of plating thickness. Properly employed, it is rapid, accurate and non-destructive.

X-ray spectrometry is analogous to optical emission spectroscopy in that the same three basic processes are involved, i.e., excitation, dispersion and detection. The instrumentation, however, is very different because the wavelengths involved in X-ray spectroscopy are shorter than those in the ultraviolet and visible regions by approximately three orders of magnitude. The wavelength range covered in most analytical applications extends from approximately 0.4–10 angstroms.

In order to achieve the high quantum energies corresponding to such comparatively short wavelengths, excitation energy is usually supplied in the form of an intense beam of primary X-radiation from an X-ray tube operated at voltages up to 50,000 or 60,000. This causes the chemical elements in the sample material to emit their characteristic X-ray wavelengths.

Separation and selection of individual wavelengths is accomplished by an X-ray monochromator consisting of two collimators and a dispersing crystal, all mounted on a goniometer. The single crystal is analogous to the ruled grating used in optical emission spectroscopy.

The intensities of the characteristic wavelengths are measured by some type of electronic radiation detector, usually a gas-filled "proportional counter" or a scintillation detector.

Detailed discussions of the theory, instrumentation and applications of X-ray spectrometry may be found in any of several texts[65,66,67].

Determination of Gold Content

The advantages and limitations of X-ray spectroscopy for the analysis of liquids and solutions in general have been summarised by Gunn[68]. Most of these principles are directly applicable to the determination of gold in plating solutions.

The most direct approach involves simply transferring a sample of the plating solution into a suitable liquid sample holder and measuring the intensity of one of the stronger characteristic gold wavelengths in the L-series, usually AuLα1. Factors other than gold concentration, however, may affect the gold intensity. These comprise what are generally termed "matrix effects" and are a consequence of variations in the concentrations of ions having significant X-ray absorption coefficients. These would include, for example, potassium and carbonate ions whose concentrations gradually increase as a result of periodic additions of potassium cyanide to cyanide-type electrolytes.

Mohrnheim[69] investigated the effect of "addition salts" on the intensity of the AuLα1 line in cyanide plating solutions. As a possible solution to the matrix problem, he proposed the use of the ratio of the intensity of the AuLα1 emission to that of the AuLβ1.

A more effective expedient involves the use of a "reference intensity" which is affected to approximately the same degree as the intensity of a selected characteristic emission. This was proposed by Andermann and Kemp[70] who used as a reference intensity the background or scattered "white" radiation at a selected wavelength. They found that whereas the absolute intensities were affected by variations in tube voltage, tube current, sample position and particle size, the ratios of the element wavelength intensities to that of the reference wavelength were substantially unaffected.

Bertin and Longobucco[71] investigated a number of reference intensities and intensity ratios. They concluded that, for the determination of gold in plating solutions, a ratio of the form $(I_L - I_B)/I_B$ gave the best results. In this expression, I_L represents the intensity of the AuLγ1 emission and I_B is the intensity of the background radiation near the AuLγ1 line.

It is desirable to be able to use the same procedure and analytical curve for the determination of gold in plating solutions of any type, e.g., cyanide, phosphate or citrate. Here the use of an intensity ratio alone may not be sufficient to overcome the effects of such large changes in matrix composition. The writer[53] has found the use of a "masking matrix" to be an effective means for completely suppressing matrix effects when used in conjunction with an intensity ratio. The masking matrix solution must, of course, be chemically compatible with all types of plating solutions and should be added in a sufficient amount so that expected variations in bath composition will have a negligible effect on the intensity ratio. Three parts, by volume, of a potassium citrate solution (250 g/l) added to one part, by volume, of the plating bath sample satisfies both of these requirements.

X-ray spectrometry provides one of the most rapid means for determining the gold content of plating solutions. A recent statistical study, moreover, indicates that a significant saving in time can be achieved while maintaining a degree of precision comparable to those of the gravimetric and electrogravimetric methods[53].

Determination of Metallic Additives and Impurities

It has been found that metallic additives, in their usual concentration ranges, may be readily determined by X-rays pectrometry using the same sample preparation and excitation conditions used for the determination of the gold content[53].

Very low concentrations of impurities usually cannot be determined directly in plating solutions although very good sensitivity can be achieved in many cases if a chemical concentration step precedes the final determination by X-ray. A number of such methods are described in the following section dealing with the analysis of gold deposits.

ANALYSIS OF GOLD DEPOSITS

In addition to controlling the compositon of the plating solution it is also desirable to know whether or not metallic additives are being codeposited in the intended proportions. It is also important to know whether impurity elements have found their way into the deposit where they may seriously impair the "noble" properties of the plate.

Direct Analysis of Deposit

In general, a direct, non-destructive quantitative analysis of the deposit is not feasible. Impurity elements, if present in high enough concentrations, can be identified by X-ray spectrometry although care must be taken not to confuse emissions from the basis metal(s) with those of the gold deposit.

Analysis of Solutions Resulting from Dissolution of Deposit

It is usually easy to isolate the gold deposit by dissolving the basis metal(s) in nitric acid. The gold can then be dissolved in aqua regia (approximately 1 : 4 : 8 parts, by volume, of nitric acid, hydrochloric acid and water, respectively). The resulting solution is placed in a liquid sample cell and examined by X-ray spectrometry. This procedure can readily be made quantitative by preparing standard solutions whose compositions correspond to those which would be obtained by dissolving gold containing definite amounts of additive or impurity elements.

The same type of sample solution can be analysed on an optical emission spectrograph using a rotating disc accessory or some other type of liquid excitation device. Alternatively, such a solution can be analysed for specific elements on an atomic absorption instrument. In this case, it might be desirable to use a sample solution more dilute than that which would be used for X-ray or spectrographic analysis.

The sensitivities of any of these methods can be increased by first bringing the aqua regia solution to fumes with sulphuric acid to precipitate the gold.

Determination of Minor and Trace Constituents After Pre-Concentration

The need for greater sensitivity than can be obtained by the methods outlined in "Analysis of Gold Electrolytes" can arise when only a limited amount of sample is available or when very low concentrations of impurities are in question.

In such cases recourse can be made to some method of collecting the small amounts of impurities from a rather dilute solution and concentrating them upon a small area where they can be analysed by X-ray spectrometry. For example, Hirano and Ujihira[72] have reported a method for the X-ray determination of trace amounts of arsenic in which the arsenic is co-precipitated with a measured amount of ferric iron by the addition of ammonium hydroxide.

One of the best techniques for pre-concentration is that based on the ion exchange principle. The adaptation of this principle to X-ray analysis was described by Grubb and Zemany[73]. Campbell, Spano and Green[74] reported the results of a rather thorough investigation into the use of ion exchange resin loaded papers as a pre-concentrated step for X-ray analysis. They determined the effects of pH, salt concentration, competing ions, exchange capacity and detection limits for 21 cations and 10 anions. After buffering the solution to the optimum pH, it is drawn several times by suction through a small disc of resin loaded paper which is then used as the X-ray specimen.

Miles, Doremus and Valent[75] report an X-ray method for the determination of trace elements in which they claim very rapid equilibration with loose granular resins (DOWEX 1-X8 or DOWEX 50W-X8). *The filtered, damp resin is packed into a cavity 3 mm deep and 21 × 31 mm in area in an aluminum plate for X-ray analysis. Conditions for the determination of 12 elements are given.

Probably the ultimate in X-ray trace analysis is a procedure described by Luke[76] in which the element(s) of interest are concentrated on an ion exchange resin disk only 0.125 inch in diameter. In order to analyse such a small area, Luke used a fully focused, curved crystal X-ray milliprobe. It is claimed that as little as 0.01 micrograms of the more sensitive metals can be determined by this means.

REFERENCES

1 Armet R. C., "Electroplating Laboratory Manual", Draper, Teddington, England (1965).
2 Grassby R. K., Gill J. A., Bradford G., *Electroplg Metal Finish.*, **19,** 12, 432 (1966).
3 Operating Instructions, Sel-Rex Corp., Nutley, New Jersey (1967).
4 Miller G. A., *Plating*, **53,** 100 (1966).
5 Smith G. F., "Mixed Perchloric, Sulphuric and Phosphoric Acids and their Applications in Analysis", 2nd Edn, G. F. Smith Chemical Co., Columbus, Ohio, U.S.A.
6 Rochat R. J., *Plating*, **36,** 817 (1949).
7 Lemons K. E., *Plating*, **52,** 307 (1965).
8 Sand H. J. S., "Electrochemistry and Electrochemical Analysis", Vol's I & II, Blackie, London (1940).
9 Lingane J. J., "Electroanalytical Chemistry", Interscience, New York (1958).
10 Langford K. E., Parker J. E., "Analysis of Electroplating and Related Solutions", Draper, Teddington, England (1971).
11 Mukherji A. K., *Anal. chim. Acta*, **23,** 4, 325 (1960).
12 "Standard Methods of Chemical Analysis", N. H. Furman (Ed.), Vol. I (1962).
13 Vogel A. I., "Quantitative Inorganic Analysis", 3rd Edn, Wiley, New York (1961).
14 Heller H. A., unpublished work.
15 Hall N., "Metal Finishing Guidebook Directory", Metals and Plastics Publications, Inc., Westwood, New Jersey (1973).
16 Stathis E. C., Gatos H. C., *Ind. Engng. Chem.* (*analyt.*), **18,** 6, 801 (1946).
17 Wagner H., *Metallwaren Ind. Galvano Tech.*, **22,** 464 (1929).
18 Nell K., *Plating*, **35,** 350 (1948).
19 Natarajan S. R., Lalitha K. S., *Metal Finish.*, **67,** 10, 79 (1969).
20 Silverman L., Zentner V., Pettinger F., Civitate, R. F., *Metal Finish.*, **67,** 9, 66 (1969).

* DOWEX, Registered Trademark of the Dow Chemical Co.

21 Unger P., Haynes H. G., *Analyst*, **71,** 839, 141 (1946).
22 Willard H. H., Young P., *J. Am. chem. Soc.*, **52,** 2, 132 (1930).
23 Smith G. F., Duke F. R., *Ind. Engng Chem.* (*analyt.*), **15,** 1, 120 (1943).
24 Berka A., Hilgard S., *Mikrochim. Acta*, (**1–2**), 174 (1966).
25 "Electroplaters' Process Control Handbook", D. G. Foulke & F. E. Crane (Eds), Reinhold, New York (1963).
26 Snell F. D., Snell C. T., "Colorimetric Methods of Analysis", 3rd Edn, Vol. 1, Van Nostrand, New York (1948).
27 Sandell E. B., "Colorimetric Determination of Traces of Metals", Interscience, New York (1959).
28 Baldwin P. C., *Plating*, **53,** 1451 (1966).
29 "Gold Recovery, Properties and Applications", E. M. Wise (Ed.), Van Nostrand, New York (1964).
30 Pantani F., Piccardi G., *Anal. chim. Acta.*, **22,** 3, 231 (1960).
31 McBryde W. A. E., Yoe J. H., *Analyt. Chem.*, **20,** 11, 1094 (1948).
32 Murphy J. W., Affsprung H. E., *Analyt. Chem.*, **33,** 12, 1658 (1961).
33 McNulty B. J., Woollard L. D., *Anal. chim. Acta.*, **13,** 2, 154 (1955).
34 Clabaugh W. S., *J. Res.*, *Natn. Bur. Stand.*, **36,** 119 (1946).
35 Cotton T. M., Woolf A. A., *Anal. chim. Acta.*, **22,** 2, 192 (1960).
36 Moeller T., *Ind. Engng. Chem.* (*analyt.*), **15,** 2, 270 (1943).
37 Stevens F., Fischer G., McArthur, D., "Analysis of Metal Finishing Effluents", Draper, Teddington, England (1968).
38 Harley J. H., Wiberley S. E., "Instrumental Analysis", Chaps. 1–4, Wiley, New York (1954).
39 "Analytical Absorption Spectroscopy", M. G. Mellon (Ed.), Wiley, New York (1950).
40 Schilt A. A., Jaselskis B., "Treatise on Analytical Chemistry", I. M. Kolthoff & P. J., Elving (Eds), Part I, Vol. 5, Interscience, New York (1964).
41 Conrad F. J., Kenna B. T., *Plating*, **52,** 1286 (1965).
42 Doherty N. F., *Plating*, **44,** 971 (1957).
43 Kolthoff I. M., Lingane J. J., "Polarography", 2nd Edn, Interscience, New York (1952).
44 Meites L., "Polarographic Techniques", 2nd Edn, Interscience, New York (1965).
45 Knotowicz A. E., Tatoian G., *Plating*, **47**, 645 (1960).
46 Craft A. H., Schumpelt K., *Plating*, **48,** 277 (1961).
47 Harrison G. R., Lord R. C., Loofbourow J. R., "Practical Spectroscopy", Prentice-Hall, New York (1948).
48 Brode W. R., "Chemical Spectroscopy", 2nd Edn, Wiley, New York (1943).
49 Sawyer R. A., "Experimental Spectroscopy", 2nd Edn, Prentice-Hall, New York (1951).
50 Duffendack O. S., Wiley F. H., Owens J. S., *Ind. Engng. Chem.* (*analyt.*), **7,** 3, 410 (1935).
51 Baer W. K., Hodge E. S., *Appl. Spectrosc.*, **14,** 6, 141 (1960).
52 McLain E. F., Wade W. H., *Plating*, **52,** 765 (1965).
53 Heller H. A., *Plating*, **56,** 277 (1969).
54 Walsh A., *Spectrochim. Acta.*, **7,** 2, 108 (1955).
55 Jones W. G., Walsh A., *Spectrochim. Acta.*, **16,** 3, 249 (1960).
56 Allan J. E., *Spectrochim. Acta.*, **18,** 3, 259 (1962).
57 Russell B. J., Shelton J. P., Walsh A., *Spectrochim. Acta.*, **8,** 4, 317 (1957).
58 Davis D. J., *Analyst*, **86,** 1026, 730 (1961).
59 Elwell W. T., Gidley J. A. F., "Atomic Absorption Spectrophotometry", 2nd Edn, Pergamon, London (1966).
60 Mikhailova T. P., Baranov S. V., Aleksandrov V. V., Sasov V. N., Rezepina V. A., Izv. Sib. Otd. Akad. Nauk USSR, Ser. Khim. Nauk **2,** 107 (1970).
61 Strelow F. W. E., Feast E. C., Mathews P. M., Bothma C. J. C., Vanzyl C. R., *Analyt. Chem.*, **38,** 1, 115 (1966).
62 Groenewald T., *Analyt. Chem.*, **40,** 6, 863 (1968).
63 Yudelevich G. A., Vall G. A., Torgov V. G., Korda T. M., *Zh. Anal. Khim.*, **25,** 5, 870 (1970).
64 Whittington C. M., Willis J. B., *Plating*, **51,** 767 (1964).
65 Liebhafsky H. A., Pfeiffer H. G., Winslow E. H., Zemany P. D., "X-ray Absorption and Emission in Analytical Chemistry", Wiley, New York (1960).
66 Birks L. S., "X-ray Spectrochemical Analysis", Interscience, New York (1959).

67 "Handbook of X-rays", E. F. Kaelble (Ed.), Part 4, McGraw-Hill, New York (1967).
68 Gunn E. L., A.S.T.M., Special tech. Publ. No. 349, Am. Soc. for Testing & Materials, Philadelphia (1964).
69 Mohrnheim A. F., *Plating*, **50,** 725 (1963).
70 Andermann G., Kemp J. W., *Analyt. Chem.*, **30,** 10, 1306 (1958).
71 Bertin E. P., Longobucco R. J., *Metal Finish.*, **60,** 3, 54 (1962).
72 Hirano S., Ujihira Y., *Bunseki Kagaku*, **12,** 747 (1963).
73 Grubb W. T., Zemany P. D., *Nature*, **176,** 221 (1955).
74 Campbell W. J., Spano E. F., Green T. E., *Analyt. Chem.*, **38,** 8, 987 (1966).
75 Miles M. J., Doremus E. H., Valent D., *Norelco Reptr*, **XIII,** 1, 32 (1966).
76 Luke C. L., *Analyt. Chem.*, **36,** 2, 318 (1964).

Chapter 34

QUALITY CONTROL

D. G. FOULKE & F. H. REID

With the change in interest from decorative to functional applications over the past two decades, the need for effective quality control of gold plating is increasingly apparent. This is important not only because of the intrinsic value of the metal, and often of the component being plated, but also as a means of ensuring that as far as possible coatings of the required physical characteristics and reliability may be consistently produced under the same nominal operating conditions. Present day gold plating is not a mere flash on handbag frames and other decorative items; it is currently used to coat control components which direct missiles, planes, communication systems and business. Consistent performance of coatings in such critical applicational areas is dependent on the degree of control exercised over each of the factors which may contribute towards the final condition of the deposit.

It is lack of appreciation of the importance of overall quality control that may well explain conflicting reports in the literature concerning the effects of operating parameters on the physical properties of deposits.

The factors involved might be summarised under the following scheme of main headings, each of which will be discussed individually in the ensuing text:

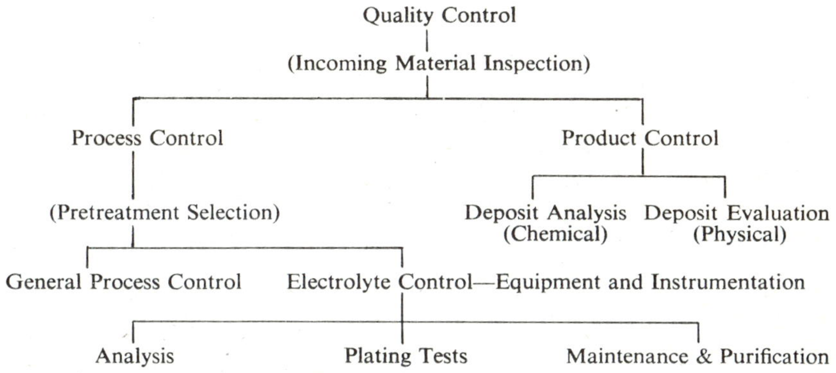

INCOMING MATERIAL INSPECTION

If any quality control procedure is to achieve its object it is clear that control of the starting basis metal must be as thorough as that of the plating process.

Surfaces to be plated should therefore be inspected for general smoothness and for the presence of heavy scale or heat treatment film, surface defects such as scratches, pits, pores or foreign inclusions, and their suitability for plating assessed in relation to the specified requirements of the final coating.

The nature and if possible the exact composition of the basis metal should also be ascertained, since this may dictate the use of special pretreatment schedules with possible adverse effects on surface finish or dimensions of components in critical cases.

Any doubt concerning the presence of inclusions or porosity in the metal should be resolved wherever possible by microscopic examination of sections which may disclose sub-surface features likely to be exposed by etching or other pretreatment, with undesirable effects.

If the surface has been heavily work-hardened during fabrication, annealing may be desirable before plating.

Following an examination of this type, if any doubt should remain as to the practical feasibility of plating the surface to specification, a full discussion with the supplier of the material or components should take place before proceeding any further.

PROCESS CONTROL

PRETREATMENT SELECTION

Choice of the correct pretreatment for the basis metal concerned is essential to achieve good adhesion of the coating. Pretreatment procedures are comprehensively discussed in Chapter 12. Their effectiveness in any particular case should be confirmed in pre-production tests and the condition of the various solutions involved should be checked by regular processing of test pieces.

GENERAL PROCESS CONTROL

Environment

The electronics, missile, atomic energy and spacecraft industries stress reliability. However, there are spectacular contrasts. For example, one can find in the building of an electronics plant a room provided with thoroughly filtered air, 10 megohm water for rinsing, workers with special clothing—donned in an intermediate room—all to ensure high reliability. In the next building is the plating shop, with plenty of air coming through the window, carrying with it insects, dust, pollen and spores of algae and fungi. Rinses are poorly engineered and ventilation normally inadequate. Parts are actually being fabricated here to be assembled in the same high reliability missile manufactured in the "clean room" next door.

Ideally, quality control begins with environment. Past experience has shown that airborne dust can create a considerable amount of trouble, even to the extent of affecting the final deposit.

Airborne dust and dirt should therefore be removed from the air brought into the plating shop and the air should be properly distributed, but this is rarely the case. Tanks should be properly vented to maintain an atmosphere which will not attack the physical environment and cause paint peelings and corrosion products to fall into the processing tanks.

WATER SUPPLY

Thorough rinsing of components between processing stages and before entering the gold bath is essential to avoid carry-over of possible contaminants This aspect is discussed by Ostrow and Nobel (see Chapter 13) and by Congreve (see Chapter 18), but a few comments may be made here in the specific context of quality control, and with particular reference to post-plate rinsing.

The first rinse, in which the majority of the gold salt drag-out is concentrated, is sometimes returned to the plating tank to replace evaporation losses, as a means of reclaiming the gold content. This practice, common in base metal plating, is very questionable in the case of high quality gold plating, where removal of accumulated impurities by continuous or overnight dummying is not feasible. Carbonates from alkaline cyanide baths, degraded organics from acid and neutral baths, and sulphates from gold sulphite electrolytes build up in the drag-out tank and may cause deleterious effects such as staining or discolouration if fed back into the plating solution.

Problems may also arise due to the use of hard water in a rinse. In one specific instance a case of dull plating from a bright cyanide gold solution was traced to the presence of a hard water film on the reclaim rinse, the trouble being overcome when this was replaced by deionised water.

Many of the buffer salts used in gold plating are based on organic acids and it is therefore not uncommon to see algae and fungal growths appearing in strike solutions and rinses, which may lead to troublesome effects, particularly in exacting applications in the semiconductor industry. In critical cases this problem has been countered by irradiating rinse water by ultra-violet light, or by the use of water treated by a two bed deioniser system and followed by a mixed bed unit. The latter, however, is an exceptional case.

For many rinses water is passed through carbon beds before entering the rinse tank. The use of 10 megohm water, common in many clean room operations, is not generally necessary, but simple deionised water may well be inadequate as a final rinse in applications calling for extreme cleanliness of the final coating.

In general it is preferable to recover gold from the drag-out rinse by continuous circulation of the solution through an ion exchange column, rather than to return solution to the plating tank. Here again it may be noted that trouble is sometimes encountered due to algae growth in the ion exchange column.

If rinses are returned to the plating tank, they should be based on deionised water, maintained scrupulously clean with respect to adventitious contamination, and added to the bath through a filter. This should preferably contain a carbon pack to remove suspended matter and any organic breakdown products.

ELECTROLYTE CONTROL

Analysis

Methods for analysis of the electrolytes are comprehensively dealt with by Heller (see Chapter 33). It is only necessary here to emphasise the importance of correct sampling, i.e., thorough stirring of the electrolyte and a reasonably accurate assessment of its volume, if the analytical results are to be useful.

The frequency of determinations of any particular component or electrolyte parameter must be fixed by experience, but will clearly depend on the intensiveness and general conditions of bath operation. As a guide to replenishments the aim should be to determine critical components at intervals which will ensure that their concentrations will not have changed between checks by more than 10% of the nominal. In individual circumstances the lack of analytical instrumentation or skill may be a limiting factor, but it must be recognised that if this is so, effective quality control is virtually impossible.

Plating Tests

Analysis must be supplemented by plating tests to check electrolyte parameters such as cathode efficiency, throwing power, and brightness-current density relationships, or others of particular importance in relation to specific applications, e.g., internal stress. The latter may equally be regarded as a product control procedure and is included later under this heading.

HULL CELL

The application of the Hull Cell to evaluating the appearance of gold deposits is severely restricted for functional coatings, primarily due to the fact that such deposits are plated at relatively low current densities (0–10 A/ft^2) whereas typical Hull Cell results cover a 0–70 A/ft^2 current density range. Secondly, all gold plating solutions because of their low metal ion content are agitated and such agitation is difficult to reproduce in a Hull Cell. Furthermore, results from such a test can be misleading since appearance is not the criterion by which functional properties such as porosity, hardness, etc., are gauged. The Hull Cell is therefore limited in its usefulness to decorative plating where only the appearance of the deposit merits consideration.

HARING AND BLUM CELL

This type of plating test, which is commonly employed to quantify throwing power, utilises two cathodes at different distances from a common anode. The

Haring-Blum Cell finds little practical use for gold plating solutions since it is difficult to reproduce identical agitation at the two cathode surfaces and because the rate of depletion of gold ions in the volume of electrolyte between anode and cathode differ considerably.

BENT CATHODE

The Bent Cathode test is the plating test most widely employed to check and correct gold solutions. Such tests are usually carried out using 1 litre of solution and 2 in. × 3 in. panels bent either semi-circular or at an angle of 45 or 90°. Each investigator, however, invariably has his own preferred test specimen shape.

The advantage of the bent panel is that it gives some idea of the appearance and thickness distribution, i.e., throwing power, to be expected under standard test conditions, which are usually the bath working conditions. The fact that such tests can be carried out quickly and isolated from the actual plating tank is initially advantageous. The deposition of discoloured or filmed deposits in recessed areas, or around the edge of the test panel (high current density areas) is an indication that all is not well with the plating bath. Invariably the cause is due to a parameter such as gold concentration, pH, brightener content, etc., being outside the recommended range, or that the solution is contaminated either by metals or organic decomposition products.

Electrolyte Purification

Procedures for the treatment of gold solutions contaminated by organic and certain metallic impurities are discussed by Ostrow and Nobel (see Chapter 13). Corrective measures of this kind should be carried out on a 1 litre sample of the electrolyte before applying them to the main bath, and it should be noted that in cases of severe metallic contamination it will often be more economical to replace the solution than to embark upon purification treatments if there is any doubt at all about their efficacy.

EQUIPMENT AND INSTRUMENTATION

Although this subject is dealt with by Ostrow and Nobel in Chapters 13 and 14, it is pertinent under the heading of "Quality Control", to underline a few salient points which are frequently overlooked or completely neglected.

On the equipment side, the tanks, bus bars, anode and cathode connections, etc., should be kept scrupulously clean in order to avoid contamination of the electrolyte and to ensure that certain areas of bath bus bar do not have a higher electrical resistance. Cleanliness also includes filtration equipment, barrels and any other equipment within close proximity to the gold plating bath.

All instrumentation such as ampere-hour meters, automatic current control devices and so forth, should be thoroughly checked and standardised on a regular basis and generally serviced at frequent intervals.

PRODUCT CONTROL

Product control, as the term implies, means the assessment and evaluation of the finished article or component, as the case may be, with respect to the properties associated with the deposit. Besides the visual inspection for obvious flaws, blemishes, imperfections, colour matching and so forth, and a few properties rarely evaluated (reflectivity, emissivity, absorptivity, etc.), methods for the quantitative measurement or qualitative assessment of these characteristics are critically discussed in considerable detail in the remainder of Part V as follows:

Solderability (Chapter 25), Contact Resistance (Chapter 26), Porosity (Chapter 27), Thickness, Hardness, Stress, Ductility and Adhesion (Chapters 28–32) and Composition (Chapter 33).

CONCLUSION

Since quality control requires that product as well as process control be maintained, laboratories for this purpose should be well equipped. It is, in fact, becoming increasingly necessary nowadays that facilities available be approved by the appropriate design authority or the main contractor.

While it has been customary to apply control tests to agreed numbers of representative samples from each batch, it is now quite common, where high reliability is concerned, to specify testing of every part before assembling into equipment. 100% testing can only be justified in the case of gold and other precious metal plating when correct and reliable functioning is of such critical importance.

Control of the quality of gold metal deposits has been important in the past and is becoming even more so nowadays. The clean room stage is now currently in operation in many plants. Gold salts are being specified as 99.99% purity and other salts added to the bath are analysed and the purity specified. The gold deposit itself is subject to specifications which may require a minimum purity of 99.9%. This means that control of quality is becoming so important that not only the electrolyte and deposit must be controlled, but the environment, equipment, general housekeeping throughout the plant and history of the basis metal, must be supervised by a suitably qualified and competent technologist who recognises the need for control and understands the chemical, physical and metallurgical aspects involved.

PART 6

Applications

Chapter 35

PRINTED CIRCUITS

G. R. Strickland

With the development of printed circuit technology, gold plating was an automatic choice for the corrosion protection of conductor patterns and for the provision of a reliable electrical contact surface on edge connector contacts, printed switch segments, etc. However, although there was no difficulty in producing gold coatings of requisite hardness and electrical properties from the then available alkaline cyanide electrolytes, it was soon found that these solutions could have severe deleterious effects on both the laminate and on the adhesives used in copper foil bonding.

A number of techniques were developed to minimise these problems. The so-called "reversal" process involves the application of a compatible screening ink or photoresist to the copper-clad laminate in a negative reverse pattern, leaving the functional conductor pattern bare. The latter is then gold plated, followed by removal of the resist and etching, usually in ferric chloride solution, during which process the gold plating itself acts as a resist. Although this method avoids attack by the gold plating solution on the adhesive bond between copper and the laminate, problems arise in production due to pin-holing or local area breakdown of the organic resist film in the first stage.

An alternative method, to produce durable gold plated surfaces on selected areas of a pattern while the remainder ultimately receives a thin coating to provide short term protection, e.g., for solderability, is to apply the reversal technique only to the edge connector contact area. After plating the contact tabs the negative resist is removed and a positive pattern applied to protect the whole circuit design during etching. Following the removal of the unwanted copper the resist is then removed and the exposed copper plated by a non-electrolytic gold process such as immersion (see Chapter 10) or electroless (see Chapter 11) to a thickness of about 0.25 microns.

It is possible to avoid the need for complete immersion of the board in a plating bath by the use of brush plating techniques (see Chapter 16) to apply gold to selected contact areas, followed again by a non-electrolytic gold process to provide a thin coating, when required, on the remainder of the circuit.

Such techniques are still used, but the very rapid growth of the printed circuit industry which took place during the latter half of the 1950's made the search for really compatible gold plating solutions not only technologically imperative, but commercially attractive and led to an intensive study of processes and a pace of new electrolyte development which has few parallels in the history of electroplating.

As a result, the plater has at his disposal today a variety of processes for the production of pure or alloy gold coatings of specific physical and electrical characteristics, and a fairly thorough appreciation of the coating properties obtainable from these solutions is essential to the printed circuit manufacturer and design engineer to guide them in the choice of the optimum process to meet any particular applicational requirement. Although gold plating electrolytes have been discussed at some length elsewhere (see Chapters 3–7), it is considered desirable for the above reason, and for the sake of completeness, to preface the description of practical plating procedures with which the present chapter is mainly concerned, by a brief recapitulation of operating conditions and deposit properties pertaining to the four major types of electrolyte, with particular reference to the requirements of printed circuitry.

Gold plating systems now commercially available fall into four main categories which may be classified according to their operable pH range.

ACID GOLD ELECTROLYTES (pH 3.0–6.0)

Electrolytes in this group have extensive patent coverage which has provided the basis for a wide variety of proprietary solutions. The consistent performance of this type of solution in intensive production is one of the main reasons for its continued popularity with circuit manufacturers.

24 carat deposits are readily obtainable from solutions without alloying metal complexes and with minimal amounts of organic brighteners, operated in the pH range of 5–6 and at temperatures of 50–70°C to give cathode efficiencies approaching 90%. Alloy gold deposits may be produced from solutions containing cobalt or nickel complexes and operated in the lower pH range of 3.5–4.5 and at temperatures of 30–40°C when cathode efficiencies of 35–45% are normally achieved.

The physical and chemical properties of coatings from these electrolytes have been widely reported and once the functional requirements of a gold deposit have been specified the appropriate electrolyte can be chosen. Table 35.1 illustrates the general composition and operating parameters of acid gold electrolytes, while Table 35.2 outlines some typical deposit characteristics.

NEUTRAL GOLD ELECTROLYTES (pH 6.0–8.5)

The very low porosity of deposits obtained from this group of electrolytes at a thickness as low as 0.2 microns is used to advantage when gold is chosen as a final solderable surface over the entire conductor pattern. A number of proprietary solutions are available to provide deposits of purity ranging from the ultimate to 23+ carat and with corresponding hardness values of 70–190 KHN_{25}.

Table 35.1

Bath Composition (g/l) and Operating Conditions of Acid Gold Electrolytes

	Low Cobalt	*High Cobalt*	*Low Nickel*	*High Nickel*	*Low Organic*	*High Organic*
Gold cyanide (68 % Au)	12	12	12	12	14.8	14.8
(Equivalent Au concentration	8.2	8.2	8.2	8.2	10	10
Chelates/citrates/ phosphates	100–300	100–300	100–300	100–300	100–250	100–250
Brighteners (Class 1)	none	none	none	none	1–20 g/l	1–20 g/l
Brighteners (Class 2)	0.03–0.1	0.3–1.0	0.03–0.1	0.3–1.0	0	0
Temperature, °C	25–45	25–45	25–45	25–45	50–70	40–70
pH (electrometric)	4.0–4.5*	4.0–4.5*	4.0–4.5*	4.0–4.5*	5.0–6.0†	5.5-6.0†
SG at 20 C‡	10–18°Be	10–18°Be	10–18°Be	10–18°Be	10–12°Be	10–12°Be
Cathode current density, A/dm^2	0–1.65	0–1.65	0–1.65	0–1.65	0–1.1	0–1.32
Cathode efficiency	ca. 40%	ca. 40%	ca. 40%	ca. 40%	ca. 90%	ca. 90%
Anodes	Platinised titanium, gold, or gold plated titanium					
Anode: cathode area ratio	1:1	1:1	1:1	1:1	2:1	2:1
Filtration	Continuous through 10 μm particle retention cotton cartridge					
Cathode oscillation speed, cm/min	400–500	400–500	400–500	400–500	400–500	400–500
Limiting deposit thickness, μm	6.25	6.25	6.25	6.25	25.4	25.4

* Adjust with 25% w/v potassium hydroxide or proprietary pH adjusting salts.
† Adjust with 50% v/v ammonium hydroxide or proprietary pH adjusting salts.
‡ Adjust with propietary conductivity salts.

Table 35.2

General Physical Characteristics of Electrodeposits from Acid Gold Electrolytes

	Low Cobalt	*High Cobalt*	*Low Nickel*	*High Nickel*	*Low Organic*	*High Organic*
Gold content, %	99.9	99.0	99.9	99.0	24 carat	24 carat
Hardness (KHN_{25})	130–180	200–250	130–150	170–200	70–110	90–140
Restivity (microhm-cm) at 0°C	10.0	16.0	7.0	12.0	2.2	2.5
Contact resistance (milliohms)	0.6	0.6	0.3	0.3	0.3	0.3
Plating rate mg/A-min at 1.1 A/dm^2	45–50	30–35	45–50	30–35	120–125	120–125
Tarnish film resistance, °C	−50 to +160	−50 to +160	−80 to +200	−80 to +200	−80 to +300	−80 to +300
Deposit distribution, %	ca. 70	ca. 70	ca. 70	ca. 70	ca. 70	ca. 70

Soft, ductile and high purity gold coatings suitable for solderability as well as low pressure electrical contact usage, are obtainable from solutions without brighteners and operated at high temperatures (60–80°C) to maintain good colour characteristics at cathode efficiencies in excess of 90%. The harder 23+ carat, relatively low stress deposits, suitable for edge connector contact applications calling for good wear resistance, are produced from

Table 35.3

Bath Composition (g/l) and Operating Conditions of Neutral Gold Electrolytes

	High Purity	*23+ Carat*
Gold potassium cyanide (68% Au)	14.7–17.5	14.7–17.5
(Equivalent Au concentration)	10–12	10–12
Chelates, citrates or phosphates	150–180	150–180
Brighteners (Class 1)	none	0.1–3.0
Temperature, °C	60–80	35–45
pH (electrometric)*	6.0–8.0	7.0–8.0
SG at 20°C†	9–12°Be	9–12°Be
Cathode current density, A/dm^2	0–0.55	0–1.1
Cathode efficiency, %	90 at 70°C	80 at 40°C
Anodes	Platinised titanium, gold, or gold plated titanium	
Anode: cathode area ratio	3:1	3:1
Filtration	Continuous through 10 μm particle retention cotton cartridge	
Cathode oscillation speed, cm/min	400–500	400–500
Limiting deposit thickness, μm	25.4	6.25

* Adjust with 50% v/v ammonium hydroxide, citric acid and/or phosphoric acid.

† Adjust with proprietary conductivity salts.

Table 35.4

General Physical Characteristics of Electrodeposits from Neutral Gold Electrolytes

	High Purity	*23+ Carat*
Gold Content, %	100	99.9
Hardness (KHN_{25})	70–90	160–190
Resistivity (microhm-cm) at 0°C	2.4	2.2–2.5
Contact resistance (milliohms)	0.3	0.3
Plating rate mg/A-min at 1.1 A/dm^2	120–125	107–112
Tarnish film resistance, °C	−80 to +300	−80 to +200
Pore-free deposit thickness on carefully polished substrate, μm	0.2	1.5–2.0
Deposit distribution, %	85.5	85.5

Table 35.5
Bath Composition (g/l) and Operating Conditions of Sulphite Complex Gold Electrolytes

	Low Alloy Metal	*High Alloy Metal*
Gold	10–14	8–12
Sulphate/sulphites/chelates	170–340	170–340
Brighteners (Class 2)	0.04–0.2	0.5–2.5
Temperature, °C	45–55	45–55
pH (electrometric)*	8.5–10	8.5–10
SG at 20°C†	8–35°Be	8–35°Be
Cathode current density, A/dm^2	0.3–0.8	0.3–0.8
Cathode efficiency, %	95	95
Anodes	Platinised titanium or gold plated titanium	
Anode: cathode area ratio	4:1	4:1
Filtration	Continuous through 10 μm particle retention cotton cartridge	
Cathode oscillation speed, cm/min	400–500	400–500
Limiting deposit thickness, μm	25.4	25.4

* Adjust with 25% w/v sodium hydroxide solution.
† Adjust with proprietary conductivity salts.

Table 35.6
General Physical Characteristics of Deposits from Sulphite Complex Gold Electrolytes

	Low Alloy Metal	*High Alloy Metal*
Gold content, %	99.9+	98–99
Hardness (KHN_{25})	130–190	180–240
Contact resistance (milliohms)	0.3	0.4
Plating rate mg/A-min at 0.3 A/dm^2	123	123
Deposit distribution, %	93.5	93.5

electrolytes with organic brightener additions and operated at somewhat lower temperatures (35–45°C). Under these conditions a high cathode efficiency is maintained, but porosity of coatings is increased.

Table 35.3 illustrates the general composition and operating parameters of neutral gold electrolytes while Table 35.4. outlines some typical deposit characteristics.

COMPLEX SULPHITE GOLD ELECTROLYTES (pH 8.5–10.0)

Electrolytes from this group will produce high metal distribution values and are particularly useful when pattern plating of varying conductor track widths in conjunction with small diameter holes, and where a minimum consistent coating thickness is desirable. Proprietary solutions are available furnishing deposits which are extremely ductile and range in purity from 98–99.9 %, with corresponding hardness values of 240–130 KHN_{25}.

Table 35.5 illustrates the general composition and operating parameters of non-cyanide gold electrolytes, while Table 35.6 outlines some typical deposit characteristics.

ALKALINE GOLD SYSTEMS (pH 8.5–13.0)

The early adaptation by Rinker[1] of an alkaline cyanide system operable at room temperature gave to the circuit manufacturer at that time, a useful range of electrolytes which were at least compatible with the laminate materials, and provided deposit hardness values to suit most requirements. However, these solutions have now been largely outmoded by the subsequent development of the acid, neutral and non-cyanide systems.

PROCESS SELECTION

As indicated in the preceding section, relatively hard, wear resistant coatings for edge connector contacts are normally applied from organic acid based electrolytes containing nickel or cobalt and providing deposit hardness values in the range 125–250 KHN_{25}.

For good solderability and maximum electrical conductivity the softer high purity gold deposits from the organic acid, neutral phosphate or alkaline non-cyanide solutions without alloying complexes may be utilised, giving hardness values of 60–90 KHN_{25}. Several investigators[2,3,4,5] have stressed that deposit thickness should not exceed 1.5 microns if soldered joints of adequate strength are to be obtained without excess formation of brittle intermetallic compounds of tin and gold. This aspect is discussed by Thwaites in Chapter 19.

For the plating of conductor patterns of varying track widths, in conjunction with small diameter holes, the good ductility and excellent metal distribution afforded by the complex sulphite electrolytes are advantageous.

When the gold plated conductor surfaces are liable to be subjected to elevated temperatures in service a nickel underplate is usually applied, which offers a hard base for the gold plate and also functions well as a barrier to diffusion between the copper substrate and the gold coating. In such cases low stress nickel deposits from the modified Watts or sulphamate nickel solutions are usually employed.

In certain applications, e.g., switch segments in flush-bonded circuitry which are subjected to high contact pressures as well as elevated temperatures, gold may be unsatisfactory due to comparatively low hardness and wear resistance. In these cases harder coatings such as palladium or rhodium may be used. It is not uncommon for combinations of precious metals to be selected, often in conjunction with an underplate of nickel, to meet specific operational requirements. Thus palladium or rhodium may be used as diffusion barrier coatings under low contact resistance gold deposits, while a hard rhodium deposit required for high wear resistance may be "capped" with a thin (0.25 micron) soft gold coating for lubricating purposes, e.g., anti-galling.

Various standards authorities have now issued specifications[6] relating to the types of gold deposits used in certain applications, and a compilation[7] reviewing the properties of electrodeposited gold and gold alloys should also be helpful in choosing the most suitable process for a particular application.

PLATING TECHNIQUES

Although gold is no more difficult to apply than any other electrodeposited metal, a more thorough approach to electrolyte selection and control as well as substrate preparation, is required if the more exacting specified tolerances and deposit characteristics laid down by the various standards authorities are to be maintained. It is also imperative that once the particular gold plating solution has been selected, any modification in the production plating sequences should be carefully assessed with respect to its possible effect on coating properties.

Having established the type of solution and the deposit characteristics required, the next stage is the design of the plating installation, bearing in mind the need to have the minimum volume of solution for cost economy purposes and yet one that will cope adequately under high production runs. In order to minimise oxide formation the capacity of the precleaning and rinse tanks as well as any metal undercoating, i.e., stress-free nickel, palladium, rhodium and acid gold strike tanks, should be sufficient to permit a continuous sequence flow relative to the gold plating unit. This is especially necessary when plating over a nickel sub-plate in manually operated plating units or where the time interval between the nickel and the acid gold "strike" is in excess of 15 seconds duration which is often sufficient under certain circumstances to allow a passive oxide film to form on the nickel coating. As a precautionary measure the installation of a nickel de-passivating dip, prior to the acid gold "strike" is worth considering. This will also be of some assistance in reducing the possibility of nickel build-up in the main gold plating solution. A solution containing 35–40 g/l of potassium cyanide and operated cathodically (4.4 A/dm^2), at room temperature for 30–60 seconds, is very effective in this context.

The prime aim in the design of an efficient gold plating unit is to achieve maximum uniformity of metal distribution. The importance of this is obvious when the plating tolerances in the various specifications are considered. Examples of poor metal distribution which all too readily spring to mind are excess gold thickness located towards each end of a row of edge connector tabs, causing stress cracks, especially when the hard alloy acid gold solutions are used, and excess gold around hole pad areas, which gives rise to embrittled solder joints.

Apart from the general solution properties which play a major role in metal distribution and which can be readily controlled and maintained, there are five inter-related basic factors of both plant and circuit design to be considered if optimum deposit distribution is to be achieved. These may be simply classified under the five "A's" as follows:

1 Anode design
2 Agitation of component and solution
3 Attachment of anodes and cathodes (electrical and mechanical)
4 Art work of circuit pattern efficiently designed for plating
5 Area of cathode accurately computed.

Item 5 needs no further comment but the remaining items merit further consideration and are based on personal experience which has produced excellent results, maintaining deposit thicknesses within ± 0.125 micron.

The simple, but very effective anode design to be described, was originally intended only for the plating of edge connector contact tabs, but has since been further extended to the plating of the complete circuit pattern. The idea originated from a detailed discussion[8] based on a study on anode behaviour in acid gold solutions.[9]

Initially the gold plating unit was constructed to allow for the installation of five gold plated titanium rod anodes 1.25 cm in diameter and 2.5 metres in length so as to form four long integrated cathode compartments completely enclosed within one tank. The lid was accurately slotted to enable the printed circuit boards to be rigidly guided while traversing between the anodes at a maximum/minimum distance of 17.2 cm/3.2 cm $\pm$ 0.5 mm.

During its construction, however, the idea was conceived of making the anodes in the form of hollow tubes through which, by means of suitably drilled 0.8 mm diameter holes, fresh solution flow from the filter unit would be directed into the catholyte region surrounding the contact tabs along the entire length of the anodes. The holes were drilled so that the spray pattern from adjacent anode lanes continuously overlapped.

Plating trials were conducted with this arrangement at various cathode oscillation speeds using a 7 cm stroke. A minimum traverse speed of 220 cm per minute was established as adequate to maintain the deposit thickness within ± 0.125 micron. Contact was made to individual circuit boards by a spring loaded nickel plated cathode contact clip located at the top, and central to the now widely adopted "picture frame" circuit design. Another important factor which helped considerably in maintaining these close tolerances was subsequent adjustment of the edge connector contact tab centres to correspond with the anode centres.

Over a period of 18 months the performance of this gold plating edge connector unit has been such that although a high total work load amounting to some 80 ampere-hours per litre has passed through the cobalt acid gold electrolyte, there has been no need for solution carbon purification, since cathode efficiency and Hull Cell performance have remained unaffected by the very small rise in heavy metal ion impurities which have remained well within the acceptable concentration limits for this type of electrolyte. This is somewhat remarkable considering the total chemical additions made, amounting to some 28 kg of gold potassium cyanide (19 kg of gold), 7.5 kg of pH adjusting salts and 7.5 litres of cobalt replenisher solution, and provides a good indication of the performance that may be achieved with careful selection and refinement of basic chemicals by the supplier, and strict adherence of production personnel to process sequences laid down by the production control laboratory.

During the period mentioned it was found necessary, however, to replate the anode tubes to an approximate thickness of 5 microns every 15.5 A-hr/l to compensate for the slow dissolution of the gold during the plating operation. This was carried out by reversing the main anode/cathode terminal connections and substituting gold plated "dummy" anodes through the slots in the lid of the tank.

TYPICAL OPERATING AND PROCESSING SEQUENCES

The preparatory cleaning stages for all copper panels before the required electroplating cycles provide an ideal surface for the more compatible acid cleaners and activating solutions to be effectively used, and whether gold is applied before or after the etching cycle, the actual plating operation remains basically the same. Obviously in the latter case the processing sequence is more flexible and will permit any minor surface blemishes, i.e., slight pits, scratches or random copper particles to be removed by mechanical lapping of the copper substrate with a fine chalk or copper chloride slurry, whereas when gold is used as a final etch-resist coating, very careful inspection and in-line production control prior to and after the application of the organic plating resist is necessary.

GOLD PLATING BEFORE ETCHING

As previously discussed, all or selected parts of the circuit may require to be gold plated at this stage of manufacture and this may well involve re-cleaning and further conditioning of the exposed conductor pattern after masking and/or organic resist coatings have been applied. However, when no such intermediate resist masking operations are involved, as may be the case following the pattern plated copper or tin-nickel process, the cleaning sequence is omitted.

The following operating cycles have been chosen as illustrating the processing sequences involved once the final plating requirements have been specified.

Gold Plating of Base Laminate Copper Pattern

Cycle 1 (*After careful examination of applied plating resist-coating*)

1 Immerse in suitable acid cleaning solution* at 18–27°C for 10 minutes.
2 Tap water spray rinse.
3 Immerse in 10% w/v ammonium persulphate with $\frac{1}{2}$% v/v sulphuric acid at 18–20°C for 1 minute (inspect for any residual resist contaminating the copper pattern area).
4 Tap water spray rinse.
5 Immerse in 10% v/v sulphuric acid at 18°–20°C for 1 minute.
6 Tap water spray rinse.
7 Immerse in 10% w/v citric acid at 18°–20°C for 1 minute.
8 Deionised water spray rinse.
9 Transfer rapidly to acid gold strike solution immersing the boards in the electrolyte with a pre-set 10% total current loading and plate at 1.65 A/dm^2 for 20 seconds.
10 Deionised water spray rinse.
11 Transfer rapidly to gold plating solution, immersing the boards in the electrolyte with a pre-set 10% current loading and electroplate under the appropriate conditions to obtain the specified coating thickness.
12 Drain and transfer to deionised water drag-out tank.
13 Tap water spray rinse.
14 Transfer to resist removal section for subsequent inspection prior to etching.

Cycle 2—*Nickel Barrier Plate to Gold Plating*

Prepare copper surface as in Cycle 1 through Sequence 1–6, then proceed as follows:

1 Transfer rapidly to "stress-free" nickel plating solution immersing the boards in the electrolyte with a pre-set 10% current loading and plate at 4.4 A/dm^2 for 15 minutes to obtain an average deposit thickness of 12.7 microns.
2 Tap water spray rinse.
3 Transfer to nickel activator (35–40 g/l KCN) at 4.4 A/dm^2 (cathodic) for 30–60 seconds.
4 Transfer rapidly via deionised water rinse to acid gold "strike" solution and continue as in Cycle 1 from Sequence 9–14.

* Proprietary acid cleaning solutions are available that do not degrade the thin organic resist coatings or etch the exposed copper circuit pattern.

Cycle 3—Nickel and Rhodium Sub-Plates Prior to Gold Plating

Prepare copper surface as in Cycle 1 through Sequence 1–6 and nickel plate as in Cycle 2 through Sequence 1–3, then proceed as follows:

1 Transfer rapidly through tap water spray rinse.
2 Immerse in 10% v/v sulphuric acid at 18–20°C for 15–30 seconds.
3 Transfer rapidly through deionised water rinse to rhodium plating solution, immersing the boards in the electrolyte with a pre-set 10% current loading and plate at 2.2 A/dm^2 for 1.5 minutes to obtain an average deposit thickness of 0.5 micron.
4 Drain and transfer to deionised water drag-out tank.
5 Transfer rapidly through deionised water spray rinse to acid gold "strike" solution and continue as in Cycle 1 from Sequence 9–14.

Although most circuit manufacturers now produce plated through hole circuitry by the electroplated copper pattern method, it is sometimes convenient for the alternative panel copper plated method to be applied. When considering the following processing cycles the precleaning sequences 1–5 in Cycle 1 should be included if the latter has been adopted.

Cycle 4—Gold Plating of PTH Copper Boards

1 Activate electroplated copper surface as in Cycle 1 from Sequence 3–14.

Cycle 5—Nickel Barrier Coating Prior to Gold Plating of PTH Copper Boards

1 Activate electrodeposited copper surface as in Cycle 1 from Sequence 3–6 and follow by Cycle 2 from Sequence 1–4 inclusive.

Cycle 6—Nickel Barrier Coating and Selective Rhodium Plating of Edge Connector Contact Tabs Prior to Gold Plating of PTH Copper Boards

One convenient method which has been adopted is to include on the original circuit art design a plating bar connecting the individual edge contact tabs so that the nickel substrate can be selectively rhodium and/or gold plated after the subsequent etching operation.

1 Activate electrodeposited copper surface as in Cycle 1 from Sequence 3–6.
2 Continue Cycle 2, Sequence 1–2.
3 De-jig and place boards in a warm convector oven or blow dry using oil-free air.
4 Selectively mask using an approved adhesive tape over the edge connector tabs.
5 Re-jig and transfer boards to activator solution and proceed as for Cycle 2, Sequence 3–4.

6 Etch complete circuit panels.
7 After etching, carefully mask the area of the board above the edge connector tabs using an approved adhesive tape and again transfer exposed nickel area to activator solution and proceed as for Cycle 2, Sequence 3 and follow by Cycle 3, Sequence 1–5.

Cycle 7—Gold Plating of Tin–Nickel PTH Circuits[10]

Following the application of the tin–nickel deposit, proceed as follows:

1 Tap water spray rinse.
2 Immerse in 10% v/v hydrochloric acid at 18–20°C for 1 minute.
3 Deionised water spray rinse.
4 Transfer rapidly to acid gold "strike" solution, immersing the boards in the electrolyte with a pre-set 10% total current loading and plate at 1.65 A/dm^2 for 20 seconds.
5 Deionised water spray rinse.
6 Transfer rapidly to gold plating solution, immersing the boards in the electrolyte with a pre-set 10% current loading and electroplate under the appropriate conditions to obtain a coating thickness of 0.25 microns.
7 Drain and transfer to deionised water drag-out tank.
8 Tap water spray rinse.
9 Transfer to resist removal section for subsequent inspection prior to etching.

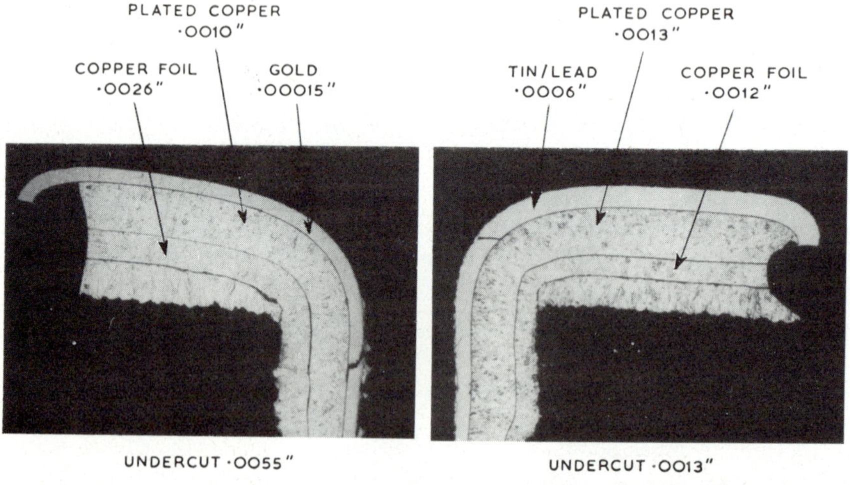

Fig. 35.1 Comparison of the degree of undercut when (a) gold (b) tin–lead are used as etch resists in ammoniacal etchants. Note the superiority of tin–lead. [Courtesy of Lea Ronal].

GOLD PLATING AFTER ETCHING

As previously discussed, any blemishes on the surface of the copper substrate can usually be removed at this stage of circuit manufacture prior to gold plating by polishing with a slurry of chalk powder. This is particularly useful for applications requiring thin, pore-free gold coatings.

The standard procedure for gold plating after the etching cycle is to include on the original circuit design art work a connecting plating bar which is common to all or part of the circuit required to be plated. During the actual etching operation the complete circuit pattern is effectively protected by either organic or metal etch-resist coatings and their subsequent removal will depend upon carefully controlled in-line solvent or chemical stripping solutions so that degradation of the base laminate does not occur.

The following operating cycles have been chosen as a way of illustrating the process sequences involved:

Cycle 8—*Selective Gold Plating of Edge Connector Contact Tabs After Etching Cycle*

(A) *PTH Pattern Plated Tin–Lead Etch Resist*

1 Following the etch inspection stage, carefully mask the area of the board above the selected contact area using an approved adhesive tape.
2 Strip the exposed tin–lead deposit on the contact area by immersing either in a hydrogen peroxide/fluoroboric acid/water mixture, or an alternative tin–lead stripping solution.
3 Double tap water spray rinse.
4 Scrub exposed copper on contact area using a slurry of precipitated chalk.
5 Tap water spray rinse.
6 Remove masking tape.
7 Dry, using a hot air, oil-free blower.
8 Re-tape carefully above the selected area.
9 Scrub exposed copper area with fresh precipitated chalk slurry.
10 Tap water spray rinse.
11 Immerse in 10% v/v hydrochloric acid at 18–20°C for 1 minute.
12 Tap water spray rinse.
13 Immerse in 10% w/v citric acid at 18–20°C for 1 minute.
14 Deionised water spray rinse.
15 Transfer rapidly to acid gold strike solution, immersing the boards in the electrolyte with a pre-set 10% total current loading and plate at 1.65 A/dm^2 for 20 seconds.
16 Deionised water spray rinse.
17 Transfer rapidly to gold plating solution, immersing the boards in the electrolyte with a pre-set 10% current loading and plate at 1.1 A/dm^2 for 20 minutes (this is assuming a cathode efficiency of 37% which will give an average deposit thickness of 5.1 microns).

18 Drain and transfer to deionised water drag-out tank.
19 Tap water spray rinse.
20 Remove masking tape, rinse and dry.

(B) *PTH Positive Acting Photoresist Coatings*

1 After etch inspection, the entire resist coating is stripped in an organic solvent mixture, followed by a hot alkaline solution soak at 80–85°C for 10 minutes.
2 Tap water spray rinse.
3 Scrub copper on contact area using a slurry of precipitated chalk.
4 Dry, using a hot air, oil-free blower.
5 Carefully mask area of the board above the selected contact area using an approved adhesive tape.
6 Scrub copper on contact area using a precipitated chalk slurry.
7 Tap water spray rinse.
8 Immerse in 10% v/v hydrochloric acid at 18–20°C for 1 minute.
9 Tap water spray rinse.
10 Continue as in Cycle 8A from sequence 13–20. The remainder of the copper plated circuit can, if required, be prepared to receive an immersion gold coating, as follows:
11 Ensure that all tape adhesive is removed by solvent wiping.
12 Scrub exposed copper circuit using a slurry of precipitated chalk.
13 Tap water spray rinse. Check for absence of water break.
14 Immerse in 1% v/v hydrochloric acid for 1 minute.
15 Tap water spray rinse.
16 Deionised water spray rinse.
17 Transfer to immersion gilding solution at 90–95°C for 15 minutes. This will give an approximate coating thickness of 0.12 micron.
18 Drain and immerse in drag-out tank.
19 Tap water spray rinse.
20 Hot deionised water soak.
21 Remove, drain and oven dry.

From the preceding examples in the section dealing with "Gold Plating Before Etching", a combination of metal sub-plate electrodeposits could be included in this section, but as the operating sequences are basically the same, Cycles 5 and 6 in particular should apply and be read in conjunction with Cycles 8A and 8B. Conversely, if base laminate copper circuits are being considered, after the etching operation has been performed, much the same operating sequences as performed in Cycle 8B should apply and be read in conjunction with Cycles 5 and 6.

REFERENCES

1 Rinker E. C., *Plating*, **40,** 861 (1953).
2 Korbelak A., Duva R., *Proc. Am. Electropl. Soc.*, **48,** 142 (1961).
3 Thompson J. M., Bjelland L. K., *Proc. Am. Electropl. Soc.*, **48,** 182 (1961).
4 Harding W. B., Pressly H. B., *Proc. Am. Electropl. Soc.*, **50,** 90 (1963).
5 Thwaites C. J., *Trans. Inst. Metal Finish.*, **43,** 143 (1965).
6 B.S. 4292; D.T.D. 938.
7 Am. Electropl. Soc. Publ. No. 58 (1970).
8 Hill R., Lea Ronal, private communication.
9 Nobel F. I., Kessler R. D., Thomson, D. W., Ostrow, R. F., Am. Electropl. Soc., Electronics Symp., New Jersey, Dec. (1966).
10 Strickland G. R., *Electroplg. Metal Finish.*, **24,** 8, 6 (1971).

Chapter 36

CONNECTORS

M. Antler

INTRODUCTION

TYPES OF CONNECTORS

An electrical contact is a junction between two or more current-carrying members which provides electrical continuity at their interfaces. Contact-containing components include (*a*) switches, relays, commutators and circuit breakers, which are designed to interrupt or establish current flow in active circuits (b) slip rings, which transmit current from a stationary to a moving frame of reference and (*c*) connectors, which are designed to enable the build-up of circuits from smaller sub-systems and their servicing, and which ordinarily are not mated or separated in active circuits.

There are many types of connectors and they can be categorised in several ways, e.g., (1) whether intended to permanently join conductors or whether designed to permit separation and rejoining (2) according to the means used to effect connection, such as by welding, soldering, severe metal deformation or light pressure (3) according to the size of the conductor and (4) whether for distribution of power or for distribution of low (signal) levels of current.

Noble metal deposits are used where freedom from insulating films is essential, such as with lightly loaded contacts in separable connectors for low energy circuits. Contacts that involve extensive plastic deformation of their surfaces, like the solderless wrap and crimp, can generally be made satisfactorily in the presence of superficial contaminant films. A separable contact usually has a wire joined to it, and a noble metal is frequently deposited on this terminating surface as well.

Separable electronic connectors are made in many configurations and sizes. Figure 36.1 shows some typical designs[18] of contacts used.

WHY USE GOLD?

Several noble metals have been used commercially on contacts for electronic connectors, but only silver and gold have widespread acceptance. Silver is, of course, limited by its tendency to tarnish in environments containing sulphur or hydrogen sulphide and is therefore not used for critical, long-life applications having open circuit potentials below a few volts, despite its low cost.

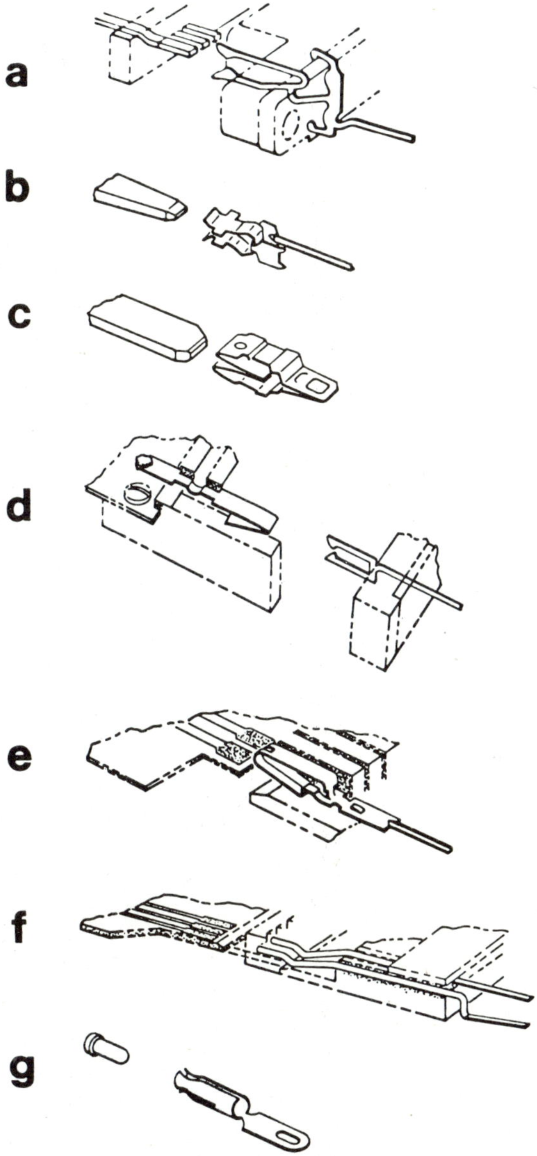

Fig. 36.1 Contact spring configurations for plug-in connectors. (a–d) blade-fork contacts (e) folded cantilever contact for printed circuit boards (f) straight cantilever contact for printed circuit boards (g) pin-socket contact.

Rhodium (and probably ruthenium) is useful above 200°C because it resists interdiffusion and sticking, a problem with gold at high temperatures when the contacts are mated for a long time. Also, it is a barrier for base elements such as copper from the substrate which otherwise would diffuse to the surface. However, these elements and palladium (the only members of the platinum group metals which are used in contacts) catalyze the formation of polymeric insulating solid films on their surfaces in atmospheres containing traces of organic pollutants[1]. This tendency is pronounced when the contacts are lightly loaded and undergo repeated engagements. The tendency of gold to catalyse such films is relatively minor and the reliability of devices with gold contacts is rarely degraded this way. Ruthenium and rhodium are more expensive than gold. The brittleness of these two metals limits them to contact elements that undergo little deformation in service. They also have a tendency to crack during plating because of internal stresses, particularly in thicknesses over 40 microinches. These fissures limit the corrosion resistance of contacts when the substrates are base metals. Palladium develops tough tarnish films in atmospheres containing sulphur dioxide at high relative humidity.

Intensive research into the properties of deposits for contact applications, and the development of many novel plating solutions and deposition techniques, have resulted in a situation today where few connector contact engineering requirements cannot be satisfactorily met by gold plate. The availability of this metal in unlimited amount is also a stimulus to its use.

Although most gold for contacts is applied by electrodeposition, cladding and welding are also used on selected portions of contact assemblies. For example, the ends of cantilever contacts of connectors for printed circuit boards having edge contacts are often coated by these methods. However, other precious metals are frequently alloyed with gold in claddings and weldments, e.g., 69% gold, 25% silver, 6% platinum; and 70% gold, 30% silver.

REQUIREMENTS OF GOLD PLATE

The requirements of gold plate for connector contacts can be described in terms of:

1 Chemical properties, which concern its ability to retain a surface free of films.

2 Mechanical properties, which involve its ability to withstand wear and deformation.

3 Joining properties, such as solderability, crimpability and performance on solderless wrap posts.

4 Process characteristics, which deal with the techniques for applying the deposits to surfaces.

CHEMICAL

The most important requirement of gold plate is that it should be free of films initially and that it retain its clean surface during the service life of the connector. For a full discussion of the factors which degrade contacts, the reader is referred to Chapters 21 and 22. The following text, however, summarises certain characteristics which determine whether the deposit will be film-free.

Porosity

Gold is ordinarily deposited on base metal underplates or substrates, as-plated porosity in the gold being the Achilles heel of the contact. In benign environments, films do not form on the surface, but in polluted atmospheres, especially at high relative humidity, insulating films may develop which cannot be broken by the contact pressure. The effect of films is most often observed on mating the contact after aging "unmade". Even a satisfactory mated contact can fail in time by ingress of atmospheric pollutants through the interstices between the members. Because contact is made at relatively few scattered asperities, the apparent or geometric touching area is mostly empty space. Therefore, low or no porosity in the gold plate is desirable, even for separable connectors that are seldom exercised while in service. Alternatively it is possible in certain circumstances, through judicious choice of underplate or by post-plate chemical inhibitors, to reduce the extent of film formation due to the environment.

Sulphides which form at pore sites on gold plated silver, copper and copper alloys are able to migrate over the surface of the gold. Figure 36.2 is a plot of distance of the advance of sulphide films at severe atmospheric conditions[7]. Since these films are insulating[10], care must be taken in designing contacts to avoid cut edges or selective plating which permits exposure of the substrate so close to the contact site that films can migrate to it. For example, it is common to chamfer the leading edge of printed circuit boards which have edge contacts after plating with gold so as to facilitate insertion into connectors. If the components serve in an environment containing sulphides, films may form on the exposed copper foil of the land and eventually migrate to the board contact. In addition, copper may transfer to the connector contact during engagement where it can form insulating films—a secondary cause of resistance failure.

The volume occupied by films which are confined to pore sites (i.e., that do not migrate) can be large. Examples are the corrosion products from gold plated nickel and copper which form in atmospheres containing sulphur dioxide at high relative humidity. On mating the connector, these solids may be dislodged and interfere with contact.

Inertness of the Bulk Deposit (Alloy Golds)

Gold deposits alloyed with significant amounts of base metals are more susceptible to film formation than those with a gold content over 99 %. One reason for this is that alloying elements may not be uniformly distributed throughout the gold. Islands rich in base metal may be on the surface[4] and if so, such deposits would be especially prone to degradation. Little is known about the intrinsic chemical stability and contact performance of high alloy gold systems.

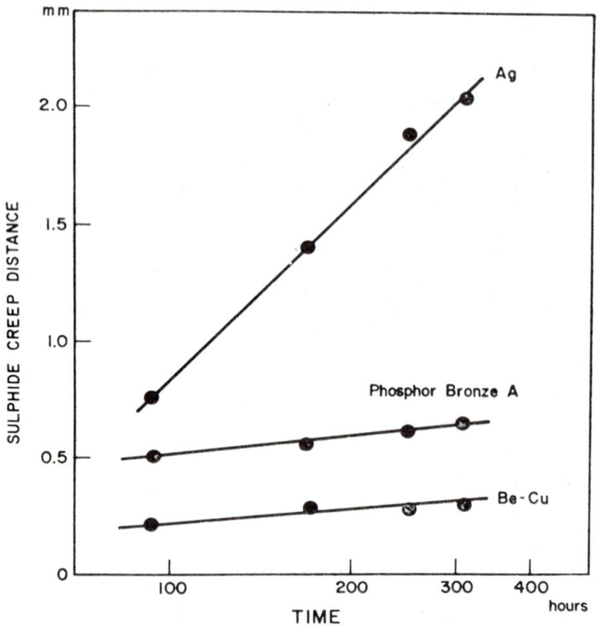

Fig. 36.2 Plot of distance of farthest advance of sulphides formed on base metals over an adjacent gold plating vs time. (Portions of the substrates were masked before plating so as to produce sharp interfaces between the deposits and the unplated areas.) The gold coatings were 200 μ-in. thick on the substrates indicated. Exposed for two weeks to 1.6 ppm of H_2S in air at 85 % RH and 32°C. Air movement over samples was 50 linear ft/min.

Thermal Stability (Diffusion)

The thermal stability of a gold deposit rests mainly on its ability to resist the formation of films by diffusion to the surface of base elements from the substrate, or of hardener metals which are codeposited with the gold. Nickel underplate is a barrier to diffusion and is widely used for contacts that serve above 125°C.[19] Diffusion rates at room temperature have recently been experimentally determined[21]. The diffusion rates were found to be surprisingly high and diffusion barriers may be necessary for contacts that are used below 125°C, particularly in the most long-lived and highest reliability applications.

Codeposited Non-metallic Material

For many years it was generally believed that gold deposits were entirely metal, either pure or alloyed with other elements. Some years ago poor solderability of gold plated contacts was related to a tendency for the deposit to "froth"[15]. When rapidly heated over a period of a few seconds to a temperature exceeding 500°C, the surfaces were observed sometimes to exude vapours or tarry materials. Although the compositions and origins of these non-metallic substances were not fully explored, it seems likely that they were either minute quantities of plating solutions trapped in the internal pores in the deposit or intrinsic polymer[14]. The anomalous low densities of some gold deposits probably originate in these impurites.

MECHANICAL

Ductility

Contacts may be formed after plating. For example, they are sometimes staked and bent prior to moulding in a plastic connector body. Socket contacts may be coined to form a partially closed entry which protects it from damage. These deposits must therefore be able to be deformed without cracking, if forming occurs at or near the contact site. Cracks have the same effect as porosity, the chemical resistance of the surface being degraded. Another related requirement of the coating is that it be able to withstand the flexing of any part, such as a spring contact, on which it is deposited.

These properties concern ductility of the deposit. Most golds that are used on contacts are not pure and hardening elements which they contain may significantly reduce their ductility[20]. Very few measurements of the ductility of gold plate have been made. In practice, tests are often performed by wrapping a plated part around mandrels of various diameters and checking for surface cracks with a 10 power loupe or by electrography[9].

Hardening elements decrease ductility and should be limited in the deposit. For example, cobalt or nickel in gold should not exceed 0.5%. Metallic impurities may also decrease ductility, iron especially being a serious offender. As a guide, it is recommended that impurity levels do not exceed 0.1% for any element in the deposit. Even so, it may be difficult to do some post-plate forming operations with hardening elements and these impurity levels. Limits can be established in particular cases by practical experience.

Friction and Wear

Low wear and friction are requirements of gold plate for contacts. Wear exposes the base substrate or underplate and is as detrimental to contact performance as plated porosity in the gold. These properties are considered further in Chapter 21.

The wear of gold in any application depends both on contact design and the intrinsic properties of the plating. Normal load and distance of sliding in mating a contact determine the rate of wear. Less well known is the effect on wear of the geometry of the contact members. Experience has shown, however, that among the most durable golds are those which contain from 0.1%–0.5% of nickel or cobalt, and have KHN_{25} hardnesses in the range, 140–250 kg/mm^2. Plating on a hard substrate or underplate (such as nickel) and contact lubricants may also improve the durability of the gold.

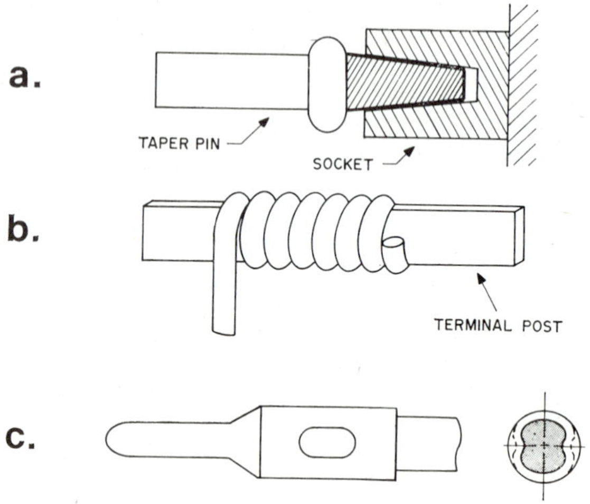

Fig.36.3 Terminations to contacts. (a) taper pin (b) solderless wrap (c) typical crimp.

Friction is important, because electronic connectors often contain large numbers of contacts, and insertion and withdrawal forces should not be excessive. In practice, the coefficient of friction depends more on the surface cleanliness of the gold deposit than on its type. Lubricants can reduce both friction and wear of the gold plate.

JOINING

The back ends of contacts for separable connectors are usually joined to wire conductors directly by soldering, crimping, or solderless wrapping. Such joints are not designed to be taken apart. Connection to wire is sometimes made indirectly by taper pins (Figure 36.3 a). The latter are high pressure contacts at the rear of the separable members in which sockets with tapered walls are mated by pins to which the conductors are crimped. Taper pins permit a few engagements.

These methods of making permanent connection do not require gold plate on the contact. Less expensive finishes, such as tin–lead alloy, are normally suitable. Since most separable contacts are quite small, barrel plating techniques are used to deposit the gold. This, however, results in the entire surface being covered. Accordingly, it has become a requirement that gold plate for separable contacts must also be suitable for permanent joints. This is not unfortunate, because only a limited thickness of insulating oxide or tarnish films can be tolerated on contacts to be permanently joined, and gold plate can keep such surfaces clean.

Solderability*

Soldering to electronic components in most cases is done with tin–lead alloys and inactive rosin fluxes. Solderability is dependent on surface cleanliness. Connector contacts having gold deposits at their back ends are, therefore, readily soldered provided that the finish has not deteriorated prior to installation of the connector by diffusion of film-forming substrate elements (such as zinc from brass or silver from silver underplatings). Porosity in gold can be troublesome for the same reason.

There are two conditions where gold can cause bad joints:

(*a*) High concentrations of codeposited organic substances may produce[22,23] weak joints. This condition does not occur with properly maintained gold plating baths.
(*b*) Tin–gold intermetallics form in the soldering process and when excessive give mechanically weak bonds. Contacts with gold deposits less than 100 microinches thick are not usually degraded in this way.

There is little difference in the solderability of gold electrodeposits which are at least 99% pure on any of the common underplatings with copper or copper alloy substrates.

Solderless Wrapping

Solderless wrap joints are made by wrapping a conductor around a hard post, normally having a rectangular cross-section with sharp edges (Figure 36.3 b). The wrapping tool applies tension and as a result the wire is severely deformed at the corners. Each turn of the wire makes four contacts, six turns commonly being used in a connection. The multiplicity of contacts in each joint and their gas tightness makes this a highly reliable connection.

It has been found[11] that solderless wrap connections to gold plated posts can be made that are able to withstand exposure (after wrapping) to a variety of corrosive laboratory environments and to temperature and humidity cycling. A satisfactory joint was defined[11] as one which (*a*) required a force to

* For complete background, refer to Chapter 19 by Thwaites.

strip the wire from the post that was equal to or greater than the tensile strength of the wire and (*b*) showed little or no increase in resistance compared to an unaged joint made to a clean post.

The platings evaluated included both "hard" (KHN_{25} = 126–175) and "soft" (KHN_{25} = 125 maximum) golds. Deposit thicknesses were 30–60 microinches on copper, silver, or nickel underplates, or 10 microinches of gold plated directly on the beryllium–copper substrate. However, if the posts were exposed to the ageing environment before wrapping, all of the platings were satisfactory, except the 10 microinch deposit which was unable to adequately protect the surfaces from corrosion. A lubricant was able to provide supplementary protection to that afforded by the gold plates.

In other tests, it was found that soft (pure) gold deposits on solderless wrap posts gave higher wire strip-off forces compared to tests using posts plated with hardened golds[16].

Crimping

Figure 36.3 c illustrates a typical crimp joint. It is made by compressing the terminal tightly onto the conductor with a special tool. Both terminal and wire are severely deformed.

Critical studies of the requirements of gold plate for crimping have not been reported. Ordinarily, crimps are made to new contacts, or to contacts that are relatively free of corrosion films. Deformation at the joint is severe and gold deposits that are sufficiently ductile to perform well at the front end of the contact may crack at the crimp. Cracking is not likely to interfere with the quality of the joint, i.e., significantly diminish the pull-out force of the wire or contribute to increase in contact resistance of the crimp joint when it is exposed to aggressive atmospheres. It is assumed, of course, that the deposits are normal in every respect. If the platings have marginal adhesion to the substrate (or to each other for multilayer systems), the deposit may flake from the barrel on crimping, and this could lead to electrical shorting of adjacent contacts in a connector if these loose particles are not removed. Cosmetically, the deposit on the outer, non-functional, surfaces of the crimp barrel may discolour at crack sites when aged.

Finally, it has been found that gold deposits sometimes transfer to the crimp dies where they build up and contribute to sticking of the contact barrel in the tool. Transfer may depend on the type of gold as well as on details of the crimping operation. A trace of lubricant on the die or on the outer surfaces of the crimp barrel can minimise this tendency.

ELECTRODEPOSITION PROCESS

Although gold plating procedures are considered in several chapters in this book, it is nevertheless relevant to emphasise several process requirements that are particularly important for connector contacts.

Throwing Power

Many contacts have deep recesses, such as the barrels of sockets, crimp receptacles and solder cups. Although it is usual to drill a small hole at the bottom of deep recesses to facilitate flow of plating solution (in crimp barrels the hole is also useful for inspection to assure the operator that the conductor is fully inserted), in practice, solution replenishment is often poor. Consequently, gold may not be uniformly distributed from the top to the bottom of the recess. Post-plate rinsing is also critical to minimise the occurrence of plating salts in these cavities. Since the functional surface can be considered to be the entire length of the contact, especially at the separable end, uniform deposits are desirable.

Gold plating solutions vary in their macrothrowing properties. Those with better characteristics are preferred, provided deposits meet composition and property requirements for good contacts.

Selective Application

Several selective plating processes have been devised to localise gold on the functional surface for cost reduction, or where two deposits are needed at different places on the contact. The most important of these methods is racking and partial immersion of the part in the plating solution. Selective masking, brush plating and jet plating (in which a stream of gold plating solution is made to impinge on a portion of the contact) are used in special cases. The application details of these processes vary with the component and are discussed elsewhere in this book.

The properties of deposits from common plating baths which are produced by some selective processes, may be different from deposits obtained by conventional racking or barrel techniques. In particular, the current density, degree of agitation during plating and other conditions are grossly different in brush and jet plating from conventional procedures. This may affect the chemical and mechanical properties of the deposit and as such, must be determined before these processes can be considered suitable for contact applications.

High Alloy Deposits

Deposits that contain 10% or more of base metals are attractive for cost reduction; the alloys not only contain less gold than conventional coatings, they are also less dense. Furthermore, some high alloy golds may possess superior mechanical characteristics compared to those low in alloy content.

High alloy golds have not been systematically explored for contacts. In addition to adequate chemical stabilities and mechanical properties, such deposits must be platable under production conditions in a reproducible manner, i.e., their solutions must produce deposits of nearly uniform compositions and properties. Unfortunately, the process requirements for alloy gold plating are stringent and often difficult to control. Since extensive process study is required before high alloy golds can be attempted in production, this has tended to limit their use.

SELECTION OF GOLD PLATINGS

FACTORS TO CONSIDER

The connector designer is concerned with specifying a finish that will provide an adequate level of surface protection against contamination by the environment in which the contact is to serve. Furthermore, it must resist deterioration during the manufacture of the connector, and later during the shipping and handling of the system of which the connector is a part. The plater, on the other hand, must select his process and equipment, determine operating conditions and institute process control, and utilise inspection procedures after plating to obtain the deposit that is required. Early collaboration of designer, plater and other personnel involved in the manufacture of the connector is desirable so that the most efficient and least costly methods are chosen.

Design for plating, in shape and dimensions of the contact, should follow well known principles. For example, care must be taken to avoid designs which in barrel plating will permit nesting or excessive tangling. Sharp corners and blind holes are undesirable.

The choice of finish itself, including the type of gold, its thickness and whether underplatings or post-gold plating surface treatments are to be used, must be made by considering (*a*) the functional requirements of the contact (*b*) the overall design of the connector (*c*) its intended service life-time and (*d*) the level of reliability in service that it is necessary to attain.

Functional requirements include mechanical, chemical, and thermal factors. Mechanical considerations involve the number of cycles of mating, sliding distance, contact load and whether the contact will be deformed in its manufacture and use. Chemical considerations include degradation due to the aggressiveness of the environment which the contact will experience, particularly by corrosive air pollutants and high relative humidity. Thermal factors relate to possible deterioration of the gold plate by diffusion of base elements from the substrate to the surface.

The overall design of the connector determines the degree to which its shell protects the contacts from exposure to the atmosphere. Tight fitting, and particularly hermetic, designs are desirable when the connector will be used in a severe environment. Hermetic designs require that the plastics, adhesives and elastomeric materials in the connector structures do not contain substances which can transfer, e.g., by outgassing and condensation, onto the contact surface, and which can corrode it or otherwise accumulate and increase contact resistance.

An important design factor involves the number of independent surfaces which join each contact. Redundancy is often desirable. Finally, the load and wipe designed into the connector enable some films that may appear on the contact surfaces to be tolerated. Although freedom from porosity in the gold plate is desirable, and even essential for the most reliable connectors and those intended for the longest service, contacts for less stringent applications can function with thin, hence porous, platings.

ROLE OF UNDERPLATE

An underplate can be defined as an electrodeposit of at least 20 microinches in thickness, to distinguish it from a strike or flash deposit. Extensive studies of the effect of an underplating on the properties of gold plated contacts have been made and it is clear that they can contribute major improvements in performance for a given thickness of gold. The purpose of an underplating is to achieve one or more of the following: (*a*) reduce the sliding wear of gold, (*b*) diminish the corrosion susceptibility of the system or (*c*) stop the tendency of the deposit to degrade by diffusion of base elements to its surface. An underplate may enable less gold to be used to achieve satisfactory behaviour.

Underplatings used for gold plated contacts are normally copper, nickel and silver. There is also limited use made of rhodium, palladium and tin–nickel. Silver at one time was the most commonly employed underplate for thin gold deposits. This practice was an historical accident and originated in the fact that before low energy circuitry became commonplace, silver was a preferred top plate. However, its tendency to form sulphides created contact reliability problems and difficulty in soldering to the back end. It was found that these problems could be partially mitigated by a gold overplate. Thus, gold was considered to protect the silver, rather than silver to upgrade the properties of the gold (see "Thin Gold", page 492).

Corrosion Resistance

The corrosion susceptibility of electroplated gold depends primarily on its porosity. If it is porous, the nature of the metal on which it is plated and the environment become controlling factors. As discussed in Chapter 23, an underplate can reduce the porosity of gold by covering substrate surface defects, such as occluded non-metallic impurities, from which most pores originate. The underplate can, therefore, be a better substrate for the gold deposit than the basis metal. Since porosity of a given mass of deposit is also directly related to the roughness of the surface on which it is plated, a levelling underplate is particularly desirable.

Pores in a deposit can be either serious or relatively unimportant, depending on the chemistry of the plate-underplate system in the environment in which it must serve. This should be considered when selecting an underplate. Thus, assuming there is some porosity in the deposit (either as-plated or produced by wear), a nickel underplate is preferred for contacts that serve in air-conditioned environments, such as those in connectors fitted to computers for business offices. The gold–nickel couple is practically inert below its critical relative humidity (70%) and this level is generally higher than normal in an air-conditioned office. On the other hand, porous gold over copper or silver will develop sulphide films below 70% relative humidity, if sulphur vapour or hydrogen sulphide are present in the atmosphere. These characteristics are considered in greater detail in Chapter 22. Table 36.1 summarises the factors to consider in choosing an underplate for a connector contact.

Wear Resistance

Hard underplatings, such as nickel, are effective in reducing the wear of gold plate, particularly with large loads, when the substrate is soft and deformable. This is fully discussed in Chapter 21. The wear relationships are complex and depend on the relative thicknesses of the deposits, their hardnesses and that of the substrate, normal load, and to a degree on contact geometry. Table 36.1 considers these factors for various underplatings.

Diffusion

Diffusion of substrate elements through gold plate to the surface to produce insulating films may limit connector life. This can occur not only at elevated temperatures, but also at room temperature in extended service. Barrier underplatings can markedly reduce diffusion, nickel being the most widely used metal for this purpose. Diffusion degradation is considered in Chapter 22, and is also referred to in Table 36.1.

THICK GOLD

It is convenient to classify gold coatings for connector contacts into "thick" and "thin". The former are plated to 100 microinches or more. However, rarely are deposits greater than 200 microinches specified for connector contacts. Thin deposits generally range from 20–60 microinches.

The justification for this classification rests historically in the attitude of the design engineer to porosity in gold plate. When the importance of porosity in controlling contact reliability became evident, users of gold plate having the most critical requirements began to over specify its thickness so as to be virtually free of pores. This practice assumed importance in the 1950's with the development of silver-hardened and later other, generally superior, alloy golds that were over 99% pure. With ordinary plating procedures on common substrates, particularly if underplates are not used, it takes 100–200 microinches to have substantially pore-free gold. The critical users were notably in

Table 36.1
Selection of Underplate for Gold Deposit

Purpose of Underplate	*Underplate*				
	Copper	*Nickel*	*Silver*	*Rhodium or Palladium*	*Tin–Nickel*
1 levelling (for thick underplates, to reduce porosity in gold plate)	yes	yes; avoid bright (brittle) nickels for substrates which will be deformed	*	no	*
2 cover substrate surface defects (to reduce porosity in gold plate)	yes	yes	*	no	*
3 diffusion barrier	yes, for Zn in brass substrate	yes, for Cu and Zn in basis metals	no (Ag diffuses rapidly through Au)	yes	*
4 wear resistance	no	yes	no	yes	*
5 corrosion resistance	no	yes, if RH is below 70%, and for service in sulphide atmospheres	yes, for salt spray environments	yes (Au top plate reduces friction and polymer formation on Rh and Pd)	yes (references 5, 6). Sn–Ni is very brittle and should not be used on substrates that will be deformed

* No published data known to author.

the computer industry during its conversion from high voltage vacuum tube and electro-mechanical relay component technology, to the technology of solid state devices, such as transistors, which function at significantly lower voltages and are less tolerant of corrosion films on connector contacts. The telephone industry also began to specify thick gold because the lifetime of telephone systems is of the order of 20–40 years, and every type of component in them must have very high reliability. On the other hand, the use of thin golds originated, as described earlier, in the attempt at rectifying the contact solderability difficulties of tarnish prone silver plate.

Thick deposits are widely used in the connector industry and, indeed, their need has been the chief stimulus to the development of selective plating processes and of cladding and welding technology which localise precious metal coatings to functional surfaces. Probably the most important application for thick gold deposits is on the edge contacts of printed circuit boards and on the mating surfaces of their connectors. Since it is difficult to design boards and connectors so as to exclude the environment from their contacts, the possibility for corrosion degradation is great.

THIN GOLD

Thin gold deposits, less than 100 microinches, are specified when a good level of corrosion resistance is necessary, but where a pore-free deposit is not required. These deposits are generally satisfactory where the environment is not severe, or where the connector is designed to effect good sealing of the contacts. Developments in underplatings, as previously described, and in post-plate treatments (discussed in the following section) which can improve reliability for thinner golds, have often made it possible to equal or surpass the performance of thick deposits used alone.

The bulk of connectors having gold plated contacts which serve general purpose applications, including those in the United States intended for aircraft and military service, are plated with a minimum of 30–50 microinches of gold, the underlayers generally being 30–150 microinches of copper or nickel. The application of silver underplate (usually at a minimum thickness of 200 microinches) is diminishing since it readily diffuses to the surface of thin gold where it can degrade contact performance. Furthermore, silver sulphide formed on the surface at pore sites in the gold readily creeps over the surface[3]. The superior properties of electroplated gold on copper or nickel are responsible for this change.

Gold is a significant factor in the cost of manufacturing connectors[17]. Typically, for thin deposits, about 10–35% of the connectors's cost is in the plating (based on 1973 gold prices)*. The value of the gold is generally greater than the expense of the plating process. This figure varies widely depending on connector design and on production levels.

* November figure of $100 per troy oz. used for the purpose of this calculation.

POST-PLATE TREATMENTS

Improvements can often be achieved in the corrosion resistance and sliding wear of gold plated contacts through corrosion inhibiting post plating treatments and by contact lubrication[12]. There is little information on field experience with these systems, particularly over long periods of time. The compositions of most post plating formulations and their methods of application are proprietary. Their commercial use is not uncommon.

Corrosion Inhibition

A number of corrosion inhibiting treatments for gold plated metals are described in the literature. In the following text, however, typical examples cited are limited to instances where the contact resistance of specimens was measured and not found to be seriously impaired.

Chromate based treatments are commonly used for thin gold plate and silver underplate on copper or copper alloy substrates. For example, gold deposits from an acid solution resist tarnishing in gaseous hydrogen sulphide or aqueous potassium sulphide after being dipped in aqueous potassium dichromate[8]. Chromating solutions protect exposed copper at the edges of gold plated edge contacts on printed circuit boards from corrosion in salt spray[13]. *N*-alkoyl or *N*-alkylaminocarboxylic acids and their salts, such as *N*-oleoyl sarcosine, are described as corrosion inhibitors in gaseous sulphur dioxide at high relative humidity and in field environments[12]. Systems evaluated included gold plated copper and its alloys, both with and without a nickel underplate.

Lubrication

The practice of contact lubrication is very old. Werner von Siemens[2] in 1860 is quoted as having said, "Metals without petroleum do not conduct at all". Since the sliding wear and high friction of contacts often seriously limits their life, lubrication of gold plated connector contacts is quite common. This point is fully covered in Chapter 21.

Octadecylamine hydrochloride is an effective solid lubricant, provided the gold plated surface has a roughness of at least 10 microinches centre line average. However, at high relative humidity in atmospheres containing sulphur dioxide, it increases the corrosion of gold plated copper when a nickel underplate is present. Lubricants with a fluid component, such as greases, do not require rough surfaces to be durable. It is possible to combine corrosion inhibitors with lubricants to achieve the advantages of reduced friction and wear, as well as protection against corrosion due to as-plated porosity in the gold and corrosion which follows sliding that, at its worst, results in wearing-through of the deposit. A typical treatment[12] involves as the inhibitor, the sodium salt of *N*-lauroyl sarcosine, applied by dipping contacts in a dilute solution in water. This is followed by immersing the parts briefly in a lubricant solution of a mixture of a 5-ring polyphenyl ether and a paraffin wax dissolved in 1,1,1-trichloroethane. The solvent evaporates from the contact after it is removed and leaves behind a residual protective film.

REFERENCES

1 Hermance H. W., Egan T. F., *Bell Syst. tech. J.*, **37**, 3, 739 (1958).
2 Holm R., "Electric Contacts Handbook", 3rd Edn, pp. 400, Springer, Berlin & New York (1958).
3 Egan T. F., Mendizza A., *J. electrochem. Soc.*, **107**, 4, 353 (1960).
4 Brenner A., "Electrodeposition of Alloys", Vol. 2, pp. 518, Academic Press, New York (1963).
5 Clarke M., Britton S. G., *Corros. Sci.*, **3**, 5, 205 (1963).
6 Clarke M., Leeds J. M., *Trans. Inst. Metal Finish.*, **43**, 2, 50 (1965).
7 Antler M., *Plating*, **53**, 1431 (1966).
8 Goethner G. A., *Metal Finish.*, **64**, 7, 58 (1966); **64**, 8, 62 (1966).
9 Noonan H. J., *Plating*, **53**, 461 (1966); see esp. Fig. 22.
10 Holm R., "Electric Contacts", 4th Edn, pp. 110, Springer, Berlin & New York (1967).
11 Auriana M., Proc. Nepcon, Long Beach, California, Jan. (1968); New York, June (1968).
12 Krumbein S. J., Antler M., I.E.E.E. Trans. on Parts, Materials, and Packaging, PMP-4, 1, 3 (1968); U.S. Pat., 3,484,209 (1969).
13 Farmer M. E., Guttenplan J. D., *Plating*, **56**, 795 (1969).
14 Munier G. B., *Plating*, **56**, 1151 (1969).
15 Ales R. E., private communication (1970).
16 Balacek F. P., private communication (1970).
17 Markstein H. W., *Electron. Packag. Prod.*, **10**, 11, 17 (1970).
18 Van Horn R. H., Proc. Electronics Components Conf., pp. 162, Washington, D.C. (1970).
19 Antler M., *Plating*, **57**, 615 (1970).
20 Baker R. G., Palumbo T. A., *Plating*, **58**, 791 (1971).
21 Pinnel M. R., Bennett J. E., Proc. Engn. Semin. on Electrical Contacts, pp. 13, Illinois Inst. of Technology, Chicago (1971).
22 Pasciak A. M., NASA TM-2290, May (1971).
23 Davis M. V., Proc. 4th Plating in the Electronic Industry Symp., pp. 210. Am. Electropl. Soc., Indianapolis (1973).

Chapter 37

SEMICONDUCTORS AND MICROELECTRONICS

E. F. DUFFEK

The rapid growth of the semiconductor industry has brought about major changes in gold plating technology. Electroplating with gold provides functional, engineered coatings that meet the wide variety of electrical, metallurgical, chemical, and mechanical requirements. Two distinct electroplating areas have emerged:

1 Electroplating of the component package piece parts and assemblies such as headers, leadframes, semiconductor die and wire bond areas, and metallised ceramics. The materials include, metals, ceramics, glass and plastics.
2 Electroplating on the active semiconductor element as an integral part of the device. The main purpose is to provide reliable, stable electrical contact to an external lead. Typical applications are wireless bonding, i.e., gold beam leads and "flip-chip" raised contacts.

Why use gold? In spite of its major cost and consequent higher component cost, gold is still required because of its many desirable properties. Metal component piece parts are plated to protect the metal during the processing, testing and life of the device. Gold facilitates chip bonding (die attachment) and wire bonding due to its softness. It also provides tarnish, oxidation and corrosion resistance, good solderability, weldability and resistance to chemical etchants and low electrical contact resistance, all of which are maintained almost indefinitely.

The time has come when technologists must really think. It does not take long to realise material costs alone. A 100 microinch (2.5 micron) deposit costs 10¢/in^2 of plated area. In the past, semiconductor components were reasonably priced with good profit margins and gold could be used rather in excess so long as the requirements were met. With the current economic climate, costs and selling prices are almost the same; the manufacturer must therefore economise in all areas to stay profitable. A most obvious place is gold plating cost reduction. Now engineers must develop technologies involving the minimum amount of this metal. Component piece parts must be made in high volume at low cost. Plating engineers and related personnel must arrive at mechanically functioning, selective, automatic equipment and achieve control of thickness distribution, metallurgical, electrochemical and functional factors on piece parts. Developments in high speed plating and mechanisation are required.

Why semiconductor packages? The semiconductor element (die, chip) for a diode, transistor, or complex integrated circuit (IC) must be protected from the hostile environment of the outside world. Moisture, ionic and particulate contamination, even air (O_2) and light contribute toward electrical degradation, corrosion of the thin metal patterns on the die and eventual failure of the device. Other considerations involving package design are:

1 Interfacing between the semiconductor die and electronic systems, i.e., bringing the miniature leads from the die to a set of large external leads, with support and isolation.
2 Compatibility with systems' requirements, i.e., heat dissipation, microwave stability and radiation hardening resistance.
3 Shape and configuration. Ease of handling for circuit board insertion.
4 High volume, low cost manufacturing.

Packaging technology for each of the various device types will be described briefly in this chapter since this will best explain the purpose and requirements of gold plating. Packages include metal cans, ceramic, glass-to-metal, single and multilayer metallised ceramic substrates, and plastic encapsulations. Device types are numerous and include diodes, transistors, integrated circuits, metal on silicon (MOS), memory arrays, hybrids, microwave and optoelectronics. The reader should consult additonal references on this subject[1-5].

GOLD PROCESSES*

PLATING SYSTEMS

Gold deposits for the semiconductor industry are Type I (99.7% Au) and Type III (99.9% Au) as described in MIL-G-45204B.[40] Hardness is specified as Grade A (90 Knoop max) since soft gold has generally been associated with purity. High purity hard golds (up to 190 Knoop) are available, but their effect on device assembly particularly wire bonding, must be evaluated in each case.

Table 37.1 summarises the bath types available. Still (rack) and barrel processes operate with a metallic gold content of about 8.2 g/l, "strike" baths at 4.1 g/l and high speed processes for selective, strip or wire at 16.4–49.2 g/l. Duplex coatings (other than a strike) are frequently used to achieve properties of two different types of gold, such as an initial deposit from an acid system

*For complete background, refer to Chapters 3–7.

for adhesion and a second plate from a neutral bath for resistance to discolouration on heating. Alloy golds are not used due to the instability of the deposit and variation in properties. The alloying of gold with low melting metals, i.e., Au–Sn, promises to find more application. Preplate layers of Cu, Ni, Ni–Co, Co, Ag and Pd are in use. Nickel is used as a diffusion barrier between copper and gold.

Acid (pH 3.0–4.5)

These systems give excellent adhesion to metals. They have been used extensively as strikes and finish plating. Unfortunately, they are the least tolerant to heavy metal and organic contaminants, these materials codepositing according to their concentration. Baths out of control drop below 60% in cathode current efficiency with problems described in the following section under "Control". Organic contamination is observed as a "reddish" deposit after heating. Batch type carbon treatment must be carried out weekly or on a regular scheduled basis.

Mild Acid (pH 4.5–6.0)

These baths are somewhat more resistant to codeposition than the lower pH systems. They require similar control measures. Current efficiencies should be about 100%.

Neutral (pH 6.0–8.5)

A higher pH and the presence of metal complexing agents allow these processes to be less subject to heavy metal codeposition and organic breakdown. The Hull cell has been found useful for the control of stabilisers, grain refiners and other components of these baths. An additional feature of these systems is their ability to control the bipolar effect which tends to dissolve gold or the basis metal anodically on non-contacting or less cathodic zones. Barrel plating distribution is therefore more uniform and narrower plating ranges, which result in cost reductions, become possible. Current efficiencies are close to 95%. These properties have led to their wide acceptance as plating solutions and more recently as strikes.

Alkaline Phosphate, Alkaline Cyanide (pH 8.5–13.0)

The optimum conditions for these systems are still under development. They are modifications of the early gold cyanide processes. The alkaline nature of these baths should allow a good tolerance toward heavy metal contamination. The best indication of the need for carbon treatment is a reddish discolouration after heating.

Non-cyanide (pH 6.0–11.0)

These systems are based on sulphite gold complexes[6] rather than cyanide complexes and offer high purity and freedom from breakdown. Iron and nickel are not appreciably complexed so codeposition is not a problem. Throwing power, levelling and waste disposal are further advantages. The baths, however, are more sensitive to operating conditions and composition control. The optimum pH is 9.5; below a pH of 8.0, the gold precipitates out. Earlier processes discoloured on heating, but this problem has now been resolved. Deposits from non-cyanide gold solutions are harder than the 90 Knoop maximum specified and must therefore be evaluated according to the user's die and wire bonding conditions. More recent solutions however, operate at lower pH's.

High Speed

Processes are generally formulated from acidic or neutral solutions with high gold contents. Their need arises from the automated, selective plating of leadframes for plastic encapsulated devices. Cathode current densities range from 50 to well over 1,000 A/ft^2.

Electroless, Immersion

These processes are not in common use due to their irreproducible behaviour and erratic adhesion to the basis metal.

CONTROL

Contaminant Codeposition

Gold plating technologists have continually strived to meet the requirements of the semiconductor industry. It is well recognised that bath maintenance, purity, filtration and cleanliness are of the utmost importance in the deposition of reproducible, high quality, pure gold. It is also agreed that processes maintained within the specified composition limits, but contaminated with certain metals, such as iron, nickel, copper, organics and so forth, will have a deleterious effect on the quality of the deposit. A decrease in the cathode current efficiency, erratic thickness distribution and not infrequently poor adhesion, are all related to these effects. The result of this sub-standard deposit, believed to be due to the codeposition of contaminants, is discolouration on heating, increased porosity, surface residues and roughness which are observed at device assembly, testing or soldering. Herein lies one of the most critical control areas and unanswered questions in gold plating technology. The plating processes exemplified by the acid systems contain complexing materials, such as phosphates and citrates. Iron, nickel, cobalt and copper readily form complexes which subsequently codeposit with gold at decreased current efficiencies. An alternative explanation for the effect of small quantities of certain impurities is not that they are codeposited, but act only to modify the properties of the gold, such as hardness and brightness. In this case, the purity of the gold deposit will not be related to the effects observed in device assembly. The change throughout the industry from low pH to neutral systems which are more tolerant of contaminants has alleviated, but not solved this problem. Organic impurities cause similar problems. A summary of maximum metal contaminant levels for various gold solutions is given in Table 37.1. Concentrations are easily determined by atomic absorption methods (see Chapter 33). Control at this level will result in high quality, pure gold deposits well suited for semiconductor assemblies. As a first approximation, about one-third of these levels will appear in the deposit-concentrations

Table 37.1

Gold Plating Systems for Semiconductor Applications

Pure Gold Type	*pH Range*	*Maximum Tolerance Level of Heavy Metals in Bath (ppm)*				*Carbon Treatment*
		Fe+Ni	*Cu*	*Pb*	*Zn*	
Low acid	3.5–4.5	50	10	4	10	Weekly
Mild acid	4.5–6.5	75	10	15	15	Weekly to Monthly
Neutral	5.0–8.0	200	25	5	25	Monthly (Hull Cell useful)
Alkaline phosphate	8.0–11.0	200	—	—	—	Monthly
Alkaline cyanide	9.0–10.0	200	—	—	—	Monthly
Non-cyanide	6.0–11.0	250	20	40	20	—
High speed	various	various				—
Electroless, immersion	various	Infrequent use				—

that have not shown problems in device assembly. Each user should determine and control their particular operating level of contaminants since the plating process, requirements and plating conditions will vary.

An alternative control procedure for iron and nickel in gold solutions is to deposit 200 microinches on a polished inert substrate and heat in air at 500°C for 5 minutes. Deposits with excess heavy metal codeposition will discolour. This test can be carried out on any system subject to heavy codeposition, but should be interpreted with caution. Test results are subject to the viewer's opinion and are not sensitive to the functional and assembly parameters of the device. This discussion also illustrates the need for research and publications on the codeposition of impurities, their control and effects on gold deposits. Contaminants will always find their way into gold solutions and practical means for their control must be available.

Cost Reductions

Gold plating cost reduction is an area of prime consideration in problem solving and control. Although barrel plating of components is a low labour effort, the control of narrow thickness ranges and quality, such as bare spots, porosity, adhesion and surface properties, still need improvement. This is discussed more fully in the section on Metal Cans.

Flatpacks, DIPs and other ceramic packages with more than 16 leads or short strips of multiple connected devices must be rack plated. This is due to their size and excessive lead bending. Package costs due to increased labour in hand racking are considerably higher than for barrel plated components. However, gold plating quality and narrow thickness ranges are easily controlled in this case.

Chemical Resistance

The chemical resistance of the packaging material is also a factor in process control. For example, the post-plating of flatpacks, cerpacks and DIPs made with low melting point glasses introduce small but critical traces of lead and zinc contamination into the gold solution and hence into the deposit. These impurities are believed to be related to stressed deposits, adhesion and solderability problems. Rigorous cleaning and activation of the metal leads are not feasible due to the chemical nature of the glass (see section on Ceramic Leadframe Packages). The processing and plating of ASTM F-15 Alloy (KOVAR), plastic encapsulated devices and various other glass-to-metal seals, bear similar considerations.

Lead Corrosion

One of the most critical control problems relating to the use of gold on semiconductor package leads is corrosion and subsequent device failure. The gold plated KOVAR system is not a good choice from the standpoint of corrosion resistance qualities. The thermodynamic tendency for corrosion is very great due to the extremely noble character of gold in contact with exposed base metal packaging materials. Pinholes, scratches and porosity cannot be completely eliminated in a practical sense from any deposit. Added to this are the thickness reduction programmes which are always in effect. The corrosion process is electrochemical in nature with gold as the cathode and KOVAR as the anode in a vigorous galvanic couple, initiated by the presence of moisture with traces of chloride ion. Further influencing factors are the intensive chemical and thermal processing of plated metals during assembly, minimal cross-sectional area of the leads, mechanical stressing of leads during testing, and printed circuit board insertion, fluxing, soldering and exposure to severe environments during service.

The following process steps have been found by experience to influence lead corrosion and its control:

1 *Metallurgical chemistry and history*[7,14]. Base metal quality control of metal constituents, reduction of carbon, non-metallic inclusions and trace metal contaminants are required for improved bulk metal quality necessary for electronic processing.

2 *Descaling and bright dipping.* These are vigorous chemical reactions intended to remove surface oxides and controlled amounts of metal. Processed leads must be smooth and bright, free of stresses, exposed inclusions and metal extrusion lines, and be within thickness tolerances. Surface finishes of 8–16 microinches are preferred. Substrate roughness fosters deposit porosity (see Chapter 23).

3 *Cleaning and activation.* Preplating steps to achieve surface cleanliness, uniform activation and smoothness.

4 *Gold striking.* Promotes adhesion, reduces porosity and contaminant control.

5 *Gold plating.* Important control parameters are bath operating conditions, cathode current efficiency, heavy metal and organic contamination, and roughness.

6 *Rinsing.* Critical deionised water rinses should be maintained at high resistivity to ensure removal of ionic residues and entrapment (hot water—1 megohm, cold water—0.1 megohm).

7 *Printed circuit board insertion.* Avoid excessive bending or stressing in mounting.

8 *Fluxing and soldering.* Fluxes must be thoroughly removed. Government standards restrict the use of acid or activated fluxes. Recent interest in the socket insertion of gold plated DIPs is related to these considerations.

MATERIALS FOR SEMICONDUCTOR PACKAGES

PACKAGE TYPES

Semiconductor devices are divided into two package groups from reliability, application and cost viewpoints:

1 Hermetic package group. Consists of metal cans, ceramic DIPs, flatpacks and other forms of glass-to-metal or ceramic envelopes with the semiconductor element completely free from the environment for an almost infinite time. These were the traditional high reliability packages, the only package forms available during the early years of the industry, starting as the metal can and then the ceramic-glass-to-metal sealed IC package.

2 Plastic encapsulated group. Brought about by the trend to lower costs and high volume production methods. Transistors, diodes and IC's are gradually being introduced into this market. Problems with moisture penetration, internal corrosion and electrical failures plague plastic package engineers. Better forms of chip protection are required. Gold plating requirements vary with each package and the package forms within that group.

METALS

Metallic leads form the interconnection between the semiconductor element and the external world. So far, all metals used in package assembly must be plated or soldered to permit device usage, i.e., electrical testing, soldering and environmental stability. Gold and tin are the most commonly plated metals on component package leads. Silver and tin–lead (electroplated or dip coated) are also used. Table 37.2 lists a selection of common metals in use or under evaluation.

Table 37.2
Metals and Alloys Used in Semiconductor Components

Type	*Application*
ASTM Alloy F-15 glass sealing metal (29% Ni, 17% Co, 54% Fe)	Metal can headers, lead pins, hybrids, ceramic packages, leadframes
Alloy 42 (42% Ni–58% Fe)	IC leads, plastic package
Alloy 52 (52% Ni–48% Fe)	IC leads, compression seal
Cu (ETP 110) and Cu alloys	IC and transistor leads, plastic packages, optoelectronic devices
CDA 194, 195 (Fe–Cu)	
CDA 116, 155 (Ag–Cu)	
CDA 638 (Al–Cu)	
CDA 706, IN 838 (90% Cu–10% Ni)	
CDA 715, IN 848, IN 732 (70% Cu–30% Ni)	
CDA 725 (9% Ni, 2% Sn, 89% Cu)	
CDA 752 (Cu–Ni–Zn, nickel silver)	
Ni (200 series)	IC leads
Steel (1010)	IC and transistor leads
Stainless steel (200, 300, 400 series)	IC leads
Cr—Steel	IC leads
Al (1100 series)	Radiation resistant devices
Clad structures	
Ag or Au on Cu–Ni, Ni or nickel silvers	IC plastic packages
Al on F–15	DIP, ceramic packages
Dumet (Cu on Fe–Ni)	Diode leads
Complex structures:	Ceramic packages, hybrids
Combination of F–15 leads,	
Ag–Cu braze, with Ni on Mo–Mn metallised ceramics	
Ni on Mo–Mn metallised ceramics	Ceramic packages

NON-METALLICS

These materials form the body of the component, selection being based on their properties, performance and cost. They all come in contact with electroplating and other processing solutions. Table 37.3 lists common package components.

Table 37.3
Non-Metallic Packaging Materials

Material	*Application*
Plastic thermosetting Silicone Epoxy Phenolic Diallylphthalate	Plastic packages, ICs, transistors, optoelectronic devices
Plastic, thermoplastic Polycarbonate ABS Acrylic Polyimide films	Plastic packages, ICs, transistors, optoelectronic devices
Ceramics Alumina (Al_2O_3) Beryllia (BeO)	Ceramic packages
Glass	
Borosilicate (hard) (SiO_2–B_2O_3)	Metal cans, diodes, ICs, flatpacks
Solder glass, low m.pt. (PbO–ZnO–B_2O_3)	ICs, DIPs, cerpacks

SEMICONDUCTORS

The heart of any semiconductor device is the miniature semiconductor element inside the package, also referred to as the chip or die. Table 37.4 lists the most common semiconductor materials used in producing microelectronic devices.

Table 37.4
Semiconductor Materials

Type	*Application*
Silicon	Device element
Germanium	Device element
Gallium arsenide	Optoelectronic devices
Gallium phosphide	Optoelectronic devices

Descriptions and properties of materials for electronics are available.[8–15]

SEMICONDUCTOR PACKAGE CONFIGURATIONS

METAL CANS

Headers

PACKAGE ASSEMBLY

One of the earliest and still very widely used package forms in the semiconductor industry is the metal can. Various standard sizes and lead counts are available from the three leaded transistor to integrated circuits with twelve leads. An exploded view of an assembly is illustrated in Figure 37.1. Materials used include ASTM F-15 glass sealing alloy for leads and eyelet, borosilicate glass for the preform and alloy F-15 (KOVAR), Alloy 42 and nickel for the can. Alloy F-15 leads are enclosed in the glass preform and assembled to the eyelet. The glass, leads, and eyelet are hermetically sealed by fusion in an oven at 1000°C. One of the leads is generally grounded to the head while the others are electrically insulated from each other and the head. This assembly which is called the "header" is descaled, bright dipped and the leads clipped to obtain the desired post height and then gold plated. Typical headers are shown in Figure 37.2. The next two operations, circuit die attachment and leadwire bonding are the heart of the device assembly process, and paramount factors in determining the quality and performance of gold deposits in semiconductor technology.

Die mounting or bonding refers to the attachment of the silicon chip onto the header (or substrate) by a combination of heat and mechanical action (scrubbing) to form an adherent metallurgical alloy between silicon and the gold substrate (Figure 37.3). Die sizes typically range from 0.020 in. × 0.020 in. to over 0.200 in. × 0.200 in. with and without gold on the backside of the wafer. A gold-silicon eutectic preform is required on headers with less than 50 microinches gold when attaching the smaller, non-gold backed die. Thickness minimums increase with increasing die size. However, as expected, the preferred direction is to use a minimum gold thickness on the header with gold backed die. The header is preheated to 450°C under a nitrogen curtain and may remain at temperature for one minute during die attachment. The thickness, quality and surface cleanliness of the gold determine the effectiveness of bonding.

Lead wire bonding provides electrical connection from the circuit die aluminium patterns to the gold plated leads of the package. Wire attachment on headers is to the post tops or sides. Gold wires, typically 1 mil diameter, are bonded by thermocompression methods at about 300°C. Ultrasonic bonding with aluminium wires and welding techniques are also used (see Chapter 20). The mechanism of wire bonding requires clean, smooth, high purity, soft gold to maintain high reliability connections. For example, engineering tests on post top header bonding have shown that excess quantities of copper in gold deposits will result in reduced wire bond yields. Requirements become

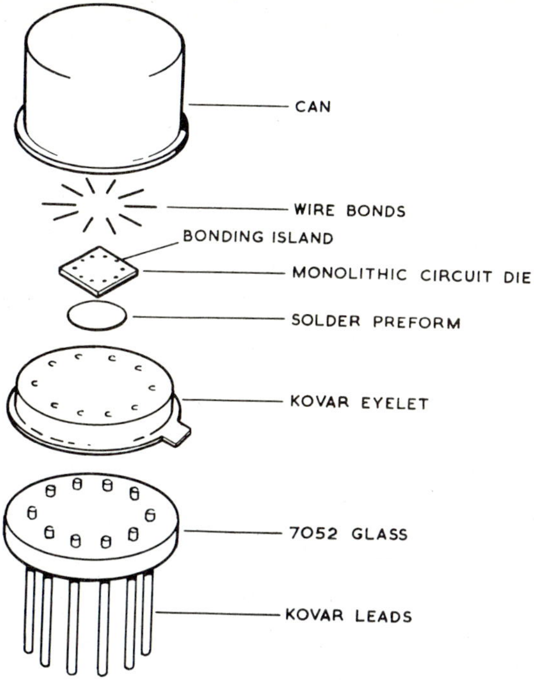

Fig. 37.1 Exploded view of TO-5 metal can assembly.

Fig. 37.2 Multi-lead TO-5, TO-18 headers.

more severe when bonding to post tops. The assembly process is completed with hermetic sealing of the metal can in an inert atmosphere.

REQUIREMENTS

Gold is needed to achieve circuit die attachment, wire bonding, soldering or welding of external leads and retention of electrical and environmental performance. The intention of MIL-G-45204B must be met. Tests designed to evaluate the gold plating and simulate functional needs of device assembly are visual, die attachment, thickness, thermal adhesion bake (5 minutes), boiling water, salt atmosphere, ductility, solderability and lead bonding. Appendix B describes these methods.

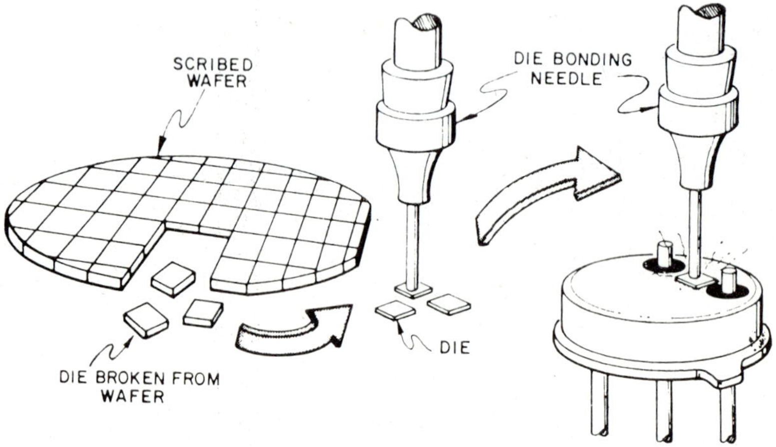

Fig. 37.3 Die bonding operation.

PLATING PROCESSES

Headers are gold plated after controlled descaling, bright dipping and cleaning. The techniques are well established having had their beginning in the vacuum tube industry. Procedure details are given in Appendix A. Gold is generally plated directly over KOVAR although nickel is occasionally deposited prior to gold.

CONTROL AND RELIABILITY

The thickness range is one of the most important parameters to be controlled since it is directly related to cost. Considerable work has been carried out in reducing the thickness spread of barrel plated headers[16–18]. Factors to be considered in thickness distribution control are:

1 Number of leads per header

2 Lengths of leads—long preferred

3 Lead position—"coned" vs straight; coned preferred

4 Barrel load and shot—generally as number of leads on header increases, shot or other types of ballast become more important

5 Processing conditions—bath control, barrel load, plating time (low current density preferred)

6 Gold thickness determination method—beta-ray backscatter, metallurgical cross-section and weigh/strip/weigh are commonly used.

Piece parts are gold plated to a minimum thickness allowing a reasonable lot tolerance percent defective (LTPD) and acceptance quality level (AQL), i.e., a specified small number of rejects[41,42]. A 5% LTPD (2% AQL) is preferred. The most difficult requirement is plating to a minimum thickness on an isolated lead. For example, a lot of 10-lead, TO-5 headers requiring a 50 microinch gold finish on an isolated lead requires a 90 microinch average thickness to pass a 5% LTPD. Stated differently, this means a sample size of 305 units from a lot of at least 3,000 piece parts (measure 2 leads/header) will allow 10 defective to pass lot and 11 defective to reject lot.

Gold thicknesses are specified in line with device package assembly and performance needs. The plating must be thick enough to prevent the basis metal from diffusing to the surface after assembly as well as be free of bare spots, porosity and cracks. Gold must be distributed as uniformly as possible since even one lead out of ten can cause failure, e.g., non-bonding. Table 37.5 lists thickness requirements for various headers, die attachment without preforms.

Table 37.5

Metal Can Gold Thickness Requirements

Thickness (μ-in.)	*Package*	*Device*
20	TO–18	Transistors—small die
35	TO–5	Transistors
50	TO–5 multilead	IC's up to 12 leads
75	All cans	High reliability components

Reliability problems related to the inherent nature of gold are intermetallic compound (Au–Al) formation[19] and external lead corrosion as discussed earlier. These problems, although not completely understood, are satisfactorily controlled in production. Further work is needed on high reliability header production, external lead corrosion control, selective plating and elimination of porosity and codeposition.

POWER DEVICES

These are also in the metal can category. Examples are shown in Figure 37.4. Typical gold thickness specifications call for 50 microinches on consumer goods and 100 microinches on high reliability products.

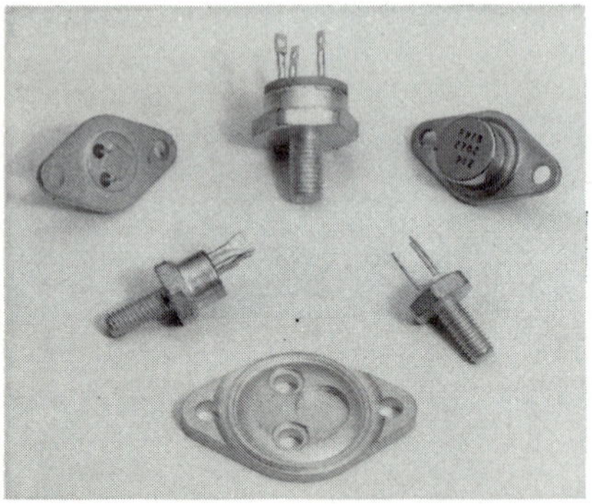

Fig. 37.4 Power devices; TO-59, 60, 61, & TO-3 base plate.

Fig. 37.5 DIP & cerpack leadframe packages. Standard process is post-plating as sealed. Units are barrel plated before lead clipped.

CERAMIC LEADFRAME PACKAGES

Flatpack and DIP

PACKAGE ASSEMBLY, HIGH AND LOW MELTING POINT GLASS SEALING

Integrated circuit devices are generally packaged as flatpacks (cerpack) or DIP's as shown in Figure 37.5. Hermetic, ceramic, glass-to-metal lead sealed packages have been the industry standard since IC's were developed. The 14 and 16 lead packages are the most common, although complex MOS and memory devices frequently have up to 50 leads. Both the flatpack and the DIP utilise Alloy F-15 or Alloy 42 (see Table 37.2) glass-to-metal structures for interconnection and isolation of the semiconductor chip. The alloy frame is known as the leadframe. Most leadframes are fabricated from strip ranging in thickness from 0.004–0.015 in. by chemical etching or stamping.

From a viewpoint of gold plating technology, there are three basic package assembly processes as leadframes types; one utilising high temperature borosilicate glass-to-metal sealing and two with low temperature solder glass-to-metal sealing. These are outlined in the following text.

High Temperature Assembly
Type 1

(*a*) Jig and fire leadframe, borosilicate glass preform and ceramic as required.
(*b*) Descale, bright dip.
(*c*) Gold plate per leadframe process.
(*d*) Die attach, wire bond.
(*e*) Lid seal, form leads.

Low Melting Point Glass Assembly
Type 2 (*refer to Figure 37.7*)

(*a*) Gold plate per leadframe process.
(*b*) Die attach, wire bond.
(*c*) Jig leadframe, glass, ceramic and fire at 525°C.

Type 3 (*refer to Figure 37.8*)

(*a*) Die attach, wire bond.
(*b*) Jig leadframe, glass, ceramics and fire at 525°C.
(*c*) Gold plate, per post-plate process.

Similar package types using thick and thin film metallised ceramic technology are discussed in a later section.

REQUIREMENTS

The plated leadframe with or without the hard glass-to-metal seal is evaluated similarly to gold plated headers. The objectives are to meet the intention of MIL-G-45204B and functional need of the particular devices. These tests include visual, die attachment, thickness, adhesion and quality bake, salt atmosphere and solderability. Methods are found in Appendix B.

The post-plated flatpacks and DIP's sealed with the soft solder glass are evaluated using applicable tests mentioned in the previous paragraph plus: (i) plating adhesion; adhesive tape pull, solderability directly after plating and baking at 275°C for 1 hour (ii) porosity (iii) surface roughness and (iv) moisture resistance. Solderability after plating has been found useful in process optimisation and gold plate adhesion. Other workers have also found it useful[20]. Porosity and surface roughness tests are performed on engineering and qualification lots to ensure reliability when in contact with corrosive environments. In addition to the above, all completed devices are thoroughly evaluated by a high reliability group.

PLATING PROCESSES

Preplated Leadframes and Hard Glass Sealed Packages (*Types* 1 *and* 2)

The combined group of metal leadframes as strips of multiple units from stamping or etched sheets of larger quantities and glass-to-metal sealed package bases are gold plated utilising processes similar to those used for headers. Smaller packages up to 16 leads are barrel plated, larger packages and leadframes by rack techniques. Generally the whole unit is electroplated although selective plating will be utilised more in the future. A typical process is described in Appendix A.

Post-Plated Packages (*Type* 3)

Post-plating has been adopted as the standard method for gold plating solder glass sealed flatpacks, cerpacks and DIPs. Considerable savings and reliability have resulted over the earlier method of preplating the entire leadframe since its inception in 1965. The general method was first developed in the writer's laboratory and is now used throughout the industry. The process outlined in the following text takes into account the chemical nature of solder glass and the control of lead and zinc in the gold.

CERPACK/DIP POST-PLATING

1 Load devices into basket, small quantities preferred.

2 Acid dip
H_2SO_4—50 to 90% by volume
Temperature—90°C
Time—1 minute

3 Water rinse (optional)

4 Acid dip
H_2SO_4—10% by volume
Temperature—25°C
Time—2 minutes

5 Water rinse (optional)

6 Bright acid dip
HNO_3—25% by volume
H_2SO_4—50% by volume
H_2O—25% by volume
Temperature—25°C
Time—10 seconds

7 Deionised water rinse.

8 Load into barrel

9 Gold strike (acid process)

10 Deionised water rinse

11 Gold plate (mild acid or neutral process)

12 Hot deionised water rinse

13 Spin dry.

CONTROL

Low levels of solution contamination must be maintained. Traces of chloride ions, lead and zinc (as well as iron and nickel) are known to be detrimental to the deposit and the insulating glass. Lead codeposition results in solderability problems whereas zinc may be related to stressed, non-adherent deposits. Figures 37.6–37.8 compare metallographic cross-sections of preplated and post-plated flatpacks.

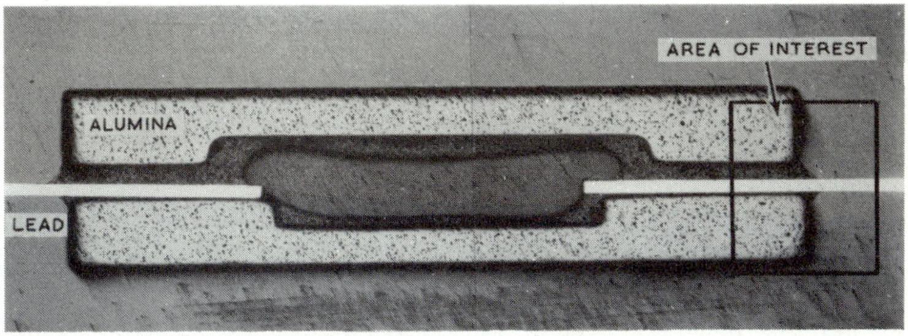

Fig. 37.6 Cerpack viewed in cross-section. Note glass-to-metal seal area.

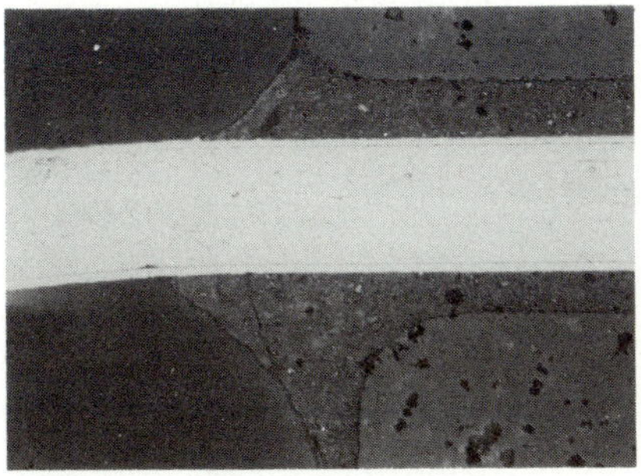

Fig. 37.7 Cerpack cross-section with preplated leadframe, 200 μ-in. gold. (200 ×).

Fig. 37.8 Cross-section of post-plated cerpack. Gold comes up to glass-to-metal seal. (200 ×).

DIODES

The familiar diode represents an early form of glass-to-metal seal technology. Diodes are used extensively in the computer industry.

Diode leads are Dumet (copper on iron–nickel). Barrel plating consists in the deposition of 100 microinches nickel and 100 microinches gold. The main control problems are attack of the glass-to-metal seal and the thickness range.

PLASTIC ENCAPSULATED PACKAGES—SELECTIVE PLATING

The need for low cost, high volume semiconductor devices in the market has put a strong emphasis on plastic encapsulated package development. Liquid cast moulded discrete units such as the transistor are well established. The current need is for ICs. Transfer moulding of selected thermoset resins offer this potential, but problems still exist. The major drawbacks have been moisture penetration to the semiconductor circuit chip, the need for chip surface passivation, wire bond failures due to differential thermal expansion of materials and ionic contamination on the surface of sensitive circuits. Plastic packaging, therefore, is used where costs are to be minimised and hermetic sealing is not required, for example, in industrial and commercial applications.

Package Assembly

Plastic IC assembly begins with leadframe plating. From a practical viewpoint only the selective plating of gold should be considered for low costs. Gold plating is required only on the die attachment and wire bond areas. Leadframe materials are Alloy 42, copper alloys and steel (Table 37.2). Moulding powders are formulations consisting of silicones, epoxies, phenolics, or diallylphthalates. Device assembly steps include mounting the die on the plated leadframe, wire bonding, chip protection with organic coatings and transfer moulding. Following the moulding operations, individual units are clipped out of the leadframe, formed, and plated with tin, tin–lead or a thin deposit of gold. Transfer moulded transistors are shown in Figure 37.9. Liquid cast moulded transistors and certain light emitting devices utilise gold plated KOVAR pins.

Selective Plating*

The need to reduce the cost of gold plating in recent years has put strong emphasis on the selective plating of leadframes. The growing acceptance of plastic encapsulated devices where gold is needed only on the die attachment and lead bond tips, an area of 0.05–0.10 in.^{2}, has made this emphasis obvious.

Required thicknesses range from 50–150 microinches since gold solder preforms are not used. Methods for selective plating of leadframes are varied and depend upon the design of the leadframe and device, and the method of masking off the restricted areas. Details of the methods are proprietary in most cases, but it is believed they include techniques such as mechanical or photoresist masking, jet plating (see Chapter 17) and contact (brush) plating (see Chapter 16). Partial immersion into the gold solution is also common where an end of the leadframe is to be gold plated as in Figure 37.9. Racks and strip plating machines are also used for this purpose.

* For a more comprehensive treatment of this topic refer to Chapter 17.

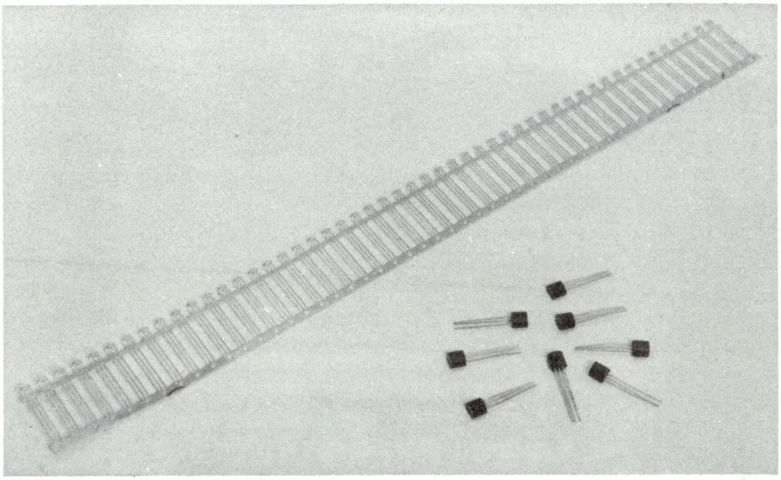

Fig. 37.9 Plastic transistor leadframe. Gold plated selectively on die and wire bond areas. Strip length is 11.5 in. Finished units in foreground.

Fig. 37.10 Selective plating line for depositing gold on leadframes for plastic semiconductor devices. From right to left in the photograph, the first five tanks are for cleaning and activation. The next three are for gold plating. [Sylvania Electric Products Inc.].

A continuous strip plating machine is shown in Figure 37.10. Although designed specifically for gold plating leadframes, it can also be used for selective plating on connectors and for controlled depth of immersion for components joined by a carrier strip. Deposits from 10–250 microinches can be achieved. Tanks are covered with lids and laminar air flow exhausted for cleanliness. Although the cost of the selectively plated gold becomes insignificant, the expense and custom built nature of each machine becomes high and is, therefore, warranted in high volume production only.

Methods for processing large sheets of leadframes with solid or liquid photoresist to restrict the plating area are also available, but the labour and processing costs generally make these useful for prototype or low volume only.

The metallurgical structure of the gold varies according to the process and plating conditions (bath type, gold content, conductivity, current density and

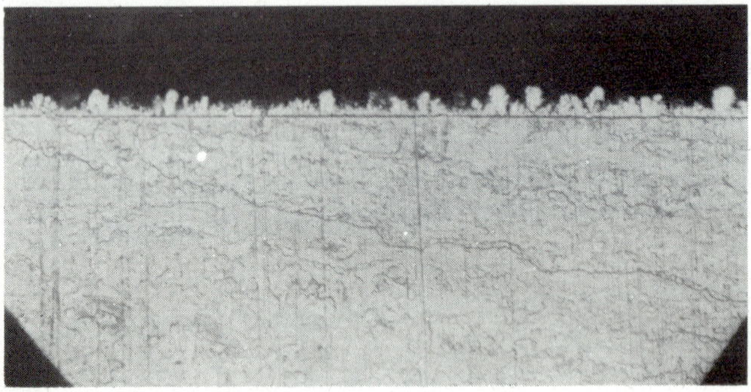

Fig. 37.11 Metallographic cross-section of die bond area on Alloy 42 leadframe, plated selectively at 800 A/ft^2. (750×).

other operating conditions). In some cases, a rough matte surface as shown by the cross-section in Figure 37.11 is preferred. So far no work has been reported on the properties of gold deposited at very high current densities. Preliminary studies have indicated that metal contaminant codeposition contents vary at the higher plating rates.

REQUIREMENTS

Gold selectively plated leadframes are evaluated by visual, die attachment, thickness, thermal adhesion bake and adhesive tape tests. Procedures are outlined in Appendix B.

PLATING PROCESS

A typical cycle is similar to that used for plating of KOVAR or Alloy 42 piece parts. For example, a continuous strip plating line uses Alloy 42, 14–16 way leadframes on a coil of 20,000 units and proceeds as follows:

1 Unreel and feed into machine

2 Vapour degrease

3 Alkaline clean—cathodic cycle

4 Water rinse

5 Acid dip
HCl—50% by volume

6 Water rinse

7 Alloy 42 activator
KCN—45 to 60 g/l
Voltage—6 (cathodic)
Temperature—65°C

8 Deionised water rinse

9 Gold strike—selectively 1–2 microinch thickness

10 Deionised water rinse

11 Gold plate—selectively
Au metal—16.4 to 32.8 g/l
Current density—150 to 1500 A/ft^2
Thickness—50 to 150 microinches

12 Drag-out rinse

13 Deionised water rinse

14 Hot dry

15 Reel on coil

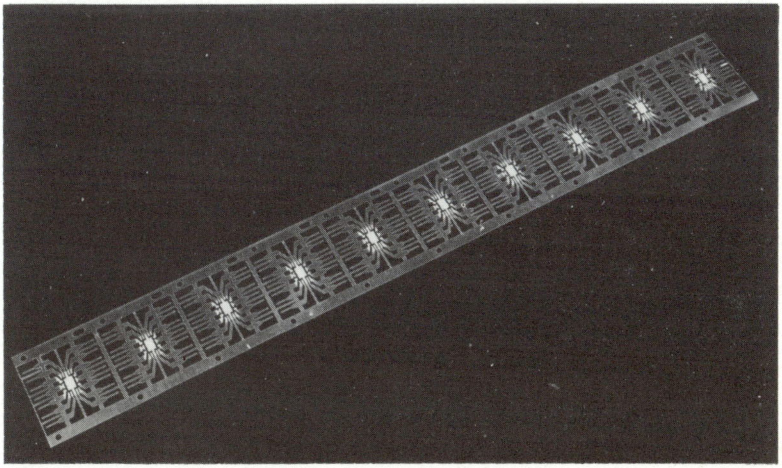

Fig. 37.12 Section of integrated circuit leadframe strip. Die and wire bond area is 0.250 × 0.250 in. plated selectively with 50 μ-in. gold.

Throughputs range from 5,000–15,000 units per hour. The plated coil is usually cut into strips of 10 each for subsequent die attachment, lead wire bonding and plastic encapsulation. A selectively plated leadframe is shown in Figure 37.12. This leaves the external leads to contend with, a frequently neglected consideration. The goal is an inexpensive lead finish such as tin or tin–lead. All units must return to processing for plating of the external leads.

This procedure is well warranted for high volume production. An alternative method, satisfactory for engineering studies and low volume production, is to preplate the entire frame with 10–20 microinches of gold (after step 9) and selectively plate gold to additional thickness where required. Processing back through the plating facility is not necessary in this case.

Leadframes (Figure 37.9) used in the production of the plastic TO-92 have an additional gold requirement. Devices are constructed with a copper alloy plated selectively on the die attachment and wire bonding areas with 100 microinches nickel and 100 microinches gold. Gold-backed silicon is bonded to the die platform at high speeds. The required die mounting must be scrubless, instantaneous and with complete wetting. Surface cleanliness and purity are critical control parameters. The copper content in the gold solution must be controlled as specified in Table 37.1. Completed units are shown in Figure 37.9.

CERAMIC PACKAGES—CONDUCTOR PATTERNED SUBSTRATES

Hermetic flatpack and DIP's utilising thick and thin film technology are also standard package forms available to the semiconductor industry. In this case, the silicon chip interconnection is to a metallised conductor pattern on ceramic rather than to a metal leadframe.

Microelectronic devices make wide use of ceramic substrates. Ceramics have two important properties: (i) high temperature strength and (ii) high thermal conductivity. These factors allow a variety of conductor patterning and silicon chip attachment methods, and permit good heat dissipation with single or multichip construction. Alumina (Al_2O_3) is the most commonly used material although others, e.g., beryllium oxide (BeO) with a higher heat dissipation than alumina are also available. Typical applications are MOS, hybrid circuitry, memory arrays, large scale integration (LSI), and single and multi-chip assembly. In each case, there is a need for a conductive pattern, usually gold, on the ceramic to support and interconnect the active and passive elements contained in a particular device or system. The conductor pattern has bonding pad areas which provide a means to attach leads suitable for connector or printed circuit board attachments. Single and multilayer substrates are in use.

Conductive patterns on ceramic substrates are made as follows:

1 Mo–Mn processes—screened or photo-etched patterns; plating before or after pattern delineation.

2 Screened conductive pastes—stainless steel screens, etched masks.

3 Evaporation, sputtering (thin films)—through metal mask or photo-etching.

4 Electroless plating—sensitised or pretreated substrates.

5 Transfer tapes.

6 Printing (letterpress).

The system selected for fabricating patterns on a substrate depends upon line definition, electrical requirements, mechanical strength and cost. For example, screened patterns are preferred if 5 mil or greater line widths and low cost are determining factors. The gold molybdenum–manganese (Mo–Mn) metallisation on an unglazed ceramic and brazed KOVAR leads is specified if the device package requires hermeticity and the highest lead strength. In the case where flip-chip or beam lead technology is to be utilised, substrate processes must be capable of yielding 2–4 mil line widths and spaces, and as such requires a more complex and more expensive photo-etching process.

Package Assembly

The classical Mo–Mn metallised ceramic system will be emphasised in this section due to the key role it plays in hermetic semiconductor package construction and gold plating technology. A typical ceramic package assembly is illustrated in Figure 37.13.

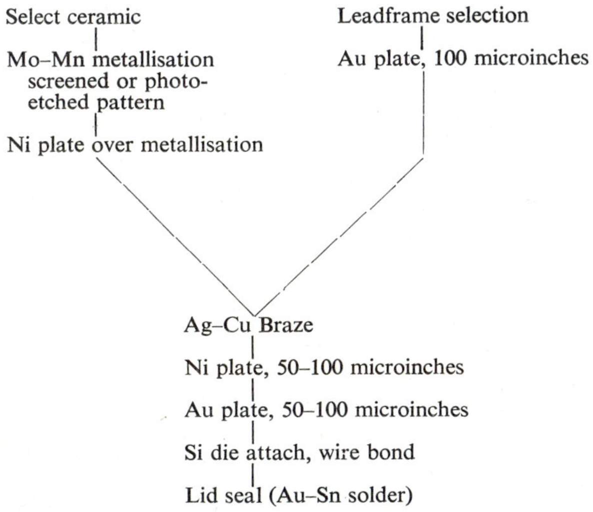

Fig. 37.13 Hermetic metallised ceramic package assembly.

Conductor patterns are gold plated Mo–Mn deposited on unglazed ceramics. Alumina is selected as the substrate material due to its good high temperature processing strength and high thermal conductivity. The Al_2O_3 selected should be high purity (96% or higher) to avoid discolouration, to ensure good electrical properties and be capable of yielding hermetic seals. The Mo–Mn is a stable, high temperature, adherent coating serving not only as the conductor, but also as the base material to which leads are brazed and lids attached. Typical bond strengths range from 10,000–15,000 $lbs/in.^2$ lead strength. The Mo–Mn technique consists of forming a thin metallic layer on a ceramic surface by applying a blended mixture of molybdenum and manganese powders, inorganic additives and an organic binder, and then firing in a hydrogen atmosphere. The mixture is applied to the ceramic by spraying, screening, dipping or painting. Patterns are generally screened on directly or over one entire surface with subsequent gold plating and etching. Other processes such as refractory metal solution metallising or sputtering are also used. Ceramic-to-metal sealing methods are recorded in the literature[10,11].

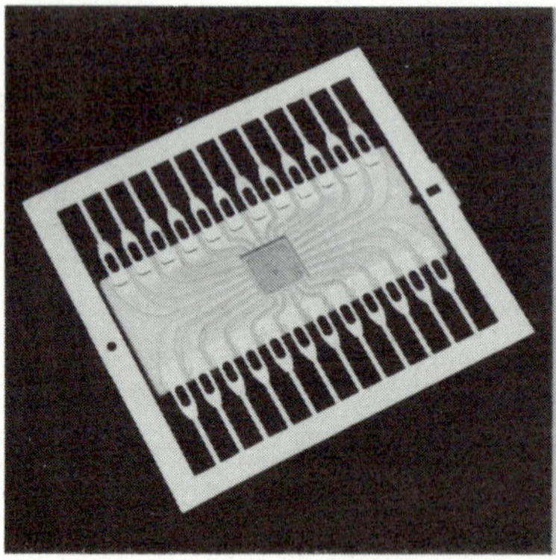

Fig. 37.14 Twenty four lead array. Package construction: thick film Mo–Mn metallisation on 1.27 in. × 0.52 in. Al_2O_3 ceramic, gold plated after leadframe braze attachment.

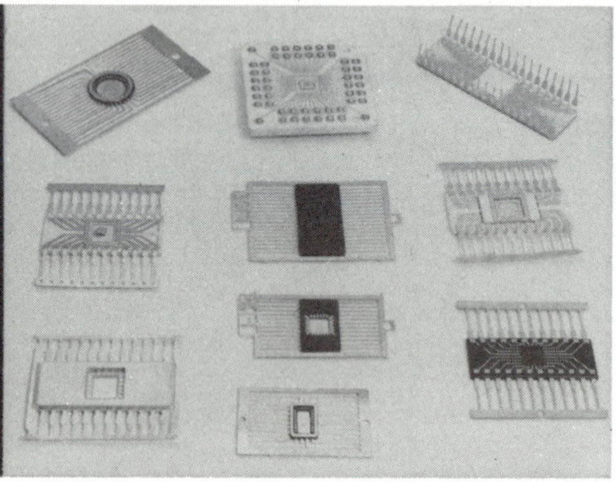

Fig. 37.15 Various 24 and 36 lead memory array and MOS packages. Assembly methods include metallised ceramic, high temperature glass-to-metal and metallised plastic (right foreground).

A second metallic coating of nickel is required to increase line conductivity and condition the Mo–Mn surface for brazing. Copper or gold plating may be deposited over the nickel surface or used in place of the nickel. A low expansion metal leadframe, e.g., KOVAR previously gold plated, is brazed to the metallised pad areas. Typical brazing materials are Ag–Cu eutectic, pure silver, gold, and copper. Ag–Cu is brazed at 780°C in hydrogen, followed by plating with nickel as a barrier layer and finally with gold. The assembly technique is continued with silicon chip attachment, wire bonding and lid sealing to gold plated surfaces. Figures 37.14 and 37.15 are typical assemblies.

Another substrate form is known as the edge-mount. In this case, there are no brazed metal leads and connection is made to the edge of the substrate similar to printed circuit boards.

Requirements

Gold plated metallised ceramics must meet requirements similar to those imposed on headers and leadframes. These include visual, die attachment, thickness, adhesion bake, electrical resistivity and lead braze. Lead tensile strengths, 90° to the surface, should be 3 lbs minimum. Gold deposits exposed on sealed assemblies are further evaluated for lead solderability, 275°C baking and resistance to moisture and salt atmosphere. Further details are given in Appendix B.

Plating Process

Gold is selected as the top metal layer owing to its good bonding, electrical and anti-tarnishing properties. A gold surface permits silicon die attachment, wire or wireless bonding as well as external lead soldering, welding and brazing.

CERAMIC SUBSTRATES

Mo–Mn metallised ceramics are readily nickel and gold plated without preparatory cleaning or activation if processed within several hours of hydrogen firing. The following procedure (basic steps only) is also used: (i) alkaline clean (ii) HCl dip (iii) nickel plate (iv) acid gold strike (v) 24 carat gold plate.

Both electroless and electrolytic systems are in use.

LEADFRAME

The basic procedure is as outlined in Appendix A.

BRAZED LEADFRAME

The following procedure is used: (i) rack (ii) alkaline clean (iii) HCl, 20–50% by volume (iv) NaCN activation, cathodic, 6 volts (v) Wood's nickel strike (vi) nickel plate, 100 microinches (vii) gold strike (viii) 24 carat gold plate, 75–100 microinches.

PROBLEMS

The Mo–Mn surface tends to be very coarse, porous and difficult to activate[25]. Plating problems are, therefore, related to adhesion, spotting out, discolouration after baking and thickness control. Rigid process control and techniques to reduce deposit porosity generally result in a good product. Multiple die attachments on a single substrate require a more effective diffusion barrier than nickel. Reliable methods for achieving non-destructive thickness measurements and uniform deposit thickness are necessary. A reliable process for plating gold–tin alloys is also needed. Lid sealing and other semi-automated bonding concepts require gold solder materials and techniques that do not necessitate preforms.

Fine-Line Patterning

THICK FILM; Mo–Mn SYSTEM

The impetus to produce narrow lines on ceramics with high yields is prompted by the advances in the field of flip-chip and beam lead technology. As integrated circuits become denser and more complex, the dimensions between chip bonding areas decrease. Conductor line widths at the point of IC chip bonding have decreased to 2–4 mils. Processes for obtaining fine line conductors involve photo-etching of Mo–Mn systems and thin films, or screen printing technology[8,21–26]. Process steps in producing fine line conductors using Mo–Mn on Al_2O_3 are shown in Figure 37.16.

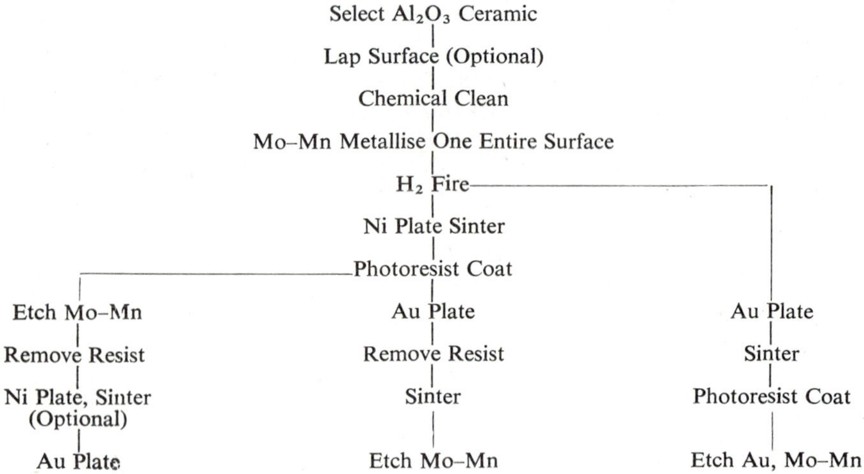

Fig. 37.16 Fine line patterning of metallised ceramics.

Gold plating steps must be carried out with mild processing when a photoresist mask is used. Photoresist residues must be removed to avoid blisters. A procedure similar to that given earlier for metallised ceramic substrates is followed. Plated substrates are sintered at 800–1000°C in hydrogen to ensure good adhesion and porosity sealing.

The Mo–Mn process fulfilled a requirement for a thick film, conductive fine line pattern on a ceramic to which silicon chips may be flip-chip bonded as well as external leads brazed. The pattern plating, photo-etch process is expensive and requires considerable skill and experience. Other methods (and their combinations) are available depending upon the requirements of the system.

THIN FILM—EVAPORATION AND SPUTTERING

Evaporated, sputtered and chemical vapour depositions films are used extensively for thin film structures. Deposition usually takes place on silicon, glass, quartz or glazed ceramics through a mask or over the entire surface.

The process involves two layers: (i) aluminium, chromium, titanium or molybdenum for adhesion to the substrate and (ii) copper, silver, gold or other precious metal for conductivity. Thin film resolution of one micron (10,000 Å) or less is possible. Gold plating is used to increase electrical conductivity and applications. Photoresist techniques are used to define the pattern. Other processes, i.e., screened metal pastes and transfer tape can also be supplemented by gold plating.

MULTILAYER SUBSTRATES

The use of multilayer ceramic substrates combined with multi-chip bonding has provided a practical means to achieve a reliable, high density, high speed package well suited for LSI, MOS memory arrays and hybrid circuit devices.

A variety of buried layer multilayer ceramic substrates have been developed. Some of these are commercially available. The basic structures are:

1 Al_2O_3/Mo–Mn/Al_2O_3/Mo–Mn, Au

2 Al_2O_3/noble metal paste/Al_2O_3/screened metal patterns

3 Al_2O_3/Al/glass/Al

4 Al_2O_3/Mo–Mn, plated Cu/glass/evaporated and plated Cu and Au[27].

The top layer metallisation is generally plated with 100 microinches of gold to provide a structure for die attachment, lead bonding and lid sealing. Double layer memory arrays and light emitting devices are shown in Figures 37.17, 37.18 and 37.19.

Hybrids

Hybrid microelectronics involves packaging of active and passive components such as IC's, transistors, resistors and capacitors into a hermetically sealed case. Various assembly methods are used including the technology for thick and thin film metallised ceramics[8,28]

1 Thick film package substrates are made using the Mo–Mn or screened paste techniques just described. Brazing or soldering is used to attach external leads.

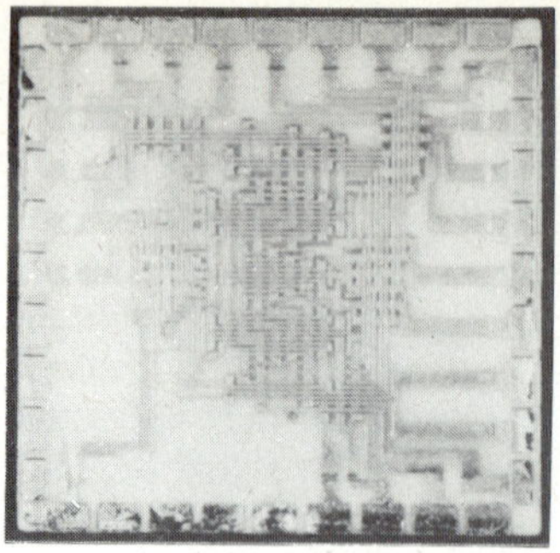

Fig. 37.17 Plated double layer ceramic. Top pattern is Cr–Au, 4 mil lines on 7 mil spaces[27].

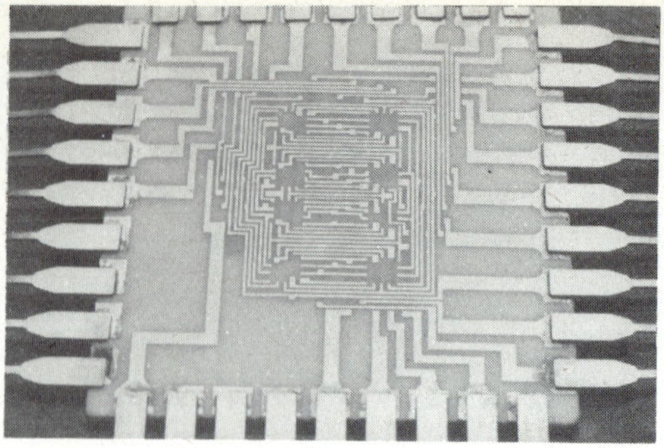

Fig. 37.18 Memory array double layer package. Top layer is Mo–Mn/Ni/Au. Fine line conductors are 4 mil with 7 mil spaces made by the photoetch process.

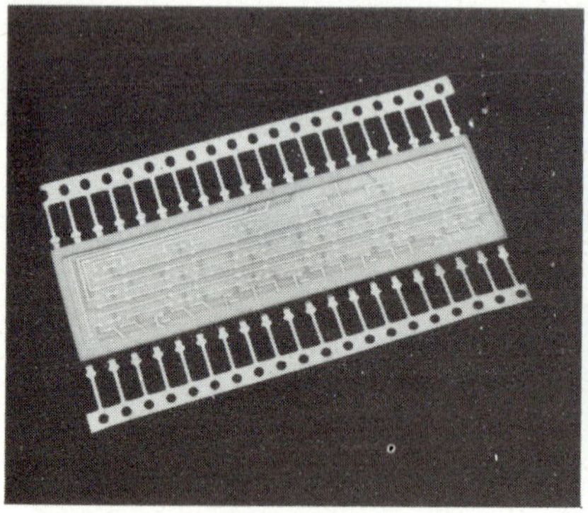

Fig. 37.19 Double layer light emitting Al_2O_3 package. Top structure is screened Mo–Mn/Ni/Au. Ceramic size is 2.0 × 0.50 in.

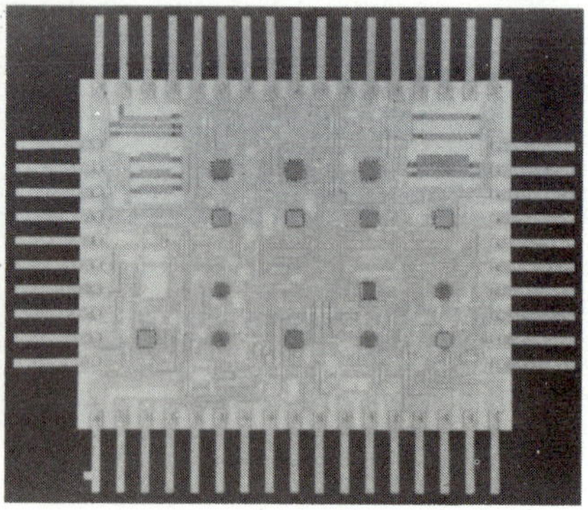

Fig. 37.20 Thin film beam lead hybrid package. Conductor patterns made by photoetch process. [Western Electric Company].

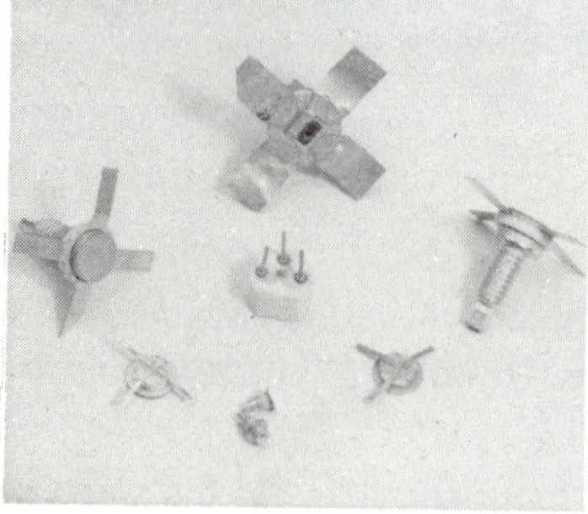

Fig. 37.21 RF power devices made with metallised Al_2O_3 and BeO.

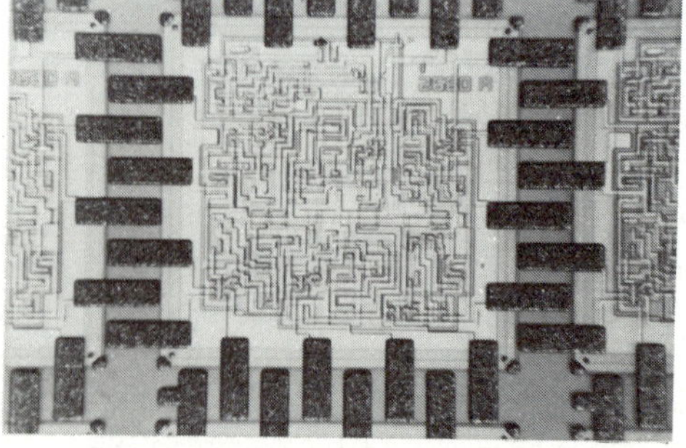

Fig. 37.22 Gold beam lead IC as plated on wafer. Beam size is 0.4 mil (10 μm) thick, 1.4 mil (35 μm) wide, and 14 mil (350 μm) long. [Western Electric Company].

2 Thin film hybrid substrates are constructed by the evaporation or sputtering of thin metal and non-metallic layers, photoetching and gold plating. The external leadframe, preplated with gold, is attached to the metallised substrate generally by thermocompression bonding or soldering. A thin film hybrid constructed with beam lead devices is shown in Figure 37.20.

RF Devices

Metallised ceramics are also used to make a variety of microwave devices as shown in Figure 37.21. Other types are Gunn diodes and coaxial microwave transistors.

PLATING ON SEMICONDUCTORS

Electroplating on semiconductor elements has been reported[29]. Early applications were deposition on the die backside for attachment and more recently on die frontside for electrical contact. In the current technology, plated metals are applied to a thin (one micron) vacuum evaporated or sputtered metal system. Recent uses devoted to elimination of costly wire bonding operations has engaged gold electroplating in some of its most stimulating applications.

Semiconductor devices require electrical contact from the semiconductor die to the external leads. The standard packaging assembly methods for which all the preceding sections are concerned, involve wire bonding—expensive assembly by hand, one at a time, using gold or aluminium wires. These wires are frail and frequently a source of reliability problems, particularly with the more complex devices. To solve these problems, the industry is developing new assembly methods and wireless bonding. These new techniques are known as "flip-chip", beam lead and face bonding, and represent the evolution of simpler, more effective and more reliable methods of producing microelectronic systems. Contact areas are known as bumps, beams, pillars and balls. The familiar package and wire leads are eliminated by turning each silicon chip face down and connecting the contact of the chip directly to the ends of a miniature leadframe or fine line conductors on a plastic or ceramic. This is a fast developing field with many potential uses for electroplating.

WIRELESS BONDING METHODS

Three interconnection techniques that use gold plating as an integral part of the device will be described. All three processes start with an insulating layer and evaporated (or sputtered) metal layers to promote adhesion and electrical contact to silicon.

Beam Leads

In this process, semiconductor devices ranging from diodes to complex IC's are fabricated in batches with electroformed terminals of gold cantilevered beyond the edges of the semiconductor chip, hence the name "beamlead". Process technology, developed initially by the Bell Telephone Laboratories, is described in the literature[30–33]. Processing involves a series of platinum and titanium sputterings, photoresist applications, two gold plating and chemical etching steps. The first gold plating stage (2 microns) is used to form the interconnecting paths on the chip and the second to increase the thickness of the beams, typically 0.4–0.68 mil (10–17 microns). Photoresist masking techniques permit very small dimensions.

In the electroforming process, an entire wafer with several thousand devices is immersed in gold plating baths[32], current passing through the areas where the metal pattern and leads are exposed by the resist pattern. A typical beam size is 0.4 mil (10 microns) thick, 1.4 mil (35 microns) wide and 14 mil (350 microns) long. Figure 37.22 illustrates an as-plated structure.

The gold beam lead structure eliminates the need for die attachment and wire bonding individual contacts to the electrical regions on the device and external lead pattern. The leads serve a structural and protective as well as an electrical function with the beam lead devices facing the substrate. The cantilevered leads extend beyond the edges of the chip which are already aligned with matching patterns on the substrate. The beam leads are bonded to these patterns by thermocompression or ultrasonic bonding, or parallel gap welding (see Chapter 20). Gold is highly ductile (type III, grade A) making a flexible lead that is not easily damaged by shock and readily conforms to the thermal expansion and contraction of the other materials. In addition, use is made of the fact that gold resists corrosion and is easy to bond or repair. Beam lead devices bonded to metallised ceramics and metal leadframes are shown in Figures 37.20 and 37.23.

Polyimide Film Package

General Electric has recently announced an assembly system that utilises silicon IC chips with electroplated pure gold raised bonding pads attached to copper-polyimide laminates[35,36]. The package leads and circuit start off as a continuous strip of 35 mm polyimide film with perforated sprocket holes along the edges for indexing. During manufacturing, 1 oz copper is laminated to one surface of the film, photo-etched and immersion tinned to leave a leadframe for chip attachment. The leadframes outer tabs are supported by polyimide, the inner leads being cantilevered over a hole that receives the silicon chip. The silicon wafer is processed with the standard aluminium pattern, coated with a glass (SiO_2) insulation layer having etched windows to the aluminium, a deposited metallic barrier, photoresist coated and plated with 0.5–1 mil gold. Thermocompression bonding joins the gold pad and the

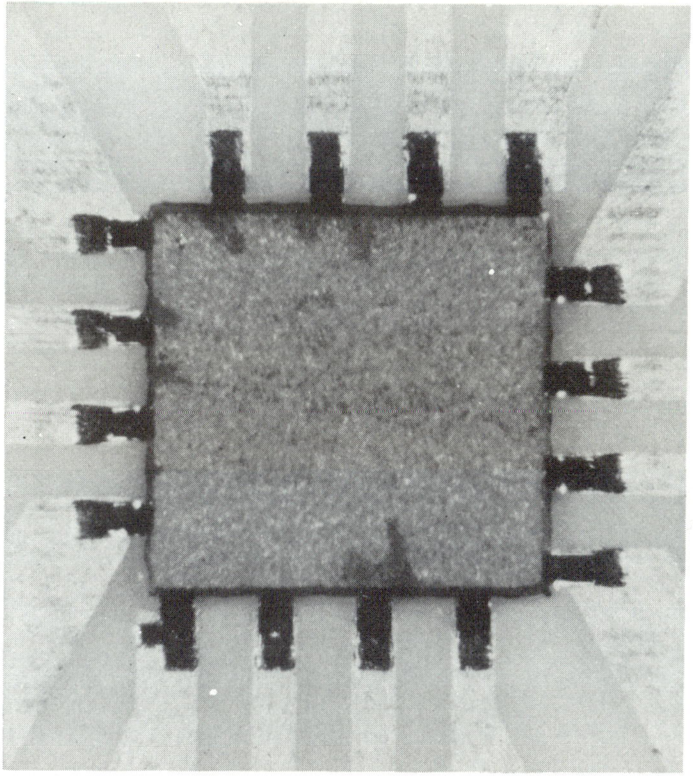

Fig. 37.23 Magnified view beam lead devices bond to metallised ceramic substrate. [Western Electric Company].

circuit leads. The silicon chip and surrounding metal pattern is then encapsulated in an epoxy resin. Figure 37.24 shows a device made using this technology.

Metal-Organic Film Structures

Dielectric films are finding new applications in microelectronics. Films of polyester, polyimide, tetrafluoroethylene, polypropylene and others clad with copper or aluminium, or sensitised and additive plated with gold and other metals, are available. Conductor patterns built on these films offer advantages in the support of fine leads, isolation of circuits and less plating area compared to a metal frame. Cost and film shrinkage are factors needing consideration. Photoresist processes for printed circuit and metal etching are employed[8,34]. A typical pattern is shown in Figure 37.25.

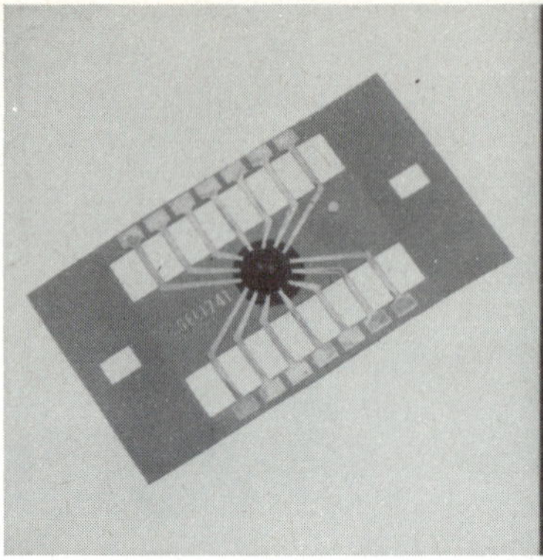

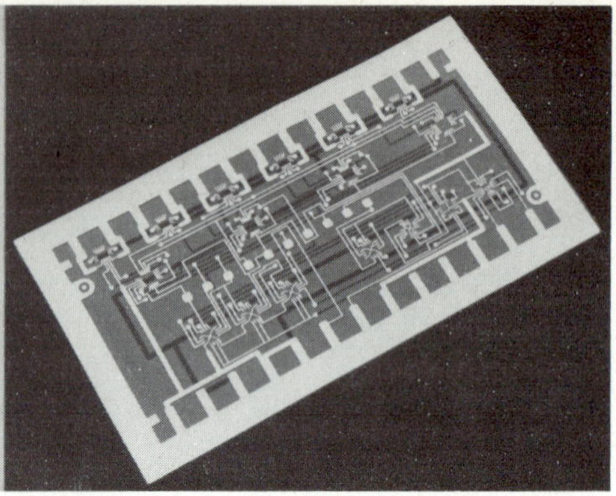

Fig. 37.24 IC package on 35 mm polyimide film. [miniMod™, General Electric Company].

Fig. 37.25 Gold plated copper circuit pattern on polyimide film. Unit size: 2.25 × 1.25 in. [3M Company].

Reflow Solder—Unibond

Development in automated assembly and low cost packaging utilising a plated reflow solder contact pad on silicon which replaces wire bonding, has also been achieved by Fairchild Semiconductor[37–39]. Unibond silicon wafers start out with standard aluminium patterns. A dielectric layer, for surface protection, a deposited metal system for adhesion and low electrical resistance, and a photoresist plating mask are applied. The next step consists of the electrodeposition of two layers; a solid pedestal metal as the first layer and a solder as the second. This provides a scheme for controlled collapse of the pad to keep the device off the substrate. Reflow solder pads are established by etching away the surrounding metal in the field down to the dielectric layer. Groups of wafers are batch processed through each step to realise low cost. A typical single wafer has about 1,000 potential devices. Various metal systems have been found useful. For example, a typical process consists in the deposition of eutectic gold–tin alloy. Modifications with a lower melting point solder, e.g., tin–lead alloy or pure gold for thermocompression bonding are also possible. Silicon wafers processed with Au–Sn reflow solder are shown in Figures 37.26 and 37.27. After scribing, each silicon IC chip is reflow soldered to a gold plated leadframe (Figure 37.28). The resultant bond has high strength, good electrical properties and is corrosion resistant since only gold alloy and the dielectric layer are exposed to the surroundings. Each unit is encapsulated by transfer moulding into a "pill" for hybrid circuit applications and, if desired, subsequently processed to a plastic DIP.

FUTURE TECHNOLOGY

The concept of plating on semiconductors for interconnections is in a

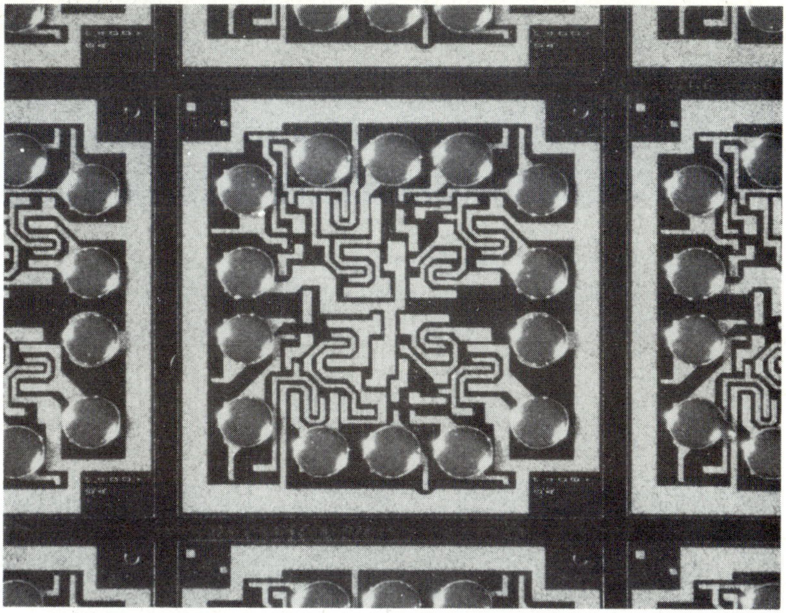

Fig. 37.26 Electroplated gold solder reflow interconnection pads on integrated circuit wafer. Pad size: 6 mils.

Fig. 37.27 Scanning electron microscope view of an electroplated raised pad; 6 mil diameter and 1 mil high. Top layer is gold. (200 ×).

Fig. 37.28 Scanning electron microscope view. Reflow solder interconnection between silicon IC chip and gold plated leadframe. (1000 ×).

dynamic state. New uses are frequently being found. For example, future applications in automated assembly will require the plating of entire circuit patterns on devices, die attachment and lead bond areas. Circuit patterns will be built up by plating to carry more power and permit wireless assembly.

CONCLUSION

Gold meets more of the requirements imposed by the semiconductor industry than any other metal. Its use will increase. There is a need for work on selective plating, gold-solder alloys and a better understanding of control problems. Future developments will show exciting applications in plating on semiconductors.

APPENDIX A

PLATING CYCLES

A typical gold plating sequence for ASTM Alloy F-15 (KOVAR) and Alloy 42 semiconductor assemblies is as follows:

1 Vapour degrease—leadframes

2 Load
barrel—headers (small packages)
rack—leadframes (large packages)

3 Alkaline cleaner
headers—soak only
leadframes—electrolytic 6 V, cathodic
Concentration—45 to 60 g/l
Temperature—70°C
Time—1 min

4 Tap water rinse

5 Acid dip
HCl—50% by volume
Temperature—60°C
Time—1 min

6 Tap water rinse

7 Cyanide activation
NaCN—45 to 60 g/l
KOH—7.5 to 15 g/l
Temperature—65°C
Voltage—6, cathodic
Time—1 to 2 min

8 Tap water rinse

9 Deionised water rinse

10 Nickel plate, 100 microinches (optional)
rinse and activate as required

11 Gold strike
Acid process
Au metal—2 to 5 g/l
pH—4.0
Current density—0 to 5 A/ft^2

12 Drag-out rinse

13 Deionised water rinse—cascade type

14 Gold plate

(*a*) Neutral process
Au metal—6 to 10 g/l
pH—7.0 to 8.0
Current density—0 to 5 A/ft^2

(*b*) Acid process
Au metal—6 to 10 g/l
pH—5.5
Current density—0 to 3 A/ft^2

(*c*) Alkaline process
Au metal—6 to 10 g/l
pH—11
Current density—0 to 2 A/ft^2

15 Drag-out rinse

16 Deionised water rinse

17 Unload to basket (headers)

18 Hot deionised water rinse

19 Centrifugal dry (headers)

20 Unload

APPENDIX B

PLATING REQUIREMENTS AND TESTS

Gold plating is evaluated according to the following tests and military standards:[40–45]

1 *Visual*

Gold plating shall be uniform in appearance, smooth, fine grained, adherent and free from exposed base metal or underplate, visible blisters, discolouration, pits, nodules, porosity and indications of "burning" on the die attachment, lead bonding and sealing areas when inspected at a magnification of 15×.

2 *Die attachment*

The semiconductor die with a gold solder preform is scrubbed on a preheated header (substrate) in a nitrogen atmosphere, at a temperature of

450–475°C for a maximum period of 60 seconds. The meniscus formed around the die must present a smooth appearance, be free of defects and foamy surfaces. At least 50% of the die must remain on the header after attempted removal by shearing. Gold plating must be free of discolouration.

3 *Thickness*

Gold plating thickness as measured by a beta-ray backscatter radiation device, cross-sectioning or weigh-strip-weigh for average thickness shall be as specified.

4 *Adhesion and quality-bake test*

Piece parts shall be placed in a forced air (nitrogen atmosphere for metallised ceramics) circulating oven at a temperature of 500 ± 10°C for five minutes. After cooling, the parts shall be inspected at a magnification of 15×. There shall be no peeling or blistering in the die attachment, bonding or sealing areas. These defects shall not exceed 1% of the remaining plated area.

5 *Boiling water test*

Piece parts shall be tested as follows for plating continuity:

Place units in a beaker of clean boiling distilled water (pH 7.0 ± 0.2) and boil for 2 hours. There shall be no evidence of rust, corrosion, or extensive discolouration visible to the naked eye. Slight discolouration is allowable at the glass-to-metal interface.

6 *Salt atmosphere*

Plated piece parts or finished assemblies shall be subjected to a salt atmosphere test as outlined in MIL-STD-883, method 1009, condition A. Evidence, when examined under a magnification of 15×, of flaking or pitting of the finish or corrosion that will interfere with the application of the device, shall be considered a failure.

7 *Ductility test—headers*

Headers with leads 0.016 to 0.019 inches in diameter shall be tested for plating ductility. One lead on each header shall be tested. Wrap (minimum 3 times) the centre half of each lead tightly around a mandrel with a diameter of 0.040 inch ±0.002 inch; the maximum spacing between each turn shall be 0.040 inch. The portion of the leads wrapped around the mandrel shall be examined using a 15× binocular microscope. The gold plating must show no evidence of flaking or cracking. There must be no exposure of the base metal.

8 *Lead solderability test*

Leads shall pass the solderability test as outlined and described in MIL-STD-202 (Method 208) and MIL-STD-883 (method 2003). The solder, flux, apparatus, procedure and evaluation criteria as therein described shall be adhered to in performing the test.

9 *Lead wire bond test*

A representative sample of bonds shall be evaluated for bond strengths. Baking of the devices may also be included. Adequate header post tops are essential to high yield bonding results. Projections, depressions and flatness must be defined and controlled.

10 *Adhesive tape adhesion test*

Federal Specification, L-T-90 Tape, "Pressure-sensitive adhesive" defines the tape to be used for testing deposit adhesion[4,5]. The tape shall be applied and removed with vigour.

11 *Moisture resistance*

Finished assemblies are tested according to MIL-STD-883 (Method 1004) or MIL-STD-202C (Method 106). Stressing involves −10 to +25°C, 90–98% relative humidity, cycling each 24 hour period for 10 days.

Acknowledgments

Appreciation is given to Fairchild Camera and Instrument Corporation for permission to publish this chapter and to all my associates in their hearty support of this topic. Special appreciation goes to Ernie Armstrong, Tom Yagi and Tim Daly for all the meaningful discussions we have had on gold plating through the years.

REFERENCES

1 Van Hoorde G. R., "Appraisal of Microelectronic Integrated Circuit Performance, Part 10—Packaging" TRW Systems, 99900-6735-RO-00, Oct. (1968).
2 Fehr G., *Solid St. Technol.*, **13**, 41, 8 (1970).
3 Batch A. T., *Electron. Packag. Prod.*, **11**, 45, 3 (1971).
4 Scrupski S. E., *Electronics*, **44**, 75, 12th April (1971); **44**, 65, 26th April (1971).
5 Cohen A., *Electron. Packag. Prod.*, **11**, 53 (1971).
6 Smith P. T., U.S. Pat., 3,057,789 (1962).

7 Colling D. A., Dowling T. J., *Electron. Packag. Prod.*, **10,** 6, 57 (1970).
8 "Handbook of Materials and Processes for Electronics", C. A. Harper (Ed.), McGraw-Hill, New York (1970).
9 "Handbook of Electronic Packaging", C. A. Harper (Ed.), McGraw-Hill, New York (1969).
10 Kohl W. H., "Handbook of Materials and Techniques for Vacuum Devices", Chap. 15, Reinhold, New York (1967).
11 Rosebury F., "Handbook of Electron Tube and Vacuum Techniques", pp. 67, Addison Wesely, Reading, Mass. (1965).
12 Harp M. B., *Electron. Packag. Prod.*, **10,** 1, 228 (1970).
13 Butt S. H., Proc. 21st Electronic Components Conf., I.E.E.E. (1971).
14 Dowling T. J., Dunn L., Colling D. A., Proc. Nepcon (1968).
15 Insulation Circuits, Directory Encyclopedia, Lake Publishing Company, Libertyville, Illinois. Yearly issue.
16 Nobel F. I., Ostrow D. O., Kessler R. B., Thomson D. W., *Plating*, **53,** 1099 (1966).
17 Nobel F. I., Thomson D. W., *Plating*, **57,** 469 (1970).
18 Nobel F. I., Kessler R. B., Thomson D. W., Am. Electropl. Soc., 3rd Plating in the Electronics Industry Symp., pp. 159, Palo Alto, California, Feb. (1971).
19 Proc. 9th Annual Reliability Physics Conf., I.E.E.E., Catalogue No. 71-C-9-PHY (1971).
20 Baker R. G., Palumbo T. A., *Plating*, **58,** 796 (1971).
21 Keys L. K., Francis F. J., Russo A. J. Herring S., Proc. 21st Electronic Components Conf., I.E.E.E. (1970).
22 Keys L. K., Proc. Wescon, (1970).
23 Keys L. K., Francis F. J., Russo A. J., Proc. ISHM Microelectronics Symp. (1970).
24 Wang C. A., Proc. 20th Electronic Components Conf., I.E.E.E. (1970).
25 Duffek E. F., *Plating*, **57,** 37 (1970); **56,** 505 (1969).
26 Surtani K. H., Proc. Nepcon (1968).
27 Duffek E. F., Proc. Nepcon (1970).
28 Mclean D. A., *Insulation Circuits,* **17,** 9, 71 (1971).
29 Duffek E. F., *Plating*, **51,** 877 (1964).
30 Lepselter M. P., *Bell Syst. tech. J.*, **44,** 233 (1966).
31 Keys L. K., Francis F. J., Russo A. J., Proc. 21st Electronic Components Conf., I.E.E.E. (1971).
32 Hodgson R. W., Szkudlapski A. H., *Plating*, **57,** 693 (1970).
33 Hodgson R. W., Proc. Nepcon (1970).
34 "Printed Circuit Handbook", C. Coombs (Ed.), Chap. 4–6, McGraw-Hill, New York (1967).
35 *Electronics*, **44,** 44, 1st Feb. (1971).
36 *Assembly Engineering*, pp. 34, Feb. (1971).
37 *Electronics*, **44,** 21, Jan. (1971).
38 *Aviat. Week*, **94,** 44, 18th Jan. (1971).
39 Duffek E. F., Blech, I., U.S. Pat., 3,480,412 (1969).
40 MIL-G-45204B, "Electrodeposited Gold Plating", 27th March (1969).
41 MIL-S-19, 500E, "General Specification for Semiconductor Devices", 1st April (1968).
42 MIL-STD-105, "Sampling Procedure and Tables for Inspection by Attributes", 29th April (1963).
43 MIL-STD-883, "Test Methods & Procedures for Microelectronics", Methods 1004, 1009, 2003, 1st May (1968); Method T5005 20th May (1968).
44 MIL-STD-202C, "Test Methods for Electronic & Electrical Component Parts", 12th Sept. (1963).
45 Federal Specification L-T-90 "Tape, Pressure-sensitive, Adhesive, Cellophane and Cellulose Acetate".

Chapter 38

REED SWITCH CONTACTS

J. F. HOULSTON

The basic reed switch comprises of two flexible, nickel-iron blades or reeds, hermetically sealed in a glass tube containing nitrogen with, in some cases, a small amount of hydrogen. The ends of the reeds are flattened and are arranged so that they overlap with a small gap between them (Figure 38.1).

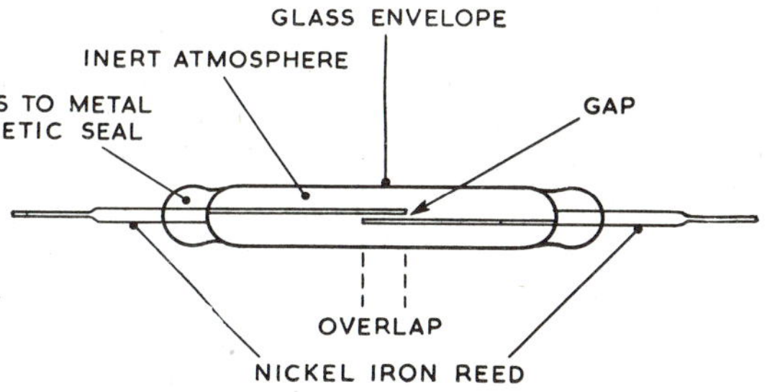

Fig. 38.1 Diagrammatic sketch of the basic reed switch.

When an external magnetising force is applied, the ends of the reeds assume opposite polarities and, if the magnetising force is strong enough, they are mutually attracted and accelerate into contact. When the force is removed the reeds spring apart to their undeflected position. A permanent magnet or current carrying coil may be used to provide the magnetising force. In the latter case the coil is usually wound around the switch and the complete assembly is known as a reed relay. Several switches may be included within one coil. The parts of the reeds which come into contact are coated with material having the appropriate properties for electrical contacts.[1]

Within the limits of its operational requirements, the switch is designed to utilise the maximum magnetic energy to promote rapid contact closure and the maximum contact pressure.[2,3,4,5] Reed design, area of overlap, the gap between the contacts, and the type and thickness of any contact material all critically affect the mechanical performance of the switch. The behaviour of the switch in life is determined largely by this mechanical performance and the properties of the contacts.

The energy required to operate the switch is related to the product of the current, in amperes, and the number of turns in the coil when the contacts have just snapped together, i.e., Ampere-Turn sensitivity (A-T). The A-T required to hold the contacts together is less than that required to initiate snap action. As the current in the coil is reduced the contacts remain together while the release A-T is exceeded. For many applications the A-T at which the switch operates and releases is critical, but these levels may be built-in by careful control of the contact gap, overlap and the thickness of the contact materials, which are usually non-magnetic.

APPLICATIONS

In general terms the reed switch fills the very wide gap between solid state switching and conventional relays. However, the reed switch does have an important advantage in that it can be operated mechanically with a magnet, or electrically with a coil.

In the former mode they are used extensively in proximity, push-button and float switches, as well as in fuel gauges and burglar alarms.

Operated electrically in a reed relay they find many applications in a wide range of electronic circuitry. A principal use is in electronic telephone exchanges where reliability and relatively high switching speeds are prime requirements. More recently the reed switch is finding increasing use in logic circuits for industrial control systems where it is used in conjunction with solid state devices.

CONTACT MATERIALS

The pressure between closed reed switch contacts depends on many design factors and for any switch is usually less than 15 g. To give the switch useable switching properties under these conditions, the contacts are coated with metals which have suitable contact characteristics.

Pure, electrodeposited gold has very low contact resistance at low contact pressures but, being soft, easily forms cold welds. For reed switches, where the separating force is also very low, pure gold is therefore relatively useless as a contact material.

This cold welding tendency can be reduced by the introduction of impurities into the gold which results in the formation of a film of surface oxides only several angstroms thick.[6,7] Provided that the hardness of the contact material is sufficient to support this film when contact is made, pure metal–metal contact and therefore adhesion, is avoided. It has been shown[5] that surface films of this thickness need not interfere with electrical conduction which takes place through the "tunnel" mechanism.

In practice there are several ways of introducing impurities into electrodeposited gold to increase the hardness of the deposit and eliminate sticking. These are outlined in the following text.

PURE GOLD (DIFFUSED)

One of the earliest methods of treating gold to prevent cold welding was to diffuse an electrodeposit of suitable thickness into the nickel-iron surface. This is accomplished by plating the reed tips with up to 4 microns of pure gold which is then diffused by heating the reeds to just over 800°C in a reducing atmosphere. The resultant surface contains gold, nickel and iron, the amounts of which depend on the deposit structure, thickness and the diffusion conditions.[8]

The presence of metals such as iron, which have unstable oxides, gives rise to contact problems. Figure 38.2 shows the typical variation of contact resistance as a function of contact pressure.

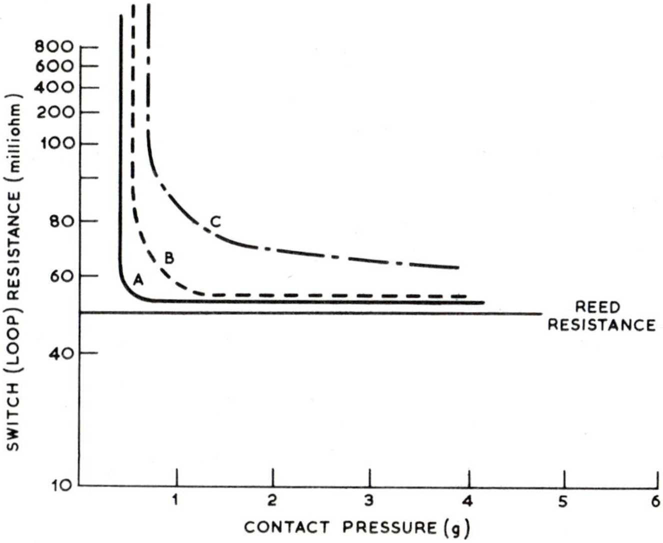

Fig. 38.2 Typical graph of contact resistance vs contact force for pure gold (diffused) reed switch contacts.
A 3 μm pure gold deposited on nickel–iron.
B 3 μm pure gold deposited on nickel–iron and diffused at 800°C for 4 min.
C 3 μm pure gold deposited on nickel–iron and diffused at 810°C for 4 min.

As the diffusion temperature increases, the variation in contact resistance increases rapidly, due to the presence of iron on the surface, and causes an initial, random, high resistance every time the switch closes. For most applications low contact resistance is not too important but, for devices switching anything but very low electrical loads, a high resistance may cause local overheating which produces rapid breakdown of the contact. Where this iron contamination is not high enough in itself to introduce initial contact resistance problems, it is possible that even small traces of oxygen will oxidise the iron present and so give rise to problems early in the life of the switch.

Increasing the diffusion time has a similar, but less critical effect.

MULTILAYER DIFFUSION

In order to overcome these difficulties many techniques have been developed in attempts to prevent iron migrating to the surface during diffusion. The best known methods involve the application of an electrodeposited "barrier" layer between the base metal and the gold. This not only slows the rate of migration of the iron to the surface but, when diffused into the gold, enhances its contact performance. Nickel, silver and copper are typical examples of metals which have been used with some success in this way.

Nickel acts as an excellent barrier and diffuses slowly enough to make the diffusion conditions less critical. Its presence in the gold readily allows the formation of a strong surface oxide which prevents cold welding. However, any gaseous or organic impurities present in the base metal or in the deposit which might otherwise have diffused through the gold and been removed, are trapped and may cause blistering. Basis metal treatment and purity of the nickel solution are therefore of prime importance.

Silver and copper diffuse rapidly into pure gold, forming alloys into which nickel and iron have much slower diffusion rates. Silver itself has some good contact properties and its presence in the gold (up to 30% by weight) has no appreciable effect on contact resistance. It does, however, allow increased solubility of oxygen in the contact surface and this could introduce contact problems under some conditions. As with single layer diffusion, plating and diffusion conditions must be rigidly controlled in order to obtain a consistent contact material.

More recent developments in plating solutions have allowed reliable deposition of gold containing the desirable metallic impurities, thereby eliminating the critical diffusion process. Nickel, cobalt and silver are the most popular additions employed.

For telecommunication applications in the United Kingdom, gold alloys containing 0.1%–0.5% by weight of nickel or cobalt have been used successfully. The factors which affect the performance of these alloys as contacts are not fully understood. Certainly, the presence in the deposits of varying amounts of organic impurities appears to be significant, tending to introduce contact resistance problems when switching electrical loads below 100 milliwatts. For loads in the range 1–5 watts, when material transfer takes place, the presence of organic impurities can increase the life of the device considerably without causing contact resistance problems.

Solution formulation and plating conditions may, therefore, be critical for any given application and it is doubtful if any particular deposit can be used successfully over a wide range of operating conditions.

Development of alloy plating solutions continues in an attempt to discover better contact materials. Recently some interesting work has been reported on gold deposits containing an even distribution of tungsten carbide particles of up to 1 micron in diameter. It is claimed[9] that this deposit has a much higher resistance to cold welding and material transfer during operation.

GOLD AS AN UNDERLAYER

Many reed switches use contacts which have been electroplated with platinum group metals such as rhodium or ruthenium. In these cases an underlayer of gold is often used to protect the base metal from attack by the strongly acid plating solutions. One micron of a low porosity, pure gold should give protection from acid attack, but the cheaper alloy golds may be used if good heat resistance in air is not required at a later stage.

As switches become smaller, the thickness of the non-magnetic layer on the contact has an increasing effect on the A-T sensitivity. In order to meet specific A-T requirements, it may be simpler and cheaper to increase the gold thickness rather than that of the platinum metal, due consideration being given to the high stress likely to be encountered with thick deposits of the latter.

ELECTROPLATING

For reliable operation it is essential that reed switch contacts remain uncontaminated. To this end the switches are assembled with ultra clean components in an atmosphere which is continuously conditioned and filtered. To minimise contamination, it is desirable that the reeds are assembled into switches soon after plating, at least the final stages of which should be carried out in a clean environment. In any case, the plating solution must be continuously filtered using 5–10 micron polypropylene filters which should suffice for most applications. Dirt of a fibrous nature from hair or clothing is particularly common and difficult to remove. All personnel in the area should wear proper clothing and headgear.

Since it is usually only necessary to plate the actual contact area of the reed, the plating is essentially a selective process. However, to avoid expensive masking operations it is normal practice to plate both sides.

The reeds are held in a vertical position with their tips immersed in the solution to the required depth. The necessary pretreatment processes may be carried out in a similar manner. It is important that the immersion depths for rinsing must be greater than for cleaning and polishing, which must in turn, be greater than for plating.

The main functions of the plating jig are as follows:

(*a*) to locate the reeds accurately with their tips immersed in the solution.

(*b*) to distribute the current equally to each reed.

The need for accurate location is highlighted for reeds with very small plated areas. Slight changes in immersion depth result in relatively large changes in surface area, making control of current density difficult. With such small surfaces, the current required by each reed is very low, often less than 1mA, and small changes in contact resistance between reed and jig produce large variations in this current. The design of a jig which minimises these effects and which is capable of being simply and, in some cases, automatically loaded remains one of the major problems in reed plating. Some manufacturers have developed reasonable systems but, as jig design is one of the critical factors in the production of predictable switches, details are seldom disclosed.

The magnetic properties of the reeds permit convenient jigging using electro- or permanent magnets to support and locate them. This is a particular asset when automatic loading and plating is involved.

For automated plating lines the choice of system depends on several considerations, not least of which are the types and numbers of variable stages involved. For plating a large quantity of a limited type of reed with a single deposit of pure or alloy gold, a continuous conveyor system is convenient. In this case there may be some advantage in pressing reeds from flat strip and leaving them connected in long lengths until after the plating stage. An important advantage of this technique is that it almost eliminates contact problems.

With the conveyer system it is difficult to vary individual process times to allow for special requirements. More flexible, and desirable for multilayer plating, is batch plating equipment which may be programmed to carry out automatically all of the stages, each of which may be individually varied. An important advantage of this type of equipment is that, as production increases, it may be progressively developed from a simple manual system.

The amount of agitation required obviously depends on the solution, but as any vertical movement of the solution surface interferes with critical level control, it can present problems. Cathode agitation is a good way of providing the necessary movement, but adequate agitation can often be achieved by using the circulatory movement of the solution. As it is pumped through suitable filters it is returned to the bath through perforated weirs so that most of the solution is directed along the surface with only sufficient disturbance below this to maintain homogeneity. The immersion depth may then be simply controlled by fine adjustment of the outlet weir.

The alignment and distance between the reeds and the anode arrangement depends on the solution, the size and shape of the contact, and the plating technique. As a general rule the proximity of the reeds to one another is limited by the surface tension of the solution which makes level control impossible if they are too close. For a single conveyer system they may be held with the contact areas directly facing anodes placed on the sides of the bath. For batch plating it is more convenient to place the anodes along the bottom of the bath. As with most precious metal plating it is usual to use expanded platinised titanium anodes.

Reed blades received from pressing may be contaminated with oil, dirt and grease. In addition to this they may possess burrs and other faults depending on the quality of the original wire or strip. To achieve the proper magnetic properties the reeds are fully annealed and degassed, after solvent degreasing, by heating in hydrogen at about 1000°C for several minutes. For more efficient removal of organic contamination it may be necessary to "wet" the hydrogen by passing it through deionised water.

Burrs and other surface irregularities may be removed mechanically prior to annealing, but as reed straightness is essential it is best carried out by electropolishing immediately before plating. This may be accomplished by treating the reeds anodically in 80% orthophosphoric acid at ambient temperature.

A typical pretreatment process for pressed reeds is as follows: (i) solvent degrease in trichlorothylene; (ii) anneal at 1000°C in wet hydrogen for 15–30 min; (iii) load into jigs; (iv) electropolish in 80% orthophosphoric acid at 40–60 A/ft^2 for 5–10 min at ambient temperature; (v) hot water rinse; (vi) activate in 20% hydrochloric acid or cathodically in 15% sodium cyanide solution; (vii) cold water rinse.

PROCESS CONTROL

The performance of the reed switch is significantly affected by the properties of the electroplated contacts. The actual mechanism of the failure of switches varies with the electrical load[10,11,12,13] and for a consistent, predictable performance over a wide range of operating conditions, a high degree of process control is necessary.

Maintenance of solution composition and control of operating conditions is an obvious requirement, particularly when depositing alloy golds. Of special interest to the reed plater is the metallic and organic contamination of the solution and the deposit.

Contamination of the solution with nickel and iron is an inevitable risk and must be kept to a minimum. One disadvantage of the bottom anode is that any reeds which fall into the bath may contact the anode with resulting rapid dissolution. Protected magnets may be positioned in the bath to remove these reeds from the anodes, but some build-up in the contamination level can be expected unless they are removed at frequent intervals.

For most solutions the initial effect of this contamination is a reduction in the cathode efficiency which may be tolerated to a degree. However, if the contaminant concentration reaches the level where codeposition occurs, contamination of the deposit itself may have serious consequences. This level will obviously vary with the solution.

Organic contamination also presents problems in the solution and in the deposit. When switching low electrical loads, i.e., with no arcing, the mechanical properties of the deposit can be critical. Stress introduced into the harder, alloy golds by organic contamination of the solution may therefore have a direct effect on the performance of the contact in these conditions. More severe contamination may cause blistering of the deposit during the sealing operation when the reed can reach temperatures in excess of 500°C.

Organic contamination of the deposit itself may come from accidental contamination or from the organic constituents of the solution[14,15]. The former source can be controlled by regular, if not continuous, treatment of the solution with activated carbon. The latter depends on the solution formulation and operating conditions, and sometimes appears as a thin tarnished layer on the plated surface which increases the contact resistance. During the heating of the reeds in the sealing process, organic contamination can be driven into the atmosphere inside the switch and have some effect on the life.

Generally, the life of reed switches increases with the thickness of the contact material. As the electrical load becomes greater, however, any improvement in the life due to this thickness becomes more insignificant. In fact, as the contact pressure decreases with the increase in non-magnetic thickness, contact resistance problems could become significant earlier in life.

For miniature switches the thickness of the non-magnetic contact has a critical effect on the sensitivity of the device and for applications demanding high sensitivity, the yield will be in proportion to the degree of thickness control.

There are several methods of measuring the plating thickness (see Chapter 28), the choice of which depends on the shape and size of the reeds, and on the contact material. For monolayer deposits on all but the smallest reeds, beta-ray backscatter techniques provide a rapid, non-destructive, reasonably accurate means of measuring deposits exceeding 1 micron. Except for measuring the thickness of the initial deposit, this method is limited for multilayer plating, particularly when the initial deposit gives much greater sensitivity than the subsequent coating. Some degree of control can be obtained by checking special test pieces which are plated with the batch of reeds.

The magnetic properties of the nickel–iron, of course, allow the use of magnetic means for measuring the thickness of the non-magnetic contact deposit. Again, the accuracy of this method is limited for deposit thicknesses of less than 2–3 microns and is not really suitable for miniature reeds. It can, however, be used to measure the total thickness of multilayer coatings.

Whichever system is used, its ultimate accuracy will depend on the calibration for which microsectioning will have to be used.

The heat involved in the assembly of switches provides a fairly severe test of the plating adhesion, but this may be more conveniently assessed by bending a sample of the reeds through 180° within the plated area. Poor adhesion will be evident if the stress area is examined under a magnification of 10 × or above.

It is extremely difficult to apply discriminating tests which will determine the performance of the electrodeposit as a contact material. With little delay between plating and assembly, the initial properties of the switch can be monitored using contact resistance measurements which will give some indication of the thickness and degree of contamination. For diffused contacts the initial contact resistance may also indicate the degree of diffusion.

A very useful and rapid method of evaluating the early contact performance is to operate the switch in a coil at about 20 Hz using a drive current of triangular wave form and displaying the dynamic contact resistance characteristics on a cathode-ray oscilloscope. Ideally, this should show a sharp rectangular form and any variation in contact resistance with contact pressure will be indicated by a tendency for the trace to curve as the switch releases. Since this test is non-destructive it may be applied to a complete batch if necessary.

Other faults may be detected by using switches at selected loads and continuously monitoring their operation for contact resistance, sticking and failure to function. For example, if a switch is operated with a load of 10 volts DC and 100 mA, the contact resistance pattern often indicates several features within the first half million operations. An immediate, random contact resistance which gradually improves, might be symptomatic of the presence of light contamination, whereas if it remains poor, heavier surface contamination, or possibly iron contamination, is indicated. A rapid, steady deterioration in contact resistance normally signifies the presence of oxygen in the atmosphere.

REFERENCES

1 Stickley B. C., Proc. 14th Ann. National Relay Conf., Oklahoma State Univ. (1966).
2 Rovnayak R. M., Proc. 18th Ann. National Relay Conf., Oklahoma State Univ. (1970).
3 Takashi S., Proc. 12th Ann. National Relay Conf., Oklahoma State Univ. (1964).
4 Cullen G. W., Proc. 19th Ann. National Relay Conf., Oklahoma State Univ. (1971).
5 Holm R., "Electric Contacts", 4th Edn, Springer, Berlin-New York (1958).
6 Holm R , Proc. 2nd Int. Res. Symp. on Electrical Contact Phenomena, Graz, Austria (1964).
7 Entwistle S. D., Craig J. A., Proc. Holm Semin. on Electrical Contact Phenomena, Illinois Inst. of Technology, Chicago (1968).
8 Ohki Y., Maruyama H., Proc. 13th Ann. National Relay Conf., Oklahoma State Univ. (1965).
9 Peiffer H. R., Marley J. E., Kobler R. J., Jacobs W., Proc. 17th Ann. National Relay Conf., Oklahoma State Univ. (1969).
10 Mitani S., Kamoshita G., Ono K., Tanii T., Proc. 16th Ann. National Relay Conf., Oklahoma State Univ. (1968).
11 Davies T. A., *Microelectronics and Reliability*, **8**, 205 (1969).
12 Hewett B. C., Charman D., Proc. 4th Int. Res. Symp. on Electrical Contact Phenomena, Univ. Coll., Swansea (1968).
13 Crawford W. M., Proc. Int. Conf. on Electromagnetic Relays, Tohuko Univ., Sendai, Japan (1963).
14 Rushton R., Maddocks R., Proc. 4th Int. Res. Symp. on Electrical Contact Phenomena, Univ. Coll., Swansea (1968).
15 Munier G. B., *Plating*, **56**, 1151 (1969).

Chapter 39

SPACE APPLICATIONS

L. MISSEL

Gold is an extremely important coating on many space vehicle components because of its resistance to change in pre- and post-launch environments. The extreme stability of the gold surface and other unique features make it attractive in space vehicles which are characterised by critical electrical contacts and optical surfaces. Surface stability contributes to high reliability performance from a very large number of functional components and minimises the repair and maintenance requirements of difficult access areas.

This chapter primarily covers applications which are unique to space and discusses them on the basis of function. First of all, however, it is necessary to describe the working environment of space and its effects.

SPACE ENVIRONMENT

Ultra-violet is the most serious radiation problem, followed by the geomagnetically trapped radiation of the Van Allen belts. The inner belt contains protons and electrons, the outer belt predominantly electrons. Cosmic-ray, auroral radiation and solar flare radiation levels are too small to be a major problem. Metals are the most resistant materials to space radiation, while organic materials and many inorganic compounds change greatly in optical and mechanical characteristics after exposure. The mechanical properties and electrical resistivity of metals can change somewhat, but optical properties tend to be stable. Gold is widely used for radiation control because of its optical stability in the space and pre-launch environments.

Thermal cycling does not occur in interplanetary travel. However, the temperature of the spacecraft tends to high or low extremes as it approaches or travels further from the sun. In the absence of an atmosphere, heat exchange to and from the spacecraft is limited to radiation transfer. Thus, control of the surface radiation characteristics is the only practical means of regulating solar energy collection, heat dissipation and temperature. Decomposition and outgassing of materials (particularly organic) and the deposition of products from these actions on gold optical surfaces may cause undesirable temperature changes due to alteration of radiation characteristics[1]. Ascent heating is a prime cause of such problems.

Reaction of the surface with high energy atomic and molecular particles results in loss of material by sputtering. Measurements of the erosion of gold surfaces on Discoverers 26 and 32 indicated an erosion rate of 0.2 ± 0.1 Å per day[2]. This can increase several orders of magnitude during a solar storm[3].

Meteoroids can cause mechanical damage, and meteoroids and cosmic dust can cause cratering, spalling, or abrasion of optical surfaces such as gold thermal control surfaces. This does not appear to be a serious problem in the absence of a meteoroid shower. The cosmic dust detector on Mariner IV registered approximately one impact per day of particles of mass greater than 10^{-13} g on its two 22 cm^2 detector plates[1].

Because of the high vacuum of space (10^{-17} torr estimated as the pressure of outer space), some metals sublime. They may then condense and short circuit electrical systems and also dull optical systems. Gold does not present this problem.

The removal of adsorbed gas layers or failure of self-healing mechanisms in vacuum causes cold welding between mating metal surfaces such as bearings or seals. Gold and some other low strength metals perform well under such conditions. Mechanical properties of materials are affected by the partial loss of the surface film of gas which is normally present on materials in the atmosphere[3].

GENERAL GOLD PLATING CONSIDERATIONS

Plating for spacecraft requires higher standards and better control than ordinary plating. Not only can defective plating seriously impair or even abort a flight if overlooked, but defective gold is difficult to strip without damaging sensitive parts. Problems with gold quality have been reported for the solar array bearings of the OGO (Orbiting Geophysical Observatory)[4] and with gold plated module pins[5]. Both problems were corrected by a change in the plating process. Also, NASA* delayed the launch of Ranger VI because loose flakes of plated gold inside the glass casing of diodes required replacement of the diodes.

The purest, softest and least stressed gold should be specified unless it is incompatible with function. Alloying, which increases hardness and stress, decreases its optical reliability, heat and tarnish resistance. The inherent internal stress in hard alloyed or unalloyed gold plating decreases corrosion protection while increasing the tendency to flake or crack, but hardness is often mandatory for wear applications. Alloys are deposited from less efficient baths with lower throwing power than the baths for the pure metal. Alloying decreases electrical conductivity and increases contact resistance. In addition, alloy golds are more susceptible to discolouration than the pure metal at elevated temperatures in air[6].

---

*National Aeronautics and Space Agency.

Most of the gold exposed to space radiation for thermal control of spacecraft is on aluminium substrates. Direct deposition on aluminium by such methods as physical vapour deposition (cathode sputtering, vacuum evaporation) or electroless deposition leads to surface blistering when the surface is exposed to low energy protons as occurs in the lower Van Allen Belt. Copper undercoats and normal plating practices prevent this condition[7]. Gold deposits in electronic gear are ordinarily not exposed to radiation.

The silver undercoat/gold topcoat plating system has become so widely known that it has been reported in newspaper accounts of the launching of Ranger Moon-probe, Mariner Venus-probe and other space vehicles. Gold deposits, particularly when applied over a silver undercoat, diffuse rapidly when exposed to moderately elevated temperatures. The use of a protective shroud during ascent heat pulse and adequate protection against re-entry thermal shock are mandatory. The popularity of silver undercoats is based on easy buffability, but it increases tarnish and thermal protection problems. A nickel layer makes a satisfactory diffusion barrier between copper and gold. Electroless nickel, however, should be avoided where long term optical stability of surfaces exposed to moderate heat is required since absorptivity increases, possibly due to phosphorus diffusion.

PROTECTION AND HANDLING

Gold excells in its ability to resist corrosion and tarnish before launch and to retain its integrity into space. All efforts to obtain a carefully controlled surface are wasted if it is not maintained up to and after launch. The most demanding surface requirements are in thermal control* and in lubrication applications. Although gold is highly resistant to atmospheric corrosion, surface contamination can cause changes in the thermal control characteristics regulating vehicle temperature, particularly thermal emission, resulting in bearing failures. Ultra-violet radiation in space can intensify the effects of residual organic contamination while breaking it down.

Matte gold deposited from a high purity acid or neutral bath is a prime choice for most space applications except for some lubrication or contact applications. It is "self-inspecting" in that even traces of pollutants alter its appearance significantly. The brighter golds do not show contamination as readily. Some sources of the problem are poor rinsing, bare hands, contaminated tools and rags, oil, dirt spots, dust, adhesive residues left from identification or inspection markings or stickers, and other pollutants from the ambient environment.

Ascent heating causes outgassing, especially from volatilisation or pyrolysis of wire, cable insulation, or potting compounds. This can be followed by recondensation on gold and other thermal control surfaces. This may cause

---

* Absorption of solar energy and emission of thermal energy which regulates temperature.

uncontrollable changes in radiation characteristics[1]. The failure of the Explorer 10 payload has been traced to foreign material condensing on the protective shroud during ascent heating. This contamination doubled absorption of solar radiation by the surface, causing excessive heating[3].

High density, high purity golds deposited from acid or neutral baths show no detectable changes in radiation properties after long protected storage. However, lower density unalloyed gold deposited from cyanide baths usually shows an increase in radiation absorption characteristics after several months storage. Less pure golds are not as resistant to discolouration during storage. The metal is also very soft and easily smeared. The thermal control characteristics of areas burnished by contact with other objects are not significantly altered[3], but mechanical damage to the coating may reduce corrosion resistance in the area.

Solvents used to clean gold such as methyl ethyl ketone, trichloroethylene, water or alcohol must be residue-free because amounts of residue invisible to the eye are enough to significantly alter thermal control properties or cause premature bearing failure under space conditions.

Gold films are sensitive and must be protected against mishandling. Any contact should be with clean, frequently replaced, lint-free, white cotton or nylon gloves. The preferred protection is sealing the item in a clean, non-contaminating plastic envelope by welding. When a part is too large, sheet plastic can be wrapped around it and secured at the edges with a tape which leaves no adhesive residue upon removal. Effective protection eliminates the need for cleaning, with all the dangers of damage inherent during handling.

Gold coatings for lubrication are often too thin to protect the underlying substrate against corrosion. Oils may be used for protection prior to launch, provided that oils used with gold over silver undercoatings are free of sulphur. Loss of oil by evaporation in space is no problem since by the time the part is spaceborne the need for corrosion protection is past, the necessary lubrication being provided by the soft metal film.

ELECTRONICS

Plating of electronic equipment is the major gold plating application in space technology, both in amount and diversity, this metal having a number of properties which make its use in electronics invaluable. These include unique capabilities in many common and specialised joining applications, high and stable electrical conductivity, resistance to tarnish or corrosion during long storage, retention of low and stable contact resistance, and resistance to radiation. Good gold is good insurance against electrical failure due to pre-launch storage. Its uses include electrical contacts, chassis, housing assemblies, heat sinks, brackets, conductors, shields, terminal boards, terminals, waveguides, horns and circuit boards[8].

The uses of gold in fabricating electronic equipment and in electronic joining applications are covered elsewhere in this book, as are thermal considerations further on in this chapter under "Thermal Control". However, antenna systems which are unique and highly diversified will be discussed in the following paragraphs.

Antennae

Antennae are constructed from a wide variety of materials including plastics and metals. Gold is useful in providing a film-free, radio frequency carrying, surface layer. Surface finishes are very important above 100 MHz, affecting both response and electrical losses. High frequency currents flow in the outer skin of a conductor with the thickness of this current carrying layer decreasing as frequency increases. If a plated top layer is above the minimum thickness required to carry the current of the plating it is the electrically functioning part of the structure. Therefore, the surface plating must be clean and untarnished and of a thickness of at least the electrical skin depth at the operating frequency.

An interesting method of gold application developed specifically to meet space industry needs is the LOCKSPRAY* gold plating process[9,10] in which an ambient temperature gold solution and reductant from a twin-headed spray gun react at the work surface to produce a pure, high quality gold film (see Chapter 15). More than a hundred unfurlable antennae in various designs and in sizes up to 30 ft^2 have been made from aerosol gold plated synthetic fabrics.

The Lincoln Experimental Satellites[14] had gold plated aluminium ground planes, almost 4 feet in diameter and 0.020 inch thick, of which 75% was cut away to lighten the structure. Gold was used to assure retention of electrical properties. The plating sequence consisted of: (i) deposit 0.5 mil electroless nickel over double zincate, (ii) bake 30 min at 175°C, (iii) activate in 70% sulphuric acid at 82°C, (iv) acid copper strike, (v) copper plate 0.5–1 mil from high speed cyanide bath, 55°C, 20 A/ft^2, 15 min, (vi) immerse in 10% hydrochloric acid, (vii) gold strike, pH 4.1, 85°C, (viii) gold plate 70–100 micro-inches, 24 carat hard acid bright gold, 8.2 g/l, pH 4.3, 32°C, 10 A/ft^2, 10 min with semi-conforming anodes of about 80% of part size to minimise edge build-up, (ix) bake at 125°C, 30 min.

The Delta launch vehicle dome[11,12] was 2 feet in diameter and made of glass-phenolic. It was gold plated to provide and assure retention of good electrical properties. After deposition of electroless copper on the resin surface, 0.5 mil of high purity, low stress gold was deposited from a neutral bath (pH 5.7–6.0).

Instrumented scale models of missiles for ballistic range blast tests required helical antennae sturdy enough to withstand test firing[15]. Two antennae, each approximately 3 feet long, 0.25 inch wide and 0.015 inch deep with the metal helices flush with the cylindrical epoxy surface, were required for each

* LOCKSPRAY—Trade Mark, Lockheed Missiles & Space Co.

Table 39.1
Typical Applications of Gold Plating on Antennae[3]

Vehicle	*Antenna*
Application Technology Satellite	Beryllium rod, copper undercoat, gold plated
Environmental Research Satellite	Gold plated steel tape
Pioneer Deep Space Probes	Epoxy-glass laminate, silver undercoat, gold plated
Telstar Communications Satellite	Gold plated beryllium wire VHF helix antenna
Delta Launching Vehicle[11,12]	Copper undercoat, gold plated plastic dome
Ranger Moon Landing Spacecraft[13]	Gold plated stainless steel antenna feed
Lincoln Experimental Satellite[14]	Gold plated aluminium ground plane

model. Strength was obtained by installing metal screws in tapped holes in the antennae grooves to anchor the electroplated antenna. The epoxy surface was made conductive by application of a thin film of silver-epoxy lacquer, after which copper was electroformed and machined to fill the grooves and provide a flush surface. The copper was plated with 100 microinches of high purity acid gold (pH 4.4, 60°C, 3 A/ft^2) for corrosion protection and preservation of electrical properties. Application of oil base lubricants in machining the copper interfered even with the use of ultrasonic cleaning (trichloroethylene or detergents) prior to gold plating. Sodium nitrite corrosion inhibitor in a water solution as the machining coolant was compatible with good gold plating.

THERMAL CONTROL

Some of the most important contributions of gold plating to the space field have been made in the area of thermal control in order to maintain components within operating temperature limits. Plated metals are very attractive for thermal control purposes because of their opacity to thermal radiation, only a few microinches having the same thermal characteristics as bulk metal[3]. Plating, therefore, is an attractive way to obtain radiation control for a minimum weight expenditure.

In the absence of an atmosphere, heat exchange to and from a vehicle is limited to radiation. Consequently, radiation control will regulate temperature under the equilibrium conditions attained relatively soon after exposure to the thermal environment of space.

Passive thermal control, using only surfaces with known reliable radiation characteristics, has been employed for essentially all unmanned flights[3] as well as manned flights. This type of regulation has the advantage of not requiring the added equipment or moving parts which are needed for active systems. Active systems do not diminish the importance of passive thermal control coatings, but includes them in a better temperature regulating system. Metals operate well in passive systems because, unlike organic materials and many inorganic compounds, their radiation characteristics are stable in space.

Table 39.2
Typical Applications of Gold in Thermal Control[3]

Vehicle	*Application*
Ariel International Scientific Satellite	60% of total surface area gold (1% silver) over nickel undercoat on aluminium
Environmental Research Satellite	Gold on external surfaces
Gemini Manned Vehicle	LOCKSPRAY gold inside of adapter section
Mariner Interplanetary Spacecraft	Gold plating on instruments, including telescope tube and secondary mirror support system[1]
Orbiting Solar Observatory	Gold plated spectrometers mounted on sail
Ranger Moon Landing Spacecraft	Gold plate on instruments, and exteriors and interiors of some electronic boxes
Surveyor Moon Landing Spacecraft	Exterior surfaces of vernier engines
Telstar Communications Satellite	Polished electroplated gold film
Transit Geodetic Satellite	Gold on inside surface of spherical satellite
Vela Radiation Detector	Gold plated aluminium thermal control coatings

Equilibrium temperature in space is controlled by the ratio of solar absorptivity (α)* to emissivity (ε)†. These are optical properties. Absorptivity refers to all electromagnetic radiation (x-ray, ultra-violet, visible, infra-red, radio frequency, etc.), while emissivity is restricted to the infra-red. Table 39.3 shows the relationship between optical properties and temperature.

Table 39.3
Values for α and ε for Typical Surfaces[16]‡

Surface Finish	α	ε	$\frac{\alpha}{\varepsilon}$	*T derived by using Stefan–Boltzmann equation* $T^4 = K \frac{absorptivity}{emissivity}$ *where* T = *temperature*, K = *constant*
Polished Metal	0.25	0.05	5	+145°C
Black paint	0.95	0.90	1.05	+ 11°C
White paint	0.20	0.90	0.22	− 81°C

* Absorptivity is the ratio of radiation absorbed by a body to that incident on it.

† Emissivity is the ratio of emitted heat to that emitted by a black body at the same temperature.

‡ Wide variation from these values exists, particularly with metals and white paints. All non-metals have high emissivities.

Local hot spots can be prevented by using good thermal conduction joints. Radiation exchange within the spacecraft is small unless there is a large temperature difference between a device and the spacecraft.

Metals used alone would provide very high temperatures because of the generally high absorptivity/emissivity ratios of their surfaces. Actual temperature control surfaces are usually a mosaic of materials with different radiation characteristics combined to give the calculated absorptivity/emissivity ratio.

Unlike organic materials and many inorganic compounds which change considerably, gold and other metals have stable optical properties when subjected to ultra-violet low energy protons and high energy electrons. The absence of surface films makes the thermal response of gold particularly reliable. As mission time increases, stability to radiation, particularly ultra-violet, becomes a more important coating requirement.

An easy gold plating procedure for aluminium which provides good adhesion, good resistance to diffusion, and reproducible stable absorptivity and emissivity is as follows: (i) degrease, (ii) non-etch type alkaline soak cleaner, (iii) deoxidise, (iv) double zincate (dilute zincate preferred), (v) 30 microinches minimum, low pH copper cyanide strike, (vi) mild acid dip, (vii) approximately 500 microinches sulphamate or non-brightened Watts nickel plate, (viii) mild acid dip, (ix) acid gold strike, (x) high purity matte acid or neutral gold plate, 75 microinches minimum.

Polishing of the gold after deposition makes the measurement of absorptivity more difficult because directional surface scratching causes a directional distribution of reflected radiation. Consequently, measured reflectance varies with orientation unless laborious procedures are used. Deep rolling marks cause the same effect. Experience shows that pure, matte, as-plated gold has a colour quality which enables the eye to distinguish differences in absorptivity of as little as 0.02 with gold from the same bath formulation.

Baking of samples at 190–205°C is a very severe test for adhesion and surface contamination. A good gold deposit is capable of withstanding 100 hours of baking at this temperature without macro- or micro-blistering, or change in absorptivity or emissivity.

The radiation characteristics of gold are strongly influenced by the roughness of the substrate and the parameters of the process. With a reasonably smooth substrate and high density, high purity, matte gold deposited from a neutral or acid bath at reasonable current densities, it is possible to take advantage of the following characteristics of the gold plating:

1 Stable absorptivity/emissivity (about 0.25α and 0.03ε), providing constant and predictable temperature in space.
2 Low emissivity (about 0.03), which prevents heat loss through radiation into space or thermal damage to adjacent sensitive components.
3 High infra-red reflectivity (well over 0.9), which aids in protection against external heat.

Examples described in the following text illustrate each of these characteristics.

Equilibrium Temperature Control (Stable Absorptivity/Emissivity Ratio)

Application of carefully controlled thin films over a surface is a widely used method of obtaining a controlled orbital temperature for spacecraft. This method is not suitable for vehicles with very short mission duration since they do not remain in space long enough to reach their steady state temperature.

Bare metal surfaces, such as gold, ordinarily have high absorptivity/emissivity ratios leading to excessively high orbital temperatures if used alone. Gold and other metals are therefore part of a mosaic of coatings of different optical properties.

The Surveyor moon landing craft[17] provides an example of the use of gold plating for equilibrium temperature control. Its vernier engines, which enable the craft to make a soft landing, have many exposed external aluminium and stainless steel parts, all gold-cobalt plated to a 150 microinch minimum thickness and then buffed. Minimum specified final gold thickness is 100 microinches which maintains the engines and propellant at their specified operating temperature. Pioneer III and IV moon probes used[18] gold plated aluminium as part of the external coating system to control temperature between 20–50°C. The instrument packages were also gold plated[19]. Telemetered data indicated a temperature of 38°C for the Pioneer III payload and 42°C for the Pioneer IV payload, verifying thermal design. For the Ranger[20] moon probes, the control temperature for the telemetering electronics gear was 20–50°C.

For the UK 3 (Ariel 3) satellite[21], electrolytic, electroless and vapour deposited gold were considered. A very high quality gold deposit was required for thermal control[16] and the comparatively large size of pieces to be plated caused considerable difficulty. An electroless process was first used, but this approach was abandoned due to difficulties in obtaining uniform optical characteristics over the entire surface, and poor adhesion between the plating and the aluminium substrate. Tests[21] indicated that a suitable high quality finish was possible by electrodepositing a hard, bright gold on a matte substrate obtained by vapour blasting electrodeposited nickel.

Gold plated parts included boom assemblies, boom hinge brackets, cross members, door assemblies, nose cone, and others. The parts were degreased in trichloroethylene, followed by vapour blasting to give a good keying surface for thermal control paints applied over part of the gold plated surface. After dull nickel plating the parts were baked, again vapour blasted, and plated to a 200 microinch minimum gold thickness. Adhesion of the plating to the aluminium substrate was found to be over 2000 psi, and resistance to thermal cycling was established by subjecting samples to 1000 cycles between +80°C and −80°C, *in vacuo*, with no detrimental effect on adhesion.

Virtually all items of equipment were temperature sensitive so that few would work satisfactorily outside the range −15°C to +60°C.[16] The operating temperature of the nickel-cadmium battery was −10°C to +40°C. The surface mosaic selected was:

	Gold	*Black Paint*	*White Paint*
Cone	65%	10%	25%
Base	55%	0%	45%

In actual flight, the difference between predicted and actual temperatures was about 5°C, with the gold the most stable of the thermal control coatings.

The Birmingham University experiment[16] required that the spacecraft have an electrically conducting surface with a projected area of at least 400 in.^{2} in any attitude. Because of its oxide layer, the aluminium used in the structure was not suitable and a noble metal had to be plated on the surface. Gold was chosen for this purpose since, of the noble metals, it was the least expensive non-tarnishing metal and the one for which most plating experience existed. It was also the surface successfully used on UK-1 to meet a similar requirement.

Aerosol applied golds are also used to provide equilibrium temperature control. Metal surfaces are coated with a polymer, such as an epoxy, before plating with aerosol gold[9]. This system was used to protect the Apollo Lunar Surface Experiments Package, placed on the moon during landing missions. Other parts that have utilised chemical spray gold for thermal control coatings include, TEFLON sheathed coaxial cables, dropsonde thermocouple shields and antenna housings[9].

Low Emissivity

Since only metals can have low emissivity surfaces, all thermal shielding must be properly selected bare metal. The Vanguard satellite[22], for example, used gold plating to minimise internal radiant heat transfer. Gold also served as an undercoat for the external evaporated coatings used to provide high visual reflectance and the necessary absorptivity/emissivity ratio for temperature stabilisation. In addition to the external shell, most internal parts and the housings of the instruments were gold plated to minimise radiant heat transfer. The satellite, carried in the nose of the third stage rocket, was protected against aerodynamic heating by a disposable cone. At the same time, the gold plating provided protection for the AZ31B magnesium structural metal.

After meticulous mechanical preparation, the magnesium alloy parts were gold plated as follows: (i) vapour degrease (ii) alkaline soak clean and rinse, (iii) phosphoric acid-potassium fluoride activation and rinse, (iv) zincate immersion treatment and rinse, (v) copper plate from a cyanide bath, 13–20 μm (0.0005–0.0008 in.) copper, and rinse, (vi) silver strike, (vii) bright silver plate, 8–13 μm (0.0003–0.0005 in.), rinse and dry, (viii) buff, (ix) vapour degrease, (x) alkaline soak clean and rinse, (xi) bright gold plate, 1.3 μm (0.00005 in.) and rinse, (xii) boiling water soak and dry, (xiii) final buff to mirror-bright finish, (xiv) alkaline soak clean, rinse, and dry.

Gemini provides another example of the use of aerosol gold as a low emissivity surface. All interior surfaces of the booster-adapter section (10 feet diameter by 8.5 feet high), housing retro rockets and life support equipment were aerosol gold coated to retain heat from the support equipment[9]. The size of this operation, let alone adhesion and processing problems, would make such an operation unlikely by other means.

High Infra–Red Reflectivity

The rate of heat transfer may be reduced by the use of insulating materials or infra-red reflecting surfaces, with weight and volume considerations favouring the reflecting type for missiles and space vehicles. In applications where thermal equilibrium is not attained, infra-red reflecting surfaces may be used. An important example is in heat shields which retard the spread of heat from engine blast.

Gold reflects well over 90% of incident infra-red radiation. In addition, its emissivity is only about 0.03. This combination of low heating and low emitting rates makes it an excellent shield against propulsion generated heat in applications such as protection of rocket fuel tanks, fuel lines, exposed wiring, components and instruments.

An example of gold as a thermal shield is its use on digital velocity meters for Ranger, Mariner, Echo, Nimbus, Alouette and others[23]. Digital velocity meters control motor shutdown when desired velocity is reached and set the final speed of the vehicle. Outer surfaces of the instrument package and brackets are plated with 100 microinches of gold-nickel alloy and buffed. This type of meter has been used on well over 100 orbiting vehicles.

Among other reported[12] applications are the use of a buffed, gold plated, copper clad laminate heat shield for protection of the valve in the rocket engine of the Apollo Lunar Module against heat radiated from the ablative thrust chamber. In others, gold plating keeps fuel cool and dissipates heat which might affect rocket fuel tanks and fuel lines.

Thin gold films on transparent plastics, applied by the aerosol process[9], have been used as optical filters with a luminous transmittance range from 90% to less than 1%. Visors worn by the astronauts during the Gemini extravehicular missions were aerosol plated with gold to reflect heat and for eye protection against far infra-red radiation. Transmittance of the visors was 14–17%. Special high density visors with a transmittance of 0.1% were used for an eclipse experiment on the Gemini XII mission.

Heat shields fabricated from reinforced plastics and then plated with chemical spray gold were used to conserve the thermal energy required to expel the propellant from the low thrust subliming attitude control motor of the Applications Technology Satellite. The gold coated heat shields reflect back 98% of the incident infra-red radiation to the motor.

Chemical spray gold was also applied on laminated plastic thermal insulation for an aluminium cryogenic tank containing solid carbon dioxide and argon. The system was designed for the Nimbus programme to provide cooling of an infra-red sensor for one year in outer space[9]. (For a detailed description of aerosol gold plating, the reader is referred to Chapter 15.)

The outside of the aluminium 6061 alloy Discoverer capsule, which was carried inside a protective shroud, was gold plated to control the desired internal temperature[13]. The shape of these large capsules resemble a hollow hemisphere. They are received for plating, as spun, and then highly polished inside and outside. A special semi-automatic polisher was originally employed, but it was subsequently found that careful manual polishing was preferable.

The plating cycle consists of the following: (i) caustic etch and rinse, (ii) nitric-sulphuric-ammonium bifluoride dip and rinse, (iii) zincate treatment and fast, thorough rinse, (iv) repetition of steps (ii) and (iii), (v) low pH, low cyanide, room temperature copper strike with live current entry, and rinse, (vi) pyrophosphate copper plate, 25 μm (0.001 in.), rinse and dry, (vii) polish to 0.025 μm (1 μ-in.) RMS or better, (viii) alkaline electroclean, cathodic and anodic, and rinse, (ix) hydrochloric acid dip and rinse, (x) cyanide gold strike with live current entry, (xi) bright gold-silver plate, 5 μm (0.0002 in.), 70 min at 0.1 A/dm^2 (1A/ft^2), (xii) rinse and dry. The fluoride formulation in step (ii) was used because nitric acid alone did not remove all the smut.

Both the strike and the gold plating tanks were built to conform to the shape of the capsule in order to make the current density as uniform as possible. After plating, the capsule was lightly buffed to provide a mirror finish. Very little buffing, however, was required.

LUBRICATION

Cold welding, to which most metals or metal combinations are subject, is the adhesion of atomically clean metal surfaces to each other without the application of heat. Even dissimilar metals of low mutual solubility can be cold welded to each other. It is normally prevented by the presence of oxide or other surface films which are maintained in the atmosphere by self-healing mechanisms.

This phenomenon typically becomes severe at pressures below 10^{-8} torr. In space, surface films can be removed by functional movement, vibration, evaporation or sputtering. Since self-healing does not occur in the absence of atmosphere, normal bearing action, as we know it, does not occur in space.

For some soft metals with low shear strength, particularly gold, seizure and cold welding *in vacuo* is not a serious problem[24]. Gold functions as a solid film lubricant in space applications involving moving parts. Among the components in this category are shutters, filter changers, focusing and zoom mechanisms, mechanical scanning devices, movable lens covers[1], horizon seekers, star finders, radar antennae, solar paddles and telescope pointing devices[13].

No one lubricant is best for all jobs, gold being most beneficial in applications where less resistant non-metallic lubricants (oils, greases, laminar solids, some polymers) would suffer chemical breakdown, radiation damage or be removed by evaporation. This occurs in exterior areas with such components as solar array orientation devices, louvres for active thermal control systems and antenna drives. Organic lubricants are more practical than gold when bearing surfaces are well shielded against radiation and high vacuum. Unlike many bonded solid film lubricants, gold is compatible with oils and greases. It can be used to upgrade the boundary lubrication effectiveness of marginal low volatility synthetic oils in space.

A study of materials for sliding electrical contacts[25] concluded that electroplated gold performed better in vacuum than wrought gold alloys, this being attributed to a unique surface chemistry related to the electroplating process. Results showed that the electroplates (Au-0.1% Co) consistently gave lower wear rates. No correlation of wear with friction or hardness was apparent. The authors indicated that surface chemistry and surface mechanical properties outweighed the importance of hardness in sliding contact wear.

The strength and durability of plated gold films for lubrication are determined largely by the degree of adhesion and the nature of the hard metal substrate, weakly bonded coatings rupturing easily. The coefficient of friction is generally at a minimum with a gold thickness in the order of 0.25 micron (10 microinches) and increases gradually with increasing thickness. These properties are reasonably unaffected by the presence of gases or a vacuum[24].

A study[26] was made of ball bearing combinations using thin metal lubricant films as solar paddle supports. Gold was found to be less temperature sensitive and to have a significantly better bearing surface than silver. The gold showed promise of exceeding by far the expected operating requirements at a service temperature range of −73°C to 204°C while exposed to space vacuum and radiation. Races and ball bearings were both made of tool steel, but the races were vapour blasted before plating and the balls were not. The plating procedure for races was as follows: (i) treatment with pumice slurry and rinse, (ii) 30 sec anodic etch at 3 volts in a solution containing potassium cyanide 64 g/l, potassium hydroxide 64 g/l, and potassium carbonate 38 g/l, no rinse, (iii) plate in solution containing potassium cyanide 64 g/l, potassium hydroxide 64 g/l, potassium carbonate 38 g/l, and potassium gold cyanide 2.5 g/l at 3 volts for 1 min, (iv) rinse, alcohol dip and dry. All solutions were at room temperature. The balls received the same treatment except that they were mixed with larger balls and barrel plated. This change increased etching time to 90 seconds and plating time to 3 minutes.

NASA evaluated[27] the running life potential in vacuum of fully machined retainers made of five different materials, with all balls and races composed of gold plated 440C stainless steel. Twenty four carat gold was compared with a 23.9 carat gold-nickel alloy, plated from an acid bath containing organic complexing agents to improve its hardness and brightness. Previously, gold plated bearings were used successfully in the electric field meter flown on Explorer VIII[27], where the use of a gold plated aluminium main housing, gold plated aluminium rotor and stator, and gold plated 440C stainless steel ball bearings solved the motor wear problem.

The NASA study concluded that (1) thin gold films as lubricants are effective in a vacuum environment, (2) pure gold plating is not as effective as the alloy gold plating, (3) fully machined retainers provide good performance and the use of relatively hard retainer materials significantly extends the useful life of the bearings, (4) the bearing failures tended to be catastrophic rather than gradual, and (5) a run-in period to compress and improve the gold surface is required.

Wear of the nickel hardened gold plate was found by the NASA researchers to be greatly superior to that of the pure metal. Pure gold tended to flake off the contacting surfaces of the balls and races, considerably increasing retainer wear and the accumulation of debris. The alloyed plate delayed the start of severe wear, thereby lengthening the operating life of the bearing. Longest life was obtained with the use of a silver undercoat under the alloyed gold.

Table 39.4

Typical Applications of Gold Plating in Space Lubrication[24,3]

Vehicle	*Application*
Orbiting Astronomical Observatory	Movable entrance slit
Pegasus Meteorite Detection Satellites	52100 (SAE)* steel races of output shaft bearing
Explorer Scientific Data Satellites	Ball bearings, rotors
Nimbus Weather Satellite	Slip rings
Orbiting Geophysical Observatory	(1) Balls and races of solar panel orientation devices (2) Stainless steel output gear of Wabble drive (3) Balls and races of orbital plane experimental package (4) Balls and races of shaft support (5) Balls, races and retainer of solar array drive and shaft support bearings

* Society of Automotive Engineers.

The OGO vehicle[4] required maximum freedom from outgassed molecules in space to enhance the validity of its geophysical measurements. Several lubricating system choices were investigated for mechanisms where hermetic sealing of oils or greases for lubrication was not permissible. A system was adopted which used 80–100 microinches of pure gold over a nickel strike with a maximum thickness of 40 microinches. The substrates in this case were 416 and 430 stainless steel. After plating, the gold was burnished with molybdenum disulphide. With the correction of early gold adhesion problems, flight data indicated satisfactory performance. An additional benefit was protection of the 400 series stainless steel substrates against corrosion in the pre-launch environment.

MISCELLANEOUS APPLICATIONS

In addition to major space applications, gold has numerous uses in which the metal's special properties help solve difficult functional problems. The following examples illustrate applications where it would be extremely difficult to substitute alternative materials.

1 Conventional seals used in high pressure liquid hydrogen fluid lines of rocket motors[28] failed after functional cycling. It was found that plating the 302 stainless steel or Inconel 718 lip seals with 250 microinches of 24 carat gold provided good sealing and eliminated galling under these conditions. Another example of this metal's use as a seal was on the heavily gold plated K-ring seal of the Surveyor[3].

Piping in the Apollo propulsion system required easy disassembly, making welding unacceptable[29]. The operating environment included very high pressures and temperature extremes, from that of hot propulsion gases to the cold of liquid hydrogen. Silver plating, over an Inco 718 substrate, was selected as the sealing medium for the hot gas seals. However, blistering of the silver and sticking of the seals resulted from test firing. The cause of blistering was diagnosed as oxygen diffusion, and a solution was found by plating the Inco 718 basis metal with 50 microinches of gold to act as an oxygen barrier before silver plating. The sticking was prevented by rhodium plating. A final bake for one hour at 504–515°C was required before use.

2 Another application was on vapour deposition shields used in fabricating reticles for a "star tracker" for space navigation in an unmanned probe[30]. Fabrication by chemical etching caused excessively rough pattern lines and was replaced by direct evaporation of the functional pattern. The process of making the sharp edged lines on the 0.005 in. thick beryllium-copper evaporation shields consisted of (1) photoprocessing, (2) plating the pattern on one side with 300 microinches of non-cyanide alkaline gold, (3) applying the test pattern on the opposite side, maintaining careful registry, (4) protecting the gold side, and (5) etching from the side opposite the gold.

3 An extensive use of spot welding in missile and space applications required gold plating of phosphor-bronze pins for protection and to provide constant welding parameters without the necessity of cleaning the pins just before welding[5]. Investigation of the cause of variable welding response of the pins indicated that the method of gold plating was more important than the type or composition of the gold. Stamped pins have deep scratches, pits, feather edges, gouges and embedded contamination in addition to normal tarnish. Unless properly processed, this is later reflected in pitting,

roughness, incomplete coverage, non-uniform coverage, poor plating adhesion and poor weldability.

The recommended plating process to insure good adhesion and uniform weld schedules was: (i) Thiourea type acid cleaner, 1 min, (ii) mild alkaline non-silicated cleaner 65°C, 1 min, (iii) dry, (iv) chemical polish, nitric-acetic acid type, 1 min, room temperature, (v) fluoroboric acid, 30 sec, (vi) copper cyanide strike 40 A/ft^2, 1 min, (vii) fluoroboric acid dip, (viii) gold strike, (ix) gold plate.

REFERENCES

1 Becker R. A., *Appl. Optics*, **6**, 5, 955 (1967).
2 McKeown D., 3rd Int. Symp. on Rarified Gas Dynamics, Paris, June (1962).
3 Rittenhouse J. B., Singletary J. B., "Space Materials Handbook", NASA SP-3051 (1969).
4 Heindl J. C., Belanger R. J., "OGO Solar Array Drive and Shaft Support Bearing Tests", TR 2311-6026-RUOOO, Sept. (1962).
5 Missel L., Torgeson D., Shaheen M. E., *Metal Finish.*, **63**, 12, 52 (1965).
6 Duva R., Foulke D. G., *Plating*, **55,** 1056 (1968).
7 Anderson D. L., Dahms R. G., Proc. of A.I.A.A./A.S.M.E. 7th Structures and Materials Conf., April (1966).
8 *Products Finish.*, **34,** 4, 98 (1970).
9 Levy D. J., Lockheed Missiles & Space Co., Sunnyvale, Calif., private communication.
10 Levy D. J., Delgado E. F., *Plating*, **52,** 1127 (1965).
11 *Products Finish.*, **33,** 8, 63 (1969).
12 Mock J. A., *Mater. Engng*, **68,** 4, 61 (1968).
13 Missel L., "Handbuch der Galvanotechnik", Vol. II, Carl Hanser, Berlin (1966); *Metal Finish.*, **63,** 2, 54 (1965).
14 *Products Finish.*, **33,** 5, 48 (1969).
15 Missel L., Shaheen M. E., *Metal Finish.*, **62,** 9, 61 (1964).
16 Semple E. C., *Radio Electron. Engr*, **35,** 1, 41 (1968).
17 "Gold Plated Engines will be Sent to Moon Aboard Surveyor on Exploratory Flight", Plating Progress, No. 19, Sel-Rex Corp., Nutley, New Jersey.
18 Clauss F. J. (Ed.), "Surface Effects on Spacecraft Materials" (Symp.), Wiley, New York (1960).
19 Curtis H., Scheiderman D., "Pioneer III and IV Space Probes", tech. Release 34-11, Jet Propulsion Laboratory (1960).
20 Happe R. A., "Materials in Space", tech. Release 34-143, Jet Propulsion Laboratory (1960).
21 *Metal Finish. J.*, 213 (1967).
22 Grupp G. W., *Metal Finish.*, **55,** 7, 40 (1957).
23 "Gold Plated Instruments Orbit the Earth", Plating Progress, No. 18, Sel-Rex Corp., Nutley, New Jersey.
24 Flom D. G., Haltner A. J., "Lubricants for the Space Environment", General Electric Co. R68SD6, March (1968). (presented at Materials Conf. of A.I.Ch.E., March (1968).
25 Abbott W. H., Bartlett E. S., "Research on Metallurgical Characteristics and Performance of Materials Used for Sliding Electrical Contacts", NASA CR-1447, Oct. (1969).
26 Lewis P., "Evaluation of Dry Film Lubricated Ball Bearings for Use in a Spatial Environment", Rep. 61GL48, General Electric Co. (1961).
27 Evans H. E., Flatley T. W., "Bearings for Vacuum Operation-Retainer Material and Design", NASA tech. Note D1339 (1962).
28 *Products Finish.*, **32,** 9, 64 (1968).
29 *Plating*, **57**, 119 (1970).
30 Whitfield E. C., Smith W. O., *Plating*, **56,** 1370 (1969).

Chapter 40

HEAVY GOLD PLATING—ELECTROFORMING

D. R. MASON

Although thick gold plating from the alkaline cyanide electrolyte bath developed by Wright and Elkington dates back to the 1840's and objects d'art[1] electroformed in gold were produced in the mid-1850's, development of electroforming as a fabrication technique is of relatively recent origin.

Reasons for this are both aesthetic and technical. Traditionally, the value of a jewellery article fabricated in solid gold lies not only in the intrinsic cost of the metal, but also in the craftmanship involved in its production and in the exclusivity of design. Though equal skill may be required in the preparation of a master model for electroforming, the feature of exclusivity must be lost if the technique is employed to make copies and it is this very capability that is most economically exploited in electroforming.

The other special features of the process, that it permits the economic fabrication of articles to designs that would be costly or impracticable to reproduce by conventional fabrication methods, is of less importance in the case of gold, where the fabrication cost can be incorporated in the final price and indeed adds to the value of the article in question. In this context, too, there is no special difficulty in working gold and gold alloys either in the massive form or as coatings on base metals, e.g., rolled gold. However, one disadvantage in the latter case in that the coated strip or wire must be cut at some stage of assembly thereby exposing the underlying base metal. Heavy gold plating after fabrication from the base metal has the advantage that not only may the whole surface of the article be covered, but the thickness of coating can be increased on significant surfaces to meet special conditions of corrosion or wear in service.

On the technical side, the demand for thick gold plating has been limited and the tendency has been to meet occasional requirements by *ad hoc* modification of available electrolytes without evaluating in depth such factors as electrolyte life, compatibility with matrix materials and the mechanical characteristics of the deposits. As a result, some of the early formulations for heavy gold plating have a very short operative life and are not suitable for electroforming due to stress or mechanical weakness of the coatings, which leads to distortion or disintegration when the support is removed.

Wider exploitation of the electroforming process seems likely with continued improvement in plating techniques and electrolyte formulations. The potentialities of the process are suggested by the fact that a good quality electroform may be only one-third of the weight of a similar fabricated component, yet still show better mechanical strength and excellent wear properties whilst being less affected by wide variations in temperature.

ELECTROLYTES

CHLORIDE ELECTROLYTES

Most of the early attempts at heavy gold plating utilised gold chloride electrolytes based on the solution formulated by Wohlwill[2] for gold refining:

Gold (as chloride)	25–40 g/l
Hydrochloric acid	23.8–55.0 g/l
Sodium chloride	10–30 g/l
Sulphuric acid	10–20 g/l
Temperature	23°C
Current density	8.6–11.0 A/dm^2

Although primarily a refining solution, adherent deposits can be obtained from this electrolyte on base metals provided they are given a preliminary strike in a cyanide solution. Deposits are, however, coarsely crystalline and the process is therefore not suited to electroplating.

During the course of an investigation into the structure of gold deposits, the writer obtained adherent coatings up to 25 microns in thickness from the following solution[3]:

Gold (as chloride)	12.5 g/l
Hydrochloric acid	115 g/l
Anodes	Gold or insoluble platinum or graphite

There is, however, a general lack of information concerning properties of deposits from chloride electrolytes. Although coatings appear to be generally lustrous, smooth and fine grained, there is a tendency for nodular growth, which may however, be offset by the use of low current densities and higher operating temperatures. From the viewpoint of electroforming there are several disadvantages. For instance, the throwing power is poor. Since gold is present in the trivalent state and cathode efficiency is low, the build-up of coatings is slow and requires lengthy plating times. Furthermore, the solutions are extremely corrosive and attack of the base metal substrate, or conductive films on non-metallic moulds or mandrels, may lead to poor adhesion unless a preliminary deposit of gold is applied from a cyanide solution. Finally, this type of solution can only be operated satisfactorily using solid gold anodes, which involves high capital investment and some security risk.

FERROCYANIDE-CHLORIDE ELECTROLYTES

As in the normal cyanide electrolytes, the essential constituent of this type of solution is an alkali metal gold double cyanide. Its development stems from the fact that in the early days of cyanide plating potassium ferrocyanide was cheap and of reasonable purity whereas gold cyanide was both expensive and impure. When the ferrocyanide is boiled with gold chloride the double cyanide of gold and potassium is formed and a precipitate, either ferric hydroxide or Prussian blue, is produced, depending on whether or not alkali is added. The reactions have been discussed by Beutel[4] and more recently by Mornheim[5] (see Chapter 4).

Paweck and Weiner[6] reported that thick deposits could be obtained from a solution of this type, the electrolyte being prepared by boiling for several hours prior to use and operated at a current density up to 60 A/ft^2 using rolled gold cathodes. No reference is made to specific thicknesses.

Kutnetsova[7] proposed the following formulation:

Gold (as chloride)	10 g/l
Potassium ferrocyanide	40 g/l
Potassium carbonate	40 g/l
Temperature	30–50°C
Current density	0.1 A/dm^2
Anodes	Gold

The solution was prepared by adding a hot solution of gold chloride to the boiling solution of ferrocyanide and carbonate, and boiling the mixture for a further 4–5 hours.

Another formulation, suggested by Yampolsky[8], is:

Chlorauric acid	2.5 g/l
Potassium ferrocyanide	15 g/l
Temperature	50°C
Current density	0.1–0.2 A/dm^2
Cathodic efficiency	25–30%

The solutions of Kutnetsova and Yampolsky would appear to the writer to be quite unsuitable for electroforming in view of the very low current density employed in both cases and, in the second formulation, the low metal content.

CYANIDE ELECTROLYTES

Since the early 1950's, alkaline cyanide electrolytes have been used almost exclusively for the production of thick gold deposits. From simple solutions of this type, deposits tend to become coarse and powdery, primarily due to high cathode polarisation which leads to an excessive discharge of hydrogen. Performance tends to be variable because the difference in anode and cathode efficiency can lead to significant changes in metal concentration as electrolysis proceeds, which may give rise to variations in structure or internal stress of coatings.

To overcome these problems a number of modifications have been developed, the simplest of which is the addition of a conducting salt, such as di-potassium phosphate, to reduce cathode polarisation and to promote smoother and finer grained deposits. A general electrolyte formulation which has been used extensively, with slight variations to meet particular cases, is as follows[9]:

Gold (as potassium gold cyanide)	6.8–10 g/l
Potassium cyanide (free)	31 g/l
Di-potassium phosphate	31 g/l
Potassium carbonate	15.5 g/l
Temperature	50–60°C
Current density	25 A/ft^2
Agitation	Cathode rod

Although coatings of up to 25 microns have been produced from this type of solution, no intensive use has been recorded. There is, therefore, a lack of information concerning the effects of such factors as total salt concentration and electrolyte breakdown products on the characteristics of deposits. In general, high gold concentrations are used to permit operation at the highest possible current density. The potassium cyanide concentration can vary between 7–120 g/l, this depending upon the types of addition agent employed. A low gold content, low cyanide solution is unsatisfactory for the production of heavy gold desposits since changes in the potassium cyanide : gold ratio during electrolysis may result in variation in hardness, structure and surface appearance of the coating. At low ratios, i.e., with molecular ratios of free cyanide to gold of 5:1, coarsely crystalline, dull deposits are produced with a tendency to treeing. At a ratio of 20:1, coatings are fairly bright. Ratios higher than 40:1 produce very rough, coarse and crystalline deposits which are invariably loosely adherent.

A potassium ferrocyanide electrolyte developed by Korbelak[10] and used in American practice is as follows:

Potassium gold cyanide	30 g/l
Potassium ferrocyanide	200 g/l
Potassium cyanide	7.5 g/l
Temperature	85°C
Current density	30–50 A/ft^2

Smooth gold coatings 20–30 mil thick, for the protection of nuclear reactor parts, were produced by Seegmiller and Gore[11] from an electrolyte containing turkey red oil as an additive:

Potassium gold cyanide	40 g/l
Potassium cyanide	70 g/l
Turkey red oil	0.5 ml/l
Temperature	60–65°C
Current density	4–10 A/ft^2
Anodes	Gold or stainless steel

This solution produces hard deposits (130 HV) with good ductility at relatively fast deposition rates of up to 1.5 mil per hour, and shows good throwing power. However, as in most of the alkaline cyanide electrolytes, potassium carbonate build-up limits the life of the solution to some 10 A-hr/l (74 g Au) after which some deterioration sets in.

Bauer[12] has proposed the following electrolyte and conditions for the continuous electroforming of smooth deposits up to a thickness of approximately 0.015 in:

Gold (as potassium gold cyanide)	18 g/l
Potassium cyanide	120 g/l
Potassium hydroxide	4 g/l
Potassium sulphite	4 g/l
Vanillin	4 g/l
Temperature	80°C
Current density	5–18 A/ft^2
Anodes	Gold

This process was reported to produce smooth deposits up to 0.015 in. thick, but electrolyte life was again limited by carbonate build-up. In an attempt to overcome this difficulty, potassium cyanide was replaced by lithium cyanide, when it was hoped to take advantage of the restricted solubility of lithium carbonate (Table 40.1) to limit carbonate build-up by continuous filtration. In practice, however, rough deposits were produced due to occlusion of lithium carbonate in the coatings.

Table 40.1
Solubility of Lithium Carbonate in Water[13]

Temperature (°C)	*Solubility (g/l)*
0	15.4
20	13.3
100	7.2

NEUTRAL AND ACID CYANIDE ELECTROLYTES

A general disadvantage of alkaline cyanide electrolytes for heavy plating and electroforming, is the relatively short useful life due to the formation of carbonate and other breakdown products such as ammonia, urea, formate, hydrocyanic acid and cyanate. Breakdown is accelerated by continuous use of the solutions at the high temperatures and current densities desirable for the achievement of fast deposition rates. Although several methods have been proposed for carbonate removal, such as freezing out and treatment with barium cyanide, these are not really practicable on a large scale.

Recent work by the writer has resulted in the development of a very stable neutral electrolyte* for the production of heavy deposits and for electroforming. The solution contains 24 g/l of potassium gold cyanide and the operating conditions are as follows:

pH	6.7
Temperature	35°C
Density	16°Beaumé
Current density	5–20 A/ft^2
Anodes	Platinum or platinised titanium
Deposition rate	60 μm/hr at 20 A/ft^2

The coatings produced are lustrous, fine grained and extremely ductile, with a hardness of 100–140 HV. The electrolyte does not suffer from short life as there is no accumulation of unwanted breakdown products, which is partially prevented by the low operating temperature and a high cathode efficiency of 90–100%. Thicknesses up to 0.080 in. have been produced from this solution, the coatings retaining their lustre and ductility.

An acid gold solution containing an organic alkyl or alkylene guanidine compound has been reported by Yamamura[14] to produce bright, crack-free deposits up to 15 microns thick with a cathode efficiency of 90%. The formulation and operating conditions are:

Potassium gold cyanide	30 g/l
Ethylene guanidine	10 g/l
Formic acid, 85%	250 g/l
pH	4.0
Temperature	50°C
Current density	0.2 A/dm^2
Agitation	None

NON-CYANIDE ELECTROLYTES

Although the use of sulphite gold salts for gold plating was recorded many years ago[15,16], it is only recently that they have been reintroduced as a basis for commerical electrolytes[17,18]. A typical formulation is:

Sodium gold sulphite	10 g/l
Potassium phosphate	30 g/l
Sodium sulphite	50 g/l
Arsenic trioxide	30 mg/l
pH	9–10
Temperature	50°C
Current density	1–6 A/ft^2

* ENFORM—Product of Engelhard Industries, Ltd.

This type of electrolyte produces pure deposits which are relatively hard and fully bright. In principle they are ideal for the production of thick deposits, even though the solution density tends to rise in proportion to the work load. A particular feature of the electrolyte is the high levelling power it exhibits. Thick deposits produced from electrolytes of this nature are ductile and smooth, requiring no surface finishing whatsoever. A disadvantage, however, is the limited life in comparison with many cyanide-based solutions, due to the formation of free sulphite and sulphate, leading to a rapid increase in solution density. In the production of thick deposits and electroforming, the life would be expected to be even more limited by the absence of regular drag-out to offset salt build-up.

ALLOY PLATING

There are a number of well established processes for the application of fairly thick gold alloy coatings to watchcases (see Chapter 41).

The most recent studies in the field of alloy plating are those of Wiesner and Frey[19] on copper-gold alloys. These authors evaluated and modified a number of commercial electrolytes with the object of producing electroforms up to 60 mil in thickness in order to study the properties of the deposit. Cyanide solutions gave rough deposits above 3–5 mil and the alloy composition was dependent upon critical control of the "free cyanide".

Deposits up to 20 mil thick were obtained without cracking or excessive roughness from the following formulations:

Gold (as potassium gold cyanide)	6–6.5 g/l	6–6.5 g/l
Copper (as di-sodium copper EDTA)	16–18 g/l	—
Copper (as tri-sodium copper DTPA)*	—	16–18 g/l
PO_4 (expressed as 85% H_3PO_4)	25 ml/l	25 ml/l
Sodium sulphite	—	6–8 g/l
pH	7.0–7.5	7.5–9.0
Temperature	65°C	65°C
Current density	0.6–1 A/dm^2	0.6–1 A/dm^2
Deposition rate	2.3–2.5 g/A-hr	2.3–2.5 g/A-hr
Deposition rate	0.4–0.5 mil/hr	0.4–0.5 mil/hr
Anodes	Platinum	Platinum

* Diethylenetriamine penta acetic acid.

Variation in current density (0.32–1.15 A/dm^2) produced varying deposit compositions, from 95–55% Au. Both of the electrolytes had limited life due to the accumulation of cyanide and organic decomposition products during operation.

Modification of a commercial acid gold electrolyte by adjusting the pH to 8.2–8.5 with potassium hydroxide and additions of varying amounts of copper as Na_3Cu DPTA, enabled fairly smooth crack-free deposits to be produced up to 0.060 in. thick. Smooth alloy deposits containing from 95–70% Au could be obtained by varying the current density. In all cases, however, a heat treatment for 3 hours at 300°C was necessary to permit the coatings to be cut without cracking for the preparation of test pieces for determination of mechanical properties. Wiesner and Distler[20] carried out extensive evaluations on the physical and mechanical properties of electroformed gold-copper alloys obtained from similar electrolytes to those of Wiesner and Frey[19], which were described above.

Providing heat treatment was carried out below 450°C, ultimate yield strengths above 70 kg/mm^2 could be achieved over a range of 65–83% Au. As the gold content increased above 83%, the strengths of the alloys decreased rapidly. Heat treatment above 300°C lowered the ultimate and yield strengths, but increased the elongation. Tables 40.2 and 40.3 show the effect of composition and heat treatment upon the tensile and yield strengths of electroplated gold-copper alloys.

Table 40.2

Physical Properties of Electrodeposited Copper-Gold Alloys*

Gold Content (%)	*Ultimate Tensile Strength* (*psi*)	*Yield Strength* (*psi*)	*Elongation* (%)
69.0	123,600	103,200	—
77.0	163,500	163,300	0.5
78.5	117,500	100,300	13
80.3	115,200	101,700	3
80.1	173,200	162,800	0.7
81.9	155,200	151,900	11
81.2	148,600	145,800	10
81.0	170,200	151,500	10
81.8	124,600	116,200	8
82.5	103,000	94,600	4
82.7	151,200	150,300	0.5
83.2	108,400	100,700	5
83.4	134,500	133,700	1.4
84.8	118,000	117,100	2.5
85.7	89,000	81,100	5
85.6	91,900	82,700	7
85.3	106,500	102,500	7
85.8	114,500	114,500	2.5
87.5	86,400	80,400	5
87.5	80,300	77,500	7
88.5	79,100	77,000	7

* Final Heat Treatment, 3 hours at 300°C and Furnace Cooled

MODULATED CURRENT

PERIODIC REVERSE CURRENT

Although periodic reverse current is used extensively in silver plating and fairly generally in the gold alloy plating of watchcases, there is little record of its application to heavy deposition. It is referred to by Yampolsky[8] in connection with a conventional alkaline cyanide bath containing 15–25 g/l of potassium gold cyanide, 100 g/l of potassium carbonate, with a free potassium cyanide content of 8–10 g/l. A cathodic current density of 20–40 A/ft^2 and an

Table 40.3
Physical Properties of Electrodeposited Copper-Gold Alloys*

Gold Content (%)	Ultimate Tensile Strength (psi)	Yield Strength (psi)	Elongation (%)	Prior Heat Treatment	
				Temperature (°C)	Time (hr)
69.0	121,000	109,500	—	300	3
70.5	133,200	127,000	0		
71.1	133,000	123,200	3		
71.8	121,000	119,900	10		
71.3	123,800	113,200	6		
72.1	100,000	92,600	1		
72.3	140,800	129,700	2	300	3
72.0	148,600	123,300	7	300	3
72.5	133,100	126,400	1.1	300	3
74.7	138,300	115,200	8	300	3
74.7	136,700	114,200	3	300	3
75.0	123,600	116,000	0	300	3
77.0	122,100	112,200	5.4	300	3
78.0	98,200	90,000	3		
79.7	131,800	131,350	1.5	300	3
80.2	126,700	126,700	7	300	3
80.0	126,700	98,300	1	350	3
81.9	89,400	73,600	25		
81.6	101,900	100,300	2	300	3
82.7	99,600	91,100	10	350	3
83.4	123,500	94,900	10	350	3
84.5	91,100	78,400	—	300	3
84.8	85,400	76,200	—	300	3
85.1	79,600	71,600	15		
85.8	93,500	83,200	13		
85.6	94,600	92,800	8.4	300	3
86.7	118,300	109,100	1		
86.0	90,300	80,000	8		
87.2	83,500	—	0.5		
89.4	72,400	65,900	6		
90.0	86,000	78,100	9	450	2

* Final Heat Treatment, 3 hours at 450°C and Furnace Cooled

anodic current density of 10 A/ft^2 is specified, but no information is given as to the actual time cycle employed.

The use of periodic reverse current plating to reduce internal stress has been reported by Bandaver, Krupnikova and Rodionova[21] using a simple cyanide electrolyte with 25–40 g/l of gold and 10 g/l of free potassium cyanide, operated at a temperature of 60–75°C with a current density in the range of

0.1–1.5 A/dm^2. The ratio of anodic to cathodic time is given as 15:2, but there is no reference to actual plating and deplating times used on this basis.

According to Heilmann[22], deposit thicknesses of up to 20 microns have been obtained from an electrolyte of the following formulation using the PR cycle indicated.

Gold (as potassium gold cyanide)	1–3 g/l
Copper (as potassium copper cyanide)	8–13 g/l
Cadmium (as potassium cadmium cyanide)	0.1–0.8 g/l
Silver (as potassium silver cyanide)	0.01–0.1 g/l
Potassium cyanide (free)	3–8 g/l
pH	9–11
Temperature	60–80°C
Current density (cathodic)	0.5–1.5 A/dm^2 for 4–20 sec
Current density (anodic)	1.0–3.0 A/dm^2 for 0.5–2 sec

Deposits from the electrolyte are hard, relatively bright and assay at 18 carat. The addition of silver enables fully bright deposits to be obtained without unduly affecting the colour which ranges from yellow to red according to the concentrations of copper and cadmium used.

SUPERIMPOSING AC ON DC CURRENT

This method, which has been fairly extensively evaluated in nickel plating as a means of reducing deposit stress, has been studied by Bertorelle[23] in the case of gold plating from an alkaline cyanide solution of composition:

Gold (as cyanide)	2 g/l
Potassium cyanide (free)	5 g/l
Disodium hydrogen phosphate	15 g/l

Satisfactory deposits of up to 46 microns were produced in a plating time of 60 minutes under the following conditions:

Temperature	65°C
Anode: Cathode ratio	1:5
Current density	0.5–3.0 A/dm^2 (optimum 1–1.5 A/dm^2)
Voltage	7–9 alternating 2–2.5 direct
Agitation	Continuous stirring

Bertorelle's results, which are summarised in Table 40.4, are reported in terms of applied AC and DC voltage, but no information is given on the actual ratio of AC:DC current.

Table 40.4
Characteristics of Gold Deposits Produced with Superimposed AC

No	*Voltage* *DC*	*AC*	*CD* *(A/dm²)*	*Time* *(min)*	*Thickness* *(µm)*	*Deposit Appearance*
1	2	1	0.5	30	12	adherent, semi-bright
2	2	2	0.7	30	14	adherent, semi-bright
3	2	3	1.0	35	26	adherent, semi-bright
4	2	4	1.2	15	11	adherent, semi-bright
5	2	9	1.3	30	27	adherent, semi-bright
6	2.5	4	1.3	30	22	adherent, semi-bright
7	2.5	7	1.5	60	46	adherent, semi-bright
8	2.5	8	2	30	23	crystalline,
9	2.5	9	2	60	44	crystalline, semi-bright
10	3	6	1.8	30	28	crystalline, semi-bright
11	3	9	3	35	32	crystalline, porous
12	4	9	3	40	37	crystalline, powdery

PULSED CURRENT

There are several patents by Winkler[24] on the application of pulsed current plating to the deposition of gold alloys, including gold-copper-nickel, gold-silver-copper and gold-silver-copper-nickel, to heavy coating thicknesses. The electrolytes used have a high gold content and gold or gold alloy anodes are used. Very precise analytical control is necessary to achieve the correct alloy composition.

An investigation into the effects of pulsed current plating on gold deposits, using pulses in the microsecond range, has recently been reported by Avila and Brown[25], though deposit thicknesses were limited to approximately 120 microinches. They claim that the main theoretical advantage of pulse plating stems from the elimination of hydrogen bubbles at the cathode by reduction of cathode polarisation. Best results were reported with an "on" time of 10–15 milliseconds and an "off" time of 115 milliseconds, the "off" time being more critical in terms of allowing the solution at the cathode to reach equilibrium before the next "on" pulse.

Improvements claimed for this type of process are (a) extremely dense, high conductivity deposits (b) increased plating rate (c) elimination of hydrogen embrittlement of the basis metal (d) reduced need for addition agents (e) high purity deposits (f) better resist performance.

Although the technique was restricted to relatively thin coatings in this investigation, the implications for the production of heavy deposits are obvious. However, claims for increase in plating rate have been questioned by Cheh[33] on the basis of a theoretical analysis which indicates that although the magnitude of the instantaneous current can be considerably higher in pulsed current than in direct current plating, the limiting overall deposition rate is in general lower. This conclusion was confirmed in tests on phosphate, citrate and cyanide gold solutions.

ULTRASONICS

A patented process[26] for electroplating and electroforming gold claims the use of exceptionally high current densities when a gold plating bath is placed in an ultrasonic field. The application of ultrasonics prevents polarisation at the electrodes, permitting the use of current densities of up to 108 A/dm^2. The throwing power of the plating solution is substantially increased to give uniform distribution of gold over the basis material.

The composition of the electrolyte and the optimum conditions are:

Gold (as gold potassium cyanide)	70 g/l
Potassium cyanide	40 g/l
Pentadecafluoro-octanoic acid	0.1% by wt.
pH	11.0
Temperature	65°C
Current density	4.3 A/dm^2
Ultrasonic field intensity	7–10 watts/in.^{2}
Frequency	20 kilocycles

Deposit purity exceeding 99.999% is reported, with a hardness of 125–225 KHN, which is apparently associated with a laminar structure developed under these conditions.

Improvement in hardness of electrodeposited gold by ultrasonic vibrations has also been reported by Vrobel[27].

APPLICATIONS

FUNCTIONAL

Cathode Sputtering Sources

Thick gold deposits are used as cathode sputtering sources. An example reported in the literature[28] involved the deposition of a 0.020 in. coating on the copper-clad face of a 3 in. thick mild steel slab measuring 44 × 28 in., to serve as a source for the sputtering of thin gold coatings on to a special grade glass for the windscreens of high speed aircraft. The sputtered film forms an interface layer in a laminated glass composite and is used for electrical heating to prevent icing-up. In preparing the source, some 250 ounces of fine gold was deposited in a continuous five day operation.

Slip Rings[29]

A well established industrial application is the production of electroformed slip rings. In this process lead wires are encapsulated in a plastic cylinder and locally exposed by machining grooves, which are then filled by gold plating to a thickness of not less than 0.015 in. The process is completed by machining and polishing of the composite surface to a 4 microinch finish. A large number of rings may be sited on each shaft or disc without affecting strength, accuracy or reliability of the assembly. The use of ultrasonic agitation to depolarise the electrodes in this process has enabled the number of rejects to be reduced by producing a more even and homogeneous deposit over several rings located on one assembly.

Cathode–Ray Tube Electrodes

Gold electroforming of electrodes for cathode-ray tubes for picture transmission were used in the early 1930's[30].

Insulated fine mesh screens were produced by electroforming on to copper plates which have the negative impression of the mesh ground into them. Deposition, partially or completely, fills the grooves and grinding the resultant electroform reproduces the required U-shapes, trapezoidal or triangular incisions. The copper master is dissolved, either chemically or electrochemically, and the knife edges left by grinding are deburred by immersion in nitric acid. An enamel coating provides electrical insulation. Although the majority of these screens were produced in nickel, for electrodes depending on the interaction of insulated and surface conductive non-oxidisable screens, the latter were electroformed in gold from a cyanide electrolyte. The same process was adopted and the grooved copper master dissolved in nitric acid, leaving the electroformed gold mesh.

Dental

Electroforming in gold has been used in dentistry to form the model of the tooth or tooth portion to which the crown or inlay is to be fitted. Although silver and copper are by far the most commonly used metals for electroforming in the dental industry, the use of gold and cadmium, together with the techniques employed, are recorded in an early Australian paper[31].

DECORATIVE

Coronet

In the decorative context, a particularly interesting example was the adoption of the process for the production of the coronet used at the Investiture of the Prince of Wales at Carnarvon in 1969[32]. Since the writer was closely concerned with this project, some detailed description may be of interest.

Electroforming was selected as the fabrication technique because the unusual surface texture of the original design could not be reproduced satisfactorily by conventional metal working methods. The original model was in wax, and from this an epoxy-resin mould was produced in two halves which were joined for the electroforming process. Electrical connections were made at eight locations by screws which were positioned so that they would be removed in final trimming.

Initial metallisation was carried out by scrubbing with detergent solution, followed by thorough rinsing and then simultaneously spraying silver and reducing solutions to produce a uniform film of silver.

In order to provide uniform gold distribution, a PVC anode cage was prepared of conforming shape to that of the epoxy mould. The cage was drilled with numerous holes and then lined with platinum mesh anodes.

The anode cage was fitted inside the epoxy and a clamp was used to keep the anode to cathode distance constant. The mould was immersed in a 200 litre gold electrolyte and the current increased very slowly to 8 A/ft^2. All electrical connections were run through PVC tubing so as not to impair the slow cathode movement of the growing electroform. The electrolyte was continually filtered at a rate of 140 l/hr. Initially a second pump and filter was used to transfer electrolyte directly into the mould as it was electroforming in the inverted position, but this was removed as trapped air bubbles were adhering to the gold surface and causing uneven deposition.

Replenishment of the electrolyte was made on the basis of regular analysis and the operation was continued for 2½ days, when it was calculated that the thickness of the electroform should be between 0.030–0.040 in. The mould was then removed from the electrolyte and rinsed in two drag-out tanks, followed by a tank containing hot water to soften the epoxy resin mould. The screws holding the two halves of the mould were removed and the mould slowly peeled away from the gold coronet (Figure 40.1). The gold electroform was

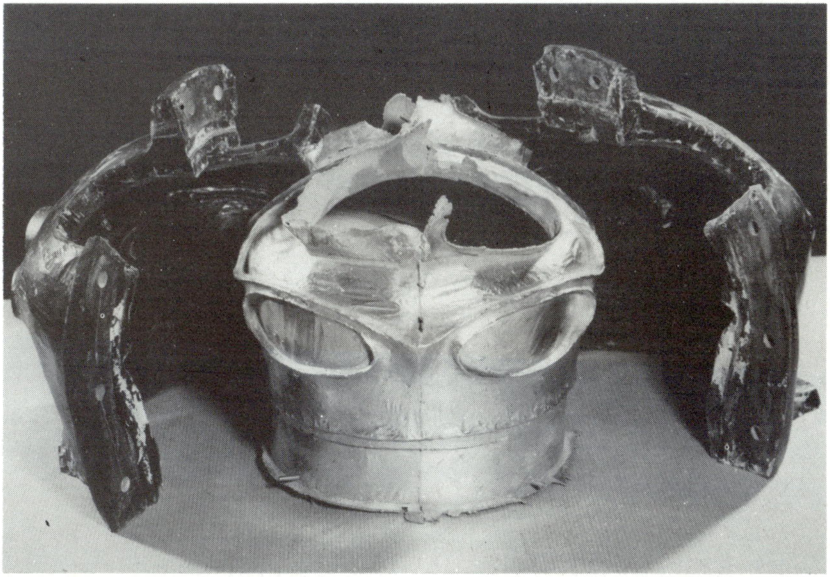

Fig. 40.1 The Prince of Wales Crown immediately after electroforming, shown here with its epoxy resin mould.

uniform, complete and the smallest of lines on the original wax coronet had been reproduced to perfection on the gold surface. A small quantity of silver from the metallising process was removed electrolytically in a sodium cyanide/sodium hydroxide electrolyte before final trimming.

Moon Models

Another recent application with which the writer was also closely concerned, was the electroforming of moon models in gold to commemorate the first moon landing of the Apollo astronauts. In this instance, PVC moulds were made from the original steel model and the electroform was produced in two halves, which were subsequently joined by cold welding and soldering. The final electroform weighed 2–3 ounces.

REFERENCES

1 "Gold-Recovery, Properties & Use", E. M. Wise (Ed.), Chapter 9, Van Nostrand, New York (1964).
2 Wohlwill E., *Z. Elektrochem.*, **4,** 379, 401 (1898).
3 "Finishing Handbook & Directory", Sawell Publications, London (1973).
4 Beutel E. B., *Z. angew Chem.* **25,** 995 (1912).
5 Mornheim A. F., *Plating*, **47,** 819 (1961).
6 Paweck R., Weiner R., *Z. Elektrochem.*, **36,** 972 (1930).
7 Kutnetsova A. N., Med Prom. SSR 12 46–8 (1958).
8 Yampolsky A. M., *Electroplg Metal Finish.*, **16,** 3, 76 (1963).
9 "Finishing Handbook & Directory", Sawell Publications, London (1973).
10 Korbelak A., "Metal Finishing Guidebook & Directory", Metals & Plastics Publications, Inc. Westwood, New Jersey (1969).
11 Seegmiller R., Gore J. K., *Proc. Am. Electropl. Soc.*, **47,** 74 (1960).
12 Bauer C. L., *Plating*, **39,** 1335 (1952).
13 Seidell A., "Solubility of Inorganic and Organic Compounds", Van Nostrand, New York (1940).
14 Yamamura K., Nagano-Ken, Hayashi, S., U.S. Pat., 3,475,290 (1969).
15 Brit. Pat., 9,431 (1842).
16 Woolrich J. S., *Dinglers polytech. J.*, **88,** 48 (1845).
17 Smith P. T., U.S. Pat., 3,057,789 (1962).
18 Shoushanian H., U.S. Pat., 3,475,292 (1969).
19 Wiesner H. J., Frey W. P., *Plating*, **56,** 527 (1969).
20 Wiesner H. J., Distler W. B., *Plating*, **56,** 799 (1969).
21 Bandaver V. V., Krupnikova E. I., Rodionova L. I., TR Gos. Nauch-Issled Inst. Splavov Obrab Tvset Metal, 31, 93 (1970).
22 Heilmann G., U.S. Pat., 3,586,611 (1971).
23 Bertorelle D. E., *Galvanotechnic*, **4,** 6, 141 (1953).
24 Winkler J., German Pat., 576,585 (1933); German Pat., 723,497 (1942).
25 Avila A. J., Brown M. J., *Plating*, **57,** 1105 (1970).
26 Litton Systems Inc., U.S. Pat., 3,427,231 (1969).
27 Vrobel L., *Trans. Inst. Metal Finish.*, **44,** 4, 161 (1966).
28 *Electroplg Metal Finish.*, **17,** 6, 217 (1964).
29 *Electroplg Metal Finish.*, **14,** 2, 69 (1961).
30 Ger. Pat., 880,678 (1935).
31 Tuckfield W. J., *Aust. J. Dent.*, (1938).
32 Mason D. R., *Electroplg Metal Finish.*, **22,** 8, (1969).
33 Cheh H. Y., *J. electrochem Soc.*, **118,** 551 (1971).

Chapter 41

DECORATIVE PLATING

B. M. Dickinson

The description "decorative" embraces all applications of gold plating of a non-technical nature and thus covers a very wide field ranging from purely ornamental uses where the thickness of gold applied may be less than 0.5 microns (usually with a clear coating of lacquer to improve corrosion and wear resistance) to those, as exemplified by high quality watchcase plating, where functional requirements dictate the use of coatings of 8–20 microns and sometimes even thicker. In the latter case the importance of adequate specifications of thickness and quality of coatings has long been recognised, but it is only comparatively recently that a general move has been made towards extending specifications to a wider range of decorative gold plating.

One difficulty in this connection stems from the problem of physically controlling the very thin coatings that may be applied to decorative articles under the description of "gilding".* The most recent British Standard Specification on gold plating, BS 4292: 1968, defines the term "gilding" as applying to coatings of less than 0.5 microns; the description "gold plating" is applicable to coatings of not less than 0.5 microns, with a gold content of at least 90%, or to coatings of not less than 2 microns with a minimum gold content of 62.5% (approx. 15 carat). Most specifications of other countries make a similar distinction between gilding and gold plating, though not necessarily at the same levels of thickness and purity. In discussions of international specifications† it has been proposed that the term "gilding" should be dropped and replaced by the generic description "gold electroplated", which would apply to coatings of a minimum thickness of 0.5 microns and a minimum gold content of 50%, so that thinner coatings would be automatically excluded if this proposal is ultimately adopted.

A further difficulty which is encountered in attempting to formulate international specifications for decorative gold plating arises from the existence in many countries of national legislation governing the quality, description and marking of gold plated wares, and it would seem that this will only be resolved in the long term. There is however, a good measure of agreement on technical aspects.

---

* Gilding is also referred to as "Gold Washing" and "Gold Flashing".

† For a more detailed treatment of international specifications, the reader is advised to study the APPENDIX by Mills.

A distinctive feature of decorative gold plating is the importance of the colour of coatings and the need for accurate colour matching when, as frequently happens, parts which may be used in conjunction with each other, e.g., watchcases and bracelets, are plated by different establishments. To achieve colour uniformity together with the necessary hardness and wear resistance of the deposit, decorative plating is based almost entirely on the use of alloy rather than pure gold deposits.

The present chapter is concerned primarily with the types of basis metal, pretreatment, plating procedures and electrolytes generally used in trade finishing, with particular reference to UK practice. Gold plating in the watch industry is dealt with in more detail in Chapter 42.

BASIS METALS

The basis metals most frequently encountered in the finishing of fancy goods are as follows:

BRASS

Extensively used in the manufacture of watch bezels, watch straps, buckles, pen nibs, pen tops, cuff-links, etc.

COPPER

Less widely employed than brass, mainly on account of its higher cost. Beryllium-copper and phosphor-bronze are sometimes used as springs in cuff-links and ear clips.

SILVER

Used in the manufacture of better class costume jewellery. Frequently, a decorative effect is obtained by plating with 2–3 microns of gold followed by diamond cutting to expose a silver design against a gold background.

GOLD

Articles fabricated from low carat gold, e.g., 9 carat are often flashed over with a higher carat gold to conceal any differences in colour between the metal used in fabrication and the brazing alloy used. The thin coating of gold also gives a more appealing colour and enhances the tarnish resistance.

MILD STEEL

Employed in the manufacture of cheap watch buckles, handbag frames, and similar goods.

ZINC-BASE ALLOYS (MAZAK, etc.)

Very widely used for the casting of intricate shapes, e.g., brooches, necklaces, cuff-link faces, watch bezels and so forth.

PLATING PROCESSES

The suggested pretreatment and plating sequences for these basis metals are summarised in Table 41.1. The same pretreatment is used for copper,

Table 41.1
Pretreatment and Plating Sequences for Various Basis Metals

Metal	*Sequence*
Brass Copper Nickel Silver	$H-W-D-W-S<$ $N-W-G(s)-W-G-D_{(R)}$ $G(s)-W-G-D_{(R)}$
Silver	$H-W-D-W-S-W-G-D_{(R)}$
Gold (low carat)	$H-W-D-W-S-W-G-D_{(R)}$
Mild steel	$H-W-D-W-H_{(A)}-W-W<$ $C_s-C(b)$ then plate as copper C_s-C Polish or burnish then plate as copper
Zinc-based alloys	$M-W-D-W-A-W-C_s-C_b$ then plate as copper
Nickel plate	$H-W-N_{(a)}-W-H_{(a)}-W-G_{(s)}-W-G-D_{(R)}$
Stainless steel	$H-W-D-W-S_{(a)}-W-W-G_s-W-G-D_{(R)}$

Code

A	= 2.5% HF or 0.5% H_2SO_4	H	= Hot soak clean
C	= Heavy copper plate	$H_{(A)}$	= 50% HCl
$C_{(b)}$	= Bright copper plate (acid or pyrophosphate)	M	= Mild alkali clean
Cs	= Cyanide copper strike	N	= Nickel plate
D	= Cathodic degrease	$N_{(a)}$	= Nickel activator*
$D_{(R)}$	= Drag-out	S	= 5–10% H_2SO_4 dip
G	= Gold plate	Sa	= Stainless steel cathodic activator (10% H_2SO_4 +5% HAc)
$G_{(s)}$	= Acid gold strike	W	= Water rinse

* Proprietary, or cathodic treatment in Cold Cleaner Solution with 4 oz/gallon of sodium cyanide.

Notes

1 Difficulty may be experienced in removing polishing compounds from deep recesses in the article. To this end it may be found advantageous to incorporate an ultrasonic tank at the hot cleaner stage, as this will greatly increase the possibility of removing the soil, without having to resort to brushing the article.

2 *Mild Steel and Zinc-Based Alloys*—As a general rule it will be found quite satisfactory to plate 5 μm copper and 5 μm nickel as the *base* for the precious metal deposit but more complete recommendations are given in BS 4292.

3 *Stainless Steel*—After activation it is essential to plate in the acid gold strike without any undue delay, as this can result in passivation of the steel.

brass and nickel-silver, which are relatively easy metals to plate and require no special precautions. Although they may be plated directly with gold, bright nickel is often used as an intermediate coating.

Mild steel is not normally plated directly with gold because adhesion of gold to the basis metal is extremely suspect, and even when this is satisfactory, corrosion through pores in the deposit will readily occur due to the large difference in potential of the plate and substrate. A substantial copper or copper/nickel undercoat is therefore applied prior to gold plating.

Care is necessary with zinc-base alloys to prevent the deposit peeling. The plating cycle indicated will give a sound and adherent deposit. Occasionally both zinc alloys and mild steel are plated directly with nickel prior to gold plating, but this is not advisable as the nickel deposit has a tendency to peel from the basis metal.

GILDING

Immersion Gilding*

This, as its description implies, involves the deposition of a thin gold coating by chemical rather than by electrolytic means. Use of this process is confined mainly to the finishing of relatively cheap mass produced articles such as shoe and watch strap buckles, although it is employed to some extent on slightly more expensive items like watch straps, brooch settings and cuff-links. Copper, brass and nickel plated articles are the most suitable for immersion gilding. Other metals and alloys must first be plated with copper or nickel.

The parts to be gilded must be thoroughly cleaned prior to immersion in the gilding solution to produce an active surface completely free from any traces of oxide or tarnish films. A suggested treatment cycle is as follows:

(i) hot soak clean (ii) water rinse (iii) cathodic degrease (iv) water rinse (v) dip in 1:1 hydrochloric acid solution containing a surfactant (vi) water rinse (vii) water rinse (viii) immersion gild (ix) drag-out (2) (x) water rinse (xi) dry (trichloroethylene or similar).

In the case of copper or brass articles, it is preferable to bright dip or chemical polish before the pretreatment cycle.

Solutions for Immersion Gilding

Two suggested formulations are as follows:

Gold (as 68% $KAu(CN)_2$)	2–4 g/l
Sodium cyanide	30 g/l
Sodium carbonate	30 g/l
Temperature	65–85°C

* The reader is referred to Chapter 10 for a fully comprehensive treatment of Immersion Plating.

Approximately 0.2 microns of gold is deposited in 4 minutes at the higher temperature. The colour is affected by the metal dissolved during the replacement reaction and for this reason the solution should not be replenished with gold salt. Some 80–85% of the metal may be deposited before the colour of the deposit becomes unacceptable.

Gold (as 68% $KAu(CN)_2$)	2–4 g/l
Dipotassium hydrogen phosphate	20 g/l
EDTA, di-potassium salt	30 g/l
pH	6.0–8.0
Temperature	90–100°C

At a pH of 6.9–7.1 and a temperature of 100°C, this solution will deposit on nickel approximately 0.25 microns in 3 minutes. At lower temperatures, deposits on nickel tend to be non-adherent. Copper and copper alloys plate more slowly, but good adherent deposits may be obtained at temperatures as low as 80°C over the whole of the stated pH range.

A minute addition of 54% potassium silver cyanide to either of these solutions will give rise to a green gold deposit, but the consistency of colour is very poor.

Colour Gilding

For many years there were no generally accepted standards for the colours of gold deposits; vague terms such as "18 carat colour", "pink" and "green" golds were used as the only definition. These terms are still widely used in gilding, but a more precise system is employed in colour matching of components plated by different establishments. This is the Swiss Standard colour table, based on a number of plates of gold alloys of standard compositions, each of which is assigned a numerical coding (see Chapter 42, Table 42.1). The most commonly used colour standards are 1N-14, 2N-18, 3N, 4N and 5N, the first corresponding to a green gold and the 4N and 5N to pink golds in the old nomenclature.

To produce this range of colours requires the use of one or more alloying additions to a pure gold deposit to give the desired shade. The following indicates the effect of various metals on the colour of deposits.

Metal	*Colour*	
Silver	Green	→ White
Zinc	Green	
Tin	Green	→ White
Nickel	Hamilton	→ White
Cobalt	Yellow	→ Pale Yellow
Copper	Pink	→ Red
Cadmium	Green	

Most gilding solutions operate at a pH above 7.0 and for this reason the adhesion of coatings, particularly on nickel, is sometimes suspect. In practice, it has been found that an acid gold strike is beneficial to the overall quality of the deposit, since this not only provides a good basis for the main gold coating, but assists in producing an even colour.

The metal concentration in gilding solutions is usually very low, with the twofold purpose of reducing metal investment cost and minimising dragout losses which could be considerable in view of the relatively short plating times and high throughputs usually involved. The solutions listed in Table 41.2 have been found to cover most of the requirements of the fancy goods trade. Due to the complexity of shapes to be plated it is customary to apply a constant voltage to the bath, rather than to calculate the work area and apply a specific current density. Colour variations may be obtained by varying the applied voltage, anode area, degree of agitation, etc., and also by the use of interrupted current. To maintain the nominal colour, however, the metal content should not be allowed to fall by more than 10% before replenishing.

Table 41.2

Bath Composition (g/l) & Operating Conditions of Typical Colour Gilding Solutions

	Bath 1 (24 *ct* *orange-yellow*)	*Bath* 2 (18 *ct* *green*)	*Bath* 3 (18 *ct* *pale pink*)	*Bath* 4 (*deep pink* –*red*)
Au(as 68% $KAu(CN)_2$)	1.0–2.0	1.0–2.0	2.0	2.0
Cu(as EDTA complex)	—	—	2.0	2.5–5.0
Ag(as 54% $KAg(CN)_2$)	—	0.125–0.4	0.05	—
K_2HPO_4	20	10	10	10
K_2EDTA	10	20	5.0	5.0
pH	6–8	6–8	6–8	6–8
Temperature, °C	35–65	35–50	35–60	35–60
Voltage	3–5	2–6	4–6	3–6
Anodes	stainless steel (EN58J or similar)	stainless steel	stainless steel	stainless steel
Time, sec	5–30	5–25	10–30	10–30

Notes

Bath No. 2 When replenishing, add 0.2 g silver for each 1 g of gold.

Bath No. 3 Copper–EDTA complex is prepared by mixing cupric carbonate with sequestric acid solution. When effervescence ceases, potassium hydroxide is added until the solution is a clear blue.

Gold and copper are replenished according to analysis. Silver content requires very little maintenance. If additions are necessary, these should be made in very small quantities only since silver concentration has a marked effect on deposit colour, particularly in low current density areas.

Care should be taken in jigging or wiring work to be gilded, particularly chain as there is a tendency for deposition of a green gold in extremely low current density areas.

Testing

Gilding deposits are of unspecified thickness, but do not exceed 0.5 microns. Tests made to assess porosity, tarnish and wear resistance, can only be empirical, and based on comparison with agreed standards.

Porosity is checked by immersion in 30% (v/v) nitric acid solution at room temperature and observing the time taken for bubbles to form on the surface.

Tarnish resistance is assessed by suspending the article under test about 1 inch above the surface of a freshly prepared 2% solution of ammonium polysulphide (50 ml) in a closed bottle or jar of 250 ml capacity, at room temperature. The time for the surface to discolour is recorded and again compared with an accepted standard.

The wearing quality of the coating is tested by holding the piece under test together with a known standard against a scratch brush (nickel silver or brass) or a leather mop and noting the time taken for the basis metal to be exposed, as a relative indication of wearability.

GOLD PLATING

Vat

As stated earlier, this definition is reserved for the thicker coatings. The most common practice in the fancy goods trade is to employ one or two solutions as "work horses", i.e., to deposit the greater part of the coating, and to achieve the required exact colour match by gilding to about 0.5 microns with a gilding solution of the appropriate composition.

The solutions listed in the following paragraphs cover most general requirements and any standard colour can be obtained by using one of these in conjunction with the gilding solutions listed in Table 41.2.

It is normal to employ an acid gold strike prior to entering the gold solution

18 *carat green*

Gold (as 68% $KAu(CN)_2$)	6–10 g/l
Silver (as 54% $NaAu(CN)_2$)	0.5–2.0 g/l
Potassium carbonate	20 g/l
Potassium cyanide	75–100 g/l
Temperature	20–30°C
Current density	0.4–0.8 A/dm^2
Anodes	Stainless steel (EN58J or similar)

This type of solution will deposit 1 micron of gold in approximately 5 minutes. Additions of wetting agents, reducing agents and grain refiners such as nickel and cobalt are added to impart the desired characteristics to the deposit for specific applications.

9 *carat colour* (23.5–23.9 *carat*)

This patented solution is based on a weak polybasic organic acid and its sodium salt, citric acid and citrates being most commonly employed. A typical formulation is:

Organic acid	80 g/l
Sodium salt	30 g/l
Gold (as $KAu(CN)_2$)	3–10 g/l
Brightener*	0.5–10 g/l
pH	3.5–4.5
Temperature	30–35°C
Current density	0.5–2.0 A/dm^2
Anodes	Platinised titanium, carbon or stainless steel.†

* Brighteners are various complexes of zinc, nickel, cobalt, indium etc.

† Stainless steel is not recommended if large quantities of inorganic ions, e.g., sulphate, are present (sometimes found in brightener).

18 *carat pink*

Gold (as $KAu(CN)_2$)	4 g/l
Copper as EDTA complex	2 g/l
Zinc as EDTA complex	2 g/l
Trisodium citrate	60 g/l
EDTA	20 g/l
pH	approximately neutral
Temperature	35°C
Current density	0.7–1.0 A/dm^2

The colour produced is in the order of 4N–5N on the standard colour table.

Jigging

Jigs should be fully insulated with contact points maintained at a minimum to avoid wastage of gold. Where an article is composed of many individual sections, such as chain, spring loaded jigs should be used to maintain tension and good contact between the joints, otherwise variations in colour may occur along the length, particularly in solutions containing silver as an alloying metal. Jig connections must be kept clean as poor contact can also lead to variation in colour.

Articles to be plated should be well spaced out. As a rule, allowance should be made for one component space between each component. Crowding of articles leads to poor deposit distribution and colour variation.

Barrel Plating*

Many cheap articles such as buckles are gilded by barrel plating, the applied voltage and other conditions being exactly the same as for rack plating, but the time being increased to $1\frac{1}{2}$–2 minutes. Solutions containing silver are generally not satisfactory since the silver tends to deposit by immersion in low current density areas. Silver may be replaced by cadmium in green gold solutions intended for barrel plating, a suitable concentration of cadmium being 0.5 g/l.

* The reader is referred to Chapter 14 for a full discussion on barrel plating.

Chapter 42

WATCH INDUSTRY APPLICATIONS

M. Massin & F. H. Reid

Applications of gold plating in the watch industry are among the most important in the so-called "decorative" field. In fact, the coatings applied to watchcases themselves may be regarded in the same category as engineering deposits since the thicknesses used can approach those employed in electroforming, and property requirements in terms of corrosion and wear resistance, with long term protection of the basis metal, are no less stringent than those pertaining to critical industrial applications.

Because in the European and other areas a major part of watch production is destined for export, specifications and legal requirements covering the plating of watchcases exist in many countries even where no general specification for decorative plating may yet have been formulated, as was the case until fairly recently in the UK. The general position in this respect is reviewed by Mills in the Appendix.

It is therefore convenient in the discussion of applications to distinguish between the coatings applied to watchcases, which are subject to fairly rigorous control, and the more purely decorative finishes applied to other components, which are more closely related to those generally used in the fancy goods trade and dealt with by Dickinson in Chapter 41.

COMPONENTS OTHER THAN CASES

In addition to bracelets, these may include mechanical parts of the movement and external features such as hands, dials and winding crowns.

In the movement itself, two types of parts are often gold plated, namely the gear wheels, and the plates and bridges. The wheels, usually of hardened brass, are plated with about 0.5 microns of pink gold by barrelling. Although the main purpose is decorative and to some extent protective, recent studies have shown that gold deposits give a very low coefficient of friction under dry sliding conditions, leading to an overall improvement in the mechanical efficiency of the gear train.

The plates and bridges are the massive parts which support the mechanical parts of the movement. They are usually produced by stamping and milling from leaded brass (39% zinc, 2% lead) in the half-hard condition. Plating may be carried out by racking or barrelling. The presence of lead in the basis metal necessitates the use of a cyanide copper strike prior to gold plating, the thickness of gold again not exceeding 0.5 microns. Pretreatment usually includes an ultrasonic cleaning stage to ensure cleanliness of the drilled holes.

The faces are stampings of brass or nickel silver on which the figures or indices showing the hours are produced by cold striking and machining. Two pins are soldered on the reverse side to provide attachment to the movement. Some faces are gold plated completely, but more commonly the face is first silver plated, the recessed area then covered by a protective lacquer, and the exposed figures finally gold plated. The masking is then removed and the whole face coated with a clear stoving cellulose lacquer.

The hands are stamped from hardened brass and carefully polished before plating by barrelling in a bright gold solution to a thickness of about 1 micron. To avoid mechanical damage to these delicate parts, special modifications of barrelling may be employed, e.g., the turbo-jet technique (see Chapter 17).

The winding crowns are turned from fluted rods of leaded brass. Some are embossed with a cap of rolled gold, but most are plated with a coating of similar thickness and colour to that used on the case. The difficulty of polishing necessitates the use of very bright plating solutions and the fluted nature of the surface calls for good throwing power. During recent years rack plating of these items which is expensive and tedious, has given way to barrel plating.

WATCHCASE PLATING

COATING REQUIREMENTS

Gold plating of watchcases was first carried out in Switzerland about 100 years ago with the object of imparting good appearance and tarnish resistance to cases in brass or nickel silver. Due to technological limitations, the majority of such plating was long based on thin (0.5–1 micron) coatings of fine gold, which achieved the desired objective to a limited extent only, in view of their restricted thickness and hardness. From this situation arose the requirement for gold alloy deposits to obtain the recognised colours of wrought gold alloys, together with greater hardness and thickness of the coatings. Such alloy deposits were first used in 1912 by a few specialist platers, and from these empirical beginnings stemmed the present day extensive gold alloy plating technology, which now permits the deposition of gold alloy coatings of almost any desired thickness and in practically all colours pertaining to wrought gold alloys.

As in all applications of gold plating, a fundamental requirement of watchcase coatings is excellent adhesion to the basis metal. This is of particular importance since the plated cases are often subjected to mechanical working operations such as, turning, drilling, milling and grinding, which constitute a very severe adhesion test on 100% of production.

The coating itself, usually a binary or ternary alloy, with a minimum thickness of 8–10 microns, should have a minimum gold content of 14 carat and be fully homogeneous in composition. The alloying metals present should be completely in the form of mixed crystals in order to achieve the necessary tarnish resistance and reproducibility of colour for a given composition.

On dissolution of the basis metal the coating should remain as a coherent shell, free from cracks or other defects associated with brittleness. Past experience has shown that coatings which show cracks or disintegrate into several pieces under this treatment, will develop unsightly tarnishing during wear due to crack formation.

To ensure good wear resistance of the coating it is desirable that the mechanical properties, such as tensile strength, ductility and hardness, should lie within certain ranges and bear the correct relationship to one another. There is no difficulty in producing very hard coatings, either by heat treatment of an "as-plated" deposit or by the use of organic additives in the plating solution, but since high hardness alone may be associated with brittleness and susceptibility to cracking, it is therefore also essential to take into account the ductility and tensile strength of the coating. Wear resistance is in fact more closely related to the metallurgical factor of "toughness", expressed by the product of tensile strength and elongation at fracture in tensile testing, good wear properties being achieved when this factor is combined with a sufficient hardness. From experience, a combination of properties within the following ranges would appear to be ideal

Tensile strength 70–90 kg/mm^2
Elongation 5% minimum
Microhardness 350–450 MHV

Coatings showing such a combination of properties may be described as tough-hard, this condition being regarded in modern tribology as desirable for minimum wear.

Finally, the deposit should have an ultrafine grain size, due consideration being given to the requirements of mechanical properties and surface brightness.

PLATING

Cheap cases which are produced from polymeric materials such as injection moulded ABS, are not considered in this chapter. It should also be noted that the term watchcase usually refers to the front of the case only, since the back will almost invariably be in stainless steel and is rarely plated.

Preparation

As received from the manufacturer the case should require no further mechanical finishing. To protect internal threads and to conserve gold, the cases are filled before plating, with a warm viscous compound based on cellulose acetobutyrate, which sets on cooling to a rubbery consistency and can be readily ejected after plating. An alternative method to this is to use conforming plastic plugs.

Plant

Installations for watchcase plating are usually small. Although these may be of manual design, some completely automatic plants are also in operation. Some installations employ rectangular tanks with transverse movement of flat racks, but it is also common to use circular plating tanks of earthenware or enamelled steel up to 60–80 litres in capacity, with anodes of stainless steel or platinised titanium arranged around the periphery. With either square or circular tanks the cases are usually jigged on cylindrical racks in groups of 50–100 according to size, provision being made for rotation of the racks during plating, with periodic reversal of direction. Current densities are low, hence current requirements are relatively small. Periodic reverse plating is commonly employed for some part of the plating period to provide a smoothing effect and to reduce internal stress in the deposit.

Procedure

Plating procedure is that normally applicable to copper base alloys. Degreasing may be carried out in two stages, comprising of an ultrasonic treatment in a chlorinated organic solvent or an aqueous medium containing a wetting agent, followed by cathodic treatment in a hot alkaline cleaner, with intermediate immersion and/or spray rinsing.

On leaded brass a thin deposit of copper from a cyanide solution is essential prior to gold plating. This may also be applied to plain brass or nickel silver as a precautionary measure to improve the adhesion of the subsequent gold deposit. If an undercoat of nickel is specified, as is particularly common in US practice, this may be applied directly to the basis metal or over a cyanide copper flash. In this case the nickel should be subjected to an effective depassivation treatment prior to gold plating. If the final plating is to be carried out in an alkaline solution, it is good practice to initially apply a thin coating of gold from an acid electrolyte to ensure good adhesion.

Electrolytes

In current practice almost all of the major types of gold plating solution are used in watchcase plating including acid, alkaline and neutral electrolytes based on aurocyanide, as well as non-cyanide solutions of the sulphite type. It is difficult to establish detailed formulations since many of the solutions are proprietary in nature, and although electrolytes of an earlier generation such as Spreter-Mermillod and Volk (see Chapter 5) are still extensively employed, these are often modified by individual users in accord with their own particular experience.

Colour control is an important aspect. An electrolyte is usually designed to deposit a gold alloy of 14–18 carat (58.5–75% gold), the balance being one or more base metals, such as copper, silver, cadmium, nickel, etc., to provide the required shade (see colour gilding, Chapters 4 and 41). Apart from their effect on colour, nickel complexes are often employed as brightening agents and antimony salts may be used in a similar role. Although organic brighteners are not uncommon, their use is generally undesirable in view of the tendency for codeposited organic material to induce stress and brittleness in deposits. In this respect there are advantages in relatively simple formulations, for example of the Volk type, which can be employed without additional brighteners, utilising periodic reverse current plating to produce an acceptable brightness.

The colour of the coating is dependent on deposit composition, which in turn is affected not only by the electrolyte composition, but also by the current density and temperature employed. Careful control of these factors is therefore essential. Due to the shape of the cases the exact calculation of area is difficult, hence the current used must be fixed to some extent on an empirical basis, based on the colour and brightness of the coating. Control of the plating operation by cathode potential would probably improve matters in this respect, but this has not yet been adopted to any significant extent.

To maintain uniformity of colour the concentration of alloying metals may be maintained by continuous dosing during plating. Even so, different batches plated under the same nominal conditions may show minor variations in tint. These are corrected if necessary, by applying a final coating from a special colour gilding solution which provides the appropriate standard shade. Problems of colour matching have been greatly alleviated by the definition of a series of colour standards from Switzerland, Germany and France, which are now in general use. The relevant specifications are NIHS 03 SO (July 1961), Switzerland; DIN 8–322 (1966), Germany; and CETEHOR 07 70 (March 1966), France. The colours, coded by a letter and one or more figures, are based on those of a series of reference alloys prepared by melting. Colour codings and corresponding alloy compositions are listed in Table 42.1.

It is important to note that these standards apply both to wrought gold and to electrodeposited alloys. In the latter case, coatings of the correct nominal composition may still not show exactly the same colour as a solid alloy due to structural differences, hence the plater must establish bath conditions empirically to obtain the required shade without reference to the exact composition of the deposit.

Post-Treatment

Deposits which are homogeneous and completely in the form of mixed crystals in the as-deposited condition, usually require no further treatment other than thorough rinsing and drying. Ultrasonics may be incorporated into one of the rinsing stages to ensure the complete removal of solution and salts from crevices and holes. Centrifugal drying is commonly used to avoid drying stains.

Table 42.1
Standard Colours in Watchcase Plating

Colour Coding	*Colour Description*	*Carat*	*Corresponding Reference Gold Alloys (Composition, parts per thousand)*				
			Gold	*Silver*	*Copper*	*Nickel*	*Zinc*
IN–14*	Pale Yellow	14	585	265	150	—	—
2N–18*	Pale Yellow	18	750	160	90	—	—
3N*	Yellow	18	750	125	125	—	—
4N*	Rose (Pink)	18	750	90	160	—	—
5N*	Red	18	750	45	205	—	—
ON†	Yellow-green	14	585	340	75	—	—
8N†	White	14	590	—	220	120	70

* Standards Common to Germany, France and Switzerland.
† Standards Common to Germany and France.

In the lower carat range some alloying constituents may be deposited from certain electrolytes, at least in part, as single metal or base metal-rich phases, the presence of which renders the coating prone to oxidation or corrosive attack in environments which would have no effect on a melted alloy of the same composition. In the as-plated condition, coatings of this type are liable to tarnish during formal testing and to experience colour changes on long term exposure to normal atmospheres, pink golds tending to develop a maroon shade and yellow golds a bluish tint. In these cases it is necessary to homogenise the coating by a diffusion heat treatment, this usually being carried out at a temperature of 300–400°C for periods ranging from several minutes to one hour so as to promote true alloy formation. To maintain maximum brightness, the treatment must be carried out in an inert atmosphere such as cracked ammonia or propane.

Heat treatment, ranging from simple immersion in boiling water to the more complex type of procedure just described in the preceding paragraph, is sometimes used to relieve stress in deposits by removing occluded gas and possibly causing partial recrystallisation. Such treatments must be carried out as soon as possible after deposition and, of course, can only be effective when the stress level is insufficient to cause cracking of the deposit in the as-plated condition.

Though not a primary objective, heat treatment may be instrumental in improving the adhesion of coatings as a result of interdiffusion at the substrate-coating interface.

TESTING OF COATINGS

Adhesion

This can be assessed by destructive tests, e.g., bending or filing, or by thermal tests on representative samples, which generally involve heating the plated case to 350–400°C for a minimum period of 5 minutes, with or without subsequent quenching in cold water. Although in principle no blistering of the coating should occur if adhesion is to be regarded as satisfactory, a small amount may be tolerated in practice. This depends on the size and distribution of blisters, since the test is quite severe and slight blistering is not necessarily indicative of failure under the relatively mild conditions of service.

Thickness

Measurement by non-destructive techniques, e.g., beta-ray backscatter, is sometimes complicated by the small dimensions of the parts. For these relatively thick deposits the microscopic method is quite satisfactory and examination of a microsection will also give a good indication of thickness distribution. It is also quite common to determine thickness by direct measurement on the shell by precision micrometer.

Gold Content

This is most accurately determined by fire assay or by chemical analysis (see Chapter 33). For control purposes however, it is still not uncommon to assess the carat of fineness of a coating by testing with acid solutions of various compositions which are designed to attack gold alloys within specified ranges of composition (Table 42.2). These may be applied directly to the

Table 42.2

Composition of Solutions used in Testing of Watchcase Coatings*

	Composition			
Fineness of Coating (parts per 1000 Au)	*HNO_3(SG 1.284) (ml)*	*$CuCl_2$Soln† (ml)*	*Water (ml)*	*HCl(conc) (ml)*
800–1000	45	3	2	
650–800	41	2	7	2–3 drops
500–650	30	1	15	
375–500	5	15	20	
	"A" 15% w/v $CuCl_2$soln		45 ml	
250–375	"A" 1.5% w/v NaCl soln		54 ml	
	"A" HNO_3(SG 1.284)		1 ml	
	Mixture "A"		10 ml	
Below 250	Water		30 ml	
	HNO_3(SG 1.18)		8–10 drops	

* As used by Central Office of Swiss Metal Control, Berne.
† Prepared by dissolving 6–7 g $CuCl_2$ in 20 ml water.

coating or to the streak produced by rubbing the coating on a touchstone, when its response may be observed individually, or by comparison with that of traces produced from reference alloys of standard compositions.

This procedure can only be used successfully by a highly experienced observer.

Continuity

The production of a coherent shell on dissolution of the basis metal is the minimum criterion of satisfactory continuity of the deposit, but the shell itself may be examined visually for cracks, pores, or other defects. The pro-

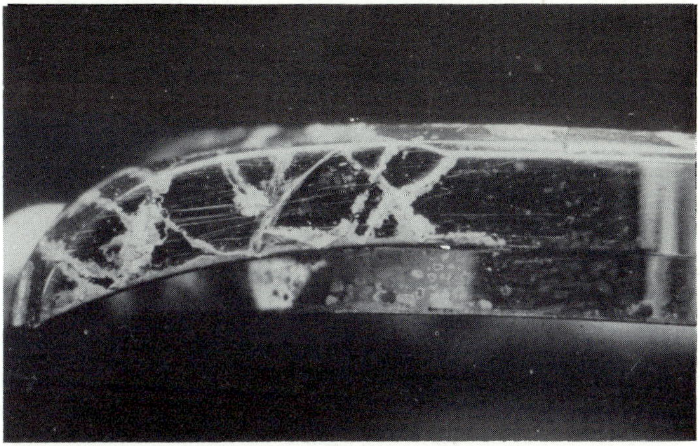

Fig. 42.1 Cracks and pores in watchcase coating revealed by exposure to a humid ammonia atmosphere for 48 hours at room temperature.

tective value of the coating can also be checked by exposing the plated case to various environments, e.g., dilute (1:2) nitric acid, ammonium sulphide vapour, humid ammonia vapour, or less aggressive reagents such as thioacetamide or acetic acid vapour. The effectiveness of ammonia vapour (48 hours at room temperature) in delineating cracks is illustrated in Figure 42.1.

Tarnish Resistance

The tarnish resistance of the coating *per se* is likely to be satisfactory if no signs of attack are noted during dissolution of the basis metal. Otherwise, either a lower carat than nominal or segregation of alloying metals may be indicated. The former can be checked by assay, the latter by exposing the plating to a reagent which is corrosive towards one or more of the alloying constituents, e.g., ammonium sulphide for silver. A simple heating test at 80°C for 48 hours will often suffice to produce tarnishing of an unsatisfactory coating.

Physical Properties

Watchcase coatings are often sufficiently thick (10 microns) to permit hardness measurement directly on the surface of the deposit, though it is always preferable to use specially prepared coatings of at least 20 microns for this purpose.

Tensile properties can only be determined quantitatively by tests on specially prepared foils. However, some qualitative indication of ductility can be gathered from simple bend tests on shells and by other signs, such as the absence of cracking at cut edges.

Comparative wear resistance of coatings may be assessed by a variety of empirical texts, which are usually made on specially prepared specimens.

Appendix

SPECIFICATIONS AND LEGAL REQUIREMENTS

R. MILLS

It is generally acknowledged that specifications play a vital role in upgrading and maintaining the quality and reliability of metallic coatings, which are applied in relatively thin layers to a variety of basis metals, or in some cases to non-metals, to confer on them such properties as corrosion resistance, wear resistance and solderability. It is essential that the thickness, porosity and mechanical properties of the coatings should be closely controlled or the plated article may not satisfactorily perform its intended function.

In no field is the importance of specifications more apparent than in the electroplating of precious metals, the high cost of which may act as an inducement to economy in the quantity of metal deposited on to an article. Electroplated coatings of gold require additional control since, like the cast and wrought metal, they can be produced in a wide range of alloys.

The oldest applications of electroplated gold coatings have been for decorative purposes in the watch, jewellery, silverware and allied trades. Watchcases and bracelets, and other items of jewellery in a cheaper range than that in which solid gold could be used, have for many years been plated with gold and gold alloys in order to simulate the appearance of the solid metal. When such plating is carried out to a high thickness specification, the article can have many years of service before the coating is worn through to reveal the basis metal. With high quality goods this period of service can, for watchcases, be greater than the useful life of the watch itself and, in the case of jewellery, greater than is necessary to satisfy the whims of fashion. On the other hand, in the absence of rigorously applied specifications or other quality controls, very thin gold coatings of low durability are frequently found.

In recent years, due largely to the rapid rise of the electronics and space industries, the use of electroplated gold coatings for engineering applications has outstripped its use for decorative purposes. In these industries gold coatings are essential for the production and manipulation of semiconductor devices and for providing highly reliable electrical contacts and heat reflectors. However, the intrinsic corrosion resistance, low contact resistance and weldability of gold are only maintained if coatings are relatively pure and possess adequate thickness and freedom from porosity. In view of the high reliability demanded in electronics and aerospace applications, when failure may have fatal consequences, the importance of high grade specifications have long been recognised.

LEGAL REQUIREMENTS

SWITZERLAND

In several countries there are legal requirements for the quality and marking of electroplated gold coatings in the interests of consumer protection. Preeminent in this field is the gold plated watchcase and Switzerland, with its heavy dependence on its watch industry, has for many years controlled gold coatings through the application of Swiss Federal law[1] by the Government Control Office in Berne.

The law applies to all exterior surfaces of watchcases and other products and refers to gold coatings which are known as "plaqué" or "doublé". The coatings must have a minimum thickness of 8 microns and be at least to the title of 350 thousandths of gold. This thickness is a minimum local thickness and the significant surface for this purpose is the entire surface with no restrictions placed on positions where thickness measurements may be made, as is the case with many European electroplating specifications. This is due to the small size of many of the component parts and the fact that they may be subjected to wear and attack by acid perspiration. However, a tolerance of 20% is permitted on the minimum thickness, effectively reducing it to 6.4 microns. The thickness of the gold coating is determined by the dissolution of the basis metal in an appropriate acid and measurement of the thickness of the shell by means of a precision micrometer recognised by the Central Bureau. In doubtful cases, and for reference purposes, the thickness of the shell must be determined by the metallographic examination of cross-sections or by any other method permitted by the Central Bureau.

Gold plated products which comply with the foregoing legal requirements may bear the designations "Plaqué", "Doublé" or a translation of these terms. Since these designations can also be used on coatings applied by a mechanical process, e.g., rolled gold, gold plated products must also bear the mark "galvanique" or the abbreviation "G" so that the complete designation must be either "Plaqué or galvanique", "Plaqué or G", "Doublé or galvanique" or "Doublé or G". No indication of title in thousandths or in carats is permitted, nor is any designation concerning the gold coating in per cent, per thousand or per kilogram, such as "Plaqué or 18 K", "14 C Gold Plated", etc. In addition, the designation must not include any adjunct liable to lead to an error concerning the title or the value of the metal.

FRANCE

France has a similar regulation covering decorative gold plating, although it is not so precise. No minimum thickness of gold coating is stipulated in the relevant article of the General Code on Taxes[2] which states that "Plaqué" or "Doublé" products should be effectively covered with a thin sheet of precious metal or should leave a shell after the basis metal has been dissolved. Products of French manufacture must be stamped using a stamp which is in the form of a perfect square in which must appear:

(a) The initials of the manufacturer and the symbol chosen by him.
(b) The words "Plaqué" or "Doublé" followed by the designation of the precious metal used (the regulation also covers silvered and platinised products) and the manufacturing process.

For imported products, the stamp has the form of a square beneath a semicircle, but containing the same information as required for products of French manufacture. Products covered with a layer of gold which is insufficient to leave a shell can only receive the designation "Gilt". They must be marked with stamps similar to those already described for products of French manufacture and those which have been imported, but the squares can only contain the initials of the manufacturer and the symbol chosen by him.

UNITED STATES

In the USA, gold coatings used in the watch industry and for general jewellery applications are covered by rules laid down by the Federal Trade Commission[3]. Their terminology "plate" or "plated" means that a sheet or shell of metal has been applied by soldering, brazing, welding or other mechanical means; it is necessary to use the words "electroplate" or "electroplated" to refer correctly to electrodeposited coatings. Thus, watchcases that have been electroplated with gold or with a gold alloy of not less than 10 Karat fineness to a thickness of not less than 0.00075 in. (19 μm) should be marked "Gold Electroplate" or "Gold Electroplated", provided that they meet stipulated tests. This marking must be immediately preceded by a correct designation of the karat fineness of a gold alloy, e.g., "16 Karat Gold Electroplate". The word "Karat" may be abbreviated to "K" or "kr" and "gold" abbreviated to "G" but the terms "electroplate" and "electroplated" must not be abbreviated.

If the thickness of the gold coating is not less than 0.0015 in. (38 μm) it may be described as "Heavy Gold Electroplate". The regulation states that this minimum thickness "shall mean that the coating of precious metal affixed to the surface of the metal stock shall be throughout the surface and at the thinnest point not less than the thickness specified after the completion of all finishing operations, including polishing, except, however, for such deviations therefrom, not exceeding 20% (minus) of the stated thickness, as may be proved by the manufacturer to have resulted from unavoidable variations in manufacturing processes and despite the exercise of due care, which deviation so proved shall be allowed if and when the quantity of precious metal is sufficient to equal the quantity necessary to provide the specified minimum thickness at all points on such watchcase including the thinnest point".

The gold coating on a watchcase must also meet the following requirements to qualify for marking:

1 *Appearance* The coating should be free from cracks, blisters, pits or other flaws.
2 *Adhesion* The watchcase should be heated to a temperature of 360–400°C and maintained at that temperature for not less than 5 minutes. The coating should show no signs of blistering, flaking, peeling or similar defects.
3 *Hardness* When subjected to a Knoop hardness test with a 25 g load the coating should achieve a rating of not less than 130.
4 *Porosity* After thorough cleaning of the surface, the watchcase should be subjected to the following tests:
   (a) Immersion in a solution of one part concentrated nitric acid (SG 1.42) and one part water at room temperature for 5 minutes.
   (b) Exposure to fumes of concentrated nitric acid (SG 1.42) in a closed vessel for 3 hours at room temperature.

At the conclusion of these tests, the gold coating should show no signs of having been attacked. Any discolouration or pitting should be considered as signs of attack. The nitric acid solution in which the watchcase was immersed should be tested for the presence of metal by making it slightly alkaline with ammonium hydroxide and by adding a solution of ammonium or sodium sulphide. The formation of a black precipitate indicates that the coating has been attacked or is porous.

With regard to jewellery, the US Federal Trade Commission has similar regulations[4] concerning the designation of electrodeposited gold coatings as those applied to watchcases. However, the requirements for articles to be marked "Gold Electroplate" or "Gold Electroplated" are considerably less onerous. The gold coating, of not less than 10 Karat fineness, must have a minimum thickness throughout which is equivalent to 0.000007 in. (0.18 μm). When the coating meets the minimum fineness, but not the minimum thickness, the marking or description may be "Gold Flashed" or "Gold Washed". Coatings of not less than 10 karat fineness, having a minimum thickness throughout equivalent to 0.0001 in. (2.5 μm) of fine gold may be given the marking or description "Heavy Gold Electroplate" or "Heavy Gold Electroplated". When coatings qualify for either of the foregoing descriptions and have been applied by the use of a special kind of electrolytic process, the designation may be accompanied by an identification of the process used, e.g., "Gold Electroplated (X Process)", "Heavy Gold Electroplated (Y Process)".

MEXICO

In Mexico[5] there is a requirement that gold plated jewellery must be clearly and visibly marked with the word "Dorado" followed by the caratage and the proportion of gold contained in the total weight of the plated object. The fineness must be not less than 417 thousands, corresponding to 10 carat, and the proportion of gold in the article must be not less than one twentieth.

UNITED KINGDOM

In the United Kingdom, there are at present no legal requirements for electroplated gold coatings as such. However, from ancient times the danger of fraudulent practice has been recognised and a series of laws have been passed with the object of controlling the work of the goldsmith. The Statute of 1403, "What things may be plated with gold or silver and what not", forbade the gilding of base metals except for church ornaments and only then if some part were left bare. In 1420 to this exception was added "knights' spurs and all the apparel that pertaineth to a baron or above that estate".

The present position with regard to electroplated gold coatings is not clear, as is recognised in the Foreword to BS* 3315:1960 "Watchcase Finishes in Gold Alloys"[6] which states; "Attention is drawn to the anomalous position law of gold alloy coverings applied to base metals. It would seem that gold alloy covered articles can be adjudged to be "manufactures of gold" and, as such, liable to compulsory hallmarking, regardless of the fact that no Assay Office would grant a hallmark to other than manufactures of "solid gold". It is hoped that this British Standard, together with the recommendations of the Board of Trade Hallmarking Committee, will form the basis of new legislation which will resolve these difficulties. In the meantime, the Assay Office advise that manufacturers contemplating the use of a gold alloy covering which is greater than one tenth of the total weight of the whole article, should consult an authorised Assay Office before proceeding".

In the intervening period, the position has changed a little, the Foreword to BS 4292, "Electroplated Coatings of Gold and Gold Alloy"[7] giving the following statement: "It should be appreciated that nothing in this standard alters the provisions of the Assay Laws about the hallmarking of wares of gold. In practice, the Assay Office would not regard an article consisting of an electroplated coating of gold or gold alloy on basis metal as a ware of gold, unless it contains more than 12.5% of gold". This higher figure of 12.5% is the one which the Department of Trade and Industry Hallmarking Committee is being recommended to adopt.

SPECIFICATIONS—DECORATIVE

UNITED KINGDOM

BS 3315:1960, "*Watchcase Finishes in Gold Alloys*"[6]

One of the earliest national specifications for gold coatings, BS 3315, covers the use of rolled gold and electroplated gold coatings on watchcases manu-

* British Standard.

factured from a non-precious metal. An essential feature of this specification is a marking clause which lays down the symbols to be used "which could, in the future, be registered with a recognised authority in much the same way as the marking applied to solid gold articles under the control of the Assay Offices".

The finished watchcase must be clearly and indelibly marked with the following information:

1 The manufacturer's name or mark.
2 A symbol indicating the method of gold coating, R for rolled gold and P for electrodeposited gold.
3 A number, followed by the symbol μ, indicating the thickness of the gold coating in microns. The minimum thickness required to satisfy the requirements of the standard is 10 microns, and the coating must be of not less than 9 carat. Tolerances on thickness are as given in Table 1.

Table 1
Tolerances on Thickness

Thickness (μm)	*Tolerance* (μm)
10 to 19	−2
20 to 39	−3
40 and above	−5

The coating thickness "shall be measured by microscopic, air gauging or electronic means or by any method capable of giving an assessment of thickness within 0.254 microns (0.00001 in.)".

The specification requires adhesion to be tested by heating the watchcase to a temperature of 360–400°C and maintaining at this temperature for not less than five minutes. The gold alloy coating shall show no sign of blistering or similar defect.

As a test for continuity, gold plated cases are required to withstand immersion for one minute in "a solution of one part of pure nitric acid (SG 1.42) to two parts of water" at room temperature.

BS 4292, "*Electroplated Coatings of Gold and Gold Alloy*"[7]

Although BS 4292 applies to gold coatings for both decorative and engineering purposes, the important requirements of thickness, purity and corrosion resistance receive separate treatment. "Gold Plated" is defined in this standard as "the finished state of an article that has been given a coating of gold or gold alloy by electrolytic means, the coating having either,

(a) an average thickness of not less than 0.5 micron and a purity of not less than 90%, or

(b) a local thickness of not less than 2 microns and a purity of not less than 62.5%."

The designation "Au" in the Classification Number for a particular coating signifies that the gold has a minimum purity of 90%, whereas the designation "G" in the Classification Number applies to coatings with a minimum purity of 62.5% (equivalent to 15 carat). Coatings which satisfy these purity requirements do not *ipso facto* comply with the standard. They may not possess the necessary corrosion resistance, which will depend not only on gold content, but also on the specific alloying elements used with the gold.

Coatings for general decorative purposes including many applications in the jewellery trade have thickness classifications as given in Table 2.

Table 2
Coatings for General Decorative Purposes

Classification Number	*Minimum Coating Thickness (μm)*
Au 0.5	0.5 (average)
Au 1	1 (average)
Au 2 or G2	2 (local)

Marking or labelling of individual articles of the above Classification Numbers is not mandatory.

Coatings for specialised applications in the jewellery trade where relatively thick coatings are required to withstand continuous wear over long periods are given the thickness classification numbers shown in Table 3.

Table 3
Coatings for Specialised Applications in the Jewellery Trade

Classification Number	*Minimum Average Thickness (μm)*	*Minimum Local Thickness (μm)*
G5	5	4
G10	10	8
G20	20	16

Such coatings are subject to mandatory marking or labelling of individual articles to indicate compliance with the standard. To quote the Marking Clause, "the plated article shall be clearly and indelibly marked with the appropriate Classification Number. If it is impracticable to mark the plated article, a label marked with the Classification Number shall be attached to the article."

Local thickness is measured at points on the significant surface that can be touched by a ball 25 mm in diameter by the microscopical examination of suitably prepared cross-sections. Prior to mounting the section in a plastics material it should be electroplated with copper or nickel, the method of deposition being such that the adhesion is good.

In the case of coatings Au 2 or G2 the average thickness is determined over any desired number of areas from the significant surface using either a given strip and weigh, a fire assay (for basis metals of tin and lead only) or a spectrophotometric analytical method, as appropriate.

In addition to the above methods, the average thickness of coatings G5, G10 and G20 may be determined by taking the average of local thickness measurements at not less than five random points on the significant surface that can be touched by a ball 25 mm in diameter.

The corrosion test laid down for all decorative gold coatings, regardless of classification or thickness, consists of exposure to hydrogen sulphide gas in a closed vessel under carefully controlled conditions for not less than 24 hours. At the end of this period, the significant surface shall be free from corrosion both in respect to the coating and to the underlying metal.

The purity of the gold coatings is determined either by a fire assay or by a spectrophotometric method. The fire assay method should be carried out only by some recognised authority skilled in the assay of precious metals.

When required by the purchaser, a burnishing test and/or an adhesive tape test must be carried out. For a sample to pass the former test, there must be no sign of blistering when an area of not more than 6 cm^2 of the significant surface is burnished for 30 seconds with a suitable burnishing tool. In preparation for the adhesive tape test, the test area is scribed to produce a grid with squares of 2 mm sides by the use of a straight edge and a hardened steel scriber which has been ground to a sharp point. Sufficient pressure is applied to cut through the gold coating to the basis metal in a single stroke. A non-transferable adhesive tape with an adhesion value of 290–310 gf per cm width is then applied to the test area with normal finger pressure, taking care to exclude all air bubbles. After an interval of 10 seconds the tape is removed by applying a steady pulling force perpendicular to the surface of the gold coating. None of the gold must be removed for the plating to pass the test.

GERMANY

*Draft Standard DIN** 8237, "*Gold Plating and Coating for Watch Cases, Requirements, Testing*"[8].

This draft standard refers to gold applied to watchcases both by electrodeposition and mechanical means, e.g., welding and hot rolling. Electrodeposited gold coatings are described by one of the following designations: "Plaqué", "Plaqué or Galvanique", "Electroplated", "Gold Plated", "Goldplattiert" and are given the code letters P or G.

---

* Deutsche Industrie Norm.

The thickness of the gold coatings must not be less than 5 microns, the preferred range of thicknesses being 5, 10, 15, 20 and 40 microns. A deviation in the stated thickness up to -20% is permissible. For coating thickness, the specified reference method is the microscopic method according to DIN 50950, "Testing of Electroplated Coatings; Microscopic Measurement of the Coating Thickness"[9]. The number of measuring points must be agreed between the purchaser and the supplier.

Another destructive method, but one which is not suitable as a referee method, can be carried out by chemical dissolution of the basis metal and the measurement of the gold shell with a micrometer. The beta-ray backscatter method is also given as a non-referee method.

The fineness of the gold plating must be a minimum of 585 thousandths, preferably 750 thousandths. The total weight of fine gold should be determined by a separation method or by wet chemical analysis.

Gold coatings are required to pass two forms of adhesion test, a file test and a bend test. When a needle-point file is applied to the coating, only a chipping deformation may occur. If the coating peels it does not pass the test. For the bend test a section of the watchcase suitable for bending is bent to approximately 90° (bending radius = 2 mm) and then bent back to its original form. If peeling or blistering of the coating occurs, the coating will be deemed to fail the test.

Gold coatings of the required minimum thickness and minimum fineness and which pass the adhesion tests, may be marked with the coating thickness in microns, the appropriate code letter and the registered trade mark of the producer.

INDIA

*IS**4252:1967 "*Electroplated Coatings of Gold for Decorative Purposes*"[10]

This standard defines the designation "gold plated" as "the finished state of an article that has been given a coating of gold or gold alloy by electrolytic means, the coating having a minimum local thickness of 2 microns and purity of not less than 60%". Four grades of gold coating are recognised as given in Table 4.

Table 4

Grades of Coatings in IS: 4252

Grade	*Minimum Average Thickness (μm)*
Au 20	20
Au 10	20
Au 5	5
Au 2	2

* Indian Standard.

The average thickness is determined by taking the average of local thickness measurements by the microscopic method given in IS3203:1965, "Methods of Testing Local Thickness of Electroplated Coatings"[11], at not less than five random points on the significant surface which are accessible for touching with a ball 25 mm in diameter.

The purity of the gold coating must not be less than 60%. Analysis is carried out on a separate gold coating, deposited on platinum or iridium at the same time as the article is plated. The coating is stripped from the specimen and the gold content is determined by the method described in IS1418:1962, "Methods for Assaying of Gold and Gold Alloys"[12].

This standard does not specify the basis metals to which the gold coatings may be applied but simply states that "the material for gold plating shall be such as to produce a coating conforming to the requirements of this standard. Coating on the basis metal may require an undercoat of nickel, copper or any other suitable metal to meet any requirements of this standard, and shall be applied according to agreement between the purchaser and the supplier."

The accelerated corrosion test which gold plated articles have to withstand consists of exposure in a gas-tight chamber at room temperature to the vapour of ammonium sulphide solution. After testing, the significant surface must be free from corrosion of the coating and of the underlying metal.

In this specification an adhesion test is mandatory and may be in the form of either a burnishing, bend, or a baking test. The bend test consists in bending a sample in a bend tester with a bending radius of 4 mm. One bend comprises bending forward to 90° and backwards to the original position, and the sample must withstand three bends without exhibiting exfoliation of the coating. To pass the baking test the sample must show no sign of blistering when heated at a temperature of 360–400°C for five minutes.

There is a mandatory marking clause in the specification; gold plated articles must be marked with the grade, purity of the coating and the name or trade mark of the manufacturer. If marking the article is impracticable, it must bear a securely attached label, giving the same information.

SPECIFICATIONS—DECORATIVE AND ENGINEERING

JAPAN

*JIS** *H*8616:1965 "*Gold Plating*"[13]

This specification makes no distinction between electroplated gold coatings intended for use in decorative or engineering applications. The standard applies to coatings at least 1 micron in thickness, which are divided into two classes, Class I covering soft gold 24 carat coatings and Class 2, hard gold coatings of 14–22 carat. Each of these classes consists of six grades as shown in Table 5.

---

* Japanese Industrial Standard.

Table 5
Classes and Grades of Coatings in JIS H8616

Class	*Grade*	*Designation*	*Description*	*Coating Thickness* (μm)
Class 1	Grade 1	GM1	24 carat (soft gold)	1
	Grade 2	GM2		2
	Grade 3	GM3		3
	Grade 4	GM4		5
	Grade 5	GM5		10
	Grade 6	GM6		20
Class 2*	Grade 1	AGM1	14–22 carat (hard gold)	1
	Grade 2	AGM2		2
	Grade 3	AGM3		3
	Grade 4	AGM4		5
	Grade 5	AGM5		10
	Grade 6	AGM6		20

* The designations of Class 2 coatings are to be prefixed by such symbols as 14K, 16K, 18K, 20K and 22K in accordance with the gold content of the coating, e.g., 14K AGM1.

Coating thickness (except for Grade 1) is determined by the microscopical examination of prepared cross-sections: the direct measurement of the thickness of all grades is carried out by means of a micrometer used on coatings separated from the basis metal by the dissolution of the latter in 50% nitric acid.

The gold content of the coating must comply with Table 6. The methods for determining the purity of the gold coating are chemical analysis and the use of a touchstone.

Table 6
Gold Content Corresponding to Caratage

Carat	*Gold Content* (%)
14K	54.2–62.5
16K	62.6–70.8
18K	70.9–79.2
20K	79.3–87.5
22K	87.6–95.8
24K	95.9 minimum

Corrosion tests specified in this standard are (a) salt spray (b) ferroxyl (c) nitric acid exposure and (d) acetamide exposure.

Adhesion is determined by heating the sample for 10 minutes at 150°C and quenching in water. Hardness is measured by the Vickers microhardness test employing a 25 g load on a gold coating not less than 30 microns in thickness on a test piece plated under the same conditions as the product. The hardness of Class 2 coatings should normally be not less than 100, but it is desirable that the actual value should be agreed between the parties concerned.

SPECIFICATIONS—ENGINEERING

UNITED KINGDOM

BS 4292, "*Electroplated Coatings of Gold and Gold Alloy*"[7]

This standard specifies requirements for gold coatings for electrical, electronic and other engineering applications in addition to those used for decorative purposes which have been dealt with in a previous section. Such coatings are classified by thickness and designated by the Classification Numbers as given in Table 7.

Table 7
Classification Numbers and Coating Thicknesses

Classification Number	*Minimum Coating Thickness (μm)*
Au 0.5	0.5 (average)
Au 1	1 (average)
Au 2	2 (local)
Au 5	5 (local)
Au 10	10 (local)
Au 20	20 (local)

Local thickness of coatings of not less than 2 microns is measured by the microsection method at any desired number of points on the significant surface that can be touched by a ball 25 mm in diameter. Since the methods used for determining the thickness of coatings less than 2 microns thick require some considerable area to be taken from the plated article, the coating thickness obtained is the average value over the area tested. The strip and weigh, and spectrophotometric analytical methods are appropriate for the determination of the average thickness of such coatings.

The gold content of coatings used for engineering purposes must not be less than 90%. Fire assay and spectrophotometric analytical methods are given for the determination of the gold content.

Adhesion is tested by either or both of two methods, the burnishing test and the adhesive tape test. Where a value for hardness is agreed between the purchaser and the supplier, the hardness is measured as microhardness using a suitable microhardness testing machine with a diamond indentor in the form of a right pyramid with a square base and an angle of 136° between opposite faces. The indentations are made on the surface of the gold coating or on cross-sections cut from the samples with a load of 10 ± 2 g. The impression is measured with a microscope objective capable of measuring to the nearest 0.25 micron.

Requirements for undercoats on ferrous metals, zinc-base alloys, and tin and lead alloys are given in the standard, but it is unlikely that the latter would be relevant to engineering applications. Ferrous metals must have an undercoat of nickel at least 10 μm thick or 8 μm of nickel on 8 μm of copper. Zinc alloys require 8 μm of nickel on 8 μm of copper. Since in electrical and electronic applications copper and copper alloys would generally be the basis metal, any undercoat necessary to meet the functional or other requirements of the standard must be the subject of agreement between the parties.

AUSTRALIA

*K*158:1967, "*Electroplated Coatings of Gold for General Engineering Purposes*"[14]

This specification classifies gold coatings into the following four types depending on the gold content:

Type	*Gold Content*
1	not less than 99.9%
2	not less than 99.0%
3	not less than 95.0%
4	less than 95.0%

The purchaser must specify the gold content of Type 4 coatings.

The gold and/or alloying elements are determined analytically, if required by the purchaser. The analysis is carried out on a coating electrodeposited on a suitable cathode (e.g., platinum, silver) at the same time and under the same conditions as the article it represents. The gold coating is stripped from the cathode and analysed by any established procedure which is acceptable to the parties concerned.

The local thickness, being the minimum thickness on the significant surface unless otherwise specified by the purchaser, should be specified according to the relevant classification given in Table 8.

The local thickness of the coating is determined by taking the average of four determinations made on the significant surface at points selected at random. The measurements are made by the microscopical examination of prepared cross-sections through the article. When it is not possible to carry out local coating thickness measurements, an average thickness requirement may be substituted by agreement. The strip and weigh method is to be used and the average thickness must comply with that specified in Table 8, unless otherwise agreed.

The porosity of the gold coating must be tested by immersion of the plated article for one minute in a solution of equal volumes of nitric acid (SG 1.42)

Table 8
Classification of Gold Coatings

Classification	*Minimum Coating Thickness* (μm)	*Minimum Coating Thickness* (*in*)
Au 0.5	0.5	0.00002
Au 1	1	0.00004
Au 2	2	0.00008
Au 2.5	2.5	0.0001
Au 5	5	0.0002
Au 10	10	0.0004
Au 20	20	0.0008
Au 30	30	0.0012
Au 40	40	0.0016
Au 50	50	0.002
Au 100	100	0.004
Au 125	125	0.005

and distilled water. There must be no indication of chemical reaction. The three corrosion tests given in the standard are not mandatory but, if required, must be specified by the purchaser. One test applicable to coatings less than 5 microns thick, consists of exposure to hydrogen sulphide gas in a closed vessel for not less than 24 hours. In a variation on this test, for coatings exceeding 5 microns, exposure to sulphur dioxide for 24 hours precedes the testing in the atmosphere of hydrogen sulphide. The third test involves suspension for not less than 3 minutes in a closed vessel over a fresh solution of ammonium sulphide contained in an open dish.

The standard requires the coating to adhere firmly to the basis metal. The purchaser is left to specify one or more of the following methods: a burnishing test, a bend test, a cutting test, a baking test and an adhesive tape test. There must be no evidence that the coating has separated from the basis metal.

Hardness testing is not mandatory, but when specified by the purchaser is determined as microhardness on the surface of the gold coating or on prepared cross-sections cut from the sample. A load of 10–20 gf is applied for 10–25 seconds using a Vickers microhardness testing machine.

UNITED STATES

ASTM B488–68, "Electrodeposited Coatings of Gold for Engineering Purposes"*[15]

This specification applies to gold and gold alloy coatings containing not less than 95.0% by weight of gold used in engineering applications where their

* ASTM—American Society for Testing and Materials.

corrosion resistance, non-tarnishing characteristics, low and stable electrical contact resistance, solderability and thermal reflectivity are advantageous. Any minimum gold content (purity) greater than 95.0% may be specified within the ranges given in Table 9. In this specification, gold content and hardness are linked with hardness (Knoop) as given in the following Table 9.

Table 9
Hardness Code for Gold Content Ranges

Purity (gold content) %	*Hardness Code*
99.9 min	A only
99.5–99.8	A, B or C
95.0–99.4	B, C or D

The hardness code in Table 9 is interpreted as follows:

Code	*Knoop Hardness Range*
A	$\not> 90$
B	91–129
C	130–200
D	> 200

No particular methods for the determination of gold content and impurity levels are specified in ASTM B488: any recognised assay or instrumental method may be used.

Hardness is measured on cross-sections of test panels plated as closely as possible to actual production conditions. For hardnesses below HK (Knoop hardness) 200, panels are plated with not less than 25 μm gold and for hardnesses above HK 200, not less than 13 μm gold. A protective backing of at least 13 μm of either copper or nickel is deposited on to the gold before microsectioning. A microhardness determination is made using a Knoop machine with a load of 25 g. At least five indentations must be made on the approximate centre line of the coating at least three indentation lengths apart, and the mean value taken.

Coatings are designated by their thickness in microinches as given in Table 10.

The recommended methods for measuring the thickness of gold coatings less than 0.0001 in. (2.5 μm) thick are the beta-ray backscatter and the X-ray fluorescence methods. Coatings exceeding 0.0001 in. in thickness should be measured by either of these two methods or else with a magnetic gauge or by microscopic examination of cross-sections.

It is left to the purchaser to specify any undercoats which he may require, the only mandatory requirement being that if copper is specified for retarding basis metal diffusion it must have a minimum thickness of 0.00001 in. The

Table 10
Coating Designations

Code	*Thickness* (*in.*)	*Thickness* (*μm*)
F	Flash*	Flash
10	0.00001	0.25
20	0.00002	0.51
30	0.00003	0.76
50	0.00005	1.3
100	0.00010	2.5
200	0.00020	5.1
500	0.00050	12.7
1000	0.00100	25.4
AS	>0.00100	>25.4

* A flash coating must be of sufficient thickness to impart the characteristic colour of gold but need not meet any other thickness requirement.

statement is made that basis metals like zinc, aluminium and steel normally require undercoatings such as copper or nickel of sufficient thickness to impart adequate corrosion protection to the basis metal.

AMS 2422, "Gold Plating for Electronic Applications"*[16]

This specification covers electroplated gold coatings applied primarily for improving the solderability, electrical conductivity, performance and appearance of electronic and electrical components. The gold is required to have a minimum purity of 97% and, unless otherwise stated, the thickness must not be less than 0.00005 in. on all the surfaces on which gold is functionally necessary. No method of testing thickness is given.

Before plating, the roughness of the basis metal must not exceed 32 microinches, unless otherwise specified. Immediately after removal from the electroplating bath, the gold plated component must be immersed in water at a temperature of at least 180°F for 15 minutes. Steel parts having a hardness greater than Rockwell C 40 are then subjected to further embrittlement relief not more than 30 minutes after completion of the hot water immersion by heating in air, preferably in a circulating air furnace to 375±10°F and holding at this temperature for 3 hours, or by heating to 275±10°F and holding at this temperature for 5 hours. Copper-base alloy parts must be subjected to additional embrittlement relief not more than 30 minutes after completion of the hot water treatment by heating in air to 275±10°F and holding at that temperature for 2 hours.

Adhesion is assessed either by a bend test or by heating the part to 350±10°F after which there must be no evidence of detachment of the coating from the basis metal.

* AMS—Aerospace Material Specification.

AMS 2425, "*Gold Plating for Thermal Control*"[17]

AMS 2425 deals with coatings used primarily for passive thermal control applications where low solar absorptance, low infra-red emittance and excellent corrosion resistance are required. Basis metals other than aluminium must be coated with not more than 0.0001 in. copper, 0.0004–0.0009 in. nickel and not less than 0.00008 in. gold. Aluminium is given an additional preliminary zinc coating. The thicknesses of the nickel and gold coatings must be determined as a routine inspection procedure, but no methods of measurement are stipulated. The gold must have a minimum purity of 98 %, but routine determination of purity is not required.

Requirements for the preparation of the basis metal for the removal of hydrogen embrittlement and for the adhesion of the gold coating are the same as for AMS 2422. A corrosion test is also specified: gold plated components are subjected to a continuous salt spray test in accordance with ASTM B117[18] for 48 hours.

SPECIFICATIONS—DEFENCE

Although Defence applications for electroplated gold coatings are of an engineering nature, it is considered desirable to deal with them separately, since certain Defence specifications include processing details for the benefit of Government inspectors who are often resident with a contractor.

UNITED KINGDOM

*DTD** 938, "*Gold Plating*"[19]

DTD 938 covers the requirements for electroplated gold coatings on electrical contacts and other components, primarily for the maintenance of low contact resistance and/or solderability. As in the case of other process specifications in the DTD series, DTD 938 controls the manner in which the gold is electrodeposited as well as specifying performance and acceptance tests.

The gold content of the coatings should not normally be less than 98 %, but in certain cases, with the agreement of the Design Authority, it may be reduced to a miniumum of 90 %. A spectrophotometric method of analysis for gold is specified.

Two classes of coating are recognised in the specification: Class Au 80 requiring a minimum local thickness of 0.00008 in. (2 μm) for improving and maintaining solderability and Class Au 200 requiring a minimum local thickness of 0.00020 in. (5 μm) for uses other than that covered by Class Au 80.

* Directorate of Technical Development.

Substantial undercoats of silver, nickel or in some cases copper, are mandatory when the basis metal consists of steel or aluminium alloy. The thickness of the gold and any undercoats are to be measured by the microscopical examination of prepared cross-sections using a magnification of not less than 1000× for Class Au 80 or 500× for Class Au 200.

A burnishing test for adhesion is required on both classes of coating, but only Class Au 200 coatings are subjected to a sulphur dioxide corrosion (porosity) test.

UNITED STATES

MIL-G-45204B, "*Gold Plating, Electrodeposited*"[20]

Three types of gold coatings are covered by this specification, the minimum gold content of Type I being 99.7%, of Type II 95.5% and of Type III 99.9%. There are eight classes of coating thickness as given in Table 11.

Table 11
Classes of Coating in MIL-G-45204B

Class	*Minimum Thickness (in.)*
00	0.00002
0	0.00003
1	0.00005
2	0.0001
3	0.0002
4	0.0003
5	0.0005
6	0.0015

The significant surfaces to which the minimum thicknesses apply are all surfaces of the component which can be touched by a sphere 0.75 in. in diameter together with any additional functional surfaces specified on the appropriate drawing.

Basis metals other than copper and its alloys, e.g., low alloy steel and zinc, must receive a suitable undercoat of not less than 0.001 in. (minus the thickness of the specified gold plating). Thickness is measured by methods considered to be appropriate as follows:

Method	*Thickness Limitation*
Microscopic	100 microinches, min.
Beta-ray backscatter radiation	300 microinches, max.
X-Ray fluorescence	100 microinches, max.

The hardness of the coatings are specified by references to grades A, B, C or D which relate to their Knoop hardness, respectively, of 90 maximum, 91–129 inclusive, 130–200 inclusive, and 201 and above.

The gold content (purity) is related to hardness as follows:

Purity	*Hardness (grades)*
Type I	A, B or C
Type II	B, C or D
Type III	A

For measuring hardness, coatings must have minimum thicknesses according to the hardness grade:

Grades A and B	:	0.002 in.
Grade C	:	0.001 in.
Grade D	:	0.0005 in.

To determine hardness, it may be necessary to prepare separate specimens in which case they should be approximately 0.5 in. × 1 in. × 0.04 in. plated concurrently with the articles represented plus additional plating under the same conditions to achieve the required thickness appropriate to Grades A/B, C or D. All samples are to be overplated with at least 0.00005 in. of nickel or copper and the hardness determined on prepared cross-sections using a Knoop hardness tester with a load of 25 grams. The Knoop hardness may be determined perpendicular to the plated surface if the thickness of the plate is at least ten times the depth of the Knoop indentation.

Adhesion of the coating is assessed, where practicable, by a bend test consisting of bending the sample through an angle of 180° on a diameter equal to the thickness of the specimen until fracture of the basis metal occurs. No detachment of the coating should take place when it is probed with a sharp instrument. When the sample is not suitable for the bend test it should be tested by cutting with a sharp instrument or by baking at 250–300°F for 1 hour. The latter test is not to be confused with an additional heat resistance test to which all samples may be subjected when specified. This consists of heating samples to 500 ± 25°F for not less than 30 minutes, after which time the parts must show no sign of blistering, discolouration or of a visible white or crystalline film.

Gold coatings may be required to pass a solderability test, in which case method 208 of MIL-STD-20, "Test Methods for Electronic and Electrical Component Parts"[21] should be used. This test is designed to demonstrate the ability of a gold plated metal component to be wetted by a new coating of solder or to form a suitable fillet when dip soldered in conjunction with a specially prepared solderable wire. Prior to soldering, samples are subjected to an accelerated ageing test which is claimed to simulate a minimum of six months natural ageing under a combination of various storage conditions that have different deleterious effects. This consists of suspending the samples for 60^{+1}_{-0} minutes 1.5 in. above boiling distilled water in a glass or other suitable non-metallic container, which is provided with a cover of one or more stainless steel plates capable of covering approximately 7/8 of the open area of the container so that a more constant temperature may be obtained. After ageing and where applicable, the sample is wrapped with 1.5 turns of a standard copper wire. This wire, of 0.025 in. diameter, is prepared by solvent cleaning, pickling in 10% hydrofluoroboric acid, immersing in liquid rosin based flux and dipping in molten solder for 5 seconds at 230±5°C. The test samples, with or without standard solderable wire, are immersed in liquid rosin based flux to MIL-F-14256, type W, to the minimum depth necessary to cover the surface to be tested, for a period of 5–10 seconds. The actual solderability test is performed using solder to type 5, composition Sn 60 of specification QQ-S-571, contained in a solder pot of sufficient size to contain at least 2 lb of solder. The dross and burned flux is skimmed from the surface of the molten solder, and it is then stirred with a clean, stainless steel paddle to ensure that it is at a uniform temperature of 230±5°C. Immediately prior to immersion of the samples, the solder is again skimmed. The test pieces are attached to a suitable dipping device, capable of producing immersion and emersion rates of 1±0.25 in. per second and a dwell time in the solder bath of 5±0.5 seconds. After soldering, the sample is allowed to cool in air and the residual flux removed by dipping in isopropyl alcohol. The soldering area is then examined at a magnification of 10× and, to be acceptable, at least 95% of the surface area must be covered by a continuous solder coating.

SPECIFICATIONS—ANODES

USSR

The only reference to gold anodes in the world's specification literature is a Russian standard GOST 6837. The reason for this, no doubt, is the widespread use of insoluble stainless steel and carbon anodes for electroplating rather than soluble gold and gold alloy anodes. GOST 6837 requires the chemical composition to comply with Mark ZI 999.9 of GOST 6835, "Gold Sheet" which signifies a minimum gold content of 99.95%. The surface of anodes must be clean, even and free from fissures, dross, exfoliation and traces of lubricant. Dents, scratches and trimming, and scraping marks are not permitted on the surfaces of anodes if their presence means that the sheet exceeds the tolerances permitted for thickness. Local areas of dullness or iridescence on the surface of anodes should not be a reason for their rejection. All edges must be even, and free from cracks and burns. Anodes are to be supplied in an unannealed condition.

The dimensions of anodes and the weight of a single anode must comply with the values given in Table 12.

Unless otherwise agreed, anodes are supplied with two holes, each 5–6 mm in diameter, situated over the width of the anode at a distance of 9 mm from the upper edge and 11 mm from the side edges, measured from the centre of the holes. The permitted tolerances are ±0.5 mm for the diameter of the holes and ±2.0 mm for the location of the holes. By agreement, anodes may be produced without holes or with one hole only.

Table 12
Dimensions and Weight of Gold Anodes

Thickness of Anode	*Width (mm)* 50	100	150	*Permissible Thickness Tolerances*	*Weight of a Single Anode (g)*
	Length (mm)				
0.1	100	—	—	−0.02	9.65
0.2	—	200	—	−0.03	77.2
0.3	—	200	—	−0.04	115.8
0.5	—	200	—	−0.06	193
1.0	—	—	300	−0.10	868.5
1.5	—	—	300	−0.15	1302.7
2.0	—	—	300	−0.20	1737
2.5	—	—	300	−0.20	2171.2
3.0	—	—	300	−0.25	2605.5
5.0	—	—	300	−0.30	4342.5

For the purposes of the specification, a batch of anodes consists of strips of a single dimension prepared from a single ingot. The weight of the batch must not be greater than 20 kg. Acceptance is carried out by visual examination and by measuring all the anodes in a batch. The thickness is measured at a distance of 15–30 mm from the edges. The weighing of anodes is carried out using technical scales of the first class with an accuracy of 0.1 g.

FUTURE TRENDS IN STANDARDISATION

The increasing use of electroplated gold coatings to replace other types of gold coating, e.g., rolled gold and gold capping does, in the opinion of many, point to the need for national and international standards and regulations for controlling their quality. As has been stated earlier in this appendix, certain countries have national regulations in the interests of consumer protection, whereas in other countries the legal position is at best vague and at worst unhelpful to the consumer. Progress will certainly be expected soon in improving national standards and world trade in jewellery and related products would be facilitated if international standards and regulations could be agreed. To this desirable end we may have to look to the International Organization for Standardization (ISO).

For some years ISO has had a Technical Committee dealing with metallic coatings, namely TC 107—Metallic and Other Non-organic Coatings, the Secretariat of which is held by Italy. Sub-committee 3—Electrodeposited Coatings, of which the Secretarist is held by the United Kingdom, has already been responsible for the publication of a number of ISO Recommendations and International Standards including those for electroplated coatings of nickel and chromium. Work is currently proceeding in this Sub-Committee on drafting an International Standard for electroplated gold coatings for decorative applications and a separate standard for these coatings when used for engineering purposes. It is to be hoped that, for the former document, the participating countries will be able to agree on the minimum thickness of gold which will qualify for the description of "electroplated gold coating" as distinct from "gilt", for example. Referee methods of determining the thickness, particularly at the thinner end of the range, will need to be agreed and those parts of the significant surface on which thickness measurements should be taken will require to be defined. As gold is an expensive metal it is also necessary, in the interests of the consumer, to define the percentages of alloying elements which may be added for the purpose of achieving a desired colour or a particular hardness. There should be less difficulty in arriving at internationally agreed standards for electroplated gold coatings for engineering purposes as a considerable amount of research has been carried out in recent years by the electronics industry in order to establish the optimum types and thickness of gold coatings necessary to ensure high reliability of electrical contacts. Furthermore, satisfactory tests for porosity and solderability have been devised to assess their performance.

REFERENCES

1 Loi fédérale sur le contrôle du commerce des métaux précieux et des ouvrages en métaux précieux. (Switzerland).
2 Code Général des Impôts. (France).
3 U.S. Federal Trade Commission: "Guides for the Watch Industry".
4 U.S. Federal Trade Commission: "Trade Practice Rules for the Jewelry Industry".
5 Official Standard on "Articles of Gold, Precious Metals and Gold Plated Metals".
6 B.S. 3315, "Watchcase Finishes in Gold Alloys".
7 B.S. 4292, "Electroplated Coatings of Gold and Gold Alloy".
8 Draft Standard DIN 8237, "Gold Plating and Coating for Watchcases, Requirements, Testing."
9 DIN 50950 "Testing of Electroplated Coatings; Microscopic Measurement of the Coating Thickness".
10 I.S. 4252, "Electroplated Coatings of Gold for Decorative Purposes".
11 I.S. 3023, "Methods of Testing Local Thickness of Electroplated Coatings".
12 I.S. 1418, "Methods for Assaying of Gold and Gold Alloys".
13 J.I.S. H8616, "Gold Plating".
14 Australian Standard K158, "Electroplated Coatings of Gold for General Engineering Purposes".
15 A.S.T.M. Specification B488, "Electrodeposited Coatings of Gold for Engineering Uses".
16 A.M.S. 2422, "Gold Plating for Electronic Applications".
17 A.M.S. 2425, "Gold Plating for Thermal Control".
18 A.S.T.M. B117, "Salt Spray (Fog) Testing".
19 D.T.D. 938, "Gold Plating".
20 MIL-G-45204B, "Gold Plating, Electrodeposited".
21 MIL-STD-202, "Test Methods for Electronic and Electrical Component Parts".

INDEX OF NAMES

(see also separate Index of Subjects)

N.B. *Only those authors mentioned by name in the text have been indexed. Names cited solely in the references at the ends of chapters have been omitted.*

INDEX OF SUBJECTS

(see also separate Index of Names)

Printed Circuit Troubleshooting

by H. R. Shemilt

ISBN 901150 03 7

SOME WORLDWIDE OPINIONS:

"Throughout, the author is concerned with practical solutions to common problems. Subject treatment is up-to-date and accurate. The layout of the book, with large pages, many drawings and photographs, and bold headings, is excellent" ... "a work on troubleshooting is unique in the field, and this volume goes far to fill a void. It is recommended for production technologists in the printed circuit field".

Plating and Surface Finishing (USA)

"It is a very convenient book, a ready troubleshooting reference and an indispensable tool for anyone involved in the fabrication of PC boards and in the research laboratories".

Electrochemical Progress (USA)

"This book is an excellent guidebook and will be indispensable to technical and research workers in this field".

Journal of the Metal Finishing Society (Japan)

"Since it covers all aspects of PC manufacture from artwork to testing, the book will be of use to designers, production engineers, chemists, inspectors, process workers and students, all of whom are likely to regard it as an investment".

Pulse (South Africa)

"Printed circuit technicians regard this work as a 'must', as it is a particularly useful book which is written in a crisp and condensed style. It gets to the root of the matter and distinguishes between the important and the trivial."

Product Finishing (UK)

"Half of each chapter describes the how and why and the other half gives a problem-cause-remedy table. By such listing of all the things which can go wrong, the book gives a better picture of the whole technology than one would gain from a conventional, straight-forward account. For this reason it could be invaluable training and reference material for apprentices, process workers, inspectors, design staff, chemists and metallurgists".

The Engineer (UK)

"Each chapter of this unique volume details with a specific stage of the printed-circuit process, presenting a comprehensive review of techniques and concluding with a tabular fault-finding guide which lists faults, causes and remedies. Although the scope ranges from artwork to testing specifications, particularly valuable sections cover plating and multilayer boards. Mr Shemilt's compendium will prove an excellent reference for technologists in industry involved in setting-up and running a printed circuit facility".

ASLIB Book List (UK)

"... a first class manual for the practical engineer".

Design Engineering (UK)

Please order direct from

Electrochemical Publications Ltd.
29 Barns Street, Ayr KA7 1XB, Scotland

Handbook of Thick Film Technology

By P. J. Holmes & R. G. Loasby

ISBN 901150 05 3

SOME WORLDWIDE OPINIONS

"The volume presents an excellent expose of the subjects and of the state-of-the-art, in addition to offer practical materials and solutions in the selection of components and techniques. It is an excellent reference for libraries and an indispensable tool for anyone interested in securing information on manufacturing of circuits".

Electrochemical Progress (USA)

"This book is recommended reading for those having interests in any aspect of thick film technology".

IEEE (USA)

"On the whole, this book serves a useful function and can be recommended both for the technologists working in industry and for the researcher requiring general information about the field."

Thin Solid Films (UK)

"The reviewer has often considered that more books should be written by staff at Government Research Establishments, particularly Establishments where practical experience on the subject has been obtained objectively. There are a great many books written by Universities which are mainly theoretical and, whilst these are valuable, books containing practical details are of more use to industry. This book falls in this latter category.

It is an excellent book and covers the subject of thick film technology in great detail and with great thoroughness. A slightly unusual feature is the detail in which the contents pages are given which is useful for quick reference".

Microelectronics and Reliability (UK)

"This is probably the best book which has yet appeared on the subject of hybrid microelectronics. It is well conceived and well produced".

Pulse (South Africa)

"This volume, edited by two leading authorities in the field, with individual chapters contributed by nine other research workers, all of them in close touch with production engineers and designers of electronic equipment, aims to provide full details of the practical aspects of the process, the materials available and the production of packaged circuit networks. To the practising technologists in thick film circuitry, conscious of the future growth and importance of his expertise, it will be a valuable source of information and guidance."

Gold Bulletin (South Africa)